Thermal Physics & Statistical Mechanics

Thermal Physics & Statistical Mechanics

Editor:

Olga Moreira, Ph.D.

ArclerPress

www.arclerpress.com

Thermal Physics & Statistical Mechanics

Editor: Olga Moreira, Ph.D.

708 3rd Avenue, 6th Floor
New York
NY 10017
United States of America
www.arclerpress.com

ISBN: 978-1-68094-436-5

Library of Congress Control Number: 2016951143

About the Editor

Olga Moreira, Ph.D.

Olga Moreira obtained her Ph.D. in Astrophysics from the University of Liege (Belgium) in 2010, her BSc. in Physics and Applied Mathematics from the University of Porto (Portugal). Her post-graduate travels and international collaborations with the European Space Agency (ESA) and European Southern Observatory (ESO) led to great personal and professional growth as a scientist. Currently, she is working as an independent researcher, technical writer, and editor in the fields of Mathematics, Physics, Astronomy and Astrophysics.

Contents

Section III: Applications

List of Contributors

Milivoje M. Kostic
Department of Mechanical Engineering, Northern Illinois University, DeKalb, IL 60115, USA; Tel.: +1-815-753-9975

Yelena Guryanova
H.H. Wills Physics Laboratory, University of Bristol, Tyndall Avenue, Bristol BS8 1TL, UK

Sandu Popescu
H.H. Wills Physics Laboratory, University of Bristol, Tyndall Avenue, Bristol BS8 1TL, UK

Anthony J. Short
H.H. Wills Physics Laboratory, University of Bristol, Tyndall Avenue, Bristol BS8 1TL, UK

Ralph Silva
H.H. Wills Physics Laboratory, University of Bristol, Tyndall Avenue, Bristol BS8 1TL, UK
De´partement de Physique The´orique, Universite´ de Gene`ve, 1211 Gene`ve, Switzerland

Paul Skrzypczyk
H.H. Wills Physics Laboratory, University of Bristol, Tyndall Avenue, Bristol BS8 1TL, UK

Lee Jinwoo
Department of Mathematics, Kwangwoon University, 20 Kwangwoon-ro, Nowon-gu, Seoul 139-701, Korea

Hajime Tanaka
Institute of Industrial Science, University of Tokyo, 4-6-1 Komaba, Meguro-ku, Tokyo 153-8505, Japan.

Tobias Damm
Institut fu¨r Angewandte Physik, Atominstitut, Institute of Quantum Electronics, Universita¨t Bonn, Wegelerstrasse 8, 53115 Bonn, Germany.

Julian Schmitt
Institut fu¨r Angewandte Physik, Atominstitut, Institute of Quantum Electronics, Universita¨t Bonn, Wegelerstrasse 8, 53115 Bonn, Germany.

Qi Liang
Institut fu¨r Angewandte Physik, Atominstitut, Institute of Quantum Electronics, Universita¨t Bonn, Wegelerstrasse 8, 53115 Bonn, Germany. Atominstitut, TU Wien, Stadionallee 2, 1020 Vienna, Austria (Q.L.); Institute of Quantum Electronics, ETH Zu¨rich, Auguste-Piccard-Hof 1, 8093 Zu¨rich, Switzerland (J.K.).

David Dung
Institut fu¨r Angewandte Physik, Atominstitut, Institute of Quantum Electronics, Universita¨t Bonn, Wegelerstrasse 8, 53115 Bonn, Germany.

Frank Vewinger
Institut fu¨r Angewandte Physik, Atominstitut, Institute of Quantum Electronics, Universita¨t Bonn, Wegelerstrasse 8, 53115 Bonn, Germany.

Martin Weitz
Institut fu¨r Angewandte Physik, Atominstitut, Institute of Quantum Electronics, Universita¨t Bonn, Wegelerstrasse 8, 53115 Bonn, Germany.

Jan Klaers
Institut fu¨r Angewandte Physik, Atominstitut, Institute of Quantum Electronics, Universita¨t Bonn, Wegelerstrasse 8, 53115 Bonn, Germany. Atominstitut, TU Wien, Stadionallee 2, 1020 Vienna, Austria (Q.L.); Institute of Quantum Electronics, ETH Zu¨rich, Auguste-Piccard-Hof 1, 8093 Zu¨rich, Switzerland (J.K.).

K. V. Reich
Ioffe Physical-Technical Institute, 194021 St. Petersburg, Russia

Alan L. Rockwood
Department of Pathology, University of Utah School of Medicine and ARUP Institute for Clinical and Experimental Pathology, USA

I.N. Shishkova
Low Temperature Department, Moscow Power Engineering Institute, Krasnokazarmennaya 14, Moscow 111250, Russia
Sir Harry Ricardo Laboratories, Centre for Automotive Engineering, School of Computing, Engineering and Mathematics, Faculty of Science and Engineering, University of Brighton, Brighton BN2 4GJ, UK

S.S. Sazhin
Sir Harry Ricardo Laboratories, Centre for Automotive Engineering, School of Computing, Engineering and Mathematics, Faculty of Science and Engineering, University of Brighton, Brighton BN2 4GJ, UK

S. V. Molodtsov
Institute of Theoretical and Experimental Physics, Moscow, Russia
Joint Institute for Nuclear Research, Dubna, Moscow Oblast, Russia

G. M. Zinovjev
Bogolyubov Institute for Theoretical Physics, National Academy of Sciences of Ukraine, Kiev, Ukraine

Shingo Takeuchi
The Institute for Fundamental Study "The Tah Poe Academia Institute", Naresuan University, Phitsanulok 65000, Thailand

Amritansu Ray
Department of Mathematics, Rajyadharpur Deshbandhu Vidyapith, Serampore, Hooghly 712203, West Bengal, India

S. K. Majumder
Department of Mathematics, Indian Institute of Engineering Science and Technology (IIEST), Shibpur, Howrah 711103, West Bengal, India

Koichi Nakagawa
Laboratory of Physics, School of Pharmacy and Pharmaceutical Sciences, Hoshi University, Tokyo 142-8501, Japan

Wolfgang Ehlers
Institute of Applied Mechanics (CE), University of Stuttgart, Pfaffenwaldring 7, 70569 Stuttgart, Germany
Stuttgart Research Centre for Simulation Technology (SRC SimTech), Stuttgart, Germany

Kai Häberle
Institute of Applied Mechanics (CE), University of Stuttgart, Pfaffenwaldring 7, 70569 Stuttgart, Germany
International Graduate School for Non-linearities and Upscaling in Porous Media (NUPUS), Stuttgart, Germany

Pavel Kovtun
Department of Physics and Astronomy, University of Victoria, Victoria BC, V8W 2Y2, Canada

John Ross
Department of Chemistry, Stanford University, Stanford, CA 94305, USA

Alejandro Fernández Villaverde
Bioprocess Engineering Group, IIM-CSIC, Eduardo Cabello 6, Vigo, 36208, Spain

Barbara A. Jones
Almaden Research Center, IBM, San Jose, California, United States of America

Justin Lessler
Department of Epidemiology, Johns Hopkins Bloomberg School of Public Health, Baltimore, Maryland, United States of America

Simone Bianco
Almaden Research Center, IBM, San Jose, California, United States of America

James H. Kaufman
Almaden Research Center, IBM, San Jose, California, United States of America

B. Gardas
University of Silesia, 40-007 Katowice, Poland; Theoretical Division, Los Alamos National Laboratory, Los Alamos, NM 87545, United States of America

J. Łuczka
University of Silesia, 40-007 Katowice, Poland; Silesian Center for Education and Interdisciplinary Research, University of Silesia, 41-500 Chorzów, Poland

A. Ptok
Institute of Nuclear Physics, Polish Academy of Sciences, ul. Radzikowskiego 152, 31-342 Kraków, Poland

J. Dajka
University of Silesia, 40-007 Katowice, Poland; Silesian Center for Education and Interdisciplinary Research, University of Silesia, 41-500 Chorzów, Poland

Chamil Abeykoon
IRC in Polymer Science and Technology, School of Engineering, Design and Technology, University of Bradford, Bradford BD7 1DP, UK

Adrian L. Kelly
IRC in Polymer Science and Technology, School of Engineering, Design and Technology, University of Bradford, Bradford BD7 1DP, UK

Javier Vera-Sorroche
IRC in Polymer Science and Technology, School of Engineering, Design and Technology, University of Bradford, Bradford BD7 1DP, UK

Elaine C. Brown
IRC in Polymer Science and Technology, School of Engineering, Design and Technology, University of Bradford, Bradford BD7 1DP, UK

Phil D. Coates
IRC in Polymer Science and Technology, School of Engineering, Design and Technology, University of Bradford, Bradford BD7 1DP, UK

Jing Deng
School of Electronics, Electrical Engineering and Computer Science, Queen's University Belfast, Belfast BT9 5AH, UK

Kang Li
School of Electronics, Electrical Engineering and Computer Science, Queen's University Belfast, Belfast BT9 5AH, UK

Eileen Harkin-Jones
School of Mechanical and Aerospace Engineering, Queen's University Belfast, Belfast BT9 5AH, UK

Mark Price
School of Mechanical and Aerospace Engineering, Queen's University Belfast, Belfast BT9 5AH, UK

Liqiu Wang
Department of Mechanical Engineering, The University of Hong Kong, Pokfulam Road, Hong Kong

Jing Fan
Department of Mechanical Engineering, The University of Hong Kong, Pokfulam Road, Hong Kong

Masayuki Tokita
Department of Physics, Faculty of Science, Kyushu University, 744 Motooka, Fukuoka, Fukuoka 812-8581, Japan

Preface

This book outlines the ongoing research and development in modern Thermal Physics & Statistical Mechanics (TPSM). It is organized into 3 sections: Selected Topics in TPSM, Advanced Topics in TPSM and Applications.

The first section starts by introducing the physical meaning of thermodynamic entropy, providing a historical context - from the Carnot's reversible cycles to the modern definition of entropy. It continues further by discussing specific questions and methods. The first studies are concerned with a better comprehension of the thermodynamics of quantum systems and small individual systems. This includes the analysis of quantum systems where the energy is not the only quantity being conserved as well as the reproduction of locally defined entropy for small individual systems. Bose–Einstein condensation of a two-dimensional photon gas and its thermodynamic behaviour at the phase transition are also investigated. The remaining studies are focused on understanding the processes involving electron gas in composite materials. This include the study of the local interaction between an electron and a photon gas as mean to examine the the Kapitza resistance (thermal boundary resistance) behaviour, the analysis of quantum mechanical model (Jellium model) as alternative procedure to examine the relationship between the absolute power of metal and the partial molar entropy, and the testing of new algorithm for solving a system of three Boltzmann equations for three components in a mixture. The second section is centered on advanced TPSM methods and approaches which are being motivated by the technological progress, research and development in other fields such as Nanophotonics, Quantum Computing, High Energy Physics, and Cosmology. For instance, the analysis of quark ensembles by using a four-fermion approximation motivated by the potential quark–gluon plasma production at heavy-ion colliders, the investigation of Bose–Einstein condensation at finite density in the Rindler space is based on the Unruh effect and in the experimental research of black holes evaporation. Moreover, Burg's maximum entropy, a method commonly used in data mining, can also be a promising tool for the analysis of complex systems such those encountered in TPSM. The use of different entropy measures (e.g. Boltzmann's entropy, Shannon's entropy, Burg's modified entropy) as well as the study of quantum entanglement from the perspective of thermal quantum field theory, phases transitions of a fluid material with solid deformations, polarized relativistic matter, and non-equilibrium thermodynamics are discussed in great detail in this section. The remaining content of this book focuses on applications of TPSM in interdisciplinary research, for example, for the study viral evolution, coupled SQUIDs (superconducting quantum interference devices), polymers extrusion, nanofluids of ultra-thermal conductivity and transport processes in soft materials.

Editor
Olga Moreira, Ph.D.

INTRODUCTION

Thermal physics is what describes the equilibrium distribution of particles for a medium. Thermal physics is the physics of real complex systems and is central to our industrial society with its properties of matter which allow for the use of machines and engines. It is a collective subject which includes thermodynamics, statistical mechanics, and kinetic theory. The physics subject is viewed in the perspective of energy (heat/ temperature). Thermal energy, measured in joules, refers to the total kinetic and potential energy of the component particles (moles, atoms, etc.) of an object. Heat energy will always be transferred from bodies with higher temperatures to those with lower temperatures. In an isolated system, the two bodies will eventually reach thermal equilibrium, a state at which both will have the same temperature. This all occurs on a microscopic level because it deals with the movements of individual atoms or molecules with in a substance; it cannot be seen with the naked eye. Thermal Physics concepts can be observed indirectly by observing and calculating new temperatures etc. Thermal energy and its reactions can be calculated by using formulas.

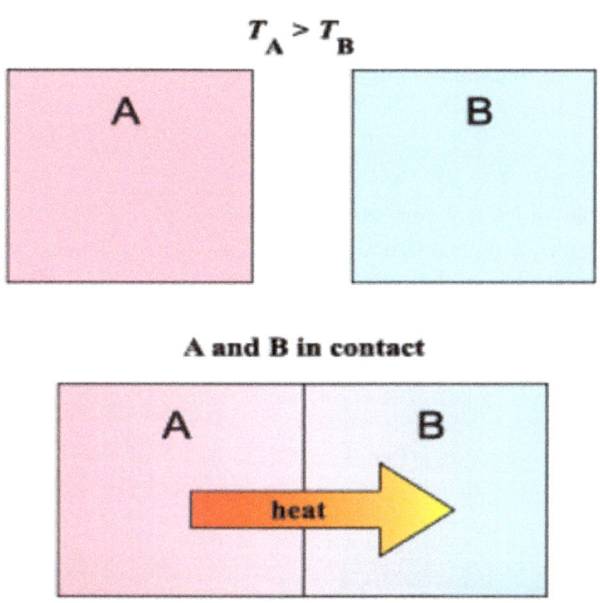

Temperature is a scalar quantity that indicates the hotness or coldness of an object and can be measured using a thermometer. Temperature is also a measure of the average kinetic energy of the molecules in a substance. If two objects have the same temperature, then their molecules have the same average kinetic energy. When the molecules are moving at a fast rate, it ultimately causes more energy creating heat. The heat can is the transfer of energy to an object, which could also be referenced as transferring kinetic energy to an object with less kinetic energy in order to achieve thermal equilibrium. The SI unit for temperature is the kelvin (K), which is based on absolute zero (the point at which the internal movement of a body's atoms or molecules is at a minimum and no heat can be removed). Absolute zero is defined as being precisely 0 K and −273.15 °C. One kelvin is equal to 273.16; the freezing point of water is 273 K (or 0°C), and the boiling point is 373 K (or 100°C).

In order to transfer units of temperature between degrees in Kelvin and degrees in Celcius, the equation below should be used:

$K = C + 273.15$

Because one degree of Celcius is roughly equal to 273 degrees Kelvin. The temperature should be plugged into the equation where it's measurement is used (ex: in the "K" if it is given in Kelvin), in order to obtain an answer.

The thermal energy departs from the sun as radiation and is then known as heat. When a person rubs his hands together rapidly, such mechanical energy raises the thermal energy content of his skin and hands. The heat is a result of this. Generally, when a substance's molecules' velocity increases, the temperature increaes as well.

The laws of thermodynamics state that energy in the heat form can be transferred between physical objects, such as in the case of a pot of water heating up after fire was put beneath it. Therefore, the fire's thermal energy is partly transmitted to the pot of water. The heat transfer between two thermodynamic systems is caused by collisions between individual molecules in the system.

Kinetic Theory: According to the Encarta dictionary kinetic theory is the behavior of gases that assumes heat is a process of energy and the internal energy of a gas is the total energy of its particles. Because of this molecules are arranged in different ways depending on the phases of the substance(solid, liquid, or gas). This theory explains the structure and nature of matter.

The assumptions of the kinetic theory are:

- All matter is made up of really small particles
- All particles are constantly in motion
- If particles collide with other particles, kinetic energy is conserved (energy is not lost or created)
- A mutual attractive force exists between particles

Solid

Solids hold their fixed positions more ridgidly and the transfer of vibrational energy is more effective (however, their large masses do not allow for large energy transfer). Examples of this are glass, plastic and gel. Solid's are found everywhere and it takes a great amount of energy to metamorphisize its state. Solid atoms are compact and move past each other slowly.

Liquid

A liquid has a fixed volume just like the solid, but unlike a solid the shape can change. The molecules are vibrating, but not completely fixed in shape. They are more compressible than solids, but much less so than gases. Liquid atoms are loosely packed together but can still move past one another. The bonds between particles can break easier than liquids, but they can also be easily reformed. The potential energy of particles in a liquid is higher than a solid because of the spacing between the particles larger than that of a solid's.

Gas

A gas will expand to fit the container in which it is put. The molecules are not fixed in position, and any forces between the molecules are weak unlike the forces with the liquids that are strong. Gases have particles that are widely spaced. Most kinetic energy is in a gas. Gasses are very compressible, as their molecules are not very well bonded and thus have much room between them that can be compressed. Gases will have a greater high potential energy than liquids because the particles of gases are farther apart than that of liquids.

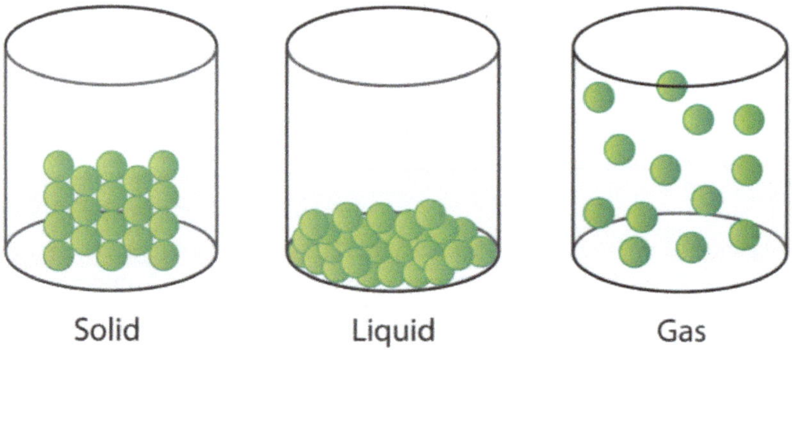

Temperature

In solids, the particles comprising the solid are held together tightly, therefore their motion is limited to vibrating back and forth in their given positions. In liquids, the particles can move back and forth across each other, but the object itself has no defined shape. In gases, the particles move throughout the volume available, interacting with each other and the walls of any container holding them. In all cases, the total thermal energy of the object is the sum total of the kinetic energies of its constituent particles.

Instead of just looking at the sum of all the individual particles' kinetic energies, you could examine the average kinetic energy of the particles comprising the object, realizing that the actual kinetic energies of individual particles may vary significantly. The average kinetic energy of the particles is directly related to the temperature of the object by the following equation:

$$K_{avg} = \frac{3}{2} k_B T$$

Examining this equation, the average kinetic energy is given in Joules, k_B is Boltzmann's Constant (1.38×10^{-23}J/K), and the temperature is given in Kelvins, the SI unit of temperature. Note that even though two objects can have the same temperature (and therefore the same average kinetic energy), they may have different internal energies.

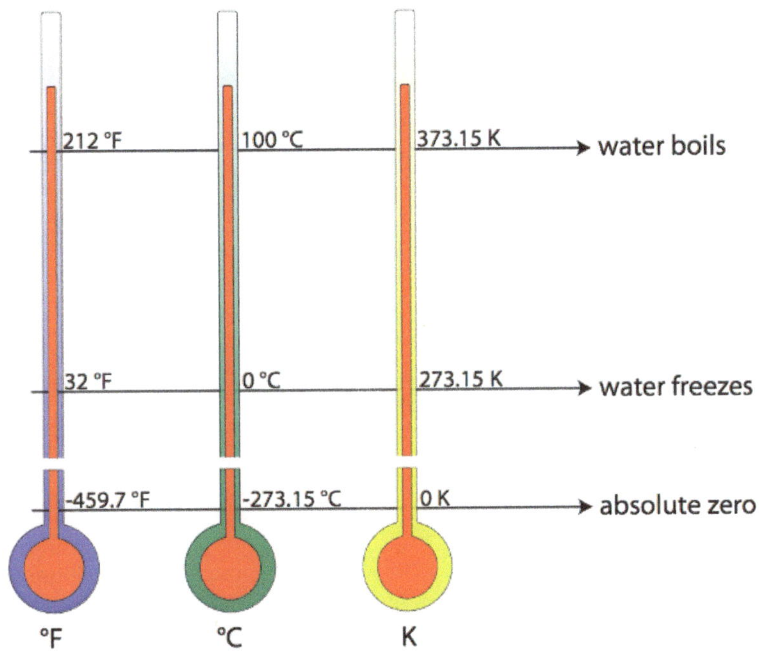

The Kelvin scale is closely related to the Celsius temperature scale, but where the Celsius scale targets the freezing point of water as 0° C, the Kelvin scale utilizes its zero at what is known as absolute zero (the point on a Volume vs. Temperature graph for a gas where the extended curve would hypothetically reach zero volume), considered a theoretical minimum temperature. There- fore, absolute zero is 0 Kelvins, which is equivalent to -273.15° Celsius. To convert from Kelvins to degrees Celsius, just add 273.15 to your temperature reading in degrees Celsius. The freezing point of water, then, is 0°C or 273.15 K, and the boiling point of water is 100°C or 373.15 K. Compare this to the Fahrenheit scale, where water freezes at 32°F, and boils at 212°F

$$T_K = T_{°C} + 273.15$$

$$T_{°F} = \tfrac{9}{5} T_{°C} + 32$$

$$T_{°C} = \tfrac{5}{9}(T_{°F} - 32)$$

Thermal Expansion

When objects are heated, they tend to expand, and when they are cooled, they tend to contract. You can use this to open glass jars with tight metal lids by running the lids under hot water. The temperature increase in the lid expands the metal lid and the glass jar, but because most metals expand more quickly than glass, the lid becomes looser, making it easier to open the jar.

This occurs because at higher temperatures, objects have higher kinetic energies, so their particles vibrate more. At these higher levels of vibration, the particles aren't bound to each other as tightly, so the object expands.

The amount an object expands can be calculated for both one-dimensional (linear) and three-dimensional (volumetric) expansion. The amount a material expands is characterized by the material's coefficient of expansion. In calculating a material's one-dimensional expansion, you can use the linear expansion formula, and the material's linear coefficient of expansion (α).

$$\Delta l = \alpha l_0 \Delta T$$

When calculating a material's three-dimensional expansion, you'll use the volumetric expansion formula, and the material's volumetric coefficient of expansion (β). Note that in most cases, the volumetric coefficient of expansion is roughly three times the linear coefficient of expansion.

$$\Delta V = \beta V_0 \Delta T$$

Heat

Heat is the transfer of thermal energy from one object to another object due to a difference in temperature. Heat always flows from warmer objects to cooler objects. The symbol for heat in physics is Q, with positive values of Q representing heat flowing into an object, and negative values of Q representing heat flowing out of an object.

When heat flows into or out of an object, the amount of temperature change depends on the material. The amount of heat required to change one kilogram of a material by one degree Celsius (or one Kelvin) is known as the material's specific heat (or specific heat capacity), represented by the symbol C.

Specific Heats of Selected Materials	
Material	**C (J/kg·K)**
Aluminum	897
Concrete	850
Diamond	509
Glass	840
Helium	5193
Water	4181

The relationship between heat and temperature is quantified by the following equation, where Q is the heat transferred, m is the mass of the object, C is the specific heat, and ΔT is the change in temperature (in degrees Celsius or Kelvins).

Phase Changes

These states include solids, liquids, gases, and plasmas. You're probably familiar with solids, liquids, and gases already. Plasmas are energetic gases that have been ionized so that they can conduct electricity (examples include stars, lightning, neon signs, etc.)

When matter changes state, its internal energy changes, so the kinetic energy of its constituent particles changes. As it is changing from one state to another, the change in energy is reflected in the bonds between the particles, and therefore the temperature of the object doesn't change. Once the state change is complete, however, changes in energy are again observed in the form of changes in temperature.

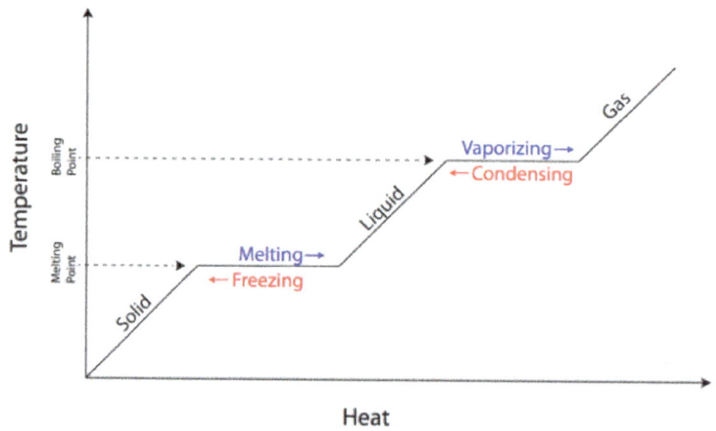

The energy required to change a specific material's state is known as the material's latent heat of transformation (L). When an object transitions from the solid to liquid phase, you use the latent heat of fusion (L_f). When an object transitions from the liquid to the gaseous phase, you use the latent heat of vaporization (L_v). You can calculate the energy required for a material to change phases using the following formula, where Q is the heat added, m is the object's mass, and L is the material's specific latent heat of transformation.

$$Q = mL$$

Latent Heats of Selected Materials				
Material	L_f (kJ/kg)	Melting Point (°C)	L_v (kJ/kg)	Boiling Point (°C)
Aluminum	399	659	10,500	2327
Helium	N/A	N/A	21	-269
Hydrogen	58	-259	455	-253
Lead	25	327	871	1750
Water	334	0	2260	100

Ideal Gas Law

In studying the behavior of gases in confined spaces, it is useful to limit ourselves to the study of ideal gases. Ideal gases are theoretical models of real gases, which utilize a number of basic assumptions to simplify their study. These assumptions include treating the gas as being comprised of many particles which move randomly in a container. The particles are, on average, far apart from one another, and they do not exert forces upon one another unless they come in contact in an elastic collision.

Under normal conditions such as standard temperature and pressure, most gases behave in a manner quite similar to an ideal gas. Heavy gases as well as gases at very low temperatures or very high pressures are not as well modeled by an ideal gas.

The Ideal Gas Law relates the pressure, volume, number of particles, and temperature of an ideal gas in a single equation, and can be written in a number of different ways.

$$PV = nRT = Nk_BT$$

In this equation, P is the pressure of the gas (in Pascals), V is the volume of the gas (in cubic meters), n is the number of moles of gas, N is the number of molecules of gas, R is the universal gas constant equal to 8.31 J/mol·K (which is also 0.08206 L·atm/mol·K), kB is Boltzmann's Constant (1.38×10^{-23} J/K), and T is the temperature, in Kelvins. To convert from molecules to moles, you can use Avogadro's Number (N0=6.02×10^{-23} molecules/mole):

$$n = \frac{N}{N_0}$$

Note that a Pascal multiplied by a cubic meter is a newton-meter, or Joule. As well, Boltzmann's constant is the ideal gas law constant divided by Avogadro's number.

Statistical Mechanics

Statistical mechanics is a branch of theoretical physics that studies, using probability theory, the average behavior of a mechanical system where the state of the system is uncertain.

A common use of statistical mechanics is in explaining the thermodynamic behavior of large systems. This branch of statistical mechanics which treats and extends classical thermodynamics is known as statistical thermodynamics or equilibrium statistical mechanics. Microscopic mechanical laws do not contain concepts such as temperature, heat, or entropy; however, statistical mechanics shows how these concepts arise from the natural uncertainty about the state of a system when that system is prepared in practice. The benefit of using statistical mechanics is that it provides exact methods to connect thermodynamic quantities (such as heat capacity) to microscopic behavior, whereas, in classical thermodynamics, the only available option would be to just measure and tabulate such quantities for various materials. Statistical mechanics also makes it possible to extend the laws of thermodynamics to cases which are not considered in classical thermodynamics, such as microscopic systems and other mechanical systems with few degrees of freedom.

Statistical mechanics also finds use outside equilibrium. An important sub-branch known as non-equilibrium statistical mechanics deals with the issue of microscopically modeling the speed of irreversible processes that are driven by imbalances. Examples of such processes include chemical reactions or flows of particles and heat. Unlike with equilibrium, there is no exact formalism that applies to non-equilibrium statistical mechanics in general, and so this branch of statistical mechanics remains an active area of theoretical research.

$$Q = mC\Delta T$$

SECTION I: SELECTED TOPICS IN TPSM

CHAPTER 1

THE ELUSIVE NATURE OF ENTROPY AND ITS PHYSICAL MEANING

Milivoje M. Kostic

Keywords: Caloric process, Carnot cycle, Clausius (in)equality, entropy, entropy generation, heat transfer, microstates number, statistical entropy, thermal energy, thermal interactions, thermal motion

INTRODUCTION

What is the underlying nature of "entropy" and why does it always increase? Why is entropy so intriguing and mysterious, unique and universal, as if it is a miraculous property of natural, material systems? How does it encompass and quantify all processes at all natural space and time scales, governed by the Second Law of Thermodynamics? And many other elusive and debatable issues, as if entropy is among the deepest unresolved mysteries in nature, defying our common sense. Entropy is the most used and often abused concept in science, but also in philosophy and society. Further confusions are produced by some attempts to generalize entropy with similar, but not the same concepts in other disciplines. Von Neumann once remarked that "whoever uses the term 'entropy' in a discussion always wins since no one knows what entropy really is, so in a debate one always has the advantage."

The historian of science and mathematician, Truesdell, explains in his essay of Method and Taste in Natural Philosophy: "Heads have split for a century trying to define entropy in terms of other things. Entropy, like force, is an undefined object, and if you try to define it, you will suffer the same fate as the force-definers of the seventeenth and eighteenth centuries: Either you will get something too special or you will run around in a circle." Trying to reason, understand, and explain fundamental concepts are daunting endeavors, and different from understanding other concepts derived from the fundamental ones. However, there is factual reality that we should be always aware of and guided by. The fundamental interactions and their inter-relations, along with subtle reasoning, may help to resolve ambiguities and close the needed loops. "Going in a circle" does not mean we are not going anywhere, as long as we succeed in comprehension of subtle "circular" interactions, including interrelated ("chicken-and-egg") causes-and-effects.

The intention of this treatise is not a review of vast entropy literature, but presentation of this author's long-contemplated reflections on physical meaning of entropy, and to put certain physical and philosophical concepts in perspective. Only two seminal references and two related publications by the author, in addition to three popular references to illustrate certain misconceptions, are cited. The classical, phenomenological thermodynamics today has unjustifiably a dubious status. Some modern physicists regard classical thermodynamics as an obsolete relic. Often, mostly due to lack of dubious comprehension, thermodynamics is considered as an engineering subject, and thus not as the most fundamental science of energy and nature. The phenomenological thermodynamics has the supremacy over other disciplines, due to its logical reasoning based on the fundamental laws and without the regard to the system complex dynamic structure, and even more complex related interactions.

There are many puzzling issues surrounding thermodynamics and the nature of heat, including subtle definitions and ambiguous meaning of very fundamental concepts. In modern times, there is a tendency by some scientists to unduly discredit thermal energy as being indistinguishable from other internal energy types. Romer [1] argues that "Heat is not a noun," and proposes to remove it from the dictionary. Ben-Naim [2] titles his book "A Farewell To Entropy," while Leff [3] in a series of articles entitled "Removing the Mystery of Entropy and Thermodynamics," argues surprisingly, that "Entropy can be introduced and understood without ever mentioning heat engines," and against the "thermal energy" concept in favor of more modern and well-defined "internal energy." Some consider the tabulated Thermodynamic Internal Energy ($u = U/m$, per unit mass) to be thermal energy, although it represents all energy types stored as kinetic and potential energy of constituent microstructure, thus thermal and

Citation: Kostic, M.M, The Elusive Nature of Entropy and Its Physical Meaning. Entropy 2014, 16, 953-967, doi:10.3390/e16020953.

mechanical elastic energy in simple compressive substances, in addition to chemical and nuclear internal energies. In more complex system structure there may be more energy types.

Entropy is related to thermal motion of a system microstructure, the latter gives rise to all thermal phenomena and related properties, namely, temperature, thermal or heat capacity, thermal energy and entropy, among others. Due to conversion of thermal energy to other energy forms, like mechanical work in heat engine, and also spontaneous and unavoidable dissipation of all other energy forms to thermal energy via so called heat generation, additional issues and often confusions arise. However, entropy is well-defined material system macro-property, precisely measured, and tabulated and/or correlated, for practical use in engineering and science. Entropy should be further reasoned, refined and explained for what it is, and not be misrepresented as something it might be or is not.

FROM CARNOT'S INGENIOUS REASONING OF REVERSIBLE CYCLES TO CLAUSIUS' DEFINITION OF ENTROPY

Sadi Carnot (1824) laid ingenious foundations for the Second Law of Thermodynamics and discovery of Entropy before the Frist Law of energy conservation was even known (Joule, 1843), and long before thermodynamic concepts were established in the second half of the nineteenth century. In historical context, it is hard to comprehend now, how Carnot then, at age 28, ingeniously and fully explained the critical concepts of reversible thermo-mechanical processes and the limits of converting heat to work at inception of the heat engines' era, when nature of heat was not fully understood. No wonder that Sadi Carnot's "Réflexions sur la puissance motrice du feu (Reflections on the Motive Power of Fire [4])," original treatise published in 1824, was not noticed at his time, when his ingenious reasoning of ideal heat engine reversible cycles, is not fully recognized, and may be truly comprehended by a few, even nowadays.

Carnot paved the way for his followers to define and prove that entropy is a state function, a material property conserved in ideal, reversible cycles (Clausius Equality—definition of entropy property), that entropy could not be destroyed since it will imply more efficient than ideal cycles, but is always generated (locally and globally, thus overall increased) due to dissipation of any and all work potentials to heat, causing generation of entropy in irreversible cycles (Clausius Inequality—definition of entropy generation); thereby, quantifying all reversible and irreversible processes and providing generalization of the Second Law of Thermodynamics. Note that Carnot erroneously assumed that the same caloric (heat) passes through the engine and extracts (produces) work by lowering its temperature, similar to how the same water flow passes through the water-wheel and produces work by lowering its elevation potential. This error, considering the knowledge at the time, in no way diminishes Carnot's ingenious reasoning and conclusions about limiting, reversible processes and its accurate limitations of heat to work conversion [5].

The consequence of a process and cycle reversibility is most ingenious and far-reaching, see elsewhere [5]. Carnot's simple and logical reasoning that mechanical work is extracted in heat engine due to the caloric (heat) passing from high to low temperature, led him to a very logical conclusion that any heat transfer from higher to lower temperature (like in a heat exchanger), without extracting possible work (like in a reversible heat engine), will be an irreversible loss of work potential [4,5]. Then he expended his logical reasoning to conclude that all reversible (ideal) heat engines, working between the same heat reservoirs, must have equal and maximum possible efficiency. Otherwise, if one with smaller efficiency is reversed and coupled with another, the impossible "perpetual motion" will be achieved. The maximum engine efficiency dependence on the reservoirs' temperatures only, is functionally expressed by Equation (1):

$$W_{Carnot} = W_{netOUT} = Q_{IN} \cdot f_C(T_H, T_L); \;\; \eta_C = \left.\frac{W_{netOUT}}{Q_{IN}}\right|_{Max} = \left.\underbrace{f_C(T_H, T_L)}_{\text{Qualitative function}}\right|_{\text{Rev.}} \tag{1}$$

What a simple and logical, ingenious reasoning! The maximum, limiting efficiency of heat engine does not depend on the medium used in the engine or its design, but only depends on (and increases with) the temperature difference between the heat source and heat sink, similar to the water wheel output dependence on the waterfall height difference.

After the First Law of energy conservation was later established (after Sadi Carnot's death), it was simple to derive the Carnot cycle efficiency, $\eta = (Q_H - Q_L)/Q_H = [f(T_H) - f(T_L)]/f(T_H)$, that led to a very important correlation deduced for a reversible Carnot cycle between any two temperature levels, say T_1 and T_2, i.e.,

$$\frac{Q(T_1)}{Q(T_2)} = \frac{Q_{\text{Ref}} \cdot f(T_1)}{Q_{\text{Ref}} \cdot f(T_2)} = \left.\frac{f(T_1)}{f(T_2)}\right|_{f(T)=\theta=T} = \frac{T_1}{T_2} = \frac{Q_1}{Q_2}, \text{i.e.}, \frac{Q}{T} = const \tag{2}$$

The above function f(T) could be arbitrary, but non-negative and increasing function of temperature, thus defining an absolute, thermodynamic temperature scale, say $f(T) = \theta = T \geq 0$ (for simplicity, but could be another non-negative and increasing function), independent of the system used for measurement (i.e., thermometer). This simple, thermodynamic temperature scale happened to be the same as the absolute temperature scale obtained using an ideal gas thermometer with constant specific heat [5].

The Carnot ratio equality, Q/T = constant (named in Carnot's honor, Equation (2)), is much more important than what it appears at first. Actually, the "Carnot Ratio Equality" is probably the most important equation in Thermodynamics and among the most important equations in natural sciences. Carnot's ingenious reasoning opened the way to the generalization of Thermodynamic reversibility and energy process-equivalency, definition of absolute thermodynamic temperature, and a new thermodynamic material property "Entropy", as well as the Gibbs Free Energy, one of the most important Thermodynamic functions for the characterization of electro-chemical systems and their equilibriums, thus resulting in the formulation of the universal and far-reaching Second Law of thermodynamics [5,6].

CLAUSIUS EQUALITY (*ENTROPY*) AND INEQUALITY (*ENTROPY GENERATION*)

Another important consequence of the Carnot ratio

equality, Equation (2), for a Carnot cycle working between the two different, but constant temperature thermal-reservoirs, T_H and $T_L < T_H$, is:

$$\frac{Q_{H,IN}}{T_H} - \frac{Q_{L,OUT}}{T_L} = \frac{Q(T_H)}{T_H} + \frac{Q(T_L)}{T_L} = 0; \quad \text{where } Q(T) = Q_{IN} = -Q_{OUT} \tag{3}$$

Or in general, a reversible cycle, working between the variable temperature thermal-reservoirs, see Figure 1, could be accomplished with an infinite number of coupled, infinitesimally small Carnot cycles, each working between two temperature reservoirs at specified temperatures and finite temperature difference. After the cycle-integration of Equation (3), for infinitely many Carnot cycles of infinitesimally small heat transfer δQ, we obtained the following equation:

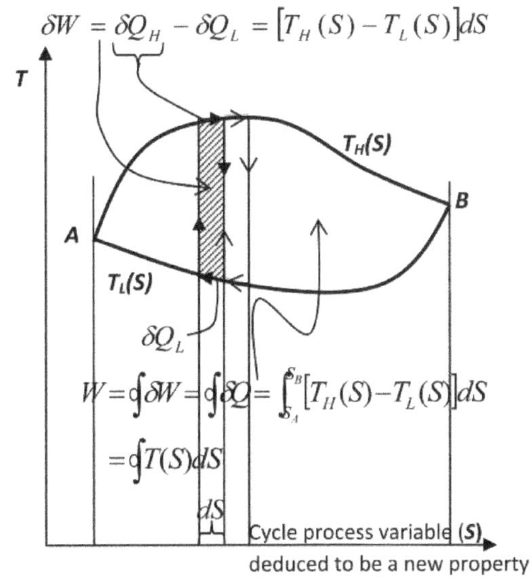

Figure 1. Variable temperature reservoirs require multi-stage Carnot cycles.

$$\int_A^B \left(\frac{\delta Q_{H,IN}}{T_H} - \frac{\delta Q_{L,OUT}}{T_L} \right) = \int_A^B \left(\frac{\delta Q(T_H)}{T_H} + \frac{\delta Q(T_L)}{T_L} \right) = \oint \frac{\delta Q(T)}{T} = 0 \tag{4}$$

Equation (4) is well-known Clausius equality (more about Clausius inequality later), which defines new fundamental property entropy, since its integral is independent on process path between any two given points A and B, namely (compare with Equation (4)):

$$\oint_{Cycle} \frac{\delta Q}{T} = \underbrace{\int_A^B \frac{\delta Q(T_H = T_{any})}{T_H}\Bigg|_{\mathrm{Rev}}}_{\text{Any Rev. Path}} + \underbrace{\int_B^A \frac{\delta Q(T_{L,as\ given})}{T_L}\Bigg|_{\mathrm{Rev}}}_{\text{Fixed Pat}} = \underbrace{\int_A^B \frac{\delta Q(T_H = T_{any})}{T_{H,Any}}\Bigg|_{\mathrm{Rev}}}_{\text{Any Rev. Path}} - \underbrace{\int_A^B \frac{\delta Q(T_{L,as\ given})}{T_{L,given}}\Bigg|_{\mathrm{Rev}}}_{\text{Fixed Pat}} = 0$$

$$S_B - S_A = \underbrace{\int_A^B \frac{\delta Q(T_H = T_{any})}{T_{Any}}}_{\text{Any Rev. Path}}\Bigg|_{\mathrm{Rev}} = \underbrace{\int_A^B \frac{\delta Q(T_{L,as\ given})}{T_{as\ given}}\Bigg|_{\mathrm{Rev}}}_{\text{Fixed Ref. Path}} = \underbrace{const\ \{\text{independent of path } Q(T = T_{any})\}}_{\text{Thus System Property}}$$

$$S_{B|Any} = S_{A|\mathrm{Ref}} + \underbrace{\int_{A|\mathrm{Ref}}^{B|Any} \frac{\delta Q(T_{Any})}{T_{Any}}\Bigg|_{\mathrm{Rev}}}_{\text{Any Path (Reversible)}} \tag{5}$$

Let us reiterate: The reversible process/cycle equivalency, deduced by Sadi Carnot, has resulted in the Clausius equality (Equation (4)). Efficiency of any reversible cycle between any two given reservoirs' temperatures cannot be smaller than any other cycle efficiency. Thus, all reversible cycle efficiencies, between the two thermal reservoirs, must be the same and maximum possible. This subtle reasoning allowed for definition of the absolute Thermodynamic temperature, and deduction, and thus proof based on Carnot's reasoning, of the Clausius equality, Equation (4), and the definition of the new property, entropy, up to an arbitrary reference value, Equation (5).

An additional consequence of Carnot's reasoning is that non-reversible (irreversible) cycles must have smaller efficiency than maximum possible reversible efficiency. Otherwise, if equal they would be reversible, and if bigger would be impossible, hence, allowing deduction (and thus proof) of the Clausius inequality based on the same Carnot's subtle reasoning. We have to keep in mind that Carnot's reasoning is based on the impossibility of making an autonomous machine (to spontaneously produce work on its own), which would be equivalent of having higher efficiency of such a cycle than a related reversible cycle, thus destroying entropy. I in turn, it is equivalent to the deduced Clausius inequality (see next), the latter in limit being an equality for reversible cycles.

Namely, as reasoned by Carnot and reemphasized above and elsewhere [5], for real irreversible (Irr) cycles, the heat engine efficiency is smaller than for reversible cycles (Rev, otherwise the Irr cycle will be reversible or impossible), i.e., for everything else being the same, then $W_{\mathrm{Irr}} < W_{\mathrm{Rev}}$. Therefore, we can reason the proof of the well-known Clausius inequality as follows:

$$\left(W_{Irr} = \oint \delta Q_{Irr} \right) < \left(\oint \delta Q_{\mathrm{Rev}} = W_{\mathrm{Rev}} \right) \tag{6}$$

Considering the inequality, Equation (6), and expending on derivation of Clausius equality, Equation (4) (recall that T is non-negative absolute temperature), we have the following:

$$\oint \frac{\delta Q}{T}\bigg|_{Irr} < \underbrace{\left(\oint \frac{\delta Q}{T}\bigg|_{\mathrm{Re}\,v} = 0\right)}_{Eq.(4)},$$

or for both, reversible and irreversible cycles:

$$\oint \frac{\delta Q}{T} \leq 0 \qquad (7)$$

Everything reasoned above is deduced from the impossibility to have a cycle with higher efficiency than a reversible cycle, and thus equivalent to "impossibility of heat to be transferred spontaneously (without any external influence) from colder to warmer body/reservoir, nor it is possible to construct any device to achieve it in an autonomous process (a process without any external influences)." A new material property, entropy, was defined using the cyclic Clausius equality, Equations (4)and (5) [6].

The cyclic Clausius inequality, Equation (7), affirms that cycle net-influx of the new quantity, entropy, within a cyclic process must be negative (i.e., must be net-outflux), since the working medium comes back to the same state (and thus the same all properties) after completing the cycle. This implies that all real, irreversible cyclic processes must generate (produce) the new property entropy, equal to the generated heat from lost cycle work potential per relevant absolute temperature, which in limit is zero, thus entropy being conserved in reversible processes. Therefore, it would be impossible to have a cyclic process that destroy entropy, since it would be equivalent to the spontaneous heat transfer from a colder to a hotter body against a thermal forcing, or to produce work from heat within a single reservoir, from within an equilibrium without a due forcing. The 'process forcing' has to be 'uniquely directional,' and so is the process energy transfer. It cannot be one way or the opposite way at will, thus defying directionality of forcing and existence of stable equilibrium. Similar reasoning has been further extended to all types of energy processes and thus establishing universality of the Second Law of entropy generation and energy degradation.

PHYSICAL MEANING OF ENTROPY: THERMAL DISPLACEMENT OF THERMAL ENERGY (DYNAMIC THERMAL-VOLUME)

During reversible, boundary heat transfer δQ_{Bry} without work interactions, thus no volume (V) change for simple compressible substance system (Sys), the system temperature within a volume will be uniform and equal to its boundary (Bry) temperature at any instant ($T_{Bry} = T_{Sys} = T$), i.e.,:

$$\begin{gathered}(\delta Q_{Bry} = T_{Bry}\delta S_{Bry})_{\mathrm{Re}\,v} = (T_{Sys}dS_{Sys} = dU_{Sys})\big|_V \text{ or } \delta Q_{\mathrm{Re}\,v} = TdS = dU \\ (\delta S_{Bry})_{\mathrm{Re}\,v} = (dS_{Sys})_{\mathrm{Re}\,v} = dS \text{ since } T_{Bry} = T_{Sys} = T \text{ [for Reversible Process]} \\ \text{However, } (\delta S_{Bry})_{Irr} \neq (dS_{Sys})_{Irr} \text{ since } T_{Bry} \neq T_{Sys} \text{ [for Irreversible Process]}\end{gathered} \qquad (8)$$

There is a tacit distinction between entropy transfer across the boundary surface δS_{Bry} (flux quantity $\delta S_{Bry} = \dot{S}_{Bry}dt$) and change (e.g., storage) of entropy property within the material system dS_{Sys} (in time dt) that occupies space volume V, and the two are the same only for reversible processes, Equation (8). We also note that process irreversibility is associated with a material system occupying a volume (it is a volumetric phenomenon) and could not be defined for an energy flux crossing a system boundary having no thickness, thus the distinction between differentials 'δ' for flux and 'd' for system.

There is an important peculiarity about heat transfer processes without any work interactions (like within heat exchangers): no heat conversion to work like in heat engine, and no other than 'thermal work potential' dissipation and heat generation, therefore the thermal energy (like original caloric) is conserved on its own. We like to name such processes, without work interactions, as "caloric processes" or caloric heat transfer. We also could define "reversible heat transfer" at infinitesimally small temperature difference, as an ideal limiting case, achieved by an ideal Carnot cycle, when the heat source and sink are at a finite temperature difference, so that work potential is extracted, instead of being dissipated as heat, like in the above caloric processes, while heat transfer takes place at infinitesimally small temperature difference at each temperature level (dT→0) [7]. Therefore, during the reversible heat transfer from higher to lower temperature reservoirs, seeFigure 2 (left), the work potential is extracted and less heat is reversibly transferred ($Q_{\mathrm{Re}\,v,2} = Q_1T_2/T_1$) to the reservoir at lower temperature (consider it to be a system, $T = T_{Sys} = T_2$), for the amount of the extracted work (as per the Carnot ratio or Carnot efficiency, and energy conservation), so the entropy is conserved since no entropy was generated in the reversible cycle (note $Q_{\mathrm{Re}\,v}/T$ = constant).

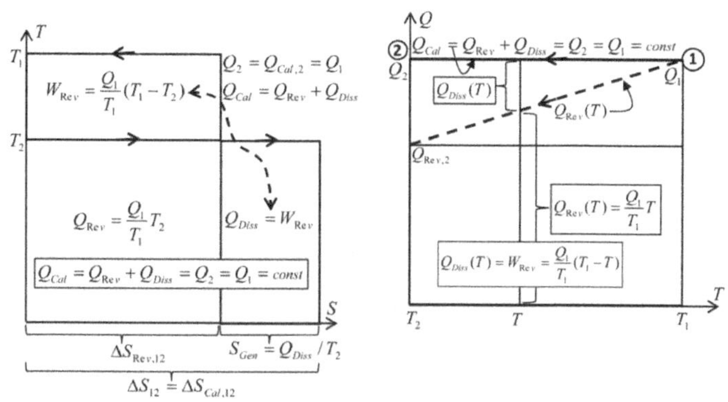

Figure 2. During a caloric heat transfer process the work potential, $W_{\mathrm{Re}\,v}$, is completely dissipated into dissipated/generated heat at a lower temperature, Q_{Diss}, which after being added to the reduced reversible heat at lower temperature, $Q_{\mathrm{Re}\,v}$, will result in conserved heat or thermal energy, $Q_{Cal} = Q_{\mathrm{Re}\,v} + Q_{Diss}$, with increased, generated entropy in the amount of dissipated work potential per relevant absolute temperature. On the right, variable system temperature $T_2 < T < T_1$ is presented.

However, during a caloric process the work potential is completely dissipated into self-generated heat at lower temperature ($Q_{Diss} = W_{Rev}$), which after being added to the equally reduced reversible heat at lower temperature ($Q_{Re\ v}$), will result in conserved heat or thermal energy ($Q_{Cal} = Q_{Re\ v} + Q_{Diss} = const$; i.e., caloric process) with increased, generated entropy in the amount of dissipated work potential per relevant absolute temperature, see Figure 2 (left).

Considering the fact that entropy is defined for reversible processes, as per the Clausius equality, it may be deduced that system entropy increase due to irreversible heat generation, is the same as if it is transferred from an internal (within) boundary, the same as entropy increase during corresponding, reversible heat transfer. Therefore, for irreversible caloric processes (without external work interactions as defined above), the entropy may be conveniently measured and calculated in the same manner as for reversible heat transfer, i.e.,

$$\Delta S = \int_{Boundary} \frac{\delta Q_{Rev}}{T_{Bry}} \text{ but also } \Delta S = \int_{System} \frac{\delta Q_{Cal}}{T_{Sys}} = \int_{System} \frac{C_{V,Sys} dT_{Sys}}{T_{Sys}} \tag{9}$$

During caloric heating (i.e., a system isochoric process without work interactions), the system thermal or heat capacity, $C_{V,Sys}$, will govern its temperature and entropy increase, regardless of the heat source temperature, the latter if higher, will cause the lesser source entropy change, resulting in source-system process irreversibility with entropy generation.

Furthermore, based on the above reasoning, since any dissipated work type may be considered as a (generated) heat source for the system, the Clausius entropy definition could be extended and generalized for any process, including work interactions, as follows:

$$\Delta S = \int_{Boundary} \frac{\delta Q_{Bry}}{T_{Bry}} + \int_{System} \frac{\delta W_{Loss}}{T_{Sys}} = \overbrace{\int_{Boundary} \frac{\delta Q_{Bry}}{T_{Bry}}}^{S_{Transfered}} + \overbrace{\int_{System} \frac{\delta Q_{Gen}}{T_{Sys}}}^{S_{Generated}} \tag{10}$$

Note that $Q_{Gen} = Q_{Diss} = W_{Loss}$ also includes the Carnot work-potential loss during 'caloric' heat transfer at finite temperature difference, so the boundary entropy transfer could be 'considered to be reversible' as in reversible processes (no 'volumetric-process' entropy generation through a boundary surface).

We could further generalize the Clausius entropy definition so that entropy change is equal to all thermal heat stored within the system, $dQ_{Sys} = \delta Q_{Bry(Re\ v)} + \delta Q_{Gen}$ per relevant, absolute system temperature T_{Sys} (at location where the heat is stored within a system), since it is irrelevant if entropy is changed due to reversible heat transfer or irreversible heat generation, i.e.,

$$\Delta S = \Delta S_{Sys} = \int_{System} \frac{dQ_{Sys}}{T_{Sys}} \tag{11}$$

Note that for reversible processes, $W_{Loss} = Q_{Gen} = 0$, then $Q_{Sys} = Q_{Bry} = Q_{Rev}$, and for caloric processes $Q_{Sys} = Q_{Cal}$. Furthermore, for reversible adiabatic process, $Q_{Sys} = Q_{Rev} + Q_{Gen} = 0$, thus isentropic process. However, for irreversible adiabatic process, there will be irreversible loss of work potential within (instead of being extracted or stored) which will dissipate and convert to heat, i.e., generate heat (and thus entropy) within, and be stored within the system, $Q_{Sys} = W_{Loss} = Q_{Gen} > 0$, thus resulting in irreversible entropy increase (Equation (11)). Accounting for all irreversibly-generated heat may be elaborate, except for reversible processes (being zero) and caloric processes (caloric heat conserved), Equation (9).

Entropy is associated with the system heat stored and not with the work transferred, only with the work lost within the system, and thus generated heat, as if it is transferred from some internal, thermal boundary within. As already stated, it is irrelevant if the heat is stored (as thermal energy) within the system by (reversible) boundary heat transfer or generated within.

For compressible ideal gas (PV = mRT) with constant specific heats C_V and $C_P = kC_V$, and the gas constant $R = C_P - C_V$, the entropy change, and thus entropy, could be determined from the Clausius definition and the First Law of energy conservation within a reference (Ref) constant, i.e.,

$$dS = m\left(C_V \frac{dT}{T} + R\frac{dV}{V}\right) = m\left(C_P \frac{dT}{T} - R\frac{dP}{P}\right)$$
$$S(T,V) - S_{Ref} = m\left[C_V \ln(T/T_{Ref}) + R\ln(V/V_{Ref})\right] \tag{12}$$

As seen from Equation (12), entropy increases with an increase of temperature and volume, thus it could be considered as "thermal dynamic-volume" or general "thermal displacement." However, during the reversible change of volume due to work transfer, temperature will change accordingly, and there will be no change of entropy, since the latter is not associated with work, but only due to heat transfer, if any. In Statistical Thermodynamics it is shown that entropy could be derived (for simple systems like mono-atomic ideal gas) as a function of thermal-motion momentum (thus function of temperature) and space position (thus function of volume) of microscopic particles of the material system.

Therefore, $dQ_{Sys} = T_{Sys} dS_{Sys} = dU_{Thermal}$, is valid in general, and it demonstrates that entropy is a thermal displacement (dynamic thermal-volume) of thermal energy, i.e., "system heat," due to temperature as a thermal potential. It is similar to physical (mechanical) volume being the displacement for mechanical energy, $dW_{Sys(IN)} = P_{Sys}(-dV_{Sys}) = -dW_{Sys(OUT)} = -P_{Sys} dV_{Sys} = dU_{Mechanical}$, due to pressure as a mechanical (elastic) potential. Thus, the differential change of internal energy, $dU = [(TdS) + (-PdV)]_{Sys} = dU_{Thermal} + dU_{Mechanical}$, to be elaborated in another manuscript, may be resolved into thermal and mechanical components. Note that change of temperature alone (without entropy change) will not cause the heat transfer (as during isentropic expansion when mechanical energy is extracted, the latter not related to entropy); the same way, the change of force (pressure) without displacement (volume change), will not cause the work transfer. The stored system heat increases the system thermal energy that is distinguished from the system internal, mechanical (elastic) energy, which will be a subject of another manuscript. For example, the heating or compressing an ideal gas with the same amount of energy will result in the same temperature and internal energies, but different states, with different volumes and entropies, and similar for other material substances.

In conclusion, entropy is a thermal displacement, an integral measure of (random) thermal energy redistribution (due to heat transfer and/or irreversible heat generation due to energy degradation) within a system mass and space, per absolute temperature level: $dS = dS_{Sys} = dQ_{Sys}/T_{Sys}$ with J/K unit. Note that adjective 'thermal' is critically important, since similar but non-thermal phenomena are not related to thermodynamic entropy.

PHYSICAL MEANING OF ENTROPY: A MEASURE OF THERMAL DISORDER

Statistical Thermodynamics deals with microscopic structures and dynamic interactions of microparticles with an objective to describe thermo-physical properties of a classical system as a statistical average of all possible and relevant microstates, described by positions and momenta of microparticles. In principle, we can reason and derive Classical Thermodynamics from relevant statistics of microstates' dynamics, down to quantum behavior of molecules and atoms. However, in reality, there are many obstacles, ambiguities and limitations due to complex microstructure and interactions at many different micro-, nano-, subatomic- and quantum- levels. Note that the statistical results are feasible for very few simple and idealized systems.

As already stated, the thermal random, fluctuation-motion of microparticles gives rise to thermal macro-properties, like temperature, thermal energy, heat capacity, and entropy, among others. Temperature is a measure of average kinetic energy of relevant microparticles, or the kinetic energy of thermal motion of a representative microparticle, while the thermal energy is the sum of all kinetic and potential energies related to the thermal motion of all microparticles, which also give rise to thermal heat capacity. When thermal motion ceases, virtually all thermal properties vanish at absolute zero temperature.

The entropy may also be expressed as a logarithmic measure of all possible thermal-microstates, W, that "correspond to orare consistent with" a given system macro-state. Thus, all real complexity is transferred to properly "counting" microstates W, to match the bulk entropy macro-state, the thermal displacement property, $dS = dQ_{Sys}/T = mC_{Sys}dT/T$, thus logarithmic integral function, with J/K unit, as accurately deduced by Clausius and precisely measured in laboratories for any system of interest. Since, for the classical systems (as continuum media), the number of microstates is actually infinite, thus uncountable, there is a need for proper discretization (discrete ranging/grouping) of the microstates to obtain a proper, countable set, e.g., if the positions (x) and momenta (p) of ideal gas molecules are within certain, relevant δx and δp ranges of each others, respectively.

Boltzmann first hypothesized that entropy is related to the probability of microstates corresponding to related macro-state. Regardless of what Boltzmann thought at the time, the well-known correlation:

$$S = k \log W \quad (\text{Note } k = k_B) \qquad (13)$$

was formalized later by Plank and inscribed on the Boltzmann's tombstone in his honor. The value of W, originally intended to be proportional to the Wahrscheinlichkeit (the German word for probability), reduces to "a metaphorical" number of all possible, relevant molecular microstates, W, corresponding to a given bulk macro-state. The k_B is a constant with entropy unit named after Boltzmann.

In Statistical Thermodynamics, "entropy" is also expressed as uncertainty of all molecular positions due to their thermal random motion (thermal randomness, thermal fluctuations, or thermal disorder). The adjective 'thermal' is critically important, but often overlooked in the literature. At absolute zero temperature the thermal motion ceases and the position of the molecules are fixed/certain in pure crystalline substance, defining the absolute zero entropy (the Third Law of Thermodynamics). For not purely crystalline substance (as 'randomly frozen' solid solution structures), the positions of the molecules are not uniquely determined and entropy is not exactly zero at absolute temperature (but some residual value); however, no uncertainty due to thermal motion.

The thermal-motion randomness or thermal disorder, as related to the entropy, was generalized by Gibbs as "a (logarithmic) measure" of sum of all microstates' uncertainties, i.e., probabilities p_i, i.e.,

$$S = -k_B \sum p_i \log_e(p_i) \qquad (14)$$

Gibbs' formulation is more general since it allows for non-uniform probability, p_i, of the microstates.

With an increase of thermal disorder, i.e., an increase of microparticle momenta (temperature) and space they occupy (volume), more information is required for its characterization than in more ordered systems, which could be described via related probabilities of microstates, thus generalizing disorder as Shanon's information entropy. However, the Gibbs or von Neumann quantum (based on the number of possible quantum microstates), or Shanon or other probabilistic entropy descriptions, are also statistical as Boltzmann's. Actually they all reduce to the Boltzmann expression for fully-randomized, with uniform probability of all W microstates, where $p_i = 1/W = \text{constant} << 1$, thus simply reducing Equation (14) to Equation (13).

Again, a statistical interpretation is important as a metaphor only: The sum of the logarithmic function of probabilities of all possible discrete microstates, p_i's, that could occur during the "random thermal fluctuations" of a given macrostate. The adjectives: possible, could occur, consistent random fluctuations (thus thermal), and the holistic meaning of the statement, have deep meanings, and could not be evaluated for any real system, but only scaled for the trivial ones. Entropy is generated when work potential is lost, i.e., dissipated (randomly equi-partitioned) via heat generation into the thermal energy at a given absolute temperature within the space occupied by the system, including caloric processes, and when expansion (elastic) work potential is lost, i.e., energy randomly redistributed within enlarged volume instead of being extracted as work (volume displacement against the surrounding equilibrium pressure), as demonstrated by the entropy dependence on temperature and volume for an ideal gas of N microparticles [$R = C_P - C_V = C_V(k-1) = k_B N/m$]:

$$dS = m\left(C_V \frac{dT}{T} + R\frac{dV}{V}\right) = m\left(C_P \frac{dT}{T} - R\frac{dP}{P}\right)$$
$$S(T,V) - S_{Ref} = m\left[C_V \ln(T/T_{Ref}) + R\ln(V/V_{Ref})\right]$$
$$= k_B \ln[(V/V_{Ref})(T/T_{Ref})^{1/(k-1)}]^N = k_B \ln W - S_{Ref} \quad (15)$$

Then:

$$W/W_{Ref} = [(V/V_{Ref})(T/T_{Ref})^{1/(k-1)}]^N = [(V^* T^{*1/(k-1)}]^N \quad (16)$$

Also, for an isochoric process:

$S = k_B \ln W$, then $dS = k_B d(\ln W) = k_B dW/W = mC_V(dT/T)_V$, then

$$\frac{dW}{W} = \frac{mC_V}{k_B}\frac{dT}{T}\bigg|_V = \frac{NC_V}{R}\frac{dT}{T}\bigg|_V \quad (17)$$

The above (Equations (16) and (17)) correlate the statistical number of microstates W with temperature T and volume V, macro-properties for ideal gas. Granted, there are certain benefits of simplified statistical descriptions to better comprehend the randomness of thermal motion and related physical quantities, but the limitations should be stated so the generalizations are not overstretched and the real physics overlooked, or worse discredited.

CONCLUSIONS: PHYSICAL MEANING OF ENTROPY

As already stated, the elusive nature of entropy is due to the elusive nature of heat and related thermal motion of the material structure, the latter also coupled with other than thermal phenomena, and the conversion of other energy types to thermal heat. The uniqueness and universality of entropy come from the fact that all processes in the universe, at all time and space scales are due to forced displacement of mass-energy, as the "cause-and-effect" of directional forcing of energy transfer (as driving force for all processes) from higher to lower energy density/potential. Therefore, the spontaneous processes are due to work-potential forcing, accompanied with its dissipation to heat and entropy generation, and thus the work potential reduction, in limit to zero when equilibrium is reached.

In general, for all processes, with different functional, natural or man-made material structures, including functional devices, the work potential (a measure of non-equilibrium) could be transferred in different ways, as forced energy displacement, to accomplish different functions and objectives, including life processes and evolution. The forced energy displacement is always accompanied with dissipation of any and all energy types to heat or thermal energy (including thermal energy to itself as transferred from higher to lower temperature) and thus contributes to the irreversible entropy generation at all space and time scales. The thermal macro-phenomena and related properties, including entropy, could be defined as statistical averages (of enough-large number) of micro-particles. Note that on an individual microscale the "Thermodynamics becomes Dynamics." In ideal, reversible processes the entropy is conserved, but there is no way to destroy entropy, even when heat is converted to work since entropy is not associated with work. Destruction of entropy would be equivalent to spontaneous heat transfer from lower to higher temperature, or to have heat engine with higher efficiency than ideal, reversible Carnot cycle, against the directional, thermal forcing and against the equilibrium existence in nature [5].

An individual microparticle alone (an atom or a molecule, etc.) does not have a certain temperature or possesses a defined amount of heat or thermal energy, nor a specified amount of entropy (all being thermal macro-properties of larger assembly of micro-particles), unless a microparticle is part of thermal, random chaotic interactions of large number of microparticles. On a small individual scale (without interactions with other particles), a microparticle entropy is non-existent, thus zero. However, within a collective larger structure with thermal random interactions/collisions, its entropy increases with spatial and other types of displacements, in limit to infinity (thus relativity of zero and infinity). These, thermal parameters characterize the collective physical state of atoms and molecules, be they solids, liquids or gases. The temperature, heat stored or thermal energy, and entropy are macroscopic properties reflecting the random (thermal) collective behavior of the constituents within a system. While temperature represents the average kinetic energy of all the system's particles (i.e., the kinetic energy of a representative particle), the stored heat, or thermal energy represents the total amount of kinetic- and potential thermal-energy possessed by all the particles in the system. Similarly, entropy quantifies chaotic displacements of thermal disorder within a system - the displacement of randomly disordered energy, i.e., thermal energy without a defined, but chaotic structure. Note again that adjective 'thermal' is critically important, but often overlooked! Moreover, the intermolecular phenomena are coupled and thus elusive. In addition to thermal interactions, giving rise to temperature, heat and entropy, the nature and strength of mutual attractions between atoms and molecules in solids and liquids give rise to other macroscopic properties such as hardness, elasticity and tensile strength in solids, viscosity in liquids, and similar.

Entropy is a thermal displacement (dynamic thermal-volume) of thermal energy due to temperature as a thermal potential ($dQ = TdS$), and thus associated with thermal heat and absolute temperature, i.e., distribution of thermal energy within thermal (randomly moving) micro-particles in space. It is an integral measure of (random) thermal energy redistribution (due to heat transfer and/or irreversible heat generation due to energy degradation, the latter being multifaceted and somewhat elusive) within a material system structure in space, per absolute temperature level: $dS = dQ_{Sys}/T = mC_{Sys}dT/T$with J/K unit. Entropy may be also expressed as a measure of thermal disorder, being related to logarithm of number of all thermal, dynamic microstates W (based on microparticles' position and momenta), $S = k_B \ln W$, or to the weighted sum of their logarithmic probabilities $S = -k_B \sum p_i \ln p_i$, that correspond to, or are consistent with the given thermodynamic macro-state. Note that the meanings of "all relevant adjectives" are deeply important to reflect reality, and as such, "disorder" has metaphoric description for real systems.

Some like to generalize the entropy concept as being a "measure of any disorder," and imply its destruction when order is created, as in evolution of life. However, during any process, natural or man-made, of creating or destroying any order or disorder, i.e., transforming any material structure, work potential will be always dissipated into heat, and entropy will always be produced, i.e., generated.

Expanding classical entropy concept to other types of disorder or information is 'overreaching' and could be a source of many misconceptions. A system form and/or functional order or disorder is not (thermal) energy order/disorder, and the former is not related to Thermodynamic entropy, see Figure 3.

Figure 3. Entropy refers to dynamic thermal-disorder of its micro-structure, which gives rise to temperature, heat capacity, entropy and thermal energy. It does not refer to form- nor functional-disorder: The same ordered bricks (on **left**) or randomly piled bricks (on **right**), if at the same temperature, density and mass (made of the same material), will have the same Thermodynamic entropy, regardless of its macro-configuration or other than thermal type of disorder.

Entropy is increasing from perfectly-ordered (uniquely defined) crystalline structure at zero absolute temperature (absolute zero reference) during the reversible heating (entropy transfer) and/or entropy generation during irreversible energy conversion (loss of work-potential to generated thermal energy), i.e., energy degradation or random equi-partition within a system material structure and space, per absolute temperature level. Furthermore, entropy of a system for a given state is the same, regardless whether it is reached by reversible heat transfer or irreversible heat or irreversible work dissipation.

The fundamental laws and physical phenomena could not be 'governed' by mathematical modeling and calculation outcomes as suggested by some [2,3], but the other way around. The energy, temperature, and especially entropy are subtle and elusive, but well-defined and precisely measured as physical quantities, and used as such. Any new approach should be correlated with existing knowledge, with the limitations clearly and objectively presented.

The phenomenological thermodynamics has the supremacy due to logical and holistic reasoning based on the fundamental laws and without regard to the systems' complex dynamic structures and even more complex interactions; thus its importance and supremacy should not be marginalized.

Nothing occurs locally, nor globally in the universe, without mass-energy exchange and entropy production/generation. The miracles are until we comprehend and explain them!

REFERENCES

1. Romer, R.H. Heat is not a noun. Am. J. Phys **2001**, 69, 107.
2. Ben-Naim, A. Farewell to Entropy: Statistical Thermodynamics Based on Information; World Scientific: Singapore, Singapore, 2008.
3. Leff, H.S. Removing the mystery of entropy and thermodynamics—Part II. Phys. Teach **2012**, 50, 87–90. With comments by Kostic, M., available online: http://goo.gl/mXYGv (accessed on 14 February 2014).
4. Carnot, S. Reflections on the Motive Power of Heat (English translation by R.H. Thurston); Chapman & Hall Ltd.: London, UK, 1897. Available online: http://www.kostic.niu.edu/Sadi-Carnot-Reflections.html (accessed on 14 February 2014).
5. Kostic, M. Revisiting the second law of energy degradation and entropy generation: From Sadi Carnot's ingenious reasoning to holistic generalization. AIP Conf. Proc **2011**, 1411, 327.
6. Clausius, R. The Mechanical Theory of Heat (English translation by Browne, W.R.); McMillan and Co.: London, UK, 1879. Available online: http://www.kostic.niu.edu/Clausius-Mechanical-Theory-of-Heat.html (accessed on 14 February 2014).
7. Kostic, M. Irreversibility and reversible heat transfer: The quest and nature of energy and entropy. Proceedings of the IMECE2004, Anaheim, California, USA, November 13–19, 2004; ASME Proceedings, ASME: New York, NY, USA, 2004.

CHAPTER 2

THERMODYNAMICS OF QUANTUM SYSTEMS WITH MULTIPLE CONSERVED QUANTITIES

Yelena Guryanova, Sandu Popescu, Anthony J. Short, Ralph Silva and Paul Skrzypczyk

Keywords: Physical sciences Theoretical physics

INTRODUCTION

Thermodynamics is one of the most successful theories of nature that we have. Since its inception almost 200 years ago it has survived the transition from classical mechanics to relativistic and quantum mechanics, with its conceptual basis unchanged. The realm of thermodynamics has also been considerably extended, with recent years witnessing the extension of thermodynamics from dealing with macroscopic systems to individual quantum systems and black holes. From its earliest days, thermodynamics was also generalized to deal not only with energy but also with various conserved quantities, introducing grand canonical ensembles, chemical potentials and so on. Again, its conceptual basis remained unchanged.

In more detail, the idea of the grand canonical ensemble, where not only energy, but also the number of particles, is a conserved quantity, goes all the way back to Gibbs[1]. A milestone was the work in 1957 of Jaynes[2,3], who, coming from a Bayesian perspective, suggested the generalization of thermodynamics to arbitrary conserved quantities through the principle of maximum entropy. The idea of the 'generalized Gibbs ensemble' is by now commonly used in quantum statistical mechanics (see for example, refs 4, 5, 6, 7, 8). More recently, refs [9, 10] considered Landauer erasure given access to an 'angular momentum bath' instead of a thermal bath, and demonstrated that information can be erased without an energy cost, provided the analogous cost is paid in angular momentum.

Very recently, there has been much renewed interest in the foundations of thermodynamics coming from the field of quantum information. On the one hand, the so-called 'single-shot information theory', which was developed initially to study finite-size effects in quantum cryptography, has proven useful for studying finite-size effects and fluctuations in quantum thermodynamics, which has lead to the development of 'single-shot quantum thermodynamics'11,12,13,14,15,16,17. On the other hand, inspired by so-called 'resource theories', which have proved to be very powerful for studying quantum information tasks, such as the theory of entanglement[18], purity[19] or asymmetry20, the 'resource theory of quantum thermodynamics'[21] was developed that, in combination with single-shot framework, has generated a lot of interest and already shown itself to be a fruitful approach to quantum th ermodynamics[22,23,24,25,26,27,28,29,30,31,32,33,34,35,36]. Our work proceeds along the lines of single-shot thermodynamics and resource theories, where a number of initial results concerning the thermodynamics of multiple conserved quantities were presented in refs 37, 38.

Here we present a generalized version of thermodynamics that deals with individual quantum systems and multiple—commuting or non-commuting—conserved quantities. What we will show is that unlike standard thermodynamics, where the second law constrains how much of the conserved quantity (energy) can be extracted from a non-equilibrium system, in the form of work, here there is no constraint on the amount of a single conserved quantity that can be extracted. In fact, we can extract as much of any individual conserved quantity as we like, if we supply an appropriate amount of other conserved quantities. What the second law constrains is the combination of conserved quantities that can be extracted—that is, the second law is seen to limit the trade-off of extractable quantities, which for the standard case of a single conserved quantity reduces to constraining how much can be extracted. At the same time, this generalized version of thermodynamics suggests that it may be worthwhile revisiting the basic concepts of the subject. Indeed, to understand the above phenomena we present an alternative viewpoint, which reinterprets some of the standard thermodynamics quantities, and is perhaps more natural and compelling. Closely related, independent work was performed by Lostaglio et al.[39] and Halpern et al.[40]. In contrast

Citation: Guryanova, Y. et al. Thermodynamics of quantum systems with multiple conserved quantities. Nat. Commun.7:12049 doi: 10.1038/ncomms12049 (2016).

to here, these works focus more on the nature of the thermal state itself when there are multiple conserved quantities.

RESULTS

Overview

Here we consider the standard general framework of thermodynamics that consists of a thermal bath, an external system out of equilibrium with respect to the bath and a number of batteries, in which we will store various conserved quantities, which are extracted from the system and bath. In our case, following Jaynes[2,3], we take the 'thermal bath' to be simply a collection of particles, each described by a generalized thermal state

$$\tau = e^{-(\beta_1 A_1 + \ldots + \beta_k A_k)} / \mathcal{Z} \tag{1}$$

where A_i are various conserved quantities, β_i are the associated inverse temperatures and is the generalized partition function. Two things are important to note: first, the quantities A_i may or may not commute, and even when they commute they may or may not be functionally dependent on one another. Second, and most importantly, energy need not be one of the conserved quantities or indeed play any role. Since energy is the generator of time evolution, such a thermal bath may not arise naturally by thermal equilibration, but have to be created externally (for example, if the Hamiltonian is zero, then no evolution occurs). Yet, as we will see, the thermodynamic flavor of the theory remains.

The batteries are systems that can each store one of the conserved quantities A_i. In our paper we will consider the batteries either explicitly or implicitly, as explained later. The system can be an individual quantum particle. Finally, the actions that we allow to be performed must conserve either exactly or on average each of the quantities A_i, which is the content of the first law.

A central result of standard thermodynamics—the content of the second law—is that if we have access only to a single thermal bath, it is impossible to extract energy, in the ordered form of work, out of it, that is $W := \Delta E^{batt} \leq 0$, where ΔE^{batt} is the change in the average energy of the battery. We show that, in our case, there is no limit on how much of any single conserved quantity A_i can be extracted, even though we have access only to a single generalized bath. More precisely, there is no limit on $W_{A_i} := \Delta A_i^{\text{batt}}$. There is, however, a global limit.

In particular, to each conserved quantity we can associate an entropic quantity $\beta_i A_i$ (the entropic nature of this quantity will be explained later). We will show that these quantities can be almost perfectly interconverted for one another inside the bath. As a result, because of the first law (conservation of A_i between bath and battery) the only constraint on the W_{A_i} given just a thermal bath is that

$$\sum_i \beta_i W_{A_i} \leq 0. \tag{2}$$

In standard thermodynamics the second law also says that if we have access to a system out of equilibrium with respect to the bath, then we can extract work, but we are limited by the change in free energy of the system, $W \leq -\Delta F_s$. In our case, we define an entropic quantity, the 'free entropy' of the system relative to the generalized bath, $\tilde{F}_s$,

$$\tilde{F}_s := \sum_i \beta_i \langle A_i^s \rangle - S_s, \tag{3}$$

where S_s is the system entropy and show that

$$\sum_i \beta_i W_{A_i} \leq -\Delta \tilde{F}_s. \tag{4}$$

We will show, with a minimal number of assumptions, that we can implement any trade-offs between conserved quantities satisfying equation (2) using the bath, and extract any combination of W_{A_i} satisfying equation (4) from a system, up to an arbitrarily small deficit because of the finite nature of the protocols. In particular, if all the conserved quantities commute, we will give explicit protocols that works for both implicit and explicit batteries, assuming exact conservation of the A_i. For more general non-commuting quantities, we will obtain the same results for implicit batteries, or explicit batteries with average conservation, but leave open the question of how to deal with strict conservation of non-commuting quantities when considering explicit battery systems.

The Generalized Thermal State

Here we consider in more detail the generalized thermal state given in equation (1)[2,3].

We begin by recalling that there are two ways to define the thermal state—by maximizing the von Neumann entropy, given appropriate constraints, or by minimizing the free energy. We start with the former. Consider a system in state σ with Hamiltonian H and average energy $\langle H \rangle := \text{tr}[H\sigma] = \bar{E}$. There are many states σ that have this particular average energy; the thermal state is the state that maximizes the entropy $S(\sigma) = -\text{tr}(\sigma \ln \sigma)$, subject to the average energy constraint. Solving the maximization problem we get $\tau(\beta) = \frac{1}{\mathcal{Z}} e^{-\beta(\bar{E})H}$, where is the partition function and the inverse temperature β is implicitly determined by the average $\bar{E}$.

In our framework we need the generalization of this idea to the case of multiple conserved quantities. In particular, we consider k quantities $A_i, i \in \{1, \cdots, k\}$ and place no restrictions on the relations between them: they may or may not commute; when they commute they may or may not be functionally dependent on one another. An example of two commuting and functionally dependent quantities are the Hamiltonian H and angular momentum L, where $H = L^2/2I$. In this case the average of one does not uniquely determine the average of the other; however, the range of admissible values is constrained, that is, such that $|\langle L \rangle| \leq \sqrt{2I\langle H \rangle}$. An example of two non-commuting conserved quantities is L_x and L_y. The generalized thermal state $\tau(\beta_1, \ldots, \beta_k)$ is then the state, which maximizes the entropy S subject to the constraint that the conserved quantities A_i have average value $\langle A_i \rangle = \overline{A_i}$. It is found to be

Definition 1. Generalized thermal state

$$\tau(\beta_1, \ldots, \beta_k) = \frac{e^{-\sum_i \beta_i A_i}}{\mathcal{Z}} \tag{5}$$

where, β_i is the inverse temperature conjugate to A_i, and the generalized partition function is $\mathcal{Z} = \text{tr}(e^{-\sum_i \beta_i A_i})$.

Note that, in general, each β_i is a function of all of the averages $\overline{A_i}$. In the case that the A_i commute, the proof is a simple generalization of the standard proof. For non-commuting observables, the proof is more involved[41].

The second way to define the thermal state (when only energy is conserved) is to fix the inverse temperature $\beta = \frac{1}{T}$ and ask for the density matrix that minimizes the free energy $F(\rho) = \langle H \rangle - TS(\rho)$. The state that solves this optimization has exactly the Gibbs form $\tau(\beta) = e^{-\beta H}/\mathcal{Z}$. Since β is given, the average energy is now implicitly defined, in contrast to the case above, where the average energy was given and the inverse temperature derived.

The idea is to do the same in the case of multiple conserved quantities, and recover the generalized thermal state via a generalized free energy. However, in the standard definition of free energy the temperature is the constant multiplying the entropy. Since we have no notion of multiple entropies, we are not afforded a way of coupling all the inverse temperatures. This is easily overcome if instead of the free energy we define $\tilde{F}(\rho) = \beta\langle H \rangle - S(\rho)$, and it is trivial to generalize this quantity to the case of multiple conserved observables.

Definition 2. Free Entropy. The free entropy of a system ρ is

$$\tilde{F}(\rho) = \sum_i \beta_i \langle A_i \rangle - S(\rho), \tag{6}$$

The free entropy is always defined with respect to a set of inverse temperatures β_i. The generalized thermal state is then the state that minimizes $\tilde{F}$ with fixed β_i.

Conceptual Viewpoint

As noted in the introduction, the effects presented in this paper suggest that it may be worthwhile to revisit the basic concepts of thermodynamics. A key aspect of this is the conceptual shift from the free energy to the free entropy.

First, we would like to emphasize that the change from the usual free energy to the free entropy is not a simple mathematical manipulation, but marks a fundamental conceptual difference. Indeed, in the standard approach to considering multiple conserved quantities, such as when considering the grand canonical ensemble, one introduces the chemical potential μ such that the free energy becomes

$$F(\rho) = \langle H \rangle + \mu \langle N \rangle - TS(\rho) \tag{7}$$

where N is the particle number operator. In this way, energy is singled out as the privileged quantity, with the chemical potential acting as the 'exchange rate' between particle number and energy (and in the same way temperature acts as the exchange rate between entropy and energy). We argue that there is no reason to single out the energy, or any other quantity for that matter. In fact, it is possible to conceive of situations in which everything is degenerate in energy, and thus where energy plays absolutely no role. We are thus lead to introduce the free entropy, which naturally and uniquely treats all quantities on an equal footing.

A second argument for considering free entropy over the free energy is that the latter might give one incorrect intuition. Indeed, in the standard treatment, the free energy puts bounds on how much of the conserved quantity (energy) can be extracted, and one may be tempted to think that even when we have multiple conserved quantities, thermodynamics is about the bounds that constrain the extraction of individual quantities. However, as we will show, this is not the case, and there are no such bounds. The only limitation is on the trade-off between the conserved quantities, and this is precisely governed by the free entropy. It is only in the standard case of a single conserved quantity that one can choose to consider the free energy, or the free entropy, with both constraining the amount of work that can be extracted.

We also note that the thermal state is the state that minimizes the free energy only when the temperature is positive; if the temperature is negative the thermal state (at negative temperature) is instead the state that maximizes the free energy. On the other hand, for all temperatures (positive or negative), the thermal state always minimizes the free entropy.

Finally, we note that it is the difference in free entropy between ρ_s and τ_s that captures the thermodynamic usefulness of the system. This difference is exactly equal to the relative entropy between these two states, $\tilde{F}(\rho_s) - \tilde{F}(\tau_s) = D(\rho_s \| \tau_s)$, where $D(\rho \| \sigma) = \text{tr}(\rho \log \rho - \rho \log \sigma)$. This is demonstrated explicitly in the Methods. This highlights the entropic nature of $\tilde{F}$.

Following on from the introduction of the free entropy, one can go a step further. Since the quantities $\beta_i \langle A_i \rangle$ all appear alongside the entropy in the definition of the free entropy, it suggests that they might be thought of as 'entropic quantities'. Note that this is true even in the standard case, where energy is the only conserved quantity; there one might think of $\beta\langle H \rangle$ as an entropic quantity.

Importantly, the sum of these entropic quantities of the batteries is the object onto which the second law of thermodynamics applies. It says that the increase in this sum is constrained by the decrease in the free entropy of the system relative to the bath equation (4).

The Set-Up

The set-up is similar in spirit to that of previous lines of work[25,42]. We consider the interaction of generalized thermal baths with quantum systems and batteries (Fig. 1). There are a number of conserved quantities, A_1 to A_k, which may or may not commute or functionally depend on each other. The generalized thermal bath consists of an unbounded collection of systems, each of which is in a generalized thermal state as defined by equation (5). Any given protocol will involve only a finite set of systems in the bath, whose combined thermal state can be written as $\tau_b(\beta_1,\ldots,\beta_k)$. We also want to consider an additional quantum system ρ_s that is both initially uncorrelated from and out of equilibrium with respect to the generalized bath, that is, $\rho_{sb} = \rho_s \otimes \tau_b(\beta_1,\ldots,\beta_k)$ and $\rho_s \neq \tau_s(\beta_1,\ldots,\beta_k)$. The main question we ask is how much of each of the conserved quantities can be extracted from the system (in conjunction with the bath, and stored in an associated battery).

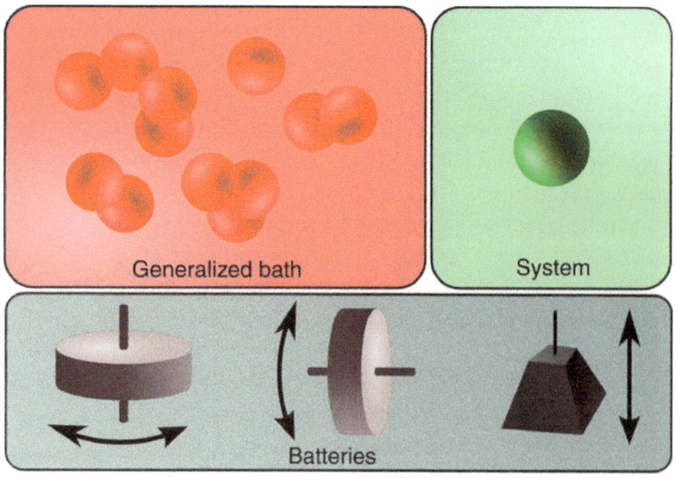

Figure 1: Schematic diagram of the set-up.

We have a generalized bath, a collection of particles each in the generalized thermal state (1). We have a system that is out of equilibrium with respect to the generalized bath. Finally, we have the collection of batteries, one for each conserved quantity. In the example depicted, the three conserved quantities are the angular momentum in the x direction, angular momentum in the z direction and the energy, with the three corresponding batteries being turntables spinning around the z and x axes, respectively, and a weight. We can perform any interaction between the components as long as each of the conserved quantities is conserved (the first law).

In the interest of being clear, we proceed by concentrating on a scenario with only two conserved quantities, A and B, since this already captures the majority of the physics contained in the general case of k conserved quantities.

In order to talk about the extraction of the conserved quantities, there are two ways in which one can proceed: by either including battery systems implicitly, or explicitly, in the formalism. In the former case, one allows the global amount of each quantity stored in the system and bath to change, and defines the changes as the amount of 'A-type work' and 'B-type work' that have been extracted from (or done on) the global system. The idea is that because of global conservation laws, when the A_i of the system and bath changes, this change is compensated by a corresponding change to the external environment (the implicit battery).

In the latter case, one introduces explicit battery systems, which by definition only accept a single type of work (that is, an A-type battery and a B-type battery). Here by definition the amount of A stored in the A-type battery is the A-type work, and similarly for B. We enforce that the global amount of A and B stored in the system, bath and battery is constant, either strictly (the entire distribution is conserved) or on average.

In the main text we consider the case of implicit batteries. We do this since dealing with implicit batteries simplifies the considerations and allows us to focus on what is arguably the most important part of the protocols, namely the interaction between system and bath. Obviously, it is preferable to have the full protocol including batteries explicitly. In doing so, there are many subtleties, which also arise in the case of standard thermodynamics. In particular, we need to impose 'no cheating' conditions that make sure that we do not make illegitimate use of batteries as sources of free entropy[25,42,43]. The danger stems from the fact that the batteries are systems out of equilibrium with respect to the bath.

Table 1: Summary of the results contained in the paper

	Commuting	Non-commuting
Implicit batteries		
Second law	✓	✓
Protocol	✓	✓
Explicit batteries (strict conservation)		
Second law	✓	✓
Protocol	✓	?
Explicit batteries (ave. conservation)		
Second law	✓	✓
Protocol	✓	✓*

The second law equations (9) holds in all instances.
*Designates that the result holds only for explicit batteries with continuous spectra.

More concretely, when considering implicit battery systems the class of allowed transformations consists of all global unitary transformations Uon the system and bath. After such a transformation, the global state is $\rho'_{sb} = U(\rho_s \otimes \tau_b(\beta_A, \beta_B))U^\dagger$ with the reduced state of the system and bath given by the reductions, $\rho'_s = \mathrm{tr}_b[\rho'_{sb}]$ and $\rho'_b = \mathrm{tr}_s[\rho'_{sb}]$, respectively. We define the A-type and B-type work to be

$$\begin{aligned} \Delta W_A &= -\Delta A_s - \Delta A_b \\ \Delta W_B &= -\Delta B_s - \Delta B_b \end{aligned} \qquad (8)$$

where $\Delta A_s = \mathrm{tr}[(A_s(\rho'_s - \rho_s))]$, $\Delta A_b = \mathrm{tr}[(A_b(\rho'_b - \tau_b))]$ and analogously for ΔB_s and ΔB_b. In addition, note that if our protocol involves multiple bath systems then $A_b = \sum_i A_b^{(i)}$, where $A_b^{(i)}$ acts non-trivially only on bath system i, that is, A_b is the sum of the local A for each system (and analogously for B_b). In equation (8) we are equating the average change of A and B, because of the unitary transformation, with the amount of A-type and B-type work that has been extracted from the system and bath. As such, our framework automatically incorporates the first law of thermodynamics for each of the conserved quantities.

Finally, an additional unrelated problem, but which often plays an important role, concerns the precise structure of the bath. In usual treatments, we may consider particles in the bath that have any energy-level spacing, such that their occupation probabilities can match any probabilities in the external system. This is used to construct efficient protocols. When considering other quantities than energy, we may be faced with quantities whose spectrum is fixed, such as angular momentum. In addition, extra constraints or relationships may exist between the different conserved quantities. This results in additional difficulties. To address these, and to remain as general as possible, we will consider baths with a minimal amount of accessible structure in terms of the eigenvalues of the conserved quantities.

The Second Law

Of great interest to us is the particular form that the second law of thermodynamics takes in the present setting. In the classical thermodynamic setting, the second law states that if one only has access to a thermal bath, then no work can be extracted, and that the maximal amount of work that can be extracted from a non-equilibrium system interacting with a thermal bath is bounded by the change in its free energy.

In our framework of multiple conserved quantities, we will see that the second law constrains the different combinations of conserved quantities that can be extracted from the system. In particular, we will show below that in the present framework, the amount of A-type work and B-type work that can be extracted is constrained such that

$$\beta_A \Delta W_A + \beta_B \Delta W_B \leq -\Delta \tilde{F}_s, \quad (9)$$

where $\Delta \tilde{F}_s = \tilde{F}(\rho'_s) - \tilde{F}(\rho_s)$. In the case where there is no system, or when the system is left in the same state, $\rho'_s = \rho_s$, then $\Delta \tilde{F}(\rho_s) = 0$, and we obtain as a corollary

$$\beta_A \Delta W_A + \beta_B \Delta W_B \leq 0. \quad (10)$$

Equations (9) and (10) constitute the second law when one has multiple conserved quantities (with and without a system).

To prove the second law, equation (9), we will need to use two further formulae, as well as the first laws, equation (8). First, since we restrict to unitary transformations, the total entropy of the global system remains unchanged, $S(\rho'_{sb}) = S(\rho_{sb})$, and from the fact that the system and bath are initially uncorrelated, along with sub-additivity, we have

$$\Delta S_s + \Delta S_b \geq \Delta S_{sb} = 0 \quad (11)$$

where $\Delta S_s = S(\rho'_s) - S(\rho_s)$, and analogously for ΔS_b and ΔS_{sb}. Second, since the bath starts in the thermal state $\tau(\beta_A, \beta_B)$, which is a minimum of the free entropy (by definition), its free entropy cannot decrease during the protocol; thus,

$$\Delta \tilde{F}_b = \beta_A \Delta A_b + \beta_B \Delta B_b - \Delta S_b \geq 0 \quad (12)$$

Now, eliminating all quantities on the bath, by substituting from the first laws, equation (8), and from equation (11), we finally arrive at

$$-\beta_A(\Delta A_s + \Delta W_A) - \beta_B(\Delta B_s + \Delta W_B) + \Delta S_s \geq 0 \quad (13)$$

which, after re-arranging and identifying terms, is straightforwardly seen to be equation (9), as desired. Thus, the first law, in conjunction with the lack of initial correlations (and sub-additivity), and the extremality of the generalized thermal state imply in a direct manner that systems obey a second law of the form given. We note that the proof does not rely on any particular properties of A and B, which need not even commute.

At this point it is worth briefly returning to the issue of implicit versus explicit batteries. If explicit batteries are included, then the unitary operations have to be extended to act on the system, bath and explicit batteries. Crucially, equation (11), and as a consequence the second law, equation (9), can be shown to hold when we are careful to avoid cheating via batteries. In the remaining we will study to what extent we can saturate equations (9) and (10), depending upon the properties of the conserved quantities (whether they commute or not), whether we consider implicit or explicit batteries and whether we consider strict or average conservation.

Commuting Observables

We will now specialize to the case of commuting observables, where we have access to joint eigenstates, and show how the second law can be saturated, both in terms of trading resources, and when extracting resources from a non-equilibrium system.

In order to remain as general as possible, we want to assume as little as possible about the structure of the generalized thermal bath. What we will require is that there exists a system in the bath (of which we can take arbitrarily many copies) with $d \geq 3$ states, $|a_i, b_i\rangle$, for $i \in \{0, 1, 2, \ldots d-1\}$, which are the joint eigenstates of A_b and B_b such that $A_b|a_i, b_i\rangle = a_i|a_i, b_i\rangle$ and $B_b|a_i, b_i\rangle = b_i|a_i, b_i\rangle$. We then need only three requirements. First, that the three eigenvalues of each observable are distinct, $a_0 \neq a_1 \neq a_2$ and $b_0 \neq b_1 \neq b_2$ (note that one consequence of this is to rule out the case in which either A or B is proportional to the identity, and thus trivially cannot be changed). Second, that the observables should be sufficiently different. In particular, that

$$\frac{a_1 - a_0}{b_1 - b_0} \neq \frac{a_2 - a_0}{b_2 - b_0}, \quad (14)$$

which amounts to saying that A and B should not be affinely related to each other, in which case they should not be thought of as different quantities. Third, that in the thermal state the joint eigenstates should not have the same populations. In particular, it must be that

$$x := \beta_A(a_1 - a_0) + \beta_B(b_1 - b_0) \neq 0,$$
$$y := \beta_A(a_2 - a_0) + \beta_B(b_2 - b_0) \neq 0. \quad (15)$$

If both x and y simultaneously vanish, then all three states have the same populations, in which case the system looks maximally mixed in this subspace. When trading quantities inside the bath, this will be the only problematic case. However, when we come to processing non-equilibrium systems, we will require simultaneously $x \neq 0$ and $y \neq 0$ in order for the bath to have enough structure to allow us to approach reversibility. We will also see that, depending on how close to reversible we want to be, we will have to exclude a small set (non-dense and of measure zero) of joint values for x and y, which are rationally related, as will be explained later. Below we outline the main ideas, and present all the details in the Methods.

We start by considering the task of trading resources within the generalized bath. That is, we consider the situation where we only have access to a generalized bath (and no external system). We will show that we can perform a unitary transformation such that, first, its free entropy changes by an arbitrarily small amount,

$$\Delta \tilde{F}_b = \beta_A \Delta A_b + \beta_B \Delta B_b \leq \epsilon \quad ((16)$$

where we used that $\Delta S_b = 0$ by definition for unitary transformations. Second, the change ΔA_b or ΔB_b can be made arbitrarily large, that is

$$\Delta A_b \geq \eta \quad \text{or} \quad \Delta B_b \geq \eta \tag{17}$$

with the other appropriately constrained by equations (16); (generally having large magnitude but opposite sign). If the above two conditions can be satisfied, we say that we can exchange A for B in an essentially reversible manner.

To show that this is possible, we proceed in two steps. We provide an explicit protocol that exchanges a suitably chosen two-dimensional subspace within the bath, and calculate the change in $\Delta\tilde{F}_b$, ΔA_b and ΔB_b that this produces. We then show that by repeating this protocol a sufficient number of times we achieve equations (16) and (17).

The explicit protocol takes the bath as n copies of $\tau(\beta_A, \beta_B)$, and a two dimensional subspace that consists of states that differ in population by Δq, and differ in the number of systems in the state $|a_1, b_1\rangle$ by Δn_1. Then, as we show in the Methods, by interchanging the population of two such states, we can achieve

$$0 < \Delta\tilde{F}_b \leq y\Delta q, \tag{18}$$

$$\frac{|\Delta A_b|}{\Delta\tilde{F}_b} = O(\Delta n_1), \quad \frac{|\Delta B_b|}{\Delta\tilde{F}_b} = O(\Delta n_1), \tag{19}$$

where the sign of ΔA_b or ΔB_b can be chosen arbitrarily, with the other quantity generally having the opposite sign, in accordance with equation (16). Hence, $\Delta\tilde{F}_b$ can be made as small as desired by making Δq arbitrarily small, while independently the relative change $|\Delta A_b|/\Delta\tilde{F}_b$ can be made as large as desired by increasing Δn_1.

Finally, by repeating the above protocol a sufficient number of times, one can trade arbitrary amounts of the conserved quantities from a generalized bath by sacrificing an arbitrarily small amount of free entropy. In particular, to achieve $\Delta A_b^{\text{total}} \geq \eta$ with $\Delta\tilde{F}_b^{\text{total}} \leq \epsilon$, one can perform the protocol above $(\epsilon/\Delta\tilde{F}_b)$ times, with $\Delta A_b/\Delta\tilde{F}_b \geq \eta/\epsilon$.

We now move onto the task of extracting resources from a single quantum system. In the above we showed that arbitrarily good interconversions can be enacted, given access only to a generalized bath. We now move on to the scenario of having a quantum system out of equilibrium with respect to the bath. Our goal is to show that we can saturate the second law given by equation (9) arbitrarily well—that is, that we can extract conserved quantities from a non-equilibrium system such that $\beta_A\Delta W_A+\beta_B\Delta W_B$ is as close as desired to the system's decrease in free entropy.

Let us consider that we have a state ρ_s, which in terms of its eigenstates and eigenvalues is given by $\rho_s=\sum_i p_i|\psi_i\rangle\langle\psi_i|$, and by convention we take the eigenvalues to be ordered, $p_{n+1}\leq p_n$. In general, the eigenbasis of the state will not coincide with the joint eigenbasis of the conserved quantities A and B. The first step is to pre-process the system, to bring it to a diagonal form in this basis. As we show in the Methods, it is always possible to do so, without even utilizing the bath, such that $\Delta\tilde{F}_s = \beta_A\Delta A_s + \beta_B\Delta B_s$, that is, in a fully reversible way that saturates equation (9). Note that in the case of explicit battery systems this pre-processing step is more complicated, nevertheless, the ideal pre-processing can still be arbitrarily well approximated. Now, having bought the system to diagonal form, we want to consider a protocol that moves a small population δp between two eigenstates, which have populations p_0 and p_1, respectively. We can implement such changes by finding two levels in the bath whose ratio of populations are close to p_0/p_1, and then swapping the 2D subspaces of the system and bath. Therefore, to carry out any such transformation requires a finely spaced set of different population ratios in the bath.

In the Methods we show that, except for a non-dense and measure zero set of x and y (as defined in equation (15)), we can find two levels in the bath whose ratio of populations is sufficiently close to p_0/p_1, such that after applying the swap operation between the appropriate two-dimensional subspaces, we achieve

$$\beta_A\Delta W_A + \beta_B\Delta W_B = -\Delta\tilde{F}_s + O(\delta p^2) \tag{20}$$

that is, up to a correction of order $O(\delta p^2)$, the combination of conserved quantities extracted, which themselves are order $O(\delta p)$, matches the change in free entropy of the system. Thus, by composing $O(1/\delta p)$ of such transformations, we can implement a protocol that transforms $\rho_s\to\tau_s(\beta_A, \beta_B)$, whereby in each stage the population changes between two states by order $O(\delta p)$, and such that

$$\beta_A\Delta W_A + \beta_B\Delta W_B = -\Delta\tilde{F}_s + O(\delta p). \tag{21}$$

Therefore, by taking δp sufficiently small we can approach the reversible regime, whereby the change in free entropy of the system matches the dimensionless combination of conserved quantities extracted. Combining this protocol with the protocol from the previous section, involving only the generalized bath and the batteries, we can obtain any combination of extracted conserved quantities.

Finally, note that the same protocol can also be used to perform efficient transformations between any two system states (where the final state is full rank), and not just to the thermal state. Our protocol also immediately gives an asymptotic protocol for the interconversion of states with no average work cost: the rate at which one can transform $\rho^{\otimes n}$ into $\sigma^{\otimes nR}$ is given by

$$R = \frac{\tilde{F}(\rho) - \tilde{F}(\tau(\beta_1, \ldots, \beta_k))}{\tilde{F}(\sigma) - \tilde{F}(\tau(\beta_1, \ldots, \beta_k))} \tag{22}$$

Here one can simply run the protocol 'forward' individually on n copies of ρ in order to obtain in the batteries $n(\beta_1\Delta W_{A1} + \ldots \beta_k\Delta W_{A_k})$, and create $\tau(\beta_1,\ldots,\beta_k)^{\otimes n}$. Then, on nR copies run the protocol 'backwards' to create $\sigma^{\otimes nR}$, having returned each battery so that finally it contains the same amount its associated quantity that it initially contained (on average).

Non-Commuting Observables

In this section we will show that when considering implicit batteries the results obtained in the previous section can easily be modified to also work for non-commuting observables. This also extends to explicit batteries with average conservation laws when the batteries have continuous spectrum. However, the same protocols do not obviously generalize to the case of explicit batteries with strict conservation. Whereas previously by virtue of the commutativity of the observables we could find a joint eigenbasis $|a_i,b_i\rangle$ that was used in our explicit protocols, that is no longer the case for non-commuting observables. Nevertheless, the generalized thermal state is diagonal in the eigenbasis of $\beta_A A+\beta_B B$,

$$e^{-\beta_A A-\beta_B B}/\mathcal{Z}=\sum_i q_i|i\rangle\langle i| \tag{23}$$

Although to each eigenstate we can no longer associate an eigenvalue for A or B, we can still associate an average value,

$$\langle a\rangle_i := \langle i|A|i\rangle$$
$$\langle b\rangle_i := \langle i|B|i\rangle \tag{24}$$

The main point is that all of our previous results hold if instead of joint eigenstates $|a_i, b_i\rangle$, with eigenvalues a_i and b_i, we use the eigenstate $|i\rangle$ with average values $\langle a\rangle_i$ and $\langle b\rangle_i$ throughout.

The only subtlety that arises is the structure we need from the bath. We still only need to use three distinct eigenstates, $|0\rangle$, $|1\rangle$ and $|2\rangle$; however, now the necessary structure relates to the average value of the conserved quantities in the eigenstates, First, it must be that

$$\frac{\langle a\rangle_1-\langle a\rangle_0}{\langle b\rangle_1-\langle b\rangle_0}\neq\frac{\langle a\rangle_2-\langle a\rangle_0}{\langle b\rangle_2-\langle b\rangle_0} \tag{25}$$

otherwise, at the level of the average values, the observables appear affinely related and therefore cannot be sufficiently distinguished to allow for trade-offs. Furthermore, we still need to be able to find eigenstates in the bath that differ in population, just as before. We can define the analogous quantities $\langle x\rangle$ and $\langle y\rangle$, and if they do not simultaneously vanish then the bath will not be maximally mixed in the subspace. Finally, in order to extract resources from systems out of equilibrium with respect to the generalized bath, there must be sufficient structure such that any ratio of populations can be approximated well enough. Again, in complete analogy to the above, if ($\langle x\rangle / \langle y\rangle$) is irrational, then we have sufficient structure. If on the other hand ($\langle x\rangle / \langle y\rangle$) is rational, we will again have to exclude a small set of values of $\langle x\rangle$ and $\langle y\rangle$ (non-dense, of zero measure), for which our results will not hold.

DISCUSSION

In this work we have studied a generalization of thermodynamics where there are multiple conserved quantities, where energy may not even be part of the story. We have been interested in what form the second law takes, and showed that it is no longer about restrictions on individual extractable quantities, but rather about the allowed ways that the conserved quantities can be traded-off for one another. Indeed, we found that we can extract as much of any individual conserved quantity as desired, as long as the other conserved quantities are appropriately consumed in the process, with the second law dictating how much of the others are necessarily consumed. In particular, we were led to introduce a dimensionless generalization of the free energy, which we termed the free entropy, that is the central quantity appearing in the second law and dictating the allowed trade-offs. Moreover, given access to any quantum system out of equilibrium with respect to the generalized bath, we showed that its free entropy change bounds the combination of conserved quantities that can be extracted.

Our results hold both for commuting and non-commuting observables, and with the desire to remain as general as possible we made only very mild assumptions about the bath. Indeed, we assumed very little about the relationship between the conserved quantities or their individual structure. The one case that remains open for future research is the case of non-commuting observables, with explicit batteries and strict conservation of the conserved quantities. Although the protocols presented for saturating the second law do not appear to generalize to this case, we do not know whether entirely different constructions will be able to achieve this goal.

METHODS

Relation between Free Entropy and Relative Entropy

Here we show that the free entropy difference between any state ρ_s and the generalized thermal state $\tau_s=\exp(-\sum_i\beta_i A_i)/Z$ is equal to the relative entropy difference between these two states. First, note that the free entropy of the thermal state is

$$\begin{aligned}\tilde{F}(\tau_s)&=\sum_i\beta_i\langle A_i\rangle_\tau+\mathrm{tr}(\tau_s\log\tau_s)\\&=\sum_i\beta_i\langle A_i\rangle_\tau-\sum_i\beta_i\langle A_i\rangle_\tau-\mathrm{tr}(\tau_s)\log\mathcal{Z}\\&=-\log\mathcal{Z},\end{aligned} \tag{26}$$

where we recall that $\mathcal{Z}=\mathrm{tr}(\exp(-\sum_i\beta_i A_i))$ is the generalized partition function. Then, it follows that

$$\begin{aligned}D(\rho_s\,||\,\tau_s)&=-S(\rho_s)-\mathrm{tr}(\rho_s\log\tau_s),\\&=-S(\rho_s)+\sum_i\beta_i\langle A_i\rangle+\log\mathcal{Z},\end{aligned} \tag{27}$$

which demonstrates the claim. Note that this result is completely analogous to the case of standard thermodynamics[44].

Trading Resources

Consider the situation where we only have access to a generalized bath (and no external system). We will show that we can perform a unitary transformation such that: (i) the free entropy $\Delta\tilde{F}_b=\beta_A\Delta A_b+\beta_B\Delta B_b$ (since $\Delta S_b=0$ by definition) changes by an arbitrarily small amount. (ii) The changes $|\Delta A_b|$ and $|\Delta B_b|$ can be made arbitrarily large.

Consider that we have a n copies of $\tau(\beta_A,\beta_B)=e^{-\beta_A A-\beta_B B}/\mathcal{Z}=\sum_i q_i|a_i,b_i\rangle\langle a_i,b_i|$. It will be convenient to label states by the number of systems found in a given eigenstate, which we shall refer to as the occupation. We denote by $|\mathbf{n},\alpha\rangle\equiv|n_0,n_1,\ldots,n_{d-1},\alpha\rangle=P_\alpha|a_0,b_0\rangle^{\otimes n_0}|a_1,b_1\rangle^{\otimes n_1}\cdots|a_{d-1},b_{d-1}\rangle^{\otimes n_{d-1}}$, where P_α is a permutation operator, permuting the bath systems, labelled by α, and $n_0+n_1+\ldots+n_{d-1}=n$. Now, we will consider only two states from the d^n which are available, corresponding to

$$\begin{aligned}|\mathbf{n},\alpha\rangle &= |n_0, n_1, n_2, n_3, \ldots, n_{d-1}, \alpha\rangle \\ |\mathbf{n}',\alpha'\rangle &= |n_0', n_1', n_2', n_3, \ldots, n_{d-1}, \alpha'\rangle\end{aligned} \quad (28)$$

that is, such that only the occupations of the first three levels differ between these states. As such, we have the constraint that $n_0+n_1+n_2=n_0'+n_1'+n_2'$. The key step in our protocol is to apply a swap operation between these two states, while leaving all other states unchanged. We assume the bath is sufficiently large so that after this step any modified systems from the bath can be discarded, and any further operations act on fresh bath systems. A direct calculation shows that the change in the average value of each quantity of interest is

$$\Delta A_b = \Delta q(a_{10}\Delta n_1 + a_{20}\Delta n_2) \quad (29)$$

$$\Delta B_b = \Delta q(b_{10}\Delta n_1 + b_{20}\Delta n_2) \quad (30)$$

$$\Delta F_b = \Delta q(x\Delta n_1 + y\Delta n_2) \quad (31)$$

where $\Delta n_k=(n'_k-n_k)$, $a_{k0}=(a_k-a_0)$, $b_{k0}=(b_k-b_0)$, x and y are as defined in equation (15), and

$$\Delta q = \left(1-\left(\frac{q_1}{q_0}\right)^{\Delta n_1}\left(\frac{q_2}{q_0}\right)^{\Delta n_2}\right)\prod_{i=0}^{d-1} q_i^{n_i} \quad (32)$$

is the difference in populations between the two states. Given $y\neq 0$, we can rewrite equation (31) as

$$\Delta\tilde{F}_b = y\Delta q\Delta n_1\left(\frac{x}{y}+\frac{\Delta n_2}{\Delta n_1}\right) \quad (33)$$

Now, for arbitrary Δn_1, we can find an integer m such that $m/\Delta n_1 < x/y \leq (m+1)/\Delta n_1$. Setting $\Delta n_2=-m$ in equation (33), we obtain

$$0 < \Delta\tilde{F}_b \leq y\Delta q. \quad (34)$$

Hence, $\Delta\tilde{F}_b$ can be made as small as desired by making Δq arbitrarily small (which can be achieved by increasing n_0). Note that it is crucial that $\Delta\tilde{F}_b \neq 0$. This is because the thermal state is the unique state that minimizes $\tilde{F}_b$. Thus, $\Delta\tilde{F}_b = 0$ implies that the bath is left completely unchanged, which in turn implies that $\Delta A_b=\Delta B_b=0$; hence, the desired transformation cannot take place. On the other hand, we find that the relative change in the conserved quantities $\Delta A_b/\Delta\tilde{F}_b$ and $\Delta B_b/\Delta\tilde{F}_b$ are

$$\begin{aligned}\frac{\Delta A_b}{\Delta\tilde{F}_b} &= \frac{a_{20}}{y}\left(1+\frac{\beta_B b_{20}}{y}\left(\frac{a_{10}}{a_{20}}-\frac{b_{10}}{b_{20}}\right)\left(\frac{x}{y}+\frac{\Delta n_2}{\Delta n_1}\right)^{-1}\right) \\ \frac{\Delta B_b}{\Delta\tilde{F}_b} &= \frac{b_{20}}{y}\left(1+\frac{\beta_A a_{20}}{y}\left(\frac{b_{10}}{b_{20}}-\frac{a_{10}}{a_{20}}\right)\left(\frac{x}{y}+\frac{\Delta n_2}{\Delta n_1}\right)^{-1}\right)\end{aligned} \quad (35)$$

In both cases, the final term satisfies $(x/y+\Delta n_2/\Delta n_1)^{-1} \geq \Delta n_1$ and can, hence, be made as large as desired by increasing Δn_1. This means that the magnitude of $\Delta A_b/\Delta\tilde{F}_b$ and $\Delta B_b/\Delta\tilde{F}_b$ will become arbitrarily large. The sign of ΔA_b will depend on the other constants, but can be modified if desired by choosing m such that $(m-1)/\Delta n_1 \leq x/y < m/\Delta n_1$ above. Note also that if $y=0$ but $x\neq 0$ we can construct an equivalent proof with the roles of x and y swapped.

Finally, by repeating the above protocol a sufficient number of times, one can trade arbitrary amounts of the conserved quantities from a generalized bath by sacrificing an arbitrarily small amount of free entropy. In particular, to achieve $\Delta A_b^{\text{total}} \geq \eta$ with $\Delta\tilde{F}_b^{\text{total}} \leq \epsilon$, one can perform the protocol above $(\epsilon/\Delta\tilde{F}_b)$ times, with $\Delta A_b/\Delta\tilde{F}_b \geq \eta/\epsilon$.

There are a number of important aspects of the above protocol: first, it relies on a minimal amount of structure in the observables A and B and the bath: it requires that there exist many copies of a bath system with a three-dimensional subspace where the action of the operators are not trivially related (by a shift and rescaling), and that the state is not maximally mixed in this subspace. Moreover, each bath system is taken to be identical, with no additional parameters necessary (that is, we do not require a family of different A_b's and B_b's, similar to the families of Hamiltonians considered in ref. 25). Second, this protocol is a only a proof-of-principle demonstration that trade-offs can be enacted. No attention was paid to the number of generalized thermal states necessary. If one cared about minimizing the resources utilized, then the above protocol would not be used, and more efficient ones would be sought. Finally, the above analysis generalizes beyond two conserved quantities to the general case of k quantities. In this case, equation (14) must hold pairwise for all quantities.

Extracting Resources from a Single Quantum System

Here we consider the scenario involving a quantum system out of equilibrium with respect to the bath. We will show that the second law given by equation (9) can be saturated arbitrarily well—that is, we can extract conserved quantities from a non-equilibrium system such that$\beta_A\Delta W_A+\beta_B\Delta W_B$ is as close as desired to the system's decrease in free entropy.

Let us consider that we have a state ρ_s, which in terms of its eigenstates and eigenvalues, is given by $\rho_s = \sum_i p_i|\psi_i\rangle\langle\psi_i|$, and by convention we take the eigenvalues to be ordered, $p_{n+1}\leq p_n$. In general, the eigenbasis of the state will not coincide with the joint eigenbasis of the conserved quantities A and B. The first step is to pre-process the system to bring it to a diagonal form in this basis. To do so we will not interact with the bath, but simply apply the unitary

$$U_s = \sum_i |a_i, b_i\rangle\langle\psi_i|, \quad (36)$$

on the system such that $\sigma_s = U_s\rho_s U_s^\dagger = \sum_i p_i|a_i,b_i\rangle\langle a_i,b_i|$. Owing to the first laws, equation (8), we have $\Delta W_A=-\Delta A_s$ and $\Delta W_B=-\Delta B_s$. Moreover, since the entropy of the system did not change, $\Delta\tilde{F}_s = \beta_A\Delta A_s + \beta_B\Delta B_s$, and we have immediately

$$-\Delta\tilde{F}_s = \beta_A\Delta W_A + \beta_B\Delta W_B \quad (37)$$

that is, in this state, unsurprisingly, we have a change in the batteries that coincides with the free entropy change of the system, and saturates equation (9). Note that in the case of explicit battery systems, the transformation in equation (36) cannot be perfectly performed at the level of the system, since $[U_s, A_s]\neq 0$ and $[U_s, B_s]\neq 0$. Nevertheless, one can implement a joint unitary on the system and batteries such that at the level of the system the transformation U_s can be approximated arbitrarily well as long as the battery systems are in appropriate states, similarly to the case of standard quantum thermodynamics[42,43].

Now, having brought the system to a diagonal form, we want to consider a transformation $\sigma_s \to \sigma_s' = \sum_i p_i' |a_i^s, b_i^s\rangle\langle a_i^s, b_i^s|$ in which only two levels of the system change their populations by a small amount. Note that here we denote the eigenvalues of A_s and B_s by a_i^s and b_i^s to differentiate them from the eigenvalues of A_b and B_b, which we will similarly denote by a_i^b and b_i^b. Namely, we would like to perform $p_1' = p_1 + \delta p, p_0' = p_0 - \delta p$. We can implement such changes by finding two levels in the bath whose ratio of populations is close to p_0/p_1 and then swapping the two-dimensional subspaces of the system and bath. To carry out any such transformation requires a finely spaced set of different population ratios in the bath. Let us now consider whether the simple bath systems we have considered so far can provide such possibilities. As before, we bring in a collection of n generalized thermal states, $\tau(\beta_A,\beta_B)$ and consider only the two states in equation (28), $|\mathbf{n}, \alpha\rangle$ and $|\mathbf{n}', \alpha'\rangle$. We shall denote the populations of these states by $q_{\mathbf{n}}$ and $q_{\mathbf{n}'}$, respectively. The global unitary transformation we will apply is the swap operator between the two-dimensional subspaces of the system and bath, and the identity everywhere else. That is, the operation that performs

$$|a_0, b_0\rangle|\mathbf{n}', \alpha'\rangle \leftrightarrow |a_1, b_1\rangle|\mathbf{n}, \alpha\rangle, \tag{38}$$

while leaving all other states unchanged. By performing this transformation, the population that is shifted between the states of the system, which coincides with the population that is shifted between the states of the bath, is $\delta p=(p_0 q_{\mathbf{n}'} - p_1 q_{\mathbf{n}})$. Now, the changes in the conserved quantities of the system and bath are found to be

$$\begin{aligned} \Delta A_s &= \delta p(a_1^s - a_0^s), & \Delta B_s &= \delta p(b_1^s - b_0^s) \\ \Delta A_b &= -\delta p(a_{\mathbf{n}'}^b - a_{\mathbf{n}}^b), & \Delta B_b &= -\delta p(b_{\mathbf{n}'}^b - b_{\mathbf{n}}^b). \end{aligned} \tag{39}$$

The change in the entropy of the system and the bath are also found to be

$$\begin{aligned} \Delta S_s &= \delta p \log\frac{p_0'}{p_1'} + O(\delta p^2) \\ \Delta S_b &= -\delta p \log\frac{q_{\mathbf{n}}}{q_{\mathbf{n}'}} + O(\delta p^2) \end{aligned} \tag{40}$$

where we use $\frac{p_0'}{p_1'}$ rather than $\frac{p_0}{p_1}$ since we can always take $p_1' \neq 0$. We want to achieve $\Delta S_s+\Delta S_b=O(\delta p^2)$, which requires that $\log(q_{\mathbf{n}}/q_{\mathbf{n}'}) = \log(p_0'/p_1') + O(\delta p)$. To see when this condition is satisfied, we first use the explicit form of the probabilities, which show that $q_{\mathbf{n}}/q_{\mathbf{n}'}=\exp(-x\Delta n_1 - y\Delta n_2)$. Hence, we require $\log(p_0'/p_1') - x\Delta n_1 - y\Delta n_2 = O(\delta p)$. Since p_0' and p_1' are arbitrary, our requirement is that $x\Delta n_1+y\Delta n_2$ should be able to come within $O(\delta p)$ of any number. As we can rescale Δn_1 and Δn_2 by an arbitrary integer, it is sufficient to obtain $0<x\Delta n_1+y\Delta n_2\leq O(\delta p)$. If x and y are not rationally related, this is always possible. However, if x/y is rational, and given in reduced form by u/v (where u and v are co-prime integers), then we need

$$0 < \frac{y}{v}(u\Delta n_1 + v\Delta n_2) \leq O(\delta p). \tag{41}$$

From number theory, one can always find Δn_1 and Δn_2 such that the term in brackets is 1 (or −1); hence, we need $|y/v|\leq O(\delta p)$. For a fixed desired accuracy ϵ (of order $O(\delta p)$), and fixed y, this rules out a finite number of x values for which x/y is a rational with a small denominator. Extending this to the x-y plane, we find that we can achieve the desired accuracy everywhere except for a non-dense set of measure zero.

Returning to the entropies, with the above in place, $\Delta S_s+\Delta S_b=O(\delta p^2)$, that is, the system and bath remain essentially uncorrelated after the transformation. Finally, we use once again the fact that the generalized thermal state is a minimum of the free entropy. This implies that the changes in population, of order $O(\delta p)$, change the free entropy only to second order, $\Delta\tilde{F}_b = O(\delta p^2)$. Putting everything together we have

$$\begin{aligned} \Delta\tilde{F}_b &= \beta_A \Delta A_b + \beta_B \Delta B_b - \Delta S_b, \\ &= -\beta_A(\Delta W_A + \Delta A_s) - \beta_B(\Delta W_B + \Delta B_s) + \Delta S_s + O(\delta p^2), \\ &= -\Delta\tilde{F}_s - (\beta_A \Delta W_A + \beta_B \Delta W_B) + O(\delta p^2). \end{aligned} \tag{42}$$

Thus, since the left-hand side is $O(\delta p^2)$, it must be that

$$\beta_A \Delta W_A + \beta_B \Delta W_B = -\Delta\tilde{F}_s + O(\delta p^2) \tag{43}$$

that is, up to a correction of order $O(\delta p^2)$, the combination of conserved quantities extracted, which themselves are order $O(\delta p)$, matches the change in free entropy of the system. Thus, by composing $O(1/\delta p)$ of such transformations, we can implement a protocol that transforms $\rho_s \to \tau_s(\beta_A,\beta_B)$, whereby in each stage the population changes between two states by order $O(\delta p)$, and such that

$$\beta_A \Delta W_A + \beta_B \Delta W_B = -\Delta\tilde{F}_s + O(\delta p). \tag{44}$$

Therefore, by taking δp sufficiently small we can approach the reversible regime, whereby the change in free entropy of the system matches the dimensionless combination of conserved quantities extracted. Combining this protocol with the protocol from the previous section, involving only the generalized bath and the batteries, we can obtain any combination of extracted conserved quantities.

It is important to stress that our assumptions only changed by a small amount relative to the previous section. In particular, we need a minimal amount of extra structure in the bath, such that it is useful to process arbitrary individual systems out of equilibrium. Furthermore, as in the previous case the protocol presented generalizes in a straightforward manner to the case of k mutually commuting conserved quantities $A_1,\ldots A_k$.

Finally, note that the same protocol can also be used to perform efficient transformations between any two system states (where the final state is full rank), and not just to the thermal state. Our protocol also immediately gives an asymptotic protocol for the interconversion of states with no average work cost: the rate at which one can transform $\rho^{\otimes n}$ into $\sigma^{\otimes nR}$ is given by

$$R = \frac{\tilde{F}(\rho) - \tilde{F}(\tau(\beta_1, \ldots, \beta_k))}{\tilde{F}(\sigma) - \tilde{F}(\tau(\beta_1, \ldots, \beta_k))} \tag{45}$$

Here one can simply run the protocol 'forward' individually on n copies of ρ in order to obtain in the batteries $n(\beta_1 \Delta W_{A1} + \ldots \beta_k \Delta W_{A_k})$ and create $\tau(\beta_1,\ldots,\beta_k)^{\otimes n}$. Then, on nR copies run the protocol 'backwards' to create $\sigma^{\otimes nR}$, having returned each battery so that finally it contains the same amount its associated quantity that it initially contained (on average).

Data Availability Statement

Data sharing not applicable to this article as no datasets were generated or analyzed during the current study.

ACKNOWLEDGEMENTS

We thank J. Hannay for many discussions while carrying out this research. We also thank S. Massar for useful comments. Y.G., P.S. and S.P. acknowledge support from the European Research Council (ERC AdG NLST). S.P. acknowledges support from a Royal Society Wolfson Merit Award. A.J.S. acknowledges support from the Royal Society and FQXi via the SVCF. R.S. acknowledges support from the Swiss National Science Foundation (Starting grant DIAQ). Y.G., A.J.S., R.S. and P.S. acknowledge support from the COST Action MP1209 Thermodynamics in the quantum regime.

REFERENCES

1. Gibbs., J. W. Elementary Principles in Statistical Mechanics Cambridge University Press (2010) Cambridge Books Online.
2. Jaynes., E. T. Information theory and statistical mechanics. Phys. Rev. 106, 620–630 (1957).
3. Jaynes., E. T. Information theory and statistical mechanics. II. Phys. Rev. 108, 171–190 (1957).
4. Rigol, M., Dunjko, V., Yurovsky, V. & Olshanii., M. Relaxation in a completely integrable many-body quantum system: an ab initio study of the dynamics of the highly excited states of 1d lattice hard-core bosons. Phys. Rev. Lett. 98, 050405 (2007).
5. Kollar, M., Wolf, F. A. & Eckstein., M. Generalized Gibbs ensemble prediction of prethermalization plateaus and their relation to nonthermal steady states in integrable systems. Phys. Rev. B 84, 054304 (2011).
6. Caux, J.-S. & Konik., R. M. Constructing the generalized Gibbs ensemble after a quantum quench. Phys. Rev. Lett. 109, 175301 (2012).
7. Dóra, B., Bácsi, Á. & Zaránd., G. Generalized Gibbs ensemble and work statistics of a quenched Luttinger liquid. Phys. Rev. B 86, 161109 (2012).
8. Pozsgay., B. The generalized Gibbs ensemble for Heisenberg spin chains. J. Stat. Mech. 2013, P07003 (2013).
9. Vaccaro, J. A. & Barnett., S. M. Information erasure without an energy cost. Proc. R Soc. A 467, 1770–1778 (2011).
10. Barnett, S. M. & Vaccaro., J. A. Beyond Landauer erasure. Entropy 15, 4956–4968 (2013).
11. Dahlsten, O. C. O., Renner, R., Rieper, E. & Vedral., V. Inadequacy of von Neumann entropy for characterizing extractable work. New J. Phys. 13, 53015 (2011).
12. del Rio, L., Åberg, J., Renner, R., Dahlsten, O. C. O. & Vedral., V. The thermodynamic meaning of negative entropy. Nature 474, 61–63 (2011).
13. Egloff, D., Dahlsten, O. C. O., Renner, R. & Vedral., V. A measure of majorization emerging from single-shot statistical mechanics. New J. Phys. 17, 073001 (2015).
14. Faist, P., Dupuis, F., Oppenheim, J. & Renner., R. The minimal work cost of information processing. Nat. Commun. 6, 7669 (2015).
15. Halpern, N. Y., Garner, A. J. P., Dahlsten, O. C. O. & Vedral., V. Introducing one-shot work into fluctuation relations. New J. Phys. 17, 095003 (2015).
16. Woods, M. P., Ng, N. & Wehner., S. The Maximum Efficiency of Nano Heat Engines Depends on More than Temperature. Preprint at http://arxiv.org/abs/1506.02322 (2015).
17. Weilenmann, M., Krämer, L., Faist, P. & Renner., R. Axiomatic Relation Between Thermodynamic and Information-Theoretic Entropies. Preprint at http://arxiv.org/abs/1501.06920 (2015).
18. Bennett, C. H., Bernstein, H., Popescu, S. & Schumacher., B. Concentrating partial entanglement by local operations. Phys. Rev. A 53, 2046–2052 (1996).
19. Horodecki, M., Horodecki, P. & Oppenheim., J. Reversible transformations from pure to mixed states and the unique measure of information. Phys. Rev. A 67, 62104 (2003).
20. Gour, G. & Spekkens., R. W. The resource theory of quantum reference frames: manipulations and monotones. New J. Phys. 10, 033023 (2008).
21. Brandão, F. G. S. L., Horodecki, M., Oppenheim, J., Renes, J. & Spekkens., R. W. Resource theory of quantum states out of thermal equilibrium. Phys. Rev. Lett. 111, 250404 (2013).
22. Janzing, D., Wocjan, P., Zeier, R., Geiss, R. & Beth., T. Thermodynamic cost of reliability and low temperatures: tightening Landauer's principle and the second law. Int. J. Theor. Phys. 39, 2717–2753 (2000).
23. Horodecki, M. & Oppenheim., J. Fundamental limitations for quantum and nanoscale thermodynamics. Nat. Commun. 4, 2059 (2013).
24. Åberg., J. Truly work-like work extraction via a single-shot analysis. Nat. Commun. 4, 1925 (2013).
25. Skrzypczyk, P., Short, A. J. & Popescu., S. Work extraction and thermodynamics for individual quantum systems. Nat. Commun. 5, 4185 (2014).
26. Brandão, F. G. S. L., Horodecki, M., Ng, N., Oppenheim, J. & Wehner., S. The second laws of quantum thermodynamics. Proc. Natl Acad. Sci. USA 112, 3275–3279 (2015).
27. Ng, N. H. Y., Mančinska, L., Cirstoiu, C., Eisert,

J. & Wehner, S. Limits to catalysis in quantum thermodynamics. New J. Phys. 17, 085004 (2015).

28. Renes., J. M. Work cost of thermal operations in quantum thermodynamics. EPJ Plus 129, 153 (2014).

29. Faist, P., Oppenheim, J. & Renner., R. Gibbs-preserving maps outperform thermal operations in the quantum regime. New J. Phys. 17, 043003 (2015).

30. Lostaglio, M., Jennings, D. & Rudolph., T. Description of quantum coherence in thermodynamic processes requires constraints beyond free energy. Nat. Commun. 6, 6383 (2015).

31. Lostaglio, M., Korzekwa, K., Jennings, D. & Rudolph., T. Quantum coherence, time-translation symmetry, and thermodynamics. Phys. Rev. X 5, 021001 (2015).

32. Ćwikliski, P., Studziski, M., Horodecki, M. & Oppenheim., J. Limitations on the evolution of quantum coherences: towards fully quantum second laws of thermodynamics. Phys. Rev. Lett. 115, 210403 (2015).

33. Lostaglio, M., Müller, M. P. & Pastena., M. Stochastic independence as a resource in small-scale thermodynamics. Phys. Rev. Lett. 115, 150402 (2015).

34. Korzekwa, K., Lostaglio, M., Oppenheim, J. & Jennings., D. The extraction of work from quantum coherence. New J. Phys. 18, 023045 (2016).

35. Alhambra, Á. M., Oppenheim, J. & Perry, C. What is the Probability of a Thermodynamical Transition? Preprint at http://arxiv.org/abs/1504.00020 (2015).

36. Perry, C., Ćwikliski, P., Anders, J., Horodecki, M. & Oppenheim, J. A Sufficient Set of Experimentally Implementable Thermal Operations. Preprint at http://arxiv.org/abs/1511.06553.

37. Halpern, N. Y. & Renes., J. M. Beyond heat baths: generalized resource theories for small-scale thermodynamics. Phys. Rev. E 93, 022126 (2016).

38. Yunger-Halpern., N. Beyond Heat Baths II: Framework for Generalized Thermodynamic Resource Theories. Preprint at http://arXiv:1409.7845 (2015).

39. Lostaglio, M., Jennings, D. & Rudolph, T. Thermodynamic resource theories, non-commutativity and maximum entropy principles. Preprint at http://arxiv.org/abs/1511.04420 (2015).

40. Halpern, N. Y., Faist, P., Oppenheim, J. & Winter, A. Microcanonical and resource-theoretic derivations of the thermal state of a quantum system with noncommuting charges. Nat. Commun. 7, 12051 (2016).

41. Liu., Y.-K. The Complexity of the Consistency and N-representability Problems for Quantum States PhD thesis Univ. California (2007).

42. Malabarba, A. S. L., Short, A. J. & Kammerlander., P. Clock-driven quantum thermal engines. New J. Phys. 17, 045027 (2015).

43. Åberg., J. Catalytic coherence. Phys. Rev. Lett. 113, 150402 (2014).

44. Donald., M. J. Free energy and the relative entropy. J. Stat. Phys. 49, 81–87 (1987).

CHAPTER 3 LOCAL NON-EQUILIBRIUM THERMODYNAMICS

Lee Jinwoo and Hajime Tanaka

INTRODUCTION

Statistical mechanics provides physical interpretations of entropy and free energy that are macro-state functions (i.e., functions defined on a domain of the phase-space points of a system, and thus inevitably non-local in character), and sets bounds on permissible processes expressed as path functions like heat and work, where a path is defined as a trajectory of macrostates of a system[1]. As a system gets smaller, the effect of fluctuations becomes significant, yet modern theory[2,3,4] provides permissible distributions of fluctuating path functions in the form of beautiful equalities[5,6,7,8,9,10]. Modern theory has identified energetics[11] and entropy production on the level of individual trajectories[12], and has linked path functions to properties of macrostates[13], as verified experimentally[14,15,16,17,18,19]. The relationship between classical and modern approaches is schematically drawn in Fig. 1.

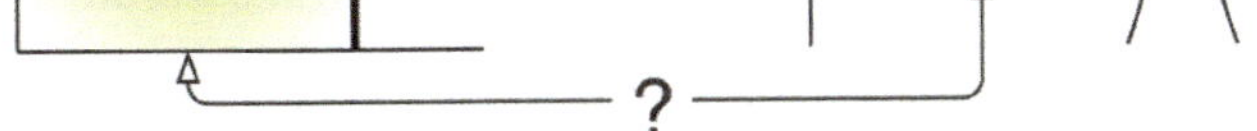

Figure 1: A universal frame of thermodynamics and a fundamental question. Thermodynamics is a theory for ensembles that are composed of a large number of fluctuating degrees of freedom. An ensemble is constrained by a macrostate λ. In equilibrium thermodynamics, λ varies in a quasi-static manner. In the modern context, the theory considers arbitrary time-varying processes. Two typical systems in both approaches are shown schematically. On the left, gas particles are fluctuating in a cylinder. The macrostate may be specified by the temperature, the volume of the cylinder, and the number of particles. On the right, a polymer chain is fluctuating under the influence of optically controlled bead connected to the end of the chain. The macrostate may be specified by the temperature, and the location of the bead applying force to the molecule. A fundamental question that we investigate is how we can justify thermodynamic descriptions of fluctuating microstates.

Citation: Jinwoo, L. & Tanaka, H. Local non-equilibrium thermodynamics. Sci. Rep. 5, 7832; DOI:10.1038/srep07832 (2015).

Despite these successes in non-equilibrium theories, however, still lacking is a unified framework analogous to classical thermodynamics[20]. It is important to note that contrary to the classical cases, a path in the modern theory is defined as a trajectory of phase-space points of a system. In general, the modern theory has been built effectively upon quantities defined locally (i.e., functions defined at each phase-space point). For example, a local form of information theoretic Shannon entropy, so called stochastic entropy, is defined as $-\ln p(x, t)$, wherep(x, t) is the phase-space density of a microstate x of a system. Such a local form is necessary for boundary conditions of a process (a path) to recover the second law of thermodynamics over a finite time interval for small scale systems[8,12]. Then we immediately encounter a serious conceptual difficulty. As a component of the second law, the entropy should be a property of an ensemble. However, it has not been clear at all how such an ensemble property as entropy could be assigned locally, or independently of a trajectory. This difficulty in the notion of locality is also closely associated with a classical view of difficulties of non-equilibrium thermodynamics; the ill-defined notion of a macrostate during dynamic evolution of a system. It is thus difficult to apply the language of equilibrium thermodynamics, most importantly the argument of counting, to non-equilibrium problems.

Without counting, for both equilibrium and non-equilibrium situations, we have no way to even imagine what entropy is. We remove this fundamental difficulty by constructing a new ensemble that is local in space and time, and is well-defined even in dynamic evolution of a system. Specifically, we count the accessible number of paths to a microstate and consider all trajectories to a microstate in non-equilibrium as an ensemble of the state. We, then, show that local entropy, similarly defined local information and local free energy are properties of the ensemble, and relate functions within the ensemble to state functions of the ensemble which have been concealed behind the integrated-out forms of previous theories.

THEORETICAL FRAMEWORK

In this article, we consider a classical stochastic system in contact with a heat bath of temperature T. To provide some intuitive grasp, we take as a simple example a polymer chain whose one end is fixed and the other end under an external control λ_τ that may vary over time $0 \le \tau \le t$(see the right panel of Fig. 1). We will refer to λ_τ as a macrostate (the state of a polymer chain) at time τ, and to the phase-space points of the polymer chain (excluding momentum variables for simplicity) as microstates. Note that, in this experiment, λ_τ is the only parameter controlling the macrostate of a polymer chain.

We define a forward process as one where an external control λ_τ is varied from λ_0 to λ_t during $0 \le \tau \le t$ in a well-defined manner, and a corresponding reverse process λ'_τ as $\lambda'_\tau \equiv \lambda_{t-\tau}$. Let l be a space-time trajectory of the system's microstates (i.e., evolving polymer configurations) during a forward process. Since l fluctuates, we repeat the process λ_τ with appropriate initial probability density p(x, 0) over all microstates x. When we consider a microstate along a path l at time τ, we will denote it as l_τ. We define path-dependent work done on the system by the external control during a forward process as follows:

$$W(l) \equiv \int_0^t \frac{\partial E(l_\tau,\tau)}{\partial \tau} d\tau, \tag{1}$$

where l is a trajectory, $E(l_\tau, \tau)$ is the energy of a microstate l_τ at time τ. Path-dependent heat transferred to the system during the same process is defined as follows:

$$Q(l) \equiv \int_0^t \frac{\partial E(l_\tau,\tau)}{\partial l_\tau} \circ dl_\tau, \tag{2}$$

where $\circ\, dl_\tau$ means either $\frac{dl_\tau}{dt}dt$ for deterministic frameworks, or a multiplication in the Stratonovich sense for stochastic frameworks. Then, for each l, by adding the heat and work during the forward conversion we have the first law of thermodynamics in the following form:

$$\Delta E = W(l) + Q(l), \tag{3}$$

where $\Delta E = E(l_t, t) - E(l_0, 0)$.

Let Λ be the set of all space-time trajectories of a system connecting times 0 and t, and Λ^x be those paths that reach a phase space point x at time t from the past (see Fig. 2). There is a natural projection π from Λ^x onto x such that $\pi(\Lambda^x) = x$, and then each microstate x can be regarded as an ensemble with conceptual rigor. This is because $\pi^{-1}(x)$ includes an infinite number of stochastic trajectories reaching x. This indicates that thermodynamic quantities can be defined for x.

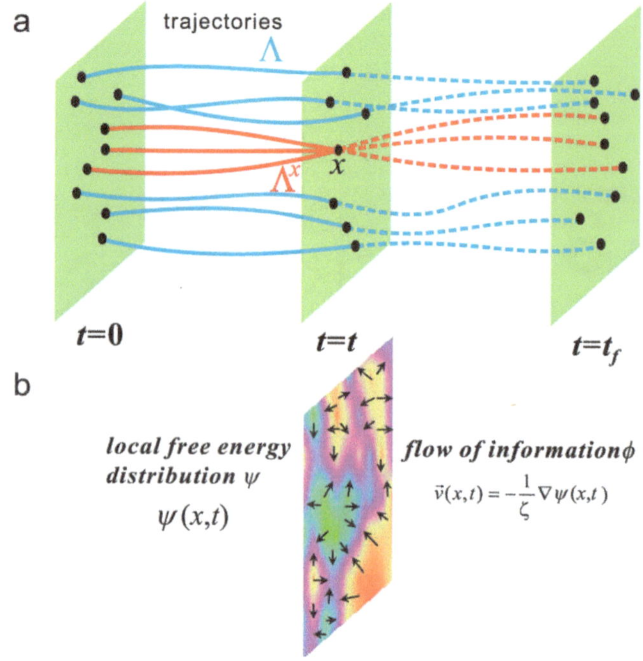

Figure 2: An evolution of a system's microstates. (a), A microstate as an ensemble of trajectories to a phase-space point. Each space in green represents the phase space of a system (excluding momentum variables for simplicity) at time t. Each trajectory shown schematically represents a stochastic evolution of the phase-space points. Among all possible trajectories Λ^x, we focus on those paths (Λ^x: thick red lines) that reach a specific point x in the phase space at time t = t. (b), Schematic figure of the distribution of ψ(x, t) at time t = t in the phase space, where the information flow is to be induced in the directions indicated by black arrows. The information flow not only causes the convective transport of the information φ(x, t), but also its divergence provides the net change of φ(x, t) (see Eq. (16)).

Our results will be built upon two assumptions. Firstly, we assume that the probability of path l ∈ Λ is represented as $p_\Lambda(l) \propto e^{-s(l)}$ for some function s. The explicit form of s depends on the type of model. For example, it would be the stochastic action with an initial condition derived by Onsager for stochastic models[21], the Hamiltonian of a system[20], or a bath at time 0 for deterministic frameworks[22]. Based on this, we develop the following idea: Since each path may not be equally probable, we assign each path a weight of the form $g(l) \equiv e^{-s(l)}$ so that less probable paths would be less counted. Let Ω(X) be the accessible number (or the total weight) of paths in a sample space X, then, we have $\Omega(X) = \sum_{l \in X} g(l)$. In this framework, every unit path, a collection of paths of which total weight is 1, is equally probable, and the probability of a microstate (e.g., a chain configuration) x at time t would be simply the ratio of the accessible number (or the total weight) of trajectories that pass x at time t to that of all possible trajectories, i.e., $\Omega(\Lambda^x)/\Omega(\Lambda)$. This new framework provides us with a basis for "counting" analogous to classical entropy.

Secondly, we assume microscopic reversibility[8,22], proven by Jarzynski in the Hamiltonian framework[22], and valid in various stochastic frameworks[2]. To state this, let us consider a forward process λ_τ ($0 \le \tau \le t$), and a space-time trajectory l of a polymer chain from a specific configuration (microstate) l_0 to another l_t during t. We also consider the time reversed conjugate l′ of the path l, i.e., $l'_\tau \equiv l_{t-\tau}$ for $0 \le \tau \le t$ under the reverse process $\lambda'_\tau = \lambda_{t-\tau}$. Here, we set the initial probability density for the reversed process p′(x, 0) as the final probability density for the forward process p(x, t) so that we have:

$$p'(l'_0,0) = p(l_t,t). \tag{4}$$

Now, the condition for microscopic reversibility[8,22] reads as follows:

$$\frac{p_\Lambda(l|l_0)}{p'_\Lambda(l'|l'_0)} = e^{-\beta Q}, \tag{5}$$

where Q is heat flowing into the system, $\beta = 1/k_BT$ (k_B: the Boltzmann constant), $p_\Lambda(l|l_0)$ is the conditional path probability in the sample space Λ, and $p'_\Lambda(l'|l'_0)$ is that for the reverse process. We note that $p_\Lambda(l|l_0) \neq p'_\Lambda(l'|l'_0)$ in general.

Given the above setup, we construct "local non-equilibrium thermodynamics", where locally-defined entropy, information, and free energy are true thermodynamic potentials for each microstate, in sharp contrast with "stochastic thermodynamics", where they are trajectory-dependent.

RESULTS

Local Functions as State Functions of Λ^x

Firstly, we prove that various local functions defined at each phase-space point x at time t are actually state functions of the newly introduced ensemble Λ^x. Let us consider a local form of information theoretic Shannon entropy (so called stochastic entropy)[8,12]. To show that this entropy is the property of the ensemble Λ^x, we define a quantity for Λ^x:

$$\phi(x,t) = \ln\Omega(\Lambda^x), \tag{6}$$

which we shall call information. As a reference value let φ_0 be the maximally attainable value of information, i.e., $\varphi_0 = \ln \Omega(\Lambda)$ (see Fig. 2a). Then, $\sigma(x, t) \equiv k_B(\varphi_0 - \varphi(x, t))$ becomes $-k_B \ln\frac{\Omega(\Lambda^x)}{\Omega(\Lambda)}$. Since the argument of the logarithm is simply the ratio of the total weight of paths in Λ^x to that of Λ, which becomes p(x, t), σ(x, t) is identical to the local Shannon entropy. Thus, stochastic entropy σ(x, t) can be viewed as a property of the ensemble Λ^x as desired. Here we emphasize that the local entropy is now defined through a means of counting. We remark that φ(x, t) may be interpreted as the information on the occurrence of the microstate (i.e., the polymer configuration), entropy σ(x, t) as lost information upon specifying a particular microstate x. Thus, the sum of information φ(x, t) and entropy σ(x, t) conserves locally. Hence, any theory for information could be interpreted using entropy and vice versa, and we will select one depending on the context. In this picture, every microstate of a system carries information, and thus we will consider information gain/consumption of each microstate but not information measurement/erasure procedures formulated in terms of mutual information[23,24,25,26]. Now we define a free-energy-like quantity as follows:

$$\psi(x,t) = E(x,t) - T\sigma(x,t). \tag{7}$$

Note that $\sigma(x, t) = k_B(\varphi_0 - \varphi(x, t))$. Thus, local free energy ψ(x, t) enables us to treat the information and energy of each microstate on an equal footing since ψ(x, t) is nothing but the sum of energy E(x, t) and information φ(x, t) adjusted by the reference value φ_0 and scaled by k_BT.

It was shown by Hummer and Szabo[14] that energy E(x, t) could be represented as weighted average of work functions over all paths in Λ^x. In detail, they derived the following relation using the Feynmann-Kac formula:

$$\left\langle e^{-\beta W}\delta(x,t)\right\rangle_\Lambda = \frac{e^{-\beta E(x,t)}}{e^{-\beta F_{eq}(\lambda_0)}}, \tag{8}$$

where the bracket indicates an average over all paths in Λ, δ(x, t) is the Dirac-delta function at a microstate x and time t, and $F_{eq}(\lambda_0)$ is the equilibrium free energy of a macrostate λ_0. Due to the end-point conditioning by δ(x, t), only the paths in Λ^x among Λ contribute to the value of the left-hand side of this relation. Thus, equation (8) shows that energy E(x, t) is to be a property of the ensemble Λ^x. Accordingly, the local free energy ψ(x, t) is also a property of Λ^x as desired. We note that the average of ψ(x, t) over microstates is known as the effective free energy of a macrostate, and relations between the effective free energy and relative entropy of macrostates have been investigated in Ref. 27.

Roles of Local Free Energy ψ(x, t) within Λ^x

Now, we derive fundamental relations for each non-equilibrium microstate, which share their mathematical forms with those of classical and modern theories.

Firstly, we show that the newly introduced local free energy ψ(x, t) is a critical quantity in determining the non-equilibrium probability of each microstate x. As above, we explicitly calculate the ratio of the total weight of paths in Λ^x to that in Λ. Then, from equation (6) and equation (7), we have the following:

$$p(x,t) = \frac{e^{-\beta E(x,t)}}{e^{-\beta\psi(x,t)}}. \tag{9}$$

This relation shows that local free energy ψ(x, t) plays a role analogous to equilibrium free energy for non-equilibrium microstates. We stress that p(x, t) is expressed solely by (micro-)state functions (i.e., true thermodynamic potentials) of the ensemble Λ^x, which is independent of specific paths to realize a microstate x. Here it may be worth noting that the structure of our theory is not dependent on the resolution used for specifying a microstate (see Methods). In the literature, equation (9) appears implicitly either from definitions of path-dependent stochastic entropy and free energy, or mixed with work functions (path functions) if an ensemble is considered[28,29]. For the latter, p(x, t) could be written in the above simple form as a property of the ensemble Λ^x only if the relation between ψ(x, t) and work functions were resolved as shown later in equation (13).

Secondly, we prove a local version of the Crooks relation. The original Crooks relation is a core equation that can generate various modern fluctuation theorems within Λ focusing on implementing a macrostateλ_t from an initial ensemble[28]. The local Crooks relation that we prove holds while implementing a single microstate, and thus it can generate various fluctuation theorems within Λ^x as will be shown below. Using equation (5) and equation (9), we have

$$\frac{p_\Lambda(l)}{p'_\Lambda(l')}=e^{\beta(\Delta E-\Delta\psi-Q)}, \quad (10)$$

where $\Delta E = E(l_t, t) - E(l_0, 0)$, and $\Delta\psi = \psi(l_t, t) - \psi(l_0, 0)$. Applying the first law as in equation (3), we have $\frac{p_\Lambda(l)}{p'_\Lambda(l')}=e^{\beta(W-\Delta\psi)}$. Now we consider path probabilities within Λ^x so that $p_\Lambda(l)=p(l_t,t)p_{\Lambda^x}(l)$ and $p'_\Lambda(l')=p'(l'_0,0)p'_{\Lambda^x}(l')$. Due to equation (4), we obtain the following localCrooks relation:

$$\frac{p_{\Lambda^x}(l)}{p'_{\Lambda^x}(l')}=e^{\beta(W-\Delta\psi)}. \quad (11)$$

This relation implies that fluctuation theorems for (realizing) a macrostate remain the same even if we focus on a single microstate. Most importantly, the local Crooks relation provides a physical interpretation for local free energy $\psi(x, t)$. To see this, we integrate equation (11) over the ensemble Λ^x, giving:

$$\left\langle e^{-\beta(W-\Delta\psi)}\right\rangle_{\Lambda^x}=1, \quad (12)$$

where the bracket indicates an average taken over all paths to x at timet. If we assume that p(x, 0) is the Boltzmann-Gibbs distribution (then, we have $\psi(x, 0) = F_{eq}(\lambda_0)$ for all x (see equation (15) below)), we may rewrite equation (12) as follows:

$$\left\langle e^{-\beta W}\right\rangle_{\Lambda^x}=e^{-\beta\Delta\psi}. \quad (13)$$

This is nothing but the local form of Jarzynski's relation, expressed for each microstate, revealing that the local free energy $\psi(x, t)$ encodes (regulates) work contents for realizing a single microstate x from an initial ensemble. This role of $\psi(x, t)$ is analogous to that of equilibrium free energy $F_{eq}(\lambda_t)$, as expected. As a corollary, we have

$$\langle W\rangle_{\Lambda^x}\geq\Delta\psi, \quad (14)$$

which shows that average efficiency of the conversion from work to ψ is less than 100%, indicating that the second law of thermodynamics holds even within the ensemble of realizations of each microstate x. It should be noted that local ensemble Λ^x is critical to reveal this role of $\psi(x, t)$: This term would otherwise be integrated out to give equilibrium free energy (see equation (15)). We remark that equation (13) is a highly desired generalization of Landauer's principle, which quantifies fluctuations in erasure processes[30,31] (see Examples and also Discussion).

Roles of Local Free Energy $\psi(x, t)$ between Λ^xs

Finally, we prove two inter-microstate relations. The first relation explains how non-equilibrium work measurement could give equilibrium free energy. Since the accessible number of microstates, $\int e^{-\beta E(x,t)}dx$, is a function of an equilibrium free energy $F_{eq}(\lambda_t)$ of a stateλ_t, equation (9) implies the following link between the instant distribution for local free energy ψ and an equilibrium free energy for the corresponding macrostate:

$$\left\langle e^{-\beta\psi(x,t)}\right\rangle_x=e^{-\beta F_{eq}(\lambda_t)}, \quad (15)$$

where the bracket indicates the average over all microstates (chain configurations). According to equation (13), non-equilibrium work measurement within Λ^x gives non-equilibrium free energy $\psi(x, t)$. Thus if the measurement is done over all paths in Λ, it corresponds to taking the average of ψ, which gives equilibrium free energy as expressed in equation (15). As a corollary, we have $\langle\psi(x, t)\rangle_x \geq F_{eq}(\lambda_t)$. In equilibrium,$\psi(x, t) = F_{eq}(\lambda_t)$ holds for all x due to the strict convexity of the exponential function[32]. This condition together with equation (9) gives the Boltzmann-Gibbs distribution, and clearly characterizes the meaning of the "least biased" in an equilibrium as presented by Jaynes[33]. Boltzmann assigned each microstate an equilibrium thermal quantity: The above equilibrium condition for $\psi(x, t)$ shows that this is valid only in equilibrium.

Next, we demonstrate the critical role of local free energy $\psi(x, t)$ in non-equilibrium relaxation. We consider a Brownian system described by the following Langevin equation, which is a minimal prototype that contains the stochasticity: $\zeta\dot{x}=-\nabla E(x,t)+\xi$, where ζ is the friction constant and ξ is the fluctuating force that satisfies the fluctuation-dissipation theorem, $\langle\xi(t)\xi(t')\rangle = 2k_BT\zeta\delta(t-t')$. Then, the probability density p(x, t) of this system obeys the Fokker-Planck equation as $\zeta\frac{\partial p(x,t)}{\partial t}=\frac{\partial(\nabla E(x,t)p(x,t))}{\partial x}+k_BT\frac{\partial^2 p(x,t)}{\partial x^2}$. From equation (9), the time evolution of local information $\varphi(x, t)$ is described by

$$\frac{\partial\phi(x,t)}{\partial t}+\vec{v}(x,t)\cdot\nabla\phi(x,t)=-\nabla\cdot\vec{v}(x,t), \quad (16)$$

where $\vec{v}(x,t)=-(1/\zeta)\nabla\psi(x,t)$. We note that this flow driven by local free energy gradient should be interpreted as information flow (not entropy flow) (see Fig. 2b). When information flows from x to x′ with a loss, we may say that entropy flows with an increase to the opposite direction, x′ to x, since the sum of them is conserved locally. Thus, equation (16) tells us that for regions of high work content, ψ spontaneously decreases through the exchange of local information and entropy until all available sources of energy are consumed so that driving force $\psi(x, t)$ becomes constant for all x, i.e.,$E(x, t) = T\sigma(x, t)$ for all x (up to an additive constant). In a non-equilibrium steady state, the relation $\vec{v}\cdot\nabla\phi+\nabla\cdot\vec{v}=0$ should be satisfied. Here we emphasize that the above equation is a closed form partial differential equation, which is for the information content between ensembles Λ^x, i.e., the quantity independent of individual paths, or (micro-)state functions. This leads to essential differences from apparently similar dynamical equations, e.g., the ordinary differential evolution equation for macroscopic Shannon-entropy[34] and the ordinary differential equation for entropy production along stochastic paths[12].

Examples

We present two examples of non-equilibrium systems where the concept of local free energy allows elegant representation of all relevant phenomena.

In the first example, we will see that local free energy $\psi(x, t)$ provides a new insight for understanding the cyclic behavior of the famous Szilard engine[35]. In step (1), a single particle in a box of volume V in contact with a heat bath of temperature T is in equilibrium. For analysis, we will split the phase-space of the particle into two: the left-half of the box and the right-half denoted as L and R, respectively, and use this coarse-grained phase-space (see Methods). Let $\beta = 1$ for a while for simplicity. Then, the free energy of the system is $F_{eq} = -\ln Z$, where $Z = \int_V e^{-E(x,t)} dx$. In step (2), a partition is inserted in the middle of the box. Before measurement, the local free energy of the left and the right would be the same, $\psi(L) = \psi(R) = -\ln Z_1 - \ln 2$, where $Z_1 = \int_{V/2} e^{-E(x,t)} dx$ (see equation (21)). Note that the local free energy is not different from the initial equilibrium free energy at this point, i.e., $\psi(L) = \psi(R) = F_{eq}$. In step (3), we measure where the particle is, and assume that it was left without loss of generality. Then, we would have $\psi(L) = -\ln Z_1$. Since ψ is a true thermodynamic potential, independent of paths realizing a stateL as shown in equation (13), we can deduce that work should be consumed upon measurement such that $\langle W \rangle_L \geq \Delta\psi = \ln 2$ on average (equation (14)). In step (4), we link a weight in accordance with the observation in step (3) and convert heat into work by the amount of ln 2. Thus, it does not violate the second law of thermodynamics.

Here we note that a similar argument also applies to Landauer›s erasure process[36]. In detail, the first and the final stage of the erasure process are identical to step (2) and step (3) of the above example of the Szilard engine, respectively. In this case, however, step (3) is achieved not by measurement but by perfectly localizing the particle to the left irrespective of its initial position (e.g., by extracting the partition separating the two halves of the box and subsequently using a piston to shift the particle to the left side of the box). In any case, consumed work is characterized by the state function $\psi(L)$, independently of paths to realize the state. We remark that fluctuations of work (not just average work) are exactly regulated by $\Delta\psi$, as expressed in equation (13).

In the second example, we illustrate how our theory converts raw data collected from fluctuating degrees of freedom into a useful set of state functions as well as the important role of local free energy $\psi(x, t)$ in non-equilibrium relaxation, both by means of Brownian dynamics simulations. We consider a Brownian particle under a tilted bi-stable potential (Fig. 3e), which has a uniform distribution at time $t = 0$. We repeated the in-silico experiment, counted the number of particles for each partitioned bin at each time, and calculated local information and free energy. Recall that information is the accessible number of paths to a point x. Due to the imposed initial condition of constant local information, the local free energy ψ initially has the same shape as the energy $E(x, t)$ over all microstates (Fig. 3a). We observe that it is ψ that drives the information flow to establish local equilibrium as an intermediate step (Fig. 3b) until global equilibrium is reached (Fig. 3d). This is an illustration of an exact mechanism for non-equilibrium relaxation (as expressed by equation (16)), which, to our knowledge, has not yet been achieved.

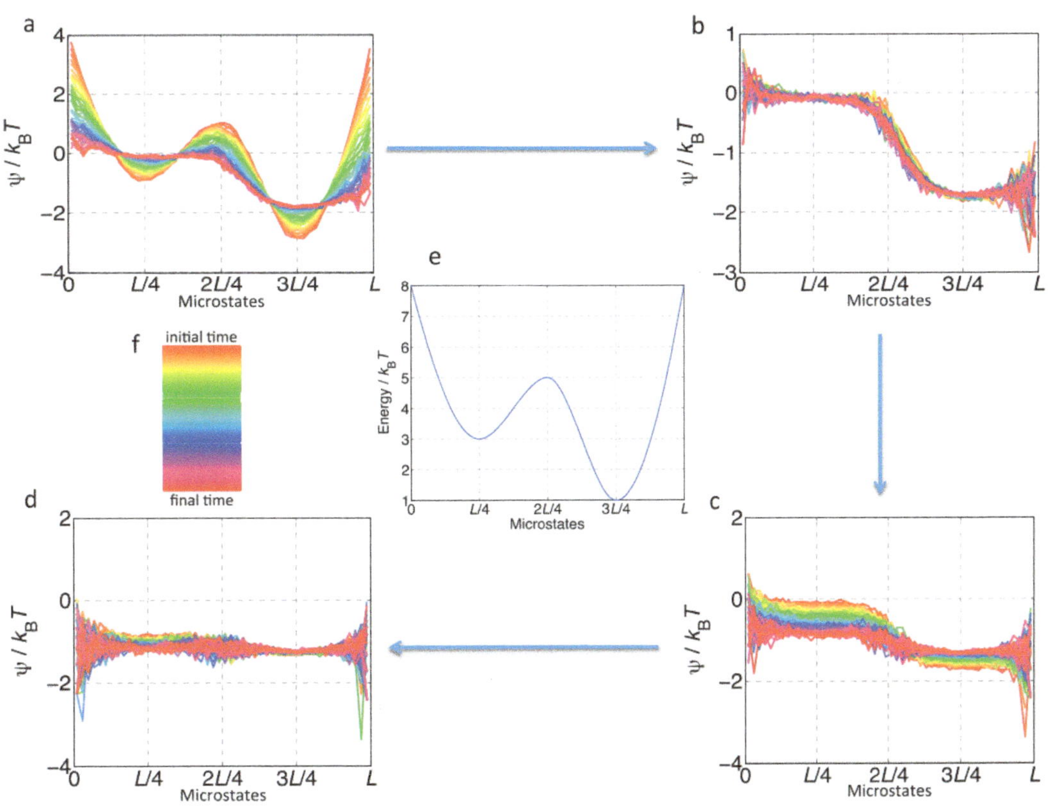

Figure 3: The mechanism of equilibration. Here, we describe a stochastic system in terms of local information and local free energy extracted from observations of stochastic trajectories, and show the exact mechanism of equilibration. A stochastic evolution in silico of a Brownian particle in a biased potential (e) is carried out 20,000 times with uniform initial distribution, and the time series of local free energy $\psi(x, t_i)$ profiles

are shown. The color code is shown in (f). The time-step taken is 0.01 in a dimensionless unit. (a–d) shows the profiles of free energy $\psi(x, t_i)$ from t_0 to t_{50}, from t_{51} to t_{100}, from t_{101} to t_{2000}, and from t_{2001} to t_{4000}, respectively. (a), Due to the energy barrier in the transition state information flows into local minimum regions. As it progresses, the barrier is lowered. (b), A local equilibrium is established where the probability of microstates is proportional to the Boltzmann factor in local regions. At this stage, the bump has disappeared. (c), A global equilibration proceeds slowly compared to the first local-equilibration process. The work potential $\psi(x, t)$ drives information to the right region. (d), A global equilibrium is established.

DISCUSSION

We have identified a new ensemble for each microstate, and revealed the important roles played by local free energy $\psi(x, t)$ of the new ensemble. Most importantly, $\psi(x, t)$ turns out to be a key quantity for generalizations of the Boltzmann-Gibbs distribution and Landauer's principle to include arbitrary fluctuations, and controls non-equilibrium relaxation. In the literature, related equations to this manuscript appear implicitly in various forms without awareness of the role of the local free energy $\psi(x, t)$. For example, the most representative form for the generalized Boltzmann-Gibbs distribution appears in the literature[28,29] as follows:

$$\frac{p(x,t)}{p_{eq}(x,t)} = \frac{e^{-\beta\Delta F}}{\langle e^{-\beta W}\rangle_{\Lambda^x}}, \tag{17}$$

where $p_{eq}(x, t)$ is the Boltzmann-Gibbs distribution, $\Delta F = F(\lambda_t) - F(\lambda_0)$ that is the difference between equilibrium free energy. This relation is a combined form of equations (9), (13), and a corollary of (15). In Ref. 37, equation (9) arises simply from the definitions of path-dependent entropy and free energy without a significant relationship between $\psi(x,t)$ and work as given in equation (13). In Ref. 31, fluctuations in Landauer's principle have been investigated using equation (17) to find Landauer's bound without awareness of the role, i.e., the exact quantification of the fluctuation, played by $\psi(x, t)$. Very recently, symmetry-breaking energetics (e.g., for a Brownian particle that is experiencing a switching from a single-well to a double-well potential and is then confined in one well resulting in loss of ergodicity) has been investigated in Ref. 38, and it has been shown that

$$\langle W\rangle_i^{(SB)} - \Delta F_i \geq k_B T \ln p_i, \tag{18}$$

where $\langle W\rangle_i^{(SB)}$ is the average work required for symmetry-breaking, $\Delta F_i = F_i - F_{eq}(\lambda_0)$, where F_i satisfies $e^{-\beta F_i} = \int_i e^{-\beta E(x,t)}dx$ and is called conformational free energy of the trapped region i of the phase-space, and p_i is the probability of trapping in the region i. If we rewrite equation (18) using the coarse-grained phase-space notation (see Methods) and equation (22), we have

$$\langle W\rangle_i^{(SB)} \geq \Delta\psi, \tag{19}$$

where $\Delta\psi = \psi(i, t) - F_{eq}(\lambda_0)$. This relation is identical to equation (14), a corollary of our main theorem, equation (13), indicating that local free energy ψ regulates fluctuations in symmetry-breaking energetics. Overall, these examples indicate the universal importance of local free energy $\psi(x, t)$ in the thermodynamic expressions.

The second issue we discuss concerns structural similarity between an equilibrium macrostate and a non-equilibrium microstate equipped with the new ensemble Λ^x. We showed that the Boltzmann-Gibbs distribution and Jarzynski's relation, which hold for (implementing) a macrostate, can be expressed in a similar manner while implementing a non-equilibrium microstate. Moreover, local functions are well-defined state functions of the new ensemble Λ^x, i.e., true thermodynamic potentials that are independent of paths of realizing a (micro-) state, playing similar roles to equilibrium macrostate functions like free energy. If we pursue this direction further, we may borrow various notions from equilibrium thermodynamics. For example, we may define non-equilibrium temperature locally as $dE(x, t)/d\sigma(x, t)$. In principle, this approach enables us to transfer all the techniques of equilibrium thermodynamics to analyse non-equilibrium local objects. We believe that there is significant value in investigating the usefulness of such an approach in non-equilibrium systems.

METHODS

Practical Definition of a Microstate

Here we briefly mention a subtlety associated with the definition of a microstate by using a polymer configuration as its example. When we try to specify a configuration of a polymer chain at time t, there is always a problem of the space-time resolution that is used to distinguish different microstates. This is particularly the case for experiments and simulations. This resolution problem can be resolved by a coarse-grained partition of the phase space. Let $\{\chi_j | j = 1, \cdots, K\}$ be a partition of the phase space with K non-overlapping subsets χ_j. We may regard χ_j as a coarse-grained microstate. The information of χ_j is defined as follows:

$$\phi(\chi_j,t) = \ln\Omega(\Lambda^{\chi_j}), \tag{20}$$

where $\Omega(\Lambda^{\chi_j}) = \sum_{x\in\chi_j}\Omega(\Lambda^x)$. Then, we apply the previous argument and obtain the non-equilibrium probability of χ_j, χ_j, $p(\chi_j,t) = e^{\phi(\chi_j,t)-\phi_0}$. In this case, we shall define the free energy of χ_j as follows:

$$\psi(\chi_j,t) = E(\chi_j,t) - T\sigma(\chi_j,t), \tag{21}$$

where $\sigma(\chi_j, t) = k_B(\varphi_0 - \varphi(\chi_j, t))$, identical to stochastic entropy, and $E(\chi_j, t)$ is defined such that $e^{-\beta E(\chi_j,t)} = \int_{x\in\chi_j} e^{-\beta E(x,t)}dx$. Then, we have

$$p(\chi_j,t) = \frac{e^{-\beta E(\chi_j,t)}}{e^{-\beta\psi(\chi_j,t)}}. \tag{22}$$

We note two things here: Firstly, the structure of our theory is invariant under the transformation of a microstate x to a coarse-grained one χ_j. Secondly, when we do not divide fluctuating degrees of freedom, our theory reduces to global theorems in a natural way. For example, let a single coarse-grained microstate χ be the set of all fluctuating degrees of freedom. Then $\psi(\chi, t)$ becomes $F_{eq}(\lambda_t)$ from equation (21). We see immediately that equations (11)–(14) in the main text reduce to the corresponding known global relations, and equations (9) and (15) reduce to trivial identities. Similarly, it may be instructive to apply the same framework to an equilibrium situation, where a system is in a stationary state with thermal fluctuations. Even in an equilibrium situation, we can think of an initial state and a state at time t. Thus we can still define $\varphi(x,t)$ and $\psi(x, t)$. Our intuition immediately tells us that $\sigma(x, t)$ and $\psi(x, t)$ should both be independent of x and t, suggesting that their averages are standard entropy and free energy, respectively.

Experimental Measurement

As an example, we consider a polymer chain (in a heat bath of temperature T) whose one end is fixed and the other end is under an external control λ_τ that varies over time $0 \le \tau \le t$ in a well-defined manner, as shown in the right panel of Fig. 1. We prepare the system initially to be sampled according to a probability density p(x, 0). Let us assume that p(x, 0) is the Boltzmann-Gibbs distribution for simplicity. We carry out the experiment by controlling λ_τ during $0 \le \tau \le t$, and measure work W done on the system. If we have additional information on the final microstate, our theory converts the fluctuating details intoφ and ψ (a microstate can be a coarse-grained one). We repeat the process N times, a large number. Let N_x be the number of the experiments whose final microstate is x. Here we put a number to pathi reaching microstate x from i = 1 to N_x and express the work done in path i as W_i. Then, we obtain the energy of x using the following equality for Λ^x:

$$\frac{1}{N}\sum_{i=1}^{N_x} e^{-\beta W_i} = \frac{e^{-\beta E(x,t)}}{e^{-\beta F_{eq}(\lambda_0)}},$$

as proven in Ref. 14. Here $F_{eq}(\lambda_0)$ is the equilibrium free energy of the initial macrostate. It is important to note that the normalization factor isN. If the summation is taken over all experiments, it reduces to the Jarzynski relation. We also proved that

$$\frac{1}{N_x}\sum_{i=1}^{N_x} e^{-\beta W_i} = \frac{e^{-\beta\psi(x,t)}}{e^{-\beta F_{eq}(\lambda_0)}},$$

giving $\psi(x, t)$. It is important that the normalization factor here is N_x. These two equalities are linked by the the relation between $\psi(x, t)$ and$F_{eq}(\lambda_t)$ in equation (15).

ACKNOWLEDGMENTS

It is a pleasure to thank C. Hyeon and H. Park in KIAS for discussions. We are also grateful to Taiki Yanagishima for critical reading of our manuscript. L.J. was supported by the National Research Foundation of Korea Grant funded by the Korean Government (NRF-2011-013-C00008), and in part by Kwangwoon University Research Year Grant in 2011. H.T. acknowledges Grants-in-Aid for Scientific Research (S) and Specially Promoted Research from the Japan Society for the Promotion of Science (JSPS), and the Aihara Project, the FIRST program from JSPS, initiated by the Council for Science and Technology Policy (CSTP).

REFERENCES

1. Callen, H. B. Thermodynamics and an Introduction to Thermostatistics (John Wiley, New York, 1985).
2. Spinney, R. & Ford, I. Nonequilibrium Statistical Physics of Small Systems: Fluctuation Relations and Beyond, chap. Fluctuation Relations: A Pedagogical Overview, 3–56 (Wiley-VCH, Weinheim, 2012).
3. Evans, D. J. & Searles, D. J. The fluctuation theorem. Adv. Phys. 51, 1529–1585 (2002).
4. Seifert, U. Stochastic thermodynamics, fluctuation theorems and molecular machines. Rep. Prog. Phys. 75, 126001 (2012).
5. Evans, D. J., Cohen, E. G. D. & Morriss, G. P. Probability of second law violations in shearing steady states. Phys. Rev. Lett. 71, 2401–2404 (1993).
6. Gallavotti, G. & Cohen, E. G. D. Dynamical ensembles in nonequilibrium statistical mechanics. Phys. Rev. Lett. 74, 2694–2697 (1995).
7. Kurchan, J. Fluctuation theorem for stochastic dynamics. J. Phys. A: Math. Gen. 31, 3719 (1998).
8. Crooks, G. E. Entropy production fluctuation theorem and the nonequilibrium work relation for free energy differences. Phys. Rev. E 60, 2721–2726 (1999).
9. Hatano, T. & Sasa, S.-I. Steady-state thermodynamics of Langevin systems. Phys. Rev. Lett. 86, 3463–3466 (2001).
10. Kawai, R., Parrondo, J. M. R. & den Broeck, C. V. Dissipation: The phase-space perspective. Phys. Rev. Lett. 98, 080602 (2007).
11. Sekimoto, K. Langevin equation and thermodynamics. Prog. Theor. Phys. Suppl. 130, 17–27 (1998).
12. Seifert, U. Entropy production along a stochastic trajectory and an integral fluctuation theorem. Phys. Rev. Lett. 95, 040602 (2005).
13. Jarzynski, C. Nonequilibrium equality for free energy differences. Phys. Rev. Lett. 78, 2690–2693 (1997).
14. Hummer, G. & Szabo, A. Free energy reconstruction from nonequilibrium singlemolecule pulling experiments. Proc. Nat. Acad. Sci. USA 98, 3658–3661 (2001).
15. Liphardt, J., Onoa, B., Smith, S. B., Tinoco, I. & Bustamante, C. Reversible unfolding of single RNA molecules by mechanical force. Science 292, 733–737 (2001).
16. Liphardt, J., Dumont, S., Smith, S., Tinoco Jr, I. & Bustamante, C. Equilibrium information from nonequilibrium measurements in an experimental test of Jarzynski's equality. Science 296, 1832–1835 (2002).
17. Trepagnier, E. H. et al. Experimental test of Hatano and Sasa's nonequilibrium steady-state equality. Proc. Nat. Acad. Sci. USA 101, 15038–15041 (2004).
18. Collin, D. et al. Verification of the Crooks fluctuation

theorem and recovery of RNA folding free energies. Nature 437, 231–234 (2005).

19. Alemany, A., Mossa, A., Junier, I. & Ritort, F. Experimental free-energy measurements of kinetic molecular states using fluctuation theorems. Nature Phys. 8, 688–694 (2012).
20. Jarzynski, C. Equalities and inequalities: Irreversibility and the second law of thermodynamics at the nanoscale. Annu. Rev. Codens. Matter Phys. 2, 329–51 (2011).
21. Onsager, L. & Machlup, S. Fluctuations and Irreversible Processes. Phys. Rev. 91, 1505–1512 (1953).
22. Jarzynski, C. Hamiltonian derivation of a detailed fluctuation theorem. J. Stat. Phys. 98, 77–102 (2000).
23. Lloyd, S. Use of mutual information to decrease entropy: Implications for the second law of thermodynamics. Phys. Rev. A 39, 5378 (1989).
24. Touchette, H. & Lloyd, S. Information-theoretic limits of control. Phys. Rev. Lett. 84, 1156 (2000).
25. Sagawa, T. & Ueda, M. Generalized Jarzynski equality under nonequilibrium feedback control. Phys. Rev. Lett. 104, 090602 (2010).
26. Sagawa, T. & Ueda, M. Fluctuation theorem with information exchange: Role of correlations in stochastic thermodynamics. Phys. Rev. Lett. 109, 180602 (2012).
27. Esposito, M. & Van den Broeck, C. Second law and Landauer principle far from equilibrium. Europhys. Lett. 95, 40004 (2011).
28. Crooks, G. E. Path-ensemble averages in systems driven far from equilibrium. Phys. Rev. E 61, 2361–2366 (2000).
29. Vaikuntanathan, S. & Jarzynski, C. Dissipation and lag in irreversible processes. Europhys. Lett. 87, 60005 (2009).
30. Dillenschneider, R. & Lutz, E. Memory erasure in small systems. Phys. Rev. Lett. 102, 210601 (2009).
31. Be´rut, A., Petrosyan, A. & Ciliberto, S. Detailed Jarzynski equality applied to a logically irreversible procedure. Europhys. Lett. 103, 60002 (2013).
32. Cover, T. M. & Thomas, J. A. Elements of information theory (Wiley-Interscience, New York: 2012).
33. Jaynes, E. T. Information theory and statistical mechanics. Phys. Rev. 106, 620–630 (1957).
34. Imparato, A. & Peliti, L. Fluctuation relations for a driven Brownian particle. Phys. Rev. E 74, 026106 (2006).
35. Leff, H. S. & Rex, A. F. (eds.) Maxwell's Demon 2 Entropy, Classical and Quantum Information, Computing (Taylor & Francis, Bristol, 2002).
36. Landauer, R. Irreversibility and heat generation in the computing process. IBM J. Res. Dev. 5, 183–191 (1961).
37. Van den Broeck, C. & Esposito, M. Ensemble and trajectory thermodynamics: A brief introduction. Physica A 418, 6–16 (2015).
38. Rolda´n, E´., Martinez, I. A., Parrondo, J. M. R. & Petrov, D. Universal features in the energetics of symmetry breaking. Nature Phys. 10, 457–461 (2014).

CHAPTER 4

CALORIMETRY OF A BOSE–EINSTEIN-CONDENSED PHOTON GAS

Tobias Damm, Julian Schmitt, Qi Liang, David Dung, Frank Vewinger, Martin Weitz and Jan Klaers

Keyword: Physical sciences Atomic and molecular physics

INTRODUCTION

Below 2.2 K, liquid helium shows peculiar hydrodynamic properties, such as a flow without viscosity, the fountain effect or the formation of vortices. This transition from a normal fluid to a superfluid has been named λ-transition, which originates from the fact that plotting the heat capacity versus temperature results in a graph resembling the greek letter λ. Soon after this discovery, it has been proposed that superfluid helium forms a macroscopic matter wave as a consequence of Bose–Einstein condensation, which describes the condensation of the ideal (interaction-free) Bose gas at low temperatures due to quantum statistics. This idea proved to be fruitful despite the fact that liquid helium is far from a system of interaction-free particles. The impressive progress in the cooling of dilute atomic gases has paved the way to realize weakly interacting Bose gases at nano-Kelvin temperatures6. Here the relation to Bose–Einstein condensation has been immediately clear. Interestingly, in contrast to liquid helium and a recent measurement of a strongly interacting atomic Fermi gas, these systems have not allowed for detailed calorimetric studies up to now. Evidence for a non-classical specific heat has been reported, but the accuracy obtained in experiments with weakly interacting atomic Bose gases has not been sufficient for an unambiguous determination of the temperature dependence of the heat capacity.

Following ultracold atomic Bose gases, other physical systems have been demonstrated to undergo Bose–Einstein condensation, for example, gases of exciton–polaritons, magnons and, in previous work of our group, photons. In contrast to a three-dimensional thermal photon gas as Planck's blackbody radiation, photons can exhibit Bose–Einstein condensation, if the thermalization process is restricted to two motional degrees of freedom. Experimentally, this situation has been realized in a microcavity enclosing a dye medium, designated as a room temperature heat bath for the photon gas. Detailed experimental studies of the thermalization and condensation process, as well as the quantum statistics of the photon condensate, have revealed the signatures of an almost ideal Bose gas.

Here we report a measurement of the calorimetric properties of a Bose–Einstein-condensed photon gas, in particular, the temperature dependence of the specific heat and entropy from the classical high temperature to the quantum-degenerate regime at low temperatures. At the phase transition, the observed specific heat shows a cusp singularity, illustrating critical behavior for a photon gas analogous to the λ-transition of liquid helium.

RESULTS

Two-Dimensional Photon Gas in a Dye Micro-cavity

In our experiment (Fig. 1a), photons are captured inside a microcavity consisting of two spherically curved mirrors while repeatedly being absorbed and re-emitted by the embedded dye medium. The cavity length is of the same order as the wavelength itself, which causes a large frequency gap between the longitudinal resonator modes (free spectral range), comparable to the emission bandwidth of the dye molecules (Fig. 1b). In this situation, the resonator becomes populated by photons of a single longitudinal mode number q only, for example, q=8. While the longitudinal mode number is frozen out, the photons may populate a multitude of transversally excited cavity modes, for example, the TEM_{8xy} sub-spectrum, which makes the photon gas effectively two-dimensional. The photon energy-momentum-relation acquires a quadratic form, resembling that of a massive particle, and a trapping potential

Citation: Damm, T. et al. Calorimetry of a Bose–Einstein-condensed photon gas. Nat. Commun. 7:11340 doi: 10.1038/ncomms11340 (2016)

for the photon gas is induced by the mirror curvature. One can show that the photon gas confined in the resonator is formally equivalent to a harmonically trapped two-dimensional gas of massive bosons, described by the dispersion

$$E \approx m(c/n)^2 + (\hbar^2/2m)(k_x^2 + k_y^2) + (m\Omega^2/2)(x^2 + y^2), \quad (1)$$

with spatial coordinates x and y, transverse wave vector components k_x and k_y, trapping frequency Ω and an effective mass $m=\hbar\omega_c(n/c)^2$, where $\hbar\omega_c$ is the photon energy in the cavity ground mode with n as the refractive index of the medium and c as the vacuum speed of light.

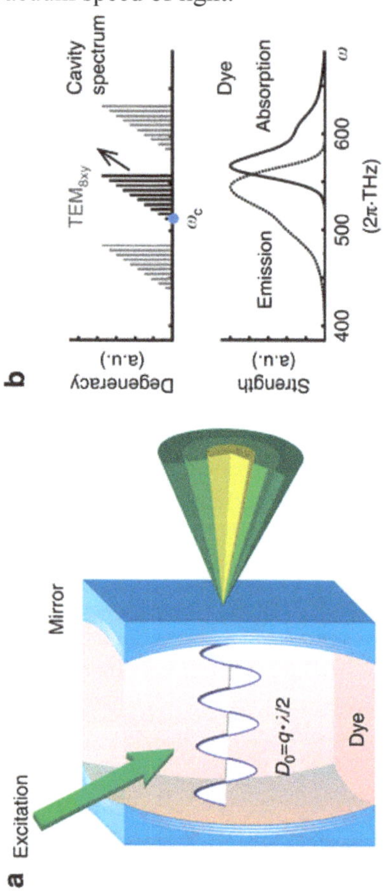

Figure 1: Bose–Einstein condensation of a two-dimensional photon gas. (**a**) Photons are captured inside a microcavity consisting of two spherically curved mirrors and get repeatedly absorbed and re-emitted by the embedded dye medium, leading to a thermalization of the photon gas to the temperature of the resonator (room temperature). (**b**) The short cavity length causes a large frequency gap between the longitudinal resonator modes (free spectral range) of order of the emission bandwidth of the dye molecules. In this situation, the resonator becomes populated by photons of a single longitudinal mode number only, here q=8. However, the photons may still populate a multitude of transversally excited cavity modes (TEM_{8xy} sub-spectrum), which effectively makes the photon gas two-dimensional. Above a critical photon number, the photon gas undergoes a Bose–Einstein condensation, leading to a massive population of the cavity ground mode (TEM_{800}). Thermodynamic information is obtained by spectroscopically analyzing the photon energy distribution across the phase transition.

Thermal equilibrium of the photon gas with the cavity environment at room temperature is achieved via repeated absorption and emission processes by the dye molecules, which establishes a thermal contact between photon gas and optical medium. Other than in a blackbody radiator, the thermalization process allows for an independent adjustment of temperature and photon number, for example, by (initial) optical pumping, which eventually goes back to a separation of energy scales of photon energy and thermal energy. In our experiment, the Bose–Einstein condensation is triggered by increasing the photon number above the saturation level at a given temperature. The corresponding critical particle number is given by

$$N_c \approx (\pi^2/3)(k_B T/\hbar\Omega)^2, \quad (2)$$

which for typical experimental parameters corresponds to $N_c \approx 90{,}000$.

One of the benefits of the given experimental system is that one can easily interpolate between equilibrium and non-equilibrium experimental conditions. Parameters, such as mirror reflectivity, dye concentration, cavity tuning and pump geometry, can be chosen such that gain and dissipation either significantly contribute to the system dynamics or effectively drop out of it. In this study, we have concentrated on the equilibrium properties of the system. Details on the corresponding experimental parameters can be found in the Methods section. In our experiments, the optical medium is pumped with a spectrally off-resonant laser source at a wavelength of λ_{exc}=532 nm, having a relatively large beam diameter of ~150 μm to keep the excitation of the medium nearly spatially homogeneous. Two acousto-optical modulators (AOMs) are used to chop the pump light to long pulses of 400ns length with a repetition rate of 400 Hz. The AOMs further control the intensity of each light pulse that allows us to adjust the average photon number with respect to the critical photon number. In our experiment, we quickly ramp the total photon number from typically 30,000 to 550,000 photons within 250 ms.

The thermodynamic properties of the system are experimentally accessible by spectroscopic means. To obtain the intracavity spectral distribution of the photons, we measure the spectral distribution outside the resonator and divide by the (wavelength-dependent) transmission coefficient of the mirrors. For this, the emitted cavity light is collimated and coupled to a monochromator (4f set-up), where it is spectrally decomposed by a diffraction grating (1,200 grooves per mm), giving an overall resolution of 0.5 nm. With this set-up, it is possible to capture the photon gas spectrum in the wavelength region starting from ≈550 nm to the cavity cutoff at λ_c=580 nm corresponding to an energy range of $\approx 4k_BT$. The latter comprises the condensate population and ≈95% of photons in excited cavity modes. The ≈5% most energetic photons of the thermal cloud are not experimentally resolvable with the present set-up.

Figure 2a shows spectra obtained at fixed temperature T=300 K for total photon numbers ranging from N≈30,000 to N≈550,000 (varying chemical potential). The observed spectra generally are in good agreement with Bose–Einstein distributions, with residual deviations at the lower photon energy part due to imperfect reabsorption. The measured energy distributions can be used to obtain full thermodynamic information of the two-dimensional photon gas. All of the derived quantities will be measured for constant volume, which here means constant (inverse squared) trapping frequency, and fixed absolute temperature T. As a first step, we have determined the condensate fraction n_0/N as a function of the reduced temperature T/T_c (Fig. 2b). From equation (2) follows that the reduced temperature T/T_c is related to the total particle number by $T/T_c=(N_c/N)^{1/2}$, with the total particle number N being obtained by integrating over the spectrum. The experimentally derived condensate fractions are slightly below the theoretical expectations shown by the solid line in Fig. 2b, describing an inverse parabola $n_0/N=1-(T/T_c)^2$. This stems from an imperfect saturation of the population in excited photon modes, which has already been observed in previous measurements, and potentially originates from a weak (thermo-optically induced) photon self-interaction.

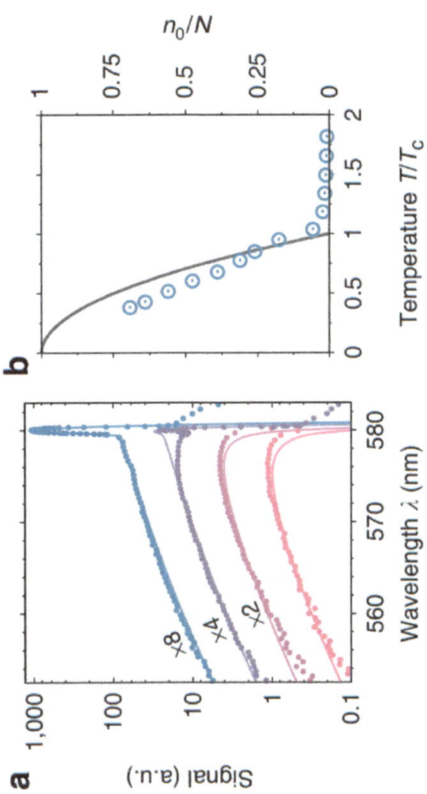

Figure 2: Spectral photon distribution and condensate fraction. (**a**) Distribution of photon energies for increasing total photon number (circles). For clarity, the spectra have been vertically shifted. The observed spectra agree well with the expected 300 K Bose–Einstein distribution functions (solid lines). (**b**) Condensate fraction n_0/N versus the reduced temperature T/T_c along with the theoretical expectation $n_0/N=1-(T/T_c)^2$. Due to an imperfect saturation of the population of the thermal cloud, the condensate fraction is observed to be systematically below the theoretically expected values.

Caloric and Entropic Properties

We next determine the average energy per photon U/N, with the zero point of the energy scale being set to the energy $\hbar\omega_c=hc/\lambda_c$ of the cavity ground state (TEM_{00} mode), corresponding to a condensate wavelength of $\lambda_c=580$ nm. On the basis of the experimentally obtained spectral photon distribution $n(\lambda)$ in the wavelength regime $\lambda\approx550$–580 nm, we extrapolate the total internal energy to be $U\approx\kappa\times\int_{550\,\mathrm{nm}}^{580\,\mathrm{nm}} n(\lambda)\, hc\, (\lambda^{-1}-\lambda_c^{-1})\, d\lambda$, with h as Planck constant and c as the vacuum speed of light. The extrapolation factor κ is uniquely determined by the assumption that the spectral distribution continues to be Boltzmann like in the experimentally not resolved wavelength regime $\lambda<550$ nm, containing the ≈5% most energetic photons of the thermal cloud. The latter sets a value of $\kappa=U/U(\lambda>550\,\mathrm{nm})\approx1.19$, with $U(\lambda>550\,\mathrm{nm})$, denoting the energy contribution of photons with wavelength $\lambda>550$ nm.

In Fig. 3a, the average energy U/N normalized to the characteristic energy at criticality k_BT_c is plotted versus the reduced temperature T/T_c, showing good agreement with the theoretical expectations for an ideal Bose gas (solid lines). At higher temperatures $T>T_c$, the energy shows a linear scaling with temperature, as expected in the classical limit, where Maxwell–Boltzmann statistics applies. In the vicinity of the condensation threshold, the energy curve changes slope, as is more clearly revealed in the graph of the heat capacity in Fig. 3b. The data points are obtained by numerically differentiating the measured energy curve with respect to temperature, following $C=\partial U/\partial T=k_B\,\partial(U/k_BT_c)/\partial(T/T_c)$. The obtained heat capacity data (circles) shows the characteristic λ-like shape as predicted theoretically (solid lines). In the high-temperature (classical) regime, C reaches a limiting value of $2k_B$ per photon. The latter stems from the four degrees of freedom, two kinetic (k_x and k_y) and two potential (x and y), which quadratically enter the photon energy of equation (1) and each contribute with $k_B/2$ to the specific heat (equipartition theorem). At criticality, the heat capacity shows a cusp with a maximum value of $C(T_c)/N=(3.8\pm0.3)k_B$, slightly below the theoretically expected value in the thermodynamic limit of $6\zeta(3)/\zeta(2)\, k_B\approx4.38k_B$. Most of this discrepancy can be explained by the finite size effects. From an exact numerical evaluation of the Bose–Einstein distribution function, one can obtain the specific heat for finite system sizes (solid lines in Fig. 3b). For the given photon numbers near $N\approx90{,}000$ at criticality, the finite size effects reduce the specific heat maximum to a value of $4.11k_B$, which agrees with the measured value within the experimental uncertainties. For $T<T_c$, the heat capacity monotonically decreases for smaller temperatures, dropping below the classical value of $2k_B$ per photon at $T/T_c\approx0.7$, and being consistent with reaching zero at T=0, as would be demanded by the third law of thermodynamics. Presently, the minimum achieved temperature is $T/T_c\approx0.4$, corresponding to condensate fractions of up to 84%, due to limited available pump power.

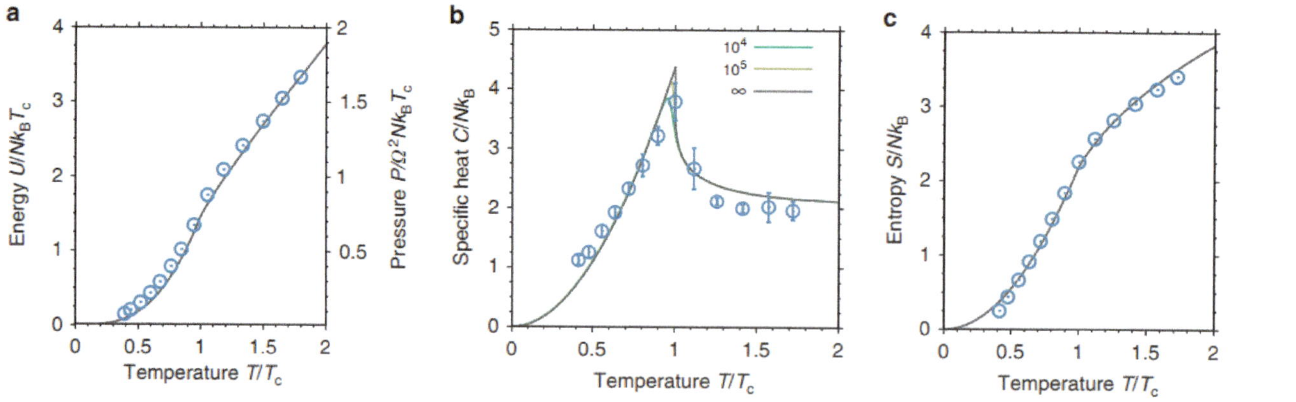

Figure 3: Caloric and entropic properties. (**a**) Internal energy (per photon) U normalized to the characteristic energy k_BT_c as a function of the reduced temperature T/T_c. In the classical high-temperature regime, energy scales linearly with temperature as is expected from Maxwell–Boltzmann-like statistics. The formation of the condensate is signalled by a change in slope close to $T=T_c$. For an ideal two-dimensional trapped Bose gas, which is realized in our experiment in good approximation, the internal energy of the photon gas can moreover be linked to its pressureP by $P=(1/2)\Omega^2U$, see axis on the right-hand side and the main text for details (Ω is the trapping frequency). Experiments are carried out at constant temperature T=300 K and photon numbers ranging from N≈30,000 to N≈550,000, whereby an increase of the photon number corresponds to a decrease in the critical temperature $T_c=T_c(N)$ of the system. s.e.m.'s are smaller than the point size. (**b**) Specific heat (per photon) versus the reduced temperature T/T_c, showing a cusp singularity at criticality (circles). In the high-temperature (classical) regime, the specific heat reaches a limiting value of $2k_B$ per photon. At criticality $T=T_c$, the heat capacity reaches a maximum value of $C(T_c)/N=(3.8\pm0.3)k_B$. The solid lines give the specific heat of the two-dimensional harmonically trapped ideal Bose gas for varying total particle numbers. The error bars indicate s.e.m. (**c**) Entropy per photon as a function of the reduced temperature T/T_c (circles), along with the theoretically expected ideal Bose gas behavior (solid line). The data are derived from a numerical integration of the specific heat curve of **b** (see text for details). The entropy monotonically decreases for decreasing temperatures, reaching a minimum value of $\approx 0.2k_B$ per photon at the lowest achieved temperature. s.e.m.'s are smaller than the point size.

The specific heat furthermore gives access to other thermodynamic quantities, as the entropy, which can be determined by the integral $S(T)=\int_0^T C(T')/T'\, dT'$, see Fig. 3c for the corresponding data. As the specific heat C(T) is not known for temperatures below $T/T_c\approx 0.4$, there is one free parameter, the constant offset $S(T/T_c\approx 0.4)$ that needs to be set to evaluate the integral and to match the experimental data (circles) to the theoretical prediction (solid line). The entropy curve monotonically decreases with decreasing temperature, reaching a minimum value of $\approx 0.2k_B$ per photon for the lowest obtained temperature. Although we presently cannot access temperatures closer to zero, the observed drop-off of the entropy curve towards lower temperatures is in accordance with the third law of thermodynamics. Note that the entropy per particle (as a function of chemical potential) also has been experimentally determined for a trapped two-dimensional atomic Bose gas, reaching entropies as low as $0.06(1)k_B$.

DISCUSSION

For a gas of non-interacting bosons, which in good approximation is realized in our experiment as demonstrated by the results given in Figs 2 and 3, one can readily link the internal energy of the gas to its pressure. In general, in the presence of a trapping potential, the pressure of a gas becomes position dependent and thus cannot serve as a global thermodynamic variable. To account for this, it has been proposed to use two global conjugated variables, harmonic volume Vand harmonic pressure P, respectively. For a harmonically trapped two-dimensional gas, the harmonic volume is defined as $V=\Omega^{-2}$, with Ω as the trapping frequency. This quantity does not have the physical units of a volume, however, it shares the same scaling with the trapping geometry as the 'true' volume of the confined gas. The harmonic pressure then follows via the usual thermodynamic relation $P=-\partial\varphi_G/\partial V$, where φ_G is the grand potential. For a two-dimensional trapped gas of non-interacting bosons, the pressure can be shown to be related to the internal energy via $P=(1/2)\Omega^2U$. Thus, Fig. 3a here does not only describe the energy as a function of temperature $U=U(T/T_c)$ (caloric equation of state) but also delivers the pressure dependence $P=P(T/T_c)$ (thermal equation of state) for a given harmonic volume V or trapping frequency Ω.

To conclude, we have determined calorimetric properties of a Bose–Einstein-condensed photon gas, in particular the temperature dependence of energy, heat capacity and entropy. Critical behavior of the photon gas is clearly demonstrated by a cusp in the specific heat curve at the condensation threshold. For the chosen experimental conditions, for example, sufficient photon reabsorption and nearly homogenous pump geometry, we do not observe significant deviations from the theoretically expected behavior of a fully equilibrated Bose gas. In comparison with other systems exhibiting Bose–Einstein condensation, the here-investigated photon gas comes closest to an ideal, that is, interaction-free, gas of bosons, allowing to match experimental results with precise, and even exact, theoretical predictions. In particular, we find that the experimentally determined specific heat of the photon gas agrees with the quantum statistical predictions down to the level of finite size effects. For the future, the spectroscopic calorimetry of a quantum-degenerate photon gas could lead to new experimental schemes for precision measurements of thermodynamic quantities as the Boltzmann constant.

METHODS

Microcavity Set-Up

In our experiment, photons are stored inside a microcavity build up from two gyro-quality mirrors. The dielectric, spherically shaped mirrors have a reflectivity in excess of $r \approx 0.99997$ in the relevant wavelength region of this experiment (λ=530–590 nm), providing a cavity finesse of order of $F \approx 10^5$ for the empty cavity. Both the mirrors share the same radius of curvature of R=1 m and are typically separated by $D_0 \approx 1.7\,\mu m$, corresponding to a free spectral range of order $\Delta\lambda_{FSR} \approx 100$ nm, which is comparable to the spectral width of the dye emission. In this situation, the dye emission is restricted to cavity modes with the longitudinal wave number q=8, effectively reducing the thermalization dynamics of the photon gas to the remaining two transversal mode numbers. The effective mass and trapping frequency introduced in equation (1) depend on the cavity geometry and typically take values ofm$\approx 6.7 \times 10^{-36}$kg and $\Omega \approx 2\pi \cdot 36.5$ GHz in our experiment. The cavity geometry is stabilized passively, by mechanical contact of the two mirrors that strongly damps fast mechanical oscillations, as well as actively utilizing a piezo translation stage that counteracts long time drifts of the resonance.

As a heat bath for the photon gas, we use a filtered solution of 10^{-3}mol l^{-1} rhodamine 6G ethylene glycol (fluorescence quantum yield $\eta \approx 0.95$, index of refraction of the solvent n=1.43). This dye fulfils the Kennard–Stepanov law $B_{21}(\omega)/B_{12}(\omega) \approx \exp(-(\omega-\omega_{ZPL})/k_B T)$, relating the Einstein coefficients of absorption $B_{12}(\omega)$ and emission $B_{21}(\omega)$ at a given frequency ω to the Boltzmann factor of that frequency (ω_{ZPL} is the zero-phonon line of the dye), which is essential for the light–matter thermalization process. The optical medium is spatially homogeneously pumped by a spectrally off-resonant laser system at a wavelength ofλ_{exc}=532 nm under an angle of $\approx 45°$ with respect to the optical axis exploiting the first reflectivity minimum at higher angles of incidence. To avoid excess population of long-lived dye triplet states and photobleaching two AOM are used to chop the pump light to pulses of 400 ns length with a repetition rate of 400 Hz.

ACKNOWLEDGEMENTS

We thank James Anglin and Henk Stoof for valuable discussions. This work has been financially supported by the DFG (We1748-17) and the ERC (INPEC).

REFERENCES

1. Tilley, D. R. & Tilley, J. Superfluidity and Superconductivity CRC Press (1990).
2. Lipa, J. A., Nissen, J. A., Stricker, D. A., Swanson, D. R. & Chui, T. C. P. Specific heat of liquid helium in zero gravity very near the lambda point. Phys. Rev. B 68, 174518 (2003).
3. London, F. The λ-phenomenon of liquid helium and the Bose-Einstein degeneracy. Nature 141, 643–644 (1938).
4. Einstein, A. Quantum theory of the monoatomic ideal gas, part II. Sitzber. Preuss. Akad. 1, 3–14 (1925).
5. Penrose, O. & Onsager, L. Bose-Einstein condensation and liquid helium. Phys. Rev. 104, 576–584 (1956).
6. Anderson, M. H., Ensher, J. R., Matthews, M. R., Wieman, C. E. & Cornell, E. A. Observation of Bose-Einstein condensation in a dilute atomic vapor. Science 269, 198–201 (1995).
7. Davis, K. B. et al. Bose-Einstein condensation in a gas of sodium atoms. Phys. Rev. Lett. 75, 3969–3973 (1995).
8. Ku, M. J. H., Sommer, A. T., Cheuk, L. W. & Zwierlein, M. Revealing the superfluid lambda transition in the universal thermodynamics of a unitary Fermi gas. Science 335, 563–567 (2012).
9. Blakie, P. B., Toth, E. & Davis, M. J. Calorimetry of Bose-Einstein condensates. J. Phys. B: At. Mol. Opt. Phys. 40, 3273–3282 (2007).
10. Ensher, J. R., Jin, D. S., Matthews, M. R., Wieman, C. E. & Cornell, E. A. Bose-Einstein condensation in a dilute gas: measurement of energy and ground-state occupation. Phys. Rev. Lett. 77, 4984–4987 (1996).
11. Shiozaki, R. F. et al. Measuring the heat capacity in a Bose-Einstein condensation using global variables. Phys. Rev. A 90, 043640 (2014).
12. Kasprzak, J. et al. Bose-Einstein condensation of exciton polaritons. Nature 443, 409–414 (2006).
13. Balili, R., Hartwell, V., Snoke, D., Pfeiffer, L. & West, K. Bose-Einstein condensation of microcavity polaritons in a trap. Science 316, 1007–1010 (2007).
14. Demokritov, S. O. et al. Bose-Einstein condensation of quasi-equilibrium magnons at room temperature under pumping. Nature 443, 430–433 (2006).
15. Klaers, J., Schmitt, J., Vewinger, F. & Weitz, M. Bose-Einstein condensation of photons in an optical microcavity. Nature 468, 545–548 (2010).
16. Klaers, J., Vewinger, F. & Weitz, M. Thermalization of a two-dimensional photonic gas in a 'white wall' photon box. Nat. Phys. 6, 512–515 (2010).
17. Schmitt, J. et al. Observation of grand-canonical number statistics in a photon Bose-Einstein condensate. Phys. Rev. Lett. 112, 030401 (2014).
18. Klaers, J., Schmitt, J., Damm, T., Vewinger, F. & Weitz, M. Statistical physics of Bose-Einstein-condensed light in a dye microcavity. Phys. Rev. Lett. 108, 160403 (2012).
19. Nyman, R. A. & Szymanska, M. H. Interactions in dye-microcavity photon condensates and the prospects for their observation. Phys. Rev. A 89, 033844 (2014).
20. Kirton, P. & Keeling, J. Nonequilibrium model of photon condensation. Phys. Rev. Lett. 111, 100404 (2013).
21. de Leeuw, A.-W., Stoof, H. T. C. & Duine, R. A. Schwinger-Keldysh theory for Bose-Einstein condensation of photons in a dye-filled optical microcavity. Phys. Rev. A 88, 033829 (2013).
22. Haugset, T., Haugerud, H. & Andersen, J. O. Bose-Einstein condensation in anisotropic harmonic traps. Phys. Rev. A 55, 2922–2929 (1997).
23. Klünder, B. & Pelster, A. Systematic semiclassical expansion for harmonically trapped ideal Bose gases. Eur. Phys. J. B 68, 457–465 (2009).
24. Kozhevnikov, A. A. Bose gas in power-like spherically

symmetric potential in arbitrary spatial dimensionality. Acta Phys. Pol. B 43, 2089–2102 (2012).

25. Marelic, J. & Nyman, R. A. Experimental evidence for inhomogeneous pumping and energy-dependent effects in photon Bose-Einstein condensation. Phys. Rev. A 91, 033813 (2015).

26. Kirton, P. & Keeling, J. Thermalization and breakdown of thermalization in photon condensates. Phys. Rev. B 91, 033826 (2015).

27. Keeling, J. & Kirton, P. Spatial dynamics, thermalization, and gain clamping in a photon condensate. Phys. Rev. A 93, 013829 (2016).

28. Tammuz, N. et al. Can a Bose gas be saturated? Phys. Rev. Lett. 106, 230401 (2011).

29. Yefsah, T., Desbuquois, R., Chomaz, L., Günter, K. J. & Dalibard, J. Exploring the thermodynamics of a two-dimensional Bose gas. Phys. Rev. Lett. 107, 130401 (2011).

30. Romero-Rochin, V. Equation of state of an interacting Bose gas confined by a harmonic trap: the role of the 'harmonic' pressure. Phys. Rev. Lett. 94, 130601 (2005).

31. de Podesta, M. et al. A low-uncertainty measurement of the Boltzmann constant. Metrologia 50, 354–376 (2013).

32. Castrillo, A. et al. The Boltzmann constant from the shape of a molecular spectral line. J. Mol. Spectrosc. 300, 131–138 (2014).

CHAPTER 5 KAPITZA RESISTANCE BETWEEN ELECTRON AND PHONON GASES IN THE 1D CASE

K. V. Reich

INTRODUCTION

In recent years, we have witnessed a growth of interest in the investigation of the processes involved in heat transfer in solid, including composite, materials. This can be traced both to present-day trends in miniaturization of electronic devices and to an ever-increasing density of electronic components per chip and impressive progress in molecular and thin-film electronics. It is these areas that provide stimuli for a search for an appropriate material to reject heat in modern electronics. Indeed, all the above systems contain in their structure nanosized regions with a high fraction of surface-bound atoms. Therefore, heat transfer in such systems differs essentially from that observed in a macroscopic case. It is known that any surface represents an obstacle to heat flow and acts as an energy barrier, which accounts for the formation of thermal resistance at the boundary between any two materials, first observed by Kapitza [1]. Reducing this resistance becomes a problem of ever-more increasing importance for nanosized systems. Recourse to high thermal-conductivity materials such as diamond, carbon nanotubes, or graphene can only alleviate but not remove the problem altogether, because the surfaces separating different components of a device always interfere with heat flow.

It thus comes as no surprise that thermal resistance is currently being actively studied in different systems, among them superlattices [2–4] and composites based both on classical semiconductors such as arsenides or nitrides [5, 6] and on nanotubes [7] and nanodiamonds [8]; surface contacts between specific materials also generate interest [9–11].

Thermal resistance attracts interest not only from experimenters but from theoreticians as well. Significantly, however, all the studies reported so far have involved only the phonon flow between two bodies, the so-called acoustic mismatch and diffuse mismatch models [12–14], even in cases where the molecular dynamics approach was invoked [15, 16]. At the same time, the case of most interest, to wit, the metal–dielectric contact, where heat is transferred between the electron gas in a metal and the phonon gas in a dielectric, has not thus far been enjoying adequate attention. This gap appears essential, because electrons are the major carriers of heat in metals, and phonons, those in a dielectric. Thus, for heat transfer to be efficient, such a mechanism should provide a major contribution to heat transfer between a dielectric and a metal [17].

Furthermore, earlier works on heat flow through a metal–dielectric surface were primarily focused on discussing a specific mechanism of interaction between electrons and phonons in a system. To cite just a few examples, analysis has been made of an interaction involving the deformation-potential interaction [18–20] or the mechanism based on the image potential of ions of the metal surface [21]. Significantly, the thermal resistance coefficient was calculated for small electron–phonon interaction coefficients only. The results obtained in these studies were only to be expected, namely, that thermal resistance decreases with increasing electron–phonon interaction coefficient. It should be stressed that the cases treated were always those of the metal–dielectric contact, while the studies attracting ever-increasing interest currently involve thermal resistance of dielectrics with nanotubes [7], which are actually semimetals.

The phenomenon itself of thermal resistance actually remains unexplored. One could consider a case, for instance, of two materials separated by an interlayer with a preset electron–phonon interaction coefficient, and ask the question of whether one could reach a low thermal resistance for large enough values of this coefficient? Another question—could one realize a situation with a large enough value of this coefficient in which the electron and the phonon would be localized close to the

Citation: K. V. Reich, "Kapitza resistance between electron and phonon gases in the 1D case," DOI: 10.1093/ptep/pts054.

interface by strong mutual interaction, so that the heat would not propagate at all? Would the thermal resistance coefficient under these conditions reach saturation or would it increase with increasing electron–phonon interaction coefficient?

To determine the thermal resistance coefficient at the boundary separating the electron from the phonon subsystems, we have considered in this work a semi-infinite electronic system coupled to the phonon subsystem at one point only (Fig. 1). In our subsequent calculations, we are going to restrict ourselves to the adiabatic case, in which the phonon frequency is assumed to be substantially smaller than the hopping integral. Nevertheless, our model is applicable to consideration of thermal current in a broad class of objects, in particular, to carbon nanotubes coated by nanodiamonds [22], as well as to Pandey chains on the surface of nanodiamonds [23, 24]. Although we treat 1D chains in this work, our findings form the basis for higher-dimensional applications in which the results are more complicated; we will present them elsewhere. Formulated in this way, the problem overlaps with another area that has been enjoying considerable interest in recent years, namely, electron–phonon interaction in thin films and nanostructures [25] and their interaction in a local region [26–29], an approach in which Holstein's model has been widely employed [30–32]. In the studies cited above, however, the results obtained were either limited to small electron–phonon interaction constants or were applicable to metals only.

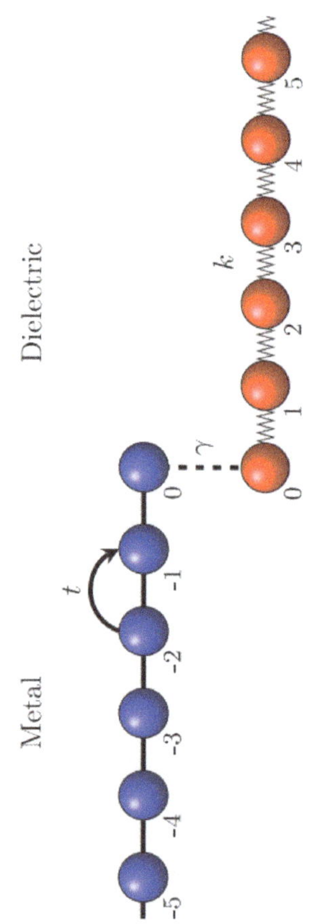

Figure 1. Schematic visualization of a semi-infinite electron system, in which the overlap integral for neighboring atoms is t, and a semi-infinite phonon system, in which elastic coupling between neighboring atoms is characterized by the elastic constant k. At site 0, local electron–phonon interaction is operative, and the electron–phonon subsystem coupling energy is γ.

All this suggests that one should consider the problem of the behavior of electrons and phonons with due allowance for local interaction between them, and the process of energy transfer between the electron and phonon gases.

MODEL

We are going to consider here a 1D problem in which the electron and phonon gases interact at one point only. The system to be studied is shown schematically in Fig. 1.

By analogy with Holstein's Hamiltonian, we write the Hamiltonian of our system with due allowance for the fact that electrons can interact with phonons at one point only, at the boundary between the electron and phonon subsystems.

$$\hat{H} = t\sum_{i=-\infty}^{0}(\hat{c}_{i+1}^{+}\hat{c}_i + \hat{c}_i^{+}\hat{c}_{i+1}) + \frac{\omega_0}{2}\sum_{i=0}^{\infty}\left(\hat{p}_i^2 + \frac{1}{2}(\hat{x}_i - \hat{x}_{i+1})^2\right) + \frac{\omega_0}{4}\hat{x}_0^2 + \gamma\hat{n}_0\hat{x}_0, \tag{1}$$

where the following notation is introduced: $\hat{c}_i^{+}, \hat{c}_i$ are the operators of creation and annihilation of an electron at the ith site. The particle number operator at the 0th site is $\hat{n_0} = \hat{c}_0^{+}\hat{c}_0$. $\hat{p}_i, \hat{x}_i$ are the operators of the momentum and displacement of the ith atom expressed in terms of the operators of creation $\hat{a}_i^{+}$ and annihilation $\hat{a}_i$ of phonons: $\hat{p}_i = i(\hat{a}_i^{+} - \hat{a}_i), \hat{x}_i = (\hat{a}_i^{+} + \hat{a}_i)$. ω_0 is the vibration frequency of atoms, and γ is the coefficient of interaction between an electron and displacement at the 0th site. We are going to assume in what follows that $\hbar = 1$. Although we are going to disregard in what follows spin degrees of freedom and direct interaction among electrons, it can be assumed in the first approximation that this will result only in a decrease of the electron–phonon interaction constant γ [33].

Our major problem lies in finding the Green's function for electrons and phonons G_{ij} = $-\langle T_\tau \hat{c}_i(\tau)\hat{c}_j^{+}(0)\rangle$, $\mathcal{D}_{ij} = -\langle T_\tau \hat{x}_i(\tau)\hat{x}_j(0)\rangle$, and the corresponding spectral functions $A_{ij}(\omega)$ = $-2\Im\mathcal{G}_{ij}(i\omega), B_{ij}(\omega) = -2\Im\mathcal{D}_{ij}(i\omega)$.

To calculate the spectral functions A and B, we shall need to derive a system of equations for the $\hat{x}, \hat{c}$ operators, which can be readily obtained from the equation of motion $\partial\hat{c}/\partial\tau = [\hat{K}, \hat{c}]$:

$$\frac{\partial\hat{c}_i}{\partial\tau} = \mu\hat{c}_0 - \delta_{i,0}\gamma\hat{x}_0\hat{n}_0 - t(\hat{c}_{i+1} + \hat{c}_{i-1})$$

$$\frac{2}{\omega_0^2}\frac{\partial^2\hat{x}_i}{\partial\tau^2} = \frac{2\gamma}{\omega_0}\hat{n}_0\delta_{i,0} + 2\hat{x}_i - (\hat{x}_{i-1} + \hat{x}_{i+1}), \tag{2}$$

where $\hat{K} = \hat{H} - \mu\hat{N}$, μ is the chemical potential, and $\hat{N}$ is the operator of the number of particles $\hat{c}_1 = \hat{x}_{-1} = 0$ are the boundary conditions for this system of equations.

Noninteracting Case

The qualitative solution of Eq. (1) for γ=0 is well known. The electron spectrum will depend on the wave vector k and the distances between neighboring atoms a: $\varepsilon(k) = 2t\cos(ka)$, with energy reckoned from the band center. Thus for the width of the electron band we obtain 4t. The phonon spectrum is given by the relation $\omega(k) = \sqrt{2}\omega_0 \sin(ka)$. In addition to the conventional, bulk states, the solution provides surface states as well for both the electrons and phonons. In the γ=0 case, a set of equations (2) allows an easy solution:

$$\mathcal{G}^0_{ij} = -\frac{1}{t}F_{ij}\left(-\frac{i\omega+\mu}{t}\right)$$

$$\mathcal{D}^0_{ij} = \frac{1}{\omega_0}F_{ij}\left(2\left(\frac{i\omega}{\omega_0}\right)^2-2\right), \quad (3)$$

where

$$F_{ij}(\Omega) = \frac{1}{\pi}\int_{-\pi}^{\pi} dx \frac{\sin(j+1)x\sin(i+1)x}{\Omega+2\cos(x)+i0}, \quad (4)$$

which can be readily verified by straightforward substitution. We are going to resort frequently in what follows to the use of this function for the values i,j=0, in which case the expression for F reduces to $F_{00} = \exp[i \arccos(\Omega/2)]$.

To the parameters of the problem introduced earlier, t, ω_0, μ, γ, we may add T, the temperature in the system. As pointed out in the Introduction, we shall restrict ourselves to the most important cases, namely, we shall assume that T, ω_0«t is the signature of the adiabatic case for temperatures low compared with the band width. Also, we shall consider below only two cases, called conventionally here "metallic", when the free electron concentration in a system is high, and the chemical potential is at the band center μ=0, and "semimetallic", if the free electron concentration is low, and the chemical potential level is at the band bottom μ=–2t. Significantly, in the case under consideration, the chemical potential μdoes not depend on the interaction parameter γ, because the infinite electron and phonon subsystems interact at one point only. This point interaction cannot affect the chemical potential of the infinite electron system. Treated in this way, chemical potential is a specified external parameter that is the same for all points of the electron subsystem.

We turn now to an analysis of the Green's functions for electrons and phonons assuming point interaction.

Electron Green's Function

To find the electron Green's function, we shall use the recursive method developed by Cini [34], which was demonstrated [31, 35, 36] to be applicable to calculation of the Green's function of the electron engaged in point interaction with phonons.

We diagonalize Hamiltonian (1) over the phonon variables:

$$\hat{H} = t\sum_{i=0}^{\infty}\varepsilon_i\hat{c}_i^+\hat{c}_i + \omega_0\sum_{i=0}^{\infty}\omega_i\hat{a}_i^+\hat{a}_i + \sum_{j=0}^{\infty}\gamma_j\hat{n}_0\hat{x}_j. \quad (5)$$

Here ωi specifies the phonon spectrum in half-space γ'j=γU_0j and U_0j is the eigenvector for the matrix of transition from representation (1) to (5); as we shall see further on, its actual form is of no significance to us here.

Consider the probability amplitude for the transition from site i to site j accompanied by emission of k phonons:

$$\Psi_{ij}(k;i\omega) = \sum_k \gamma_0\cdots\gamma_{mk}\left\langle\prod_k a_k^{m_k}\hat{c}_i\frac{1}{i\omega-\hat{H}}\hat{c}_j^+\right\rangle. \quad (6)$$

The summation here is run over all possible combinations of phonon variables, with their total number k. We readily see that the electron Green's function is directly connected with this quantity: $\mathcal{G}_{ij}(i\omega) = \Psi_{ij}(0;i\omega)$.

We have now to identify in Hamiltonian (5) the terms accounting for the noninteracting phonons and electrons $\hat{H}_0$ and their interaction $\hat{H}_1$: $\hat{H} = \hat{H}_0 + \hat{H}_1$. This opens the way to deriving recurrence relations for the quantities Ψij(k;iω) by averaging the operator identity:

$$\frac{1}{\omega-\hat{H}} = \frac{1}{\omega-\hat{H}_0} + \frac{1}{\omega-\hat{H}_0}\hat{H}_1\frac{1}{\omega-\hat{H}}.$$

Significantly, at this point, an essential simplification is introduced in Cini's method; more specifically, dispersion is assumed to be absent:ωi=ω_0. This approach is justified not only for the optical but for acoustic phonons as well. This can be partially assigned to the fact that in this way one does not neglect spatial dispersion for the coupling constant γ'. It was shown [37] that this method can be generalized by including spatial dispersion of frequencies. In this case, for the Green's function, one obtains a perturbation theory series, but it will not differ qualitatively from the results derived in the above approximation.

One can write in the above approximations the following recurrence relations for the Ψij(k;iω) functions:

$$\Psi_{ij}(k;i\omega) = \mathcal{G}^0_{ij}\delta_{k,0} + \mathcal{G}^0_{i0}(i\omega-k\omega_0)[\Psi_{0j}(k+1;i\omega)+\gamma^2k\Psi_{0j}(k-1;i\omega)]. \quad (7)$$

Significantly, here $\gamma^2 = \sum\gamma_j'^2$, a consequence of Uij being a unitary transition matrix.

The solution of this system of recurrence relations for the Green's function $\mathcal{G}_{ij}(i\omega) = \Psi_{ij}(0;i\omega)$ can be written as

$$\mathcal{G}_{ij} = \mathcal{G}^0_{ij}(i\omega) + \frac{\mathcal{G}^0_{i0}(i\omega)\mathcal{G}^0_{0j}(i\omega)}{\mathcal{G}^0_{00}(i\omega)}(C_{\mathrm{el}}(i\omega)-1), \quad (8)$$

where

$$C_{\mathrm{el}}(i\omega) = \cfrac{1}{1-\cfrac{\gamma^2\mathcal{G}^0_{00}(i\omega-\omega_0)\mathcal{G}^0_{00}(i\omega)}{1-\cfrac{2\gamma^2\mathcal{G}^0_{00}(i\omega-2\omega_0)\mathcal{G}^0_{00}(i\omega-\omega_0)}{\cdots}}}. \quad (9)$$

For small ω_0, we can expand the C_{el} function in a Taylor series. As is evident from the form of expression (9), as well as of Green's function $\mathcal{G}^0$ (3), the only small dimensionless parameter in this function is $\gamma^2\omega_0/t^3$, so that $C_{el}(i\omega,\omega_0)=C_{el}(i\omega,0)+O(\gamma^2\omega_0/t^3)$. But then the expression for C_{el} allows significant simplification. It can readily be shown that by comparing the expression for C_{el} written for ω_0=0 with that for the complementary error function erfc expressed in the form of continued fractions [38, 39] one can obtain an exact analytical expression for $\omega_0/t \to 0$. In fact, this is a main result of this paper:

$$C_{\text{el}} = -\frac{i\sqrt{\pi}}{\sqrt{2}\gamma \mathcal{G}^0_{00}(i\omega)} w\left(\frac{1}{\sqrt{2}\gamma \mathcal{G}^0_{00}(i\omega)}\right), \tag{10}$$

where $w = \exp(-x^2)\text{erfc}(-ix)$ is the Faddeeva function. We note that the Faddeeva function $w(1/x) \sim ix/\sqrt{\pi}$ for $x \to 0$ [40]. Thus, for $\gamma \to 0$, $C_{\text{el}} \to 1$ and $G(i\omega) \to G^0(i\omega)$.

As already mentioned, the simplification accepted in this method, namely, neglect of the spatial dispersion of phonons, can hardly affect seriously the final results. This effect can obviously be neglected altogether in the $\omega_0 \ll t$ approximation. Indeed, for $\omega_0 \to 0$ we also have $\omega_i < \sqrt{2}\omega_0 \to 0$. This brings us to the exact solution for the Green's function allowing for dispersion in the form $\mathcal{G}^{\text{disp}}(i\omega, \omega_1, \omega_2, \ldots) = \mathcal{G}^{\text{disp}}(i\omega, 0, 0, \ldots) + O(\omega_0/t)$. But $\mathcal{G}^{\text{disp}}(i\omega, 0, 0, \ldots) = \mathcal{G}_{ij}(i\omega, \omega_0 = 0)$. In other words, spatial dispersion of phonons would become manifest only in the next order of the smallness parameter ω_0/t.

Figure 2 plots the spectral functions at different sites A_{00}, $A_{0,-1}$, $A_{-1,-1}$ versus electron energy, as well as versus the electron–phonon interaction energy. The graphs relate to the metallic case, where the major fraction of electrons reside close to the band center. The spectral functions for the semimetallic case can be obtained by displacing the points from which the energy is reckoned to the left by 2t. Significantly, the spectral function A_{00} at the band center (at the ω=0 point) is larger than $A_{-1,-1}$, which implies the existence of a bound state on the surface. As γ increases, states with energies both lower than −2t and higher than 2t appear in the density of states A_{00} at site 0. This is a direct consequence of an electron now acquiring a probability of both emitting and absorbing a phonon. As the interaction energy γ continues to increase still more, the density at site 0 progressively approaches the Gaussian form. This correlates with the independent boson models, because at high γ the interaction between the 0th and −1th sites becomes inessential, and electrons at the 0th site behave independently. At the same time, the −1th site actually becomes the boundary of the electron subsystem, with the result that, with increasing γ, the density $A_{-1,-1}$ approaches the density A_{00} at low γ.

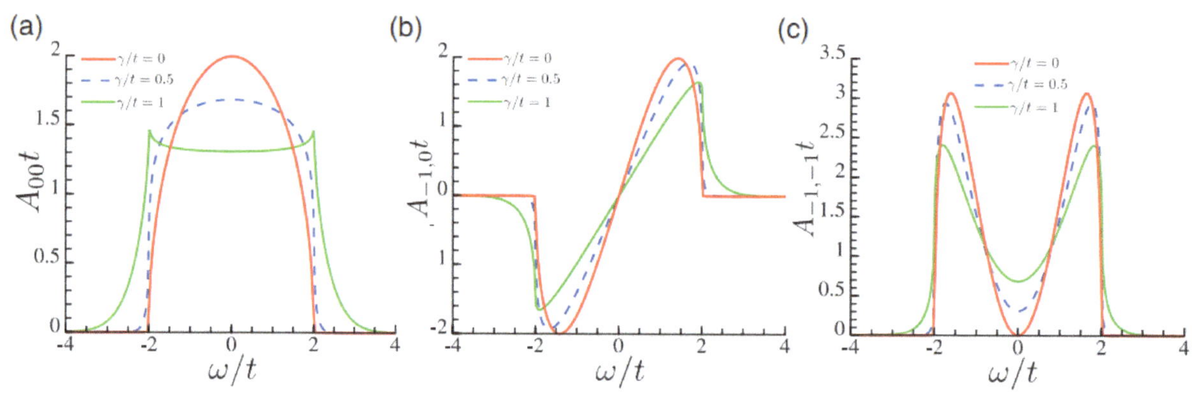

Figure 2. Spectral functions for an electron (a) A_{00}, (b) $A_{0,-1}$, (c) $A_{-1,-1}$ plotted versus its energy ω for different parameters of interaction γbetween electrons and phonons at site 0.

Phonon Green's Function

We turn now to a consideration of the behavior of phonons in a system with local electron–phonon interaction. Equations (2) for the phonon Green's function can be solved using standard techniques. But, in contrast to the uniform case [41], in the case of point interaction one can obtain an exact solution:

$$\mathcal{D}_{ij}(i\omega) = \mathcal{D}^0_{ij} + \gamma^2 \mathcal{D}^0_{i0}\Pi(i\omega)\mathcal{D}^0_{0j}. \tag{11}$$

It should be pointed out that, while this equation resembles in its form the solution of Dyson for small interaction parameters, the present solution was obtained without using the condition of smallness of the coefficient γ; indeed, the smallness parameter here acted the size of the interaction region.

We shall write here only the spectral part of the polarization operator C=−2ℑΠ because, as follows from the Lehmann relation

$$\Pi(i\omega_n) = \int \frac{d\omega}{2\pi}\frac{C(\omega)}{i\omega_n - \omega}, \tag{12}$$

if one knows it, reconstruction of the polarization operator should not present any difficulties:

$$C(\omega) = 2\int \frac{d\omega'}{2\pi} A_{00}(\omega')A_{00}(\omega' + \omega)(n_F(\omega') - n_F(\omega' + \omega)). \tag{13}$$

This function can be readily calculated as soon as we find the electron spectral function at the zero site A_{00}. In the low-temperature case of interest to us here, one can invoke Eqs. (10) and (8) to rewrite Eq. (13) in the form

$$C(\omega) = \frac{2}{\gamma^2}\int_{-\omega}^{0} d\omega' \Re w\left(\frac{1}{\mathcal{G}^0_{00}(\omega')\gamma\sqrt{2}}\right)\Re w\left(\frac{1}{\mathcal{G}^0_{00}(\omega' + \omega)\gamma\sqrt{2}}\right). \tag{14}$$

Renormalization of the Interaction Constant

Prior to continuing the discussion of the phonon properties and thermal resistance, we should first renormalize the electron–phonon interaction constant. To do this, we write the potential of effective interaction between the electron and the phonons in the form:

$$V_{ep} = \gamma^2 \mathcal{D}^0_{00} = \lambda^2 \mathcal{D}_{00}, \tag{15}$$

where λ is the normalized interaction constant.

Electron–phonon interaction in a local region (close to the boundary) exerts the strongest effect on phonons with small wavelengths, i.e., with high energies. Therefore the relation for λ will be given here for phonon energies of about $\sqrt{2}\omega_0$:

$$\lambda = \frac{\gamma}{\sqrt{1+\frac{\gamma^2}{\omega_0}\Re\Pi(\sqrt{2}\omega_0)}}. \tag{16}$$

In the "metallic" case, where the electron concentration is high μ=0, they reshape the phonon spectrum strongly; in these conditions, the effect of phonons on electrons may be safely neglected. Therefore, calculation of Eq. (14) may be performed, in the first approximation, under the assumption of $\gamma \to 0$. Recalling that $\omega_0 \ll t$, one may also assume that $\Re\Pi(\omega) \approx \Re\Pi(0)$. With these simplifications we come to $\Pi(0) \approx -0.9/t$, from which it follows immediately that for γ equal to $\gamma_c = 1.1\sqrt{\omega_0 t}$ the effective electron–phonon interaction constant grows strongly.

It is known [33] that λ relates the renormalized frequencies ω with the "bare" ones ω_0: $\lambda \approx \gamma\omega_0^2/\omega^2$. Therefore, an increase of λ can be identified with mode "softening", i.e., decrease of phonon frequency. Thus, in the metallic case, we meet with softening of high-energy phonons.

In contrast, for low-energy phonons $\Re D_{00} < 0$, which translates into reversal of the sign of γ^2/ω_0 in Eq. (16). Thus low-energy phonons "harden" with increasing λ, i.e., their frequency increases.

The dependence of the effective electron–phonon interaction constant on the λ parameter is presented in graphical form in Fig. 3(a). Figure 3(b) visualizes the behavior of the phonon spectral function at the zero site.

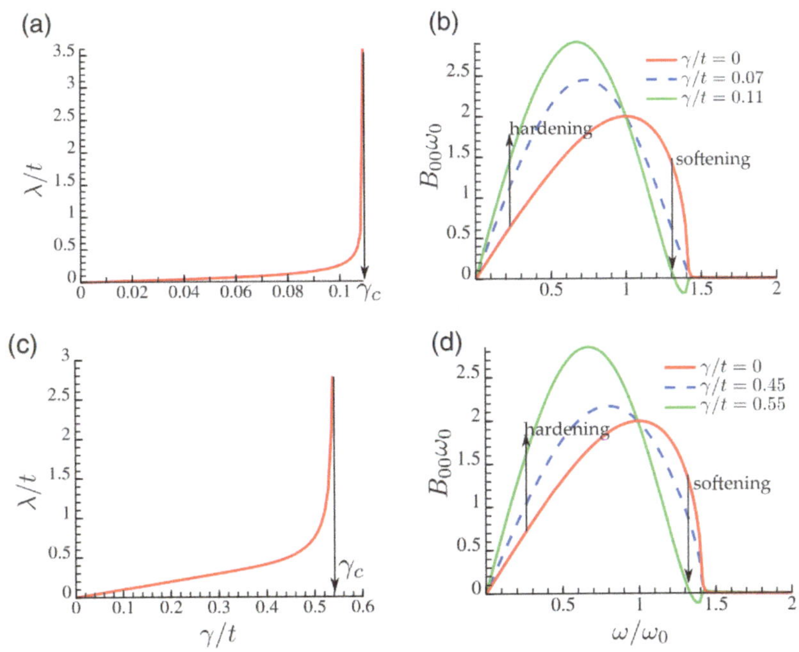

Figure 3. (a) Normalized constant of electron–phonon interaction λ plotted versus γ for the metallic case (μ=0). Close to γc, this dependence follows a strongly nonlinear pattern. (b) Spectral function of phonons at site 0 D_{00} in the metallic case drawn for different γ. Arrows identify hardening of low-energy phonons and softening of the high-energy ones. (c) Variation of the normalized electron–phonon interaction constant λ with γ in the semimetallic case (μ=−2). (d) Spectral function of phonons at site 0 in the semimetallic case drawn for different γ. The ω_0/tparameter was chosen equal to 0.01.

As seen from Eq. (15), whether low-energy phonons soften or harden can also be inferred readily from the derivative of the spectral function with respect to energy. The latter can be derived from (11):

$$\left.\frac{\partial B}{\partial \omega}\right|_0 = \frac{2}{\omega_0^2}\left(\sqrt{2} - \frac{2\sqrt{2}\gamma^2}{\omega_0}\Re\Pi(0) + \frac{\gamma^2}{2\pi}A^2(0)\right) \tag{17}$$

We immediately see that, in the metallic case, an increase of γ initiates accordingly growth of the derivative, i.e. hardening of low-energy phonons, the situation shown in Fig. 3(b). Consider now the empty band case, i.e. the case of semimetals μ=−2t. In this case, ℜΠ exhibits a more complex, nonlinear pattern of variation with γ. At small γ, ℜΠ is close to zero, because at small interaction coefficients the number of electrons at the Fermi level is small. At high interaction energies, however, as is evident from Fig. 2(a), the density of states at energies −2t increases significantly. The electron concentration and hence ℜΠ grows with increasing γ. At large γ parameters, when the number of electrons at the Fermi level becomes large, we actually come to the metallic case, with ℜΠ saturating.

The dependence of the electron–phonon interaction constant on γ for semimetals is shown in graphic form in

Fig. 3(c). Figure 3(d) visualizes the behavior of the spectral function for phonons at site 0. Just as in the metallic case, we observe here softening of the high-energy phonons and hardening of the low-energy phonons.

KUBO FORMULA FOR KAPITZA RESISTANCE

Thermal resistance at the boundary separating the electron from the phonon system can be calculated in terms of the Kubo formalism. Incidentally, while in practical calculations it is quite often more convenient to use thermal resistance, nevertheless, quite frequently it turns out more appropriate to calculate the inverse quantity, namely, surface conductance. Transition from one quantity to the other being trivial, we are going to refer in what follows to thermal surface conductance only.

We shall follow a path largely similar to that chosen in the works dealing with calculation of the resistance for phonon systems [42, 43]. We calculate the average heat flux J at point z_2 at time t resulting from application of a temperature jump ΔT at point z_1 at time zero. We first define the ratio

$$K(z_1, z_2, t) = \frac{\mathcal{J}(z_2, t)}{\Delta T(z_1)}. \quad (18)$$

The surface conductance K can be found by setting $z_1 = z_2 = 0$ for $t \to \infty$. In other words, we are interested in the dependence of the heat flux crossing the boundary on the applied temperature difference in the steady state. The average flux can be expressed in terms of the density matrix $\rho(t)$:

$$\mathcal{J}(z_2, t) = \frac{1}{S}\int d\mathbf{r}\, \mathrm{Tr}\, \rho(t)\hat{\mathbf{J}}(\mathbf{r}, t)\delta(z - z_2)\mathbf{n}, \quad (19)$$

where **n** is a vector normal to the surface, S is the surface area, and $\hat{\mathbf{J}}(\mathbf{r}, t)$ is the heat current operator.

$$\rho(t) = \frac{1}{N}\exp\left(-\int d\mathbf{r}\beta(\mathbf{r})\hat{\mathcal{H}}(\mathbf{r})\right). \quad (20)$$

Here $\hat{\mathcal{H}}$ is the Hamiltonian density , and N is the normalization constant. The local inverse temperature $\beta(\mathbf{r})$ identifies the temperature jump at point z_1: $\beta(\mathbf{r}) = 1/[T + \Delta T\Theta(z - z_1)]$ The expression for the Hamiltonian density $\hat{\mathcal{H}}$ will be presented later. It is pertinent to note here that it does not depend on temperature and is related to the current operator through the equation of continuity: $\partial\hat{\mathcal{H}}/\partial t + \nabla\hat{\mathbf{J}} = 0$.

In the first order in temperature nonuniformity $\delta\beta(\mathbf{r}) = -\beta^2\Delta T\Theta(z - z_1)$ we obtain:

$$\rho(t) = \rho_0\left(1 - \frac{1}{\beta}\int_0^\beta d\tau \int d\mathbf{r}'\delta\beta(\mathbf{r}')\left[\hat{\mathcal{H}}(\mathbf{r}', -i\tau) - \langle\hat{\mathcal{H}}(\mathbf{r}', -i\tau)\rangle_0\right]\right), \quad (21)$$

where $\rho_0 = \exp(-\beta\hat{H})/\mathrm{Tr}\exp(-\beta\hat{H})$, and $\langle\cdots\rangle_0$ specifies averaging over the equilibrium state.

In calculation of the average current (19), the first and last terms of the expression for $\rho(t)$ do not contribute; using the equation of continuity for the current operator, one comes to:

$$\mathcal{J}(z_2, t) = \frac{1}{S}\int d\mathbf{r}d\mathbf{r}'\, \mathrm{Tr}\, \rho_0\frac{1}{\beta}\int_0^\beta d\tau \int_0^t dt'\delta\beta(\mathbf{r}')\nabla\hat{\mathbf{J}}(\mathbf{r}, -i\tau')\hat{\mathbf{J}}(\mathbf{r}, t)\delta(z - z_2)\mathbf{n}. \quad (22)$$

After integrating by parts integral $\boldsymbol{r}'$ and substituting the expression thus found in expression (18), we obtain in the steady state case $t \to \infty$ and close to the boundary $z_1, z_2 \to 0$:

$$K = \frac{1}{T}\Im\frac{\pi(\omega + i0)}{\omega}\bigg|_{\omega\to 0} \quad (23)$$

where π is the correlator of the heat currents incident on the surface $\mathbf{n}\hat{\mathbf{J}}$:

$$\pi(i\omega) = \frac{1}{S}\int\int d^2\mathbf{r}'d^2\mathbf{r}\int_0^\beta d\tau e^{i\omega\tau}\langle T_\tau \mathbf{n}\hat{\mathbf{J}}(\tau, \mathbf{r})\mathbf{n}\hat{\mathbf{J}}(0, \mathbf{r}')\rangle, \quad (24)$$

where the integrals are over the whole infinite plane at z=0. The density of heat flow passing through the surface that separates the electron from the phonon gas can be readily specified by presenting Hamiltonian (1) as a sum of the energy operators at a site $\hat{h}_i$, i.e., the Hamiltonian density $\hat{H} = \sum\hat{h}_i$, where

$$\hat{h}_i = \begin{cases} t/2\left(\hat{c}_{i+1}^{+}\hat{c}_i + \hat{c}_i^{+}\hat{c}_{i+1}\right) & i < 0 \\ t/2\left(\hat{c}_{-1}^{+}\hat{c}_0 + \hat{c}_0^{+}\hat{c}_{-1}\right) + \gamma\hat{n}_0\hat{x}_0 + \omega_0/2(p_0^2 + x_0^2) - \omega_0/4x_0x_1 & i = 0 \\ \omega_0/2(p_i^2 + x_i^2) - \omega_0/4(x_i x_{i+1} + x_{i-1}x_i) & i > 0 \end{cases}. \quad (25)$$

In this case, the heat current operator $\hat{\mathbf{J}} = i\sum_{lm}\mathbf{R}_l[\hat{h}_m, \hat{h}_l] - \mu\mathbf{J}_e$, where **Rl** are the site position and **Je** are the electron current operator (see Ref. [44]). The only term proportional to the energy of interaction between the electrons and phonons γ specifies the heat flow passing through the boundary:

$$\mathbf{nJ} = i\frac{1}{2}\gamma t\left(\hat{c}_{-1}^{+}\hat{c}_0 - \hat{c}_0^{+}\hat{c}_{-1}\right)x_0. \tag{26}$$

Substituting the expression for the energy current operator into the correlation function (24) and averaging using Wick's theorem, we come to

$$\pi(i\omega) = \frac{1}{4}\gamma^2 t^2 a^2 \int_0^\beta d\tau e^{i\omega\tau}\mathcal{D}_{00}(\tau)\left(\mathcal{G}_{00}(\tau)\mathcal{G}_{11}(-\tau) + \mathcal{G}_{11}(\tau)\mathcal{G}_{00}(-\tau) - 2\mathcal{G}_{01}(\tau)\mathcal{G}_{10}(-\tau)\right). \tag{27}$$

One can readily demonstrate the validity of the following relation

$$\int_0^\beta d\tau e^{i\omega\tau}\mathcal{G}_\alpha(\tau)\mathcal{G}_\beta(-\tau)\mathcal{D}_\gamma(\tau) = \int \frac{d\varepsilon_\alpha}{2\pi}\frac{d\varepsilon_\beta}{2\pi}\frac{d\omega'}{2\pi}A_\alpha(\varepsilon_\alpha)A_\beta(\varepsilon_\beta)B_\gamma(\omega') \times \frac{n_F(\varepsilon_\beta - \omega') - n_F(\varepsilon_\alpha)}{i\omega + \varepsilon_\beta - \omega - \varepsilon_\alpha}\left(n_B(\omega') + n_F(\varepsilon_\beta)\right), \tag{28}$$

where n_F, n_B are the Fermi–Dirac and Bose–Einstein distribution functions. Expressing correlator (27) through the spectral functions and substituting it into expression (23) leads us to:

$$K = -\frac{\pi}{4T}\gamma^2 t^2 a^2 \int \frac{d\omega'}{2\pi}B_{00}(\omega')\int \frac{d\varepsilon_\beta}{2\pi}\left.\frac{\partial n_F(\varepsilon)}{\partial\varepsilon}\right|_{\varepsilon_\beta - \omega'}\left(n_B(\omega') + n_F(\varepsilon_\beta)\right) \times \left[A_{00}(\varepsilon_\beta - \omega')A_{-1,-1}(\varepsilon_\beta) + A_{-1,-1}(\varepsilon_\beta - \omega')A_{00}(\varepsilon_\beta) - 2A_{-1,0}(\varepsilon_\beta - \omega')A_{-1,0}(\varepsilon_\beta)\right]. \tag{29}$$

Figure 4 displays graphically this dependence of surface conductance on temperature for various interaction parameters γ, for both the metallic and semimetallic cases. This relation becomes substantially simplified in the cases under consideration here.

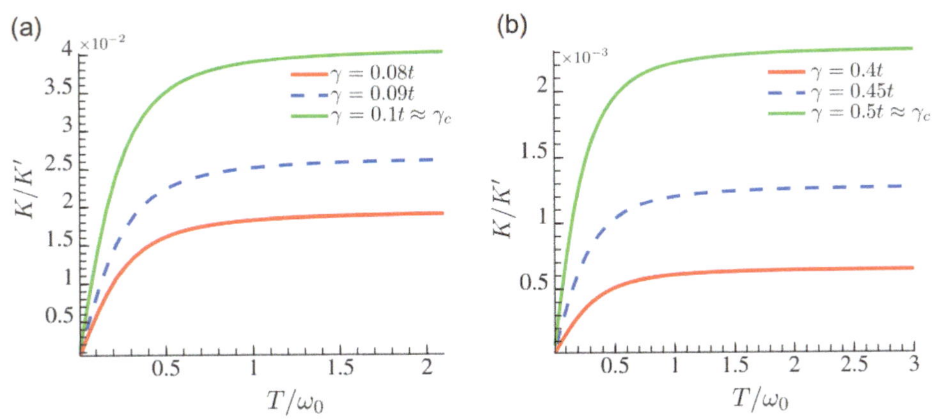

Figure 4. The surface conductance K plotted versus temperature T for various electron–phonon interaction parameters γ for the (a) metallic and (b) semimetallic cases. The graphs are drawn in units of $K'=\gamma^2a^2/(4\omega_0)$, $\omega_0/t=0.01$.

We first note that $dn_F(x)/dx=-\delta(x)$, and also $n_B(x) + n_F(x) = 1/\sinh(\beta x)$. In turn, for $\beta \to \infty$ this function behaves as $-\pi^2/2\beta^2 d\delta(x)/dx$, and for $\beta \to 0$, as $1/\beta x$. We may also note that B00 is nonzero within a narrow interval of frequencies close to ω0; therefore $A(\omega) \approx A(0)$, because $\omega_0 \ll t$. This means that terms containing A can be factored out from underneath the integral sign.

We shall now consider two limiting cases: low temperatures $\omega_0 \gg T$, for which the expression for surface conductance (29) assumes the form

$$K = \frac{\pi}{16}\gamma^2 t^2 a^2 T\left[A_{00}(0)A_{-1,-1}(0) - A_{-1,0}^2(0)\right]\left.\frac{\partial B_{00}(\omega)}{\partial\omega}\right|_0, \tag{30}$$

and high temperatures $T \gg \omega_0$:

$$K = \frac{1}{4}\gamma^2 t^2 a^2 \left[A_{00}(0)A_{-1,1}(0) - A^2_{-1,0}(0)\right] \int \frac{d\omega}{2\pi}\frac{B_{00}(\omega)}{\omega}. \tag{31}$$

As seen from the above expressions for the surface conductance, in the cases of both the metal–dielectric and semimetal–dielectric contacts the relation is linear at low temperatures and saturates at high temperatures, as observed in Fig. 4. This holds for any values of the electron–phonon interaction constants close to the boundary. For weak electron–phonon interactions, this conclusion correlates well with available experimental data [45–47] and coincides with the theoretical result obtained for metals [21].

The expression obtained earlier for the derivative (17) can be complemented with another one for the integral of the phonon spectral function:

$$\int \frac{d\omega}{2\pi}\frac{B_{00}(\omega)}{\omega} = \frac{1}{\omega_0}\left(1 - \frac{\gamma^2}{\omega_0}\Re\Pi(0) - \frac{2\sqrt{2}}{15\pi}\gamma^2 A^2(0)\right). \tag{32}$$

Thus, the system of equations produced [(8), (14), (17), (30)–(32)] provides a complete solution to the problem of the dependence of thermal resistance on the electron–phonon interaction constant.

The above approach permits one to calculate readily the dependence of thermal resistance on the effective electron–phonon interaction constant in the metallic and semimetallic cases, at temperatures both higher and lower than the characteristic phonon frequencies ω_0.

Figures 5(a) and 5(b) plot these dependences for low and high temperatures, accordingly, for the metallic case. We see immediately that they follow practically the same pattern. This was only to be expected, because the dependences on λ of the relations (32) and (17) are practically identical.

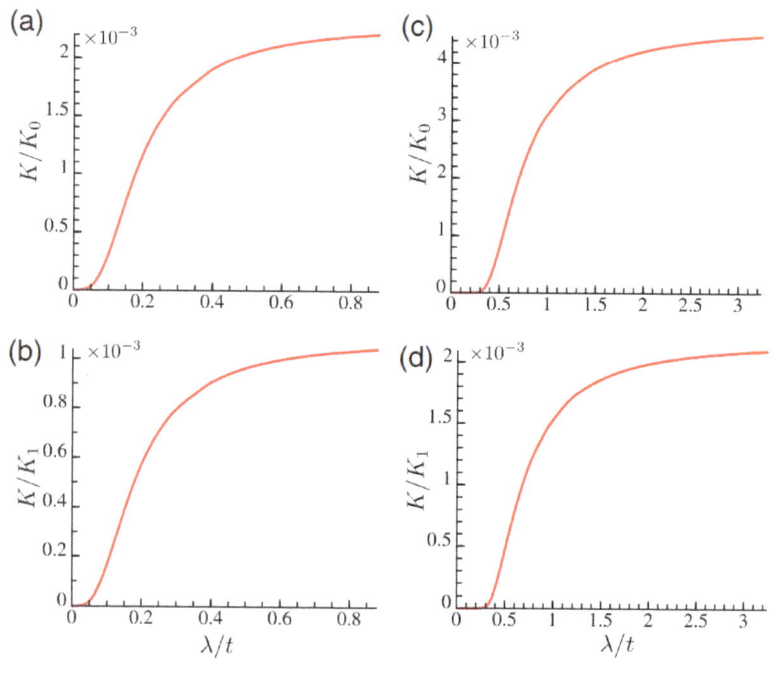

Figure 5. The surface conductance Kof the metal–dielectric boundary in the (a) low- and (b) high-temperature cases plotted versus the renormalized electron–phonon interaction constant λ, and the surface conductance K of the semimetal–dielectric boundary in the (c) low- and (d) high-temperature cases plotted versus the renormalized electron–phonon interaction constant λ. The graphs are drawn in units of $K_0=(\pi/8)(t^2a^2/\omega_0)T/\omega_0$, $K_1=t^2a^2/(4\omega_0)$, $\omega_0/t=0.01$.

While for small interaction coefficients the thermal conductivity scales as $\approx\lambda^2$, at large λ it reaches saturation. The explanation for the latter feature lies in the fact that at large interaction parameters electrons and phonons find it energetically preferable to reside close to the interaction region.

Figures 5(c) and 5(d) plot surface conductance for the semimetallic case in the low- and high-temperature domains, respectively. These curves are seen to practically coincide, for the same reasons as in the metallic case. For small parameters of interaction among electrons and phonons, the surface conductance varies in an essentially nonlinear pattern, more specifically, as $\approx \exp(-t^2/\lambda^2)$, while at large λ it saturates, as it does in the metallic case. Apart from the essentially different behavior at zero, the semimetallic case differs from the metallic one in the values of the λ parameter at which the transition to saturation becomes evident. We readily see that in the semimetallic case saturation is approached at large values of λ. This should be assigned to the fact that only at $\lambda\approx t$ does electron–phonon coupling become strong enough to result in a high enough electron concentration comparable to that in the metallic case.

CONCLUSION

Thus we have succeeded in "summing up" for the adiabatic case the solution for the electron Green's function derived by Cini's method. We have obtained an analytic expression for the density of states of electrons and phonons in a 1D system for the case of local electron–phonon interaction. A careful analysis of the phonon density of states suggests a firm conclusion that in the metallic and semimetallic cases, as the local electron–phonon interaction constant increases, the low-energy phonon modes harden, and the high-energy phonon modes soften. These effects should become manifest, as has been demonstrated, for instance, in Ref. [48], in the Raman spectra of such systems.

Basically, the approach proposed for the calculation of thermal resistance at the boundary between the electron and phonon subsystems is applicable for arbitrary parameters of a system, in both the adiabatic and nonadiabatic cases.

The exact solution obtained in the adiabatic case for Green's function for electrons and phonons made it possible to derive the relation for thermal conductance of the boundary, which covered a broad range of temperatures and electron–phonon interaction parameters, both in the metallic and semimetallic cases. This offers the possibility of answering a number of questions not treated in earlier publications [18–21], which considered only the case of weak electron–phonon interaction at the metal–dielectric boundary.

We have shown that the behavior with temperature of the thermal conductance does not qualitatively depend on the electron–phonon interaction constant. At high electron–phonon interaction constants, just as at low ones, thermal conductance grows with increasing temperature, to reach saturation at temperatures on the order of the characteristic phonon energies (see Table 1).

Table 1. Main results

Parameters	Semimetal ($\mu = -2t$)	Metal ($\mu = 0$)
$\gamma_c/\sqrt{\omega_0 t}$	5.4	1.1
phonons		
low frequency	hardening	
high frequency	softening	
K		
$T \to 0$	$\sim T$	
$T \to \infty$	saturation at $T \gtrsim 0.5\omega_0$	
$\lambda \to 0$	$\exp(-\lambda^2/t^2)$	λ^2
$\lambda \to \infty$	saturation at $\lambda \gtrsim t$	saturation at $\lambda \gtrsim 0.5t$

Earlier publications [18–21] seemed to suggest that thermal conductance should grow indefinitely with increasing electron–phonon interaction constant. We have shown that, at a strong electron–phonon interaction, thermal conductance should reach saturation.

In addition, in contrast to earlier publications, we have studied the dependence of thermal conductance on the electron–phonon interaction coefficient for both the metallic and semimetallic cases. These two cases should reveal a significant difference both in the strong dependence of thermal conductance on the electron–phonon interaction coefficient—indeed, at coefficients in excess of the critical level, thermal conductance grows sharply—and in the characteristic magnitude of this conductance.

ACKNOWLEDGEMENTS

The author owes sincere gratitude to A. Ya. Vul' for helpful advice and assistance provided in the course of the work. Financial support from the Dynasty Foundation, RFBR (No. 12-02-31518, No. 12-08-00174), and the Ministry of Education and Science of Russia (project number 8621) is gratefully acknowledged.

REFERENCES

1. P. L. Kapitza, J. Phys. USSR, 4, 181 (1941).
2. G. D. Mahan, Phys. Rev. B, 83, 125313 (2011).
3. D. G. Cahill, W. K. Ford, K. E. Goodson, G. D. Mahan, A. Majumdar, H. J. Maris, R. Merlin, and S. R. Phillpot, J. Appl. Phys., 93, 793 (2003).
4. V. Rawat, Y. K. Koh, D. G. Cahill, and T. D. Sands, J. Appl. Phys., 105, 024909 (2009).
5. A. I. Persson, Y. K. Koh, D. G. Cahill, L. Samuelson, and H. Linke, Nano Lett., 9, 4484 (2009).
6. M. Kazan, A. Bruyant, P. Royer, and P. Masri, Surf. Sci. Rep., 65, 111 (2010).
7. Z. Han and A. Fina, Prog. Polym. Sci., 36, 914 (2011).
8. S. V. Kidalov and F. M. Shakhov, Materials, 2, 2467 (2009).
9. B. A. Cola, J. Xu, and T. S. Fisher, Int. J. Heat Mass Transfer, 52, 3490 (2009).
10. D. Wang, M. T. Carlson, and H. H. Richardson, ACS Nano, 5, 7391 (211).
11. M. Battabyal, O. Beffort, S. Kleiner, S. Vaucher, and L. Rohr, Diamond Relat. Mater., 17, 1438 (2008).
12. L. Zhang, P. Keblinski, J.-S. Wang, and B. Li, Phys. Rev. B, 83, 064303 (2011).
13. M. Hu, P. Keblinski, and P. K. Schelling, Phys. Rev. B, 79, 104305 (2009).
14. K. Termentzidis, J. Parasuraman, C. A. Da Cruz, S. Merabia, D. Angelescu, F. Marty, T. Bourouina, X. Kleber, P. Chantrenne, and P. Basset, Nanoscale Res. Lett., 6, 288 (2011).
15. T. Luo and J. R. Lloyd, Int. J. Heat Mass Tran., 53, 1 (2010).
16. H. A. Patel, S. Garde, and P. Keblinski, Nano Lett., 5, 2225 (2005).
17. A. Majumdar and P. Reddy, Appl. Phys. Lett., 84, 4768 (2004)
18. A. Sergeev, Physica B, 263–264, 217 (1999).
19. A. V. Sergeev, Phys. Rev. B, 58, R10199 (1998).
20. M. L. Huberman and A. W. Overhauser, Phys. Rev. B,

50, 2865 (1994).

21. G. D. Mahan, Phys. Rev. B, 79, 075408 (2009).
22. A. Vul, K. Reich, E. Eidelman, M. L. Terranova, A. Ciorba, S. Orlanducci, V. Sessa, and M. Rossi, Adv. Sci. Lett., 3, 110 (2010).
23. K. V. Reich, JETP Lett., 94, 22 (2011).
24. A. E. Aleksenskii, A. Y. Vul', S. V. Konyakhin, K. V. Reich, L. V. Sharonova, and E. D. Eidel'man, Phys. Solid State, 54, 578 (2012).
25. M. Galperin, M. A Ratner, and A. Nitzan, J. Phys.: Condens. Matter, 19, 103201 (2007).
26. K. V. Reich and E. D. Eidelman, Europhys. Lett., 85, 47007 (2009).
27. K. V. Reich and E. D. Eidelman, Phys. Solid State, 53, 1704 (2011).
28. T. Olsen, Phys. Rev. B, 79, 235414 (2009).
29. B. Dóra and M. Gulácsi, Phys. Rev. B, 78, 165111 (2008).
30. M. Chakraborty, A. N. Das, and A. Chakrabarti, J. Phys.: Condens. Matter, 23, 025601 (2011).
31. L. K. Dash, H. Ness, and R. W. Godby, J. Chem. Phys., 132, 104113 (2010).
32. Q. Liu, J. Ye, and Y. Zhao, Phys. Chem. Chem. Phys., 12, 6045 (2010).
33. V. L. Ginzburg, D. A. Kirzhnits, A. K. Agyei, and J. L. Birman, High-Temperature Superconductivity (Consultants Bureau, New York, 1982).
34. M. Cini, Phys. Rev. B, 17, 2486 (1978).
35. M. Cini and A. D'Andrea, J. Phys. C, 21, 193 (1988).
36. H. Ness, J. Phys.: Condens. Matter, 18, 6307 (2006).
37. M. Cini and M. Cuozzo, J. Phys.: Condens. Matter, 8, 9011 (1996).
38. L. Lorentzen, Rocky Mt. J. Math., 33, 409 (2003).
39. A. Cuyt, V. Brevik Petersen, B. Verdonk, H. Waadeland, and W. B. Jones, Handbook of Continued Fractions for Special Functions (Springer, New York, 2008).
40. M. Abramowitz and I. A. Stegun, Handbook of Mathematical Functions with Formulas, Graphs, and Mathematical Tables (Dover, Mineola, NY, 1964).
41. J. Loos, M. Hohenadler, A. Alvermann, and H. Fehske, J. Phys.: Condens. Matter, 18, 7299 (2006).
42. A. Maassen van den Brink and H. Dekker, Phys. Rev. B, 51, 17842 (1995).
43. P. T. Leung and K. Young, Phys. Rev. B, 36, 4973 (1987).
44. G. D. Mahan, Many-Particle Physics (Plenum, New York, 1993).
45. R. J. Stoner and H. J. Maris, Phys. Rev. B, 48, 16373 (1993).
46. A. J. Schmidt, K. C. Collins, A. J. Minnich, and G. Chen, J. Appl. Phys., 107, 104907 (2010).
47. K. C. Collins, S. Chen, and G. Chen, J. Appl. Phys., 97, 083102 (2010).
48. C. Q. Sun, L. K. Pan, C. M. Li, and S. Li, Phys. Rev. B, 72, 134301 (2005).

CHAPTER 6

PARTIAL MOLAR ENTROPY OF ELECTRONS IN A JELLIUM MODEL: IMPLICATIONS FOR THERMODYNAMICS OF IONS IN SOLUTION AND ELECTRONS IN METALS

Alan L. Rockwood

Keywords: Jellium model, Non-equilibrium thermodynamics, Equilibrium thermodynamics, Seebeck, Thermodynamic law, Absolute thermoelectric power

INTRODUCTION

An earlier paper in this journal discussed thermoelectrochemical effects and also examined the relationship between the partial molar entropy of electrons in metals and the absolute Seebeck coefficient of the metal, also known as the absolute thermoelectric power [1]. By applying microscopic theory to the problem it was argued that the partial molar electronic entropy and the absolute thermoelectric power are essentially equivalent for all isotropic materials.

Stated briefly, the argument for the validity this proposal uses Eq. (1)

$$G(T)=S+F\sigma \qquad (1)$$

where S is the partial molar entropy of electrons in the metal, F is Faraday's constant, and σ is the absolute Seebeck coefficient of the metal. In this equation G(T) is a universal function, i.e. the same for all materials. In the limit of infinite number density both S and σ become zero for a degenerate free electron Fermi gas. This implies that G(T) is also zero. Since G(T) is a universal function, if G(T) must be zero for one isotropic material it is true for all isotropic materials, and one concludes that

$$S=-F\sigma \qquad (2)$$

for all isotropic materials. In plain language this relationship means that the partial molar entropy of electrons in a metal is essentially equivalent to the absolute Seebeck coefficient of the metal.

Earlier work also concluded that the partial molar entropy of the electron and the absolute thermoelectric power are essentially equivalent [2] and [3]. The restriction to isotropic materials has been discussed by Tykodi [2].

As noted above, the relationship summarized in Eq. (2) was derived by considering the thermodynamic properties of a free electron Fermi gas. However, it is good practice to validate a scientific result using more than one line of evidence. Therefore, the present paper takes this analysis one step further to consider the jellium model of electronic structure, which is more similar to a real metal than the free electron model. The jellium model answers certain objections that could be raised against the free electron Fermi gas model.

THEORETICAL DEVELOPMENT

Because G(T) is a universal function for all materials, the question of whether a particular method of evaluating G(T) is valid or not depends, in part, on what one means by the term "material". If one is using a theoretical construct for the test material then an important question is whether the theoretical material has properties that are sufficiently similar to real materials to consider the material as a valid test case. For example, if the argument from Ref. [1] for Eq. (2) of the present paper were to fail, it would be because the physical model upon which it was built (a strongly degenerate free electron Fermi gas) had somehow failed to capture some essential feature of real materials.

Citation: Alan L. Rockwood, Partial molar entropy of electrons in a jellium model: Implications for thermodynamics of ions in solution and electrons in metals, Electrochimica Acta, Volume 112, 1 December 2013, Pages 706-711, ISSN 0013-4686, http://dx.doi.org/10.1016/j.electacta.2013.08.190.

The free electron Fermi gas differs from real conductors in several respects.

- It does not include electron–phonon interactions.
- It does not include interactions of electrons with a periodic lattice.
- It does not include electron–electron interactions.

Although some investigators may consider that an analysis of the problem based on the free electron Fermi gas is sufficient to validate Eq. (2), (the present author included,) for other investigators the differences between the free electron gas model and real materials could open objections to using this model to validate Eq. (2). In particular one could ask "Is the degenerate free electron Fermi gas similar enough to real materials to be used to validate Eq. (2)?" Therefore, let us examine the three issues above, starting with electron–phonon interactions.

It is well known that acoustic properties and electronic structure vary between one material and another. Clearly, electron–phonon coupling must therefore vary between one material and another, and since G(T) does not depend on the material, it must mean that G(T) does not depend on electron–phonon interactions.

Next, let us consider the interactions of electrons with a periodic lattice. Lattice properties vary greatly from one conductor to another. Some conductors, such as liquid metals, do not even have a periodic lattice. Therefore, G(T) cannot depend on lattice periodicity.

There is still the possibility that G(T) could depend, in a general way, on the existence of a background of positive charge which interacts with the electrons, even if lattice periodicity is not a factor. Therefore, if one is to go beyond the free electron Fermi gas model to validate Eq. (2) one should use a model in which electrons interact with a background of positive charge. One should also include electron–electron interactions. The simplest system which includes these interactions is the uniform electron gas model, also known as the jellium model. In this model electrons interact with each other as well as a uniformly distributed positive charge.

The thermodynamics of this model can be analyzed in the light of a landmark paper by Kohn and Sham [4]. That paper builds upon earlier results from Hohenberg and Kohn [5] to derive a set of equations they describe as a Hartree–Fock method, corrected for correlation effects, and applies these results, together with a finite temperature generalization from Mermin [6] to analyze the thermodynamics of an interacting Fermi gas. In Section 3 of [4] the authors show that within the approximations employed in their paper the thermodynamics of an electron gas in a potential v(r) can be treated in much the same way as a free particle system in an effective potential shifted relative to that of a free particle system of the same number density.

At high number density the energy per electron of a homogeneous electron gas in the ground state is given by:

$$E = \frac{A_F}{r_s^2} - \frac{B_E}{r_s} + C_C \ln(r_s) - D_C \tag{3}$$

where the term containing AF is the energy of a degenerate free electron gas, the term containing BE is the exchange energy, the terms containing CC and DC represent the correlation energy, and rs is the Wigner–Seitz radius, which in the case of a jellium is the radius of a sphere whose volume is equal to the mean volume per electron. In atomic units, AF and BE have values of 2.21 and 0.916 respectively. The value of CC is variously quoted as 0.0313 or 0.0622, depending on the source, and the value of DC is variously quoted as 0.115 or 0.096, depending on the source [7], [8], [9] and [10].

If N is the total number of electrons then the total energy of the ground state is

$$E_{Total} = \frac{NA_F}{r_s^2} - \frac{NB_E}{r_s} + NC_C \ln(r_s) - ND_C \tag{4}$$

Considering the relationship between the Wigner–Seitz radius, particle number, and volume

$$\frac{4\pi}{3} r_s^3 = \frac{V}{N} \tag{5}$$

one can re-write the total energy as

$$E_{Total} = N^{5/3} A_F \left(\frac{3V}{4\pi}\right)^{-2/3} - N^{4/3} B_E \left(\frac{3V}{4\pi}\right)^{-1/3} + NC_C \ln\left(\left(\frac{3V}{4\pi N}\right)^{1/3}\right) - ND_C \tag{6}$$

Assuming that V is constant, the Fermi level at 0 K is

$$\mu_{T=0} = \frac{\partial E_{total}}{\partial N} \tag{7}$$

$$\mu_{T=0} = \frac{5}{3} N^{2/3} A_F \left(\frac{3V}{4\pi}\right)^{-2/3} - \frac{4}{3} N^{1/3} B_E \left(\frac{3V}{4\pi}\right)^{-1/3} + C_c \ln\left(\left(\frac{3V}{4\pi N}\right)^{1/3}\right) - \left(\frac{C_C}{3} + D_C\right) \tag{8}$$

$$\mu_{T=0} = \frac{5A_F}{3r_s^2} - \frac{4B_E}{3r_s} + C_c \ln(r_s) - \left(\frac{C_C}{3} + D_C\right) \tag{9}$$

The functional form of this expression is similar to that of the energy per electron (Eq. (3)) but with different constants. For example, it uses 5AF/3 in place of AF. The leading term in Eqs. (8) and (9) is the zero temperature Fermi level of the degenerate free electron Fermi gas. One can see by inspection that as rs approaches zero the free particle term dominates, and the other terms become negligible by comparison.

Given the fact that the electron density in a jellium is uniform, it seems reasonable to assume that the exchange and correlation energies of low-lying excited states should not be greatly different from those of the ground state. If this were strictly true, then the energy level spacing would be the same as the free particle spacing at the same density. In this case the density of states (which is inversely proportional to the energy level spacing) will show the same limiting form as the density of states of the free electron Fermi gas, and the analysis in Ref. [1] applies, leading to the same conclusion as before, namely that Eq. (2) of the present paper is valid.

Taking a different approach for estimating the density of states near the zero-temperature Fermi level, assume that the energy of promoting an electron from the highest occupied orbital (the zero temperature Fermi level) to the lowest unoccupied level is the same as the difference in energy between an N electron jellium and an N + 1 electron jellium. This naïve assumption probably not strictly valid, but let us see where it leads

$$\Delta E_{N,N+1} = E_{Total,N+1} - E_{Total,N} = \sim \frac{\partial \mu_{T=0}}{\partial N} \quad (10)$$

$$\frac{\partial \mu_{T=0}}{\partial N} = \frac{10A_F}{9r_s^2 N} - \frac{4B_E}{9r_s N} - \frac{C_C}{3N} \quad (11)$$

The density of states is given by the inverse of the energy level spacing

$$\text{Density of states} = D(E_F) = \left(\frac{\partial \mu_{T=0}}{\partial N}\right)^{-1} \quad (12)$$

$$D(E_F) = \frac{N}{\frac{10A_F}{9r_s^2} - \frac{4B_E}{9r_s} - \frac{C_C}{3}} \quad (13)$$

In the limit of infinite electron number density (i.e. in the limit of zero Seitz radius), the first term in the denominator swamps the other terms, and the density of states becomes that of the degenerate free electron Fermi gas. The analysis based on the degenerate free electron gas therefore applies [1]. Briefly, both the partial molar entropy and the absolute thermoelectric power are proportional to temperature, and the proportionality constants for both approach zero as the density approaches infinity. Substituting this result into Eq.(1) gives the result that G(T) is zero, implying the validity of Eq. (2).

Next, assume that the preceding equation is of the correct functional form but that the constants associated with the exchange and correlation terms (the analogs of the constants BE and CC) are different. In this case the first term in the denominator of the density of states equation still dominates as the electron gas density approaches infinity, and the density of states approaches that of the degenerate free electron Fermi gas. The analysis of Ref. [1] still applies, implying the validity of Eq. (2).

Next assume that the second term in the denominator depends not on r_s^{-1}, but on r_s^{-E}, where E is an arbitrary number not necessarily equal to one. In that case, as long as the condition

E<2 (14)

holds, the first term in the denominator still dominates as rs approaches zero, the density of states approaches that of the degenerate free electron Fermi gas, and the analysis of Ref. [1] still applies, implying the validity of Eq. (2). An analogous argument applies to the last term in the denominator.

From the above development one sees that the assumptions about the exchange and correlation energies are not stringent in order for the analysis from Ref. [1] to apply. Furthermore, it is not necessary to assume that the exponent on the Seitz radius in the first term in the denominator (the free particle-related term) has a value of two. As long as the exponent in Eq. (13) is greater than zero and also greater than the exponent in the exchange-related term (parameter E in Eq. (14)) then the density of states will approach zero in the limit of zero Seitz radius. Therefore, it is reasonable to conclude that Eq. (2) is valid, and the partial molar electronic entropy is essentially equivalent to the absolute Seebeck coefficient.

Consider yet another way of looking at the problem. One can calculate that under the assumptions leading to Eq. (13) the ratio of the heat capacity of the interacting electron gas to the heat capacity of the non-interacting electron gas is given by the ratio of the density of states under for two models:

$$\frac{C_{interacting}}{C_{non-interacting}} = \frac{D(E_F)_{interacting}}{D(E_F)_{non-interacting}} = \frac{\frac{10A_F}{9r_s^2}}{\frac{10A_F}{9r_s^2} - \frac{4B_E}{9r_s} - \frac{C_C}{3}} \quad (15)$$

$$\frac{D(E_F)_{interacting}}{D(E_F)_{non-interacting}} = \left(1 + \frac{2B_E}{5A_F} r_s - \frac{3C_C}{10A_F} r_s^2\right)^{-1} \quad (16)$$

This ratio has also been calculated to higher accuracy by Gell-Mann, who obtained an expression of slightly different functional form [11]

$$\frac{C_{interacting}}{C_{non-interacting}} = \frac{D(E_F)_{interacting}}{D(E_F)_{non-interacting}} = (1 + 0.083 r_s(-\ln r_s - 0.203) + \cdots)^{-1} \quad (17)$$

According to Gell-Mann this is an exact relationship for the first few terms of a series expansion. Gell-Mann also compared his result to an earlier approximate result obtained by Pines [12]

$$\frac{C_{interacting}}{C_{non-interacting}} = \frac{D(E_F)_{interacting}}{D(E_F)_{non-interacting}} = (1 + 0.083 r_s(-\ln r_s + 1.47) + \cdots)^{-1} \quad (18)$$

Regardless of whether one uses Eqs. (16), (17) and (18), or some other closely related equation, the qualitative result is the same. As the number density of the interacting gas approaches infinity the density of states approaches that of the degenerate free electron Fermi gas. Since the density of states of the degenerate free electron gas approaches zero one concludes that the heat capacity and entropy of the interacting electron gas (the jellium) approaches zero in the limit of infinite number density.

The partial molar electronic heat capacity and the partial molar entropy can be calculated as follows. The total heat capacity is related to the density of states by the following relationship

$$C_{total} = \frac{\pi^2 k_B^2}{3} D(E_F) T \quad (19)$$

where kB is Boltzmann's constant. The total entropy is given by

$$S_{total} = \int_{T=0}^{T} \frac{C_{total}}{T} dT = \frac{\pi^2 k_B^2}{3} D(E_F) T = C_{total} \quad (20)$$

The two expressions are the same because of the particular functional form of C_{total}.

Partial molar quantities are given by

$$\frac{\partial C_{total}}{\partial N} = \frac{\partial S_{total}}{\partial N} = \frac{\pi^2 k_B^2}{3} \frac{\partial D(E_F)}{\partial N} T \qquad (21)$$

Here we have written the partial molar quantities in terms of particle number, though it would more conventionally be expressed on a molar basis. One can express the density of states, D(EF), in the following form:

$$D(E_F)_{interacting} = D(E_F)_{non-interacting}(1 + f(r_s))^{-1} = \frac{9r_s^2 N}{10A_F}(1 + f(r_s))^{-1} \qquad (22)$$

where f(rs) is taken from Eqs. (16), (17) and (18) or a related equation that expresses the density of states in terms of the density of states of a free electron Fermi gas multiplied by a correction factor. For example

$$f(r_s) = \frac{2B_E}{5A_F} r_s - \frac{3C_C}{10A_F} r_s^2 \qquad (23)$$

or

$$f(r_s) = 0.083 r_s(-\ln r_s - 0.203) + \cdots \qquad (24)$$

or

$$f(r_s) = 0.083 r_s(-\ln r_s + 1.47) + \cdots \qquad (25)$$

etc.

By making the appropriate substitutions into the equations above and calculating the partial derivatives one can verify that, for any reasonable model for the density of states, the partial molar entropy of the interacting gas (the jellium model) approaches zero as the density approaches infinity.

DISCUSSION

Let us consider whether the assumptions and approximations in treatment of the electron gas given by Kohn and Sham [4] would affect the arguments presented in the present paper. First, recognize that the requirement to invalidate the argument presented here is quite stringent. It is only necessary that the theory give a correct qualitative description for the argument presented here to be valid. It is not required that the equations predict the correct numerical values under all conditions, but only that the theory predicts the correct limiting values at zero Seitz radius, i.e. at infinite electron density.

Addressing the question of the applicability of the approach of Kohn and Sham, their theory is based on two principle approximations. One was that the lattice potential interacting with the electron gas is slowly varying. Being based on density of positive charge that is uniform throughout the material, the jellium model automatically satisfies that requirement. It is, in fact, the reference system relative upon which Kohn and Sham based their approximations. The second approximation made by Kohn and Sham is that the electron density is high. This requirement is automatically satisfied in the present paper because it is based on the limiting case of infinite particle density. Therefore, one can justify an opinion that the general approach of Kohn and Sham provides a reasonable way analyze the present problem, recalling that Kohn and Sham proposed that the thermodynamics of a real system of interacting electrons can be treated as similar to a system of non-interacting electrons subjected to an altered potential. Nevertheless, the assumptions and approximations of Kohn and Sham, as well as those by Gell-Mann, Pines, etc. should be critically examined by the community to see if there are any assumptions that could invalidate the arguments presented in this paper.

Note particularly the fact that Kohn and coauthors introduced the approach to electronic structure calculations known today as density functional theory (DFT). One of the properties of DFT is that, although quite accurate for the calculation of the ground state, the DFT orbitals is of uncertain physical significance. The scientific community should discuss whether those uncertainties would affect the conclusions of the present paper.

However, at least partially mitigating this uncertainty is that in the present paper a number of different approaches were used to estimate the density of states. It is not necessary that any of them are numerically correct at any finite number density. It only requires that they give the same result at infinite number density. All approaches lead to the same result at infinite number density, i.e. the density of states approaches that of the degenerate free electron Fermi gas, which in turn approaches zero under those conditions. This reduces them all to the problem of the previously discussed case, the degenerate free electron Fermi gas [1]. It is also worth mentioning that part of the arguments above were based on the works of Gell-Mann and Pines, and those papers preceded the papers by Kohn and coworkers by almost a decade and are therefore not dependent on the work of Kohn and Sham.

As mentioned earlier, it is generally a good practice to validate a result using as many ways as possible. Therefore, it would be very useful to consider other physical models in addition to the free electron model and the jellium model. For example, one might consider the properties of BCS superconductors and ask what would happen in BCS superconductor theory if one were to add a small number of electrons to the conductor. Would the total entropy increase, decrease, or be unaffected? Since the absolute thermoelectric power of all known superconductors is essentially zero [13], if BCS theory predicts a value of zero for the partial molar entropy this result would be another theoretical validation of Eq. (2).

Theoretical models that include band structure would also be of interest. As discussed in Ref. [1] and references contained therein it is known that the absolute thermoelectric power of some materials is positive, and for other materials it is negative. This would imply that the partial molar electronic entropy can be either positive or negative. If the partial molar entropy calculated from theoretical band structure calculations would mimic this feature, i.e. being either positive or negative, it would be consistent with the idea that the partial molar electronic heat capacity is equivalent to the absolute thermoelectric power. One could, perhaps, go a step further and compare theoretical calculations of the partial molar heat capacity with theoretical calculations of the absolute thermoelectric power. One potential challenge to overcome would be to assure that the inevitable approximations one would need to make in order to perform the calculations would not invalidate the comparison.

The gold standard for validating that the absolute thermoelectric power is equivalent to the partial molar electronic heat capacity would be calorimetric measurements of the partial molar electronic heat capacity. Conceptually, one would measure the heat capacity of a sample both before and after a change in the number of electrons in the sample. The results would be compared to thermoelectric parameters for the same material. In principle this type of experiment is possible, but the effect is expected to be exceedingly small, and with our present level of technology the experimental difficulties probably make this experiment impractical.

If Eq. (2) is valid then it would have several implications. The first is that it would provide a missing piece of information that would allow one to determine the partial molar entropy of ions in solution using electrochemical Peltier measurements. This was implicit by a rearrangement of the equations in Ref. [14] and was more explicitly stated in Ref. [1]. Summarizing the key relationships and concepts from Ref. [1], consider a generic electrochemical process under reversible conditions

$$M_{metal} \rightarrow M^{+}_{solution} + e^{-}_{metal} \qquad (26)$$

The entropy of this process is given by two relationships. One comes from electrochemical Peltier heat measurements. The entropy of the process also equals the difference between the sum of the partial molar entropies of the products minus the partial molar entropy of the metal, Setting the experimentally entropy obtained by electrochemical Peltier measurements equal to the entropy expressed in terms of partial molar entropies, one can then use the partial molar entropy of the metal (i.e. the molar entropy obtained by third law measurements) and the partial molar entropy of electrons in the metal (obtained from the absolute thermoelectric power using Eq. (2)) to determine the partial molar entropy of ions in solution. Without a way to determine the partial molar entropy of electrons in the metal the problem of determining the partial molar entropy of ions in solution using this method is unsolvable. Eq. (2) provides the necessary information, turning an unsolvable problem into a solvable one.

A second implication is that it would imply that current thermodynamic theory might be incomplete. Briefly, Eq. (2) is a relationship between macroscopic thermodynamic functions. However, it seems that Eq. (2) cannot be derived or deduced from the currently accepted laws of thermodynamics. If Eq. (2) is valid and cannot be derived from existing thermodynamic theory then current thermodynamic theory is incomplete. If so then Eq.(2) itself could be considered to be one statement of the new principle. Ref. [1] discussed this rather ambitious proposal more fully and invited the wider scientific community to discuss it as well. If the proposed relationship between partial molar electronic entropy and absolute thermoelectric power can be validated by direct experimental measurements, as discussed earlier in this paper, the implications regarding current thermodynamic theory would not be altered.

A third implication is that it would provide greater unity between the theories of reversible and irreversible thermodynamics. It does this by equating certain quantities between the two theories.

A fourth implication is that it would provide a way to determine the electric field within a conductor placed in a temperature gradient.

This fourth implication requires some explanation. Domenicali has shown that there is a relationship between the temperature coefficient of the electrochemical potential and the electric field existing within a conductor placed in a temperature gradient [15]

$$\Delta\varnothing = \frac{1}{e}\left(\frac{\partial\mu}{\partial T} + S^*\right)\Delta T \qquad (27)$$

where $\Delta\emptyset$ is the electrostatic potential difference between two points at different temperature within the conductor, e is the absolute value of the electron charge, μ is the chemical potential of the electron, and S* is the "transported entropy" of the electron.

Domenicali expressed his quantities on a per-particle basis rather than a molar basis. However, the equation is identical if expressed on a molar basis, provided one substitutes Faraday's constant in place of e in Eq. (27), yielding:

$$\Delta\varnothing = \frac{1}{F}\left(\frac{\partial\mu}{\partial T} + S^*\right)\Delta T \qquad (28)$$

For small temperature and position differences the relationship between position and temperature is given by

$$\Delta T = \frac{\partial T}{\partial x}\Delta x \qquad (29)$$

Substituting this into Eq. (28) and rearranging yields

$$E = \frac{\Delta\varnothing}{\Delta x} = \frac{1}{F}\left(\frac{\partial\mu}{\partial T} + S^*\right)\frac{\partial T}{\partial x} \qquad (30)$$

By definition transported entropy is related to the absolute thermoelectric power by

$$S^* = -F\sigma \qquad (31)$$

Substituting this into Eq. (29) and rearranging yields

$$E = \frac{\Delta\varnothing}{\Delta x} = \frac{1}{F}\left(\frac{\partial\mu}{\partial T} - F\sigma\right)\frac{\partial T}{\partial x} \qquad (32)$$

Domenicali uses μ in a way that can be considered to be equivalent to the partial molar Gibbs free energy of a material in an environment independent of extrinsic electrostatic potentials. Its temperature coefficient therefore gives the partial molar entropy according to the following relationship

$$\frac{\partial\mu}{\partial T} = -S \qquad (33)$$

Substituting this into Eq. (32) and rearranging gives an equation for the electric field inside a homogeneous material in the presence of a temperature gradient

$$E = \frac{\Delta \varnothing}{\Delta x} = -\frac{1}{F}(S + F\sigma)\frac{\partial T}{\partial x} \tag{34}$$

All of the quantities on the right-hand side are either constants or are experimentally accessible via generally recognized methods, except for S. Eq. (34) is a general thermodynamic relationship, and if a method is found to measure S (the partial molar electronic entropy of the metal) it would enable one to determine the electric field inside a homogeneous conductor placed in a temperature gradient.

Eq. (1) is also a general thermodynamic relationship. Substituting Eq. (1) into Eq. (34)yields

$$E = \frac{\Delta \varnothing}{\Delta x} = -\frac{G(T)}{F}\frac{\partial T}{\partial x} \tag{35}$$

However, as discussed above an argument has been presented in favor of Eq. (2) which relates partial molar entropy to thermoelectric power. Eqs. (1) and (2) combine to yield

$$G(T)=0 \tag{36}$$

Substituting this into Eq. (35) yields

$$E=0 \tag{37}$$

At the present time this seems to be the only way to evaluate the electric field in a homogeneous conductor in the presence of a temperature gradient. However, this result depends on the validity of Eq. (2), which as discussed above, has been proposed as an additional thermodynamic principle.

What if Eq. (2) is wrong? In that case the electric field in a metal in the presence of a temperature gradient is given by Eq. (35) without the result that G(T) = 0. In any case, because G(T) is a universal function [1], [2] and [3], the internal electric field must be the same for all conductors in a temperature gradient.

CONCLUSIONS

The thermodynamics of a jellium model provide a second way to evaluate the general relationship between the partial molar entropy and absolute thermoelectric power of a metal. It is shown here that the density of states of a jellium model approaches the density of states of a degenerate free electron Fermi gas as the number density approaches infinity. The thermodynamics of the jellium model therefore approach those of a degenerate free electron Fermi gas as the number density approaches infinity. Therefore, since it had been previously shown that for a degenerate free electron Fermi gas the absolute thermoelectric power is essentially equivalent to the partial molar electronic entropy, the same result must be true for a jellium, and since the relationship between the two quantities is universal the same result must be true for all materials. This result, if valid, provides a way to solve certain problems that have traditionally been considered unsolvable, such as measuring the partial molar entropies of ions in solution. It also implies that current thermodynamic theory may be incomplete and that an additional thermodynamic law or principle may be needed. Eq. (2) itself could serve as this new principle. This principle also provides a way to solve another problem that has traditionally considered unsolvable, determining the electric field in a homogeneous metal placed in a thermal gradient.

ACKNOWLEDGEMENTS

Financial support was provided by the ARUP Institute for Clinical and Experimental Pathology.

REFERENCES

1. A.L. Rockwood, Comments on "The Seebeck coefficient and the Peltier effect in a polymer electrolyte membrane cell with two hydrogen electrodes" by S. Kjelstrup, P.J.S. Vie, L. Akyalcin, P. Zefeniya, J.G. Pharoah, O.S. Burnheim
2. Electrochim. Acta 99 (2013) 166., Electrochim. Acta 107c (2013) 686–690.
3. R.J. Tykodi, Thermodynamics of Systems in Nonequililbrium States, Thinker's Press, Davenport, Iowa, USA, 2002, pp. 304–306.
4. A.L. Rockwood, Relationship of thermoelectricity to electronic entropy, Phys. Rev. A 30 (1984) 2843–2844.
5. W. Kohn, L.J. Sham, Self-consistent equations including exchange and correlation effects, Phys. Rev. 140 (1965) 1133–1138.
6. P. Hohenberg, W. Kohn, Inhomogeneous electron gas, Phys. Rev. 136 (1964) B864–B871.
7. N.D. Mermin, Thermal properties ofthe inhomogeneous electron gas, Phys. Rev. 137 (1965) A1441–A1443.
8. J.A. Alonso, N.H. March, Electrons in Metals and Alloys, Academic Press, New York, 1989, pp. 58.
9. P. Nozieres, D. Pines, The correlation energy of a free electron gas, Phys. Rev. 111 (1958) 442–454.
10. N.W. Ashcroft, N.D. Mermin, Solid State Physics, Holt, Rinehart, and Winston, New York, 1976, pp. 336.
11. M. Gell-Mann, K. Brueckner, Correlation energy of an electron gas at high density, Phys. Rev. 106 (1957) 364–368.
12. M. Gell-Mann, Specific heat of a degenerate electron gas at high density, Phys. Rev. 106 (1957) 369–372.
13. D. Pines, Solid State Physics, vol. 1, Academic Press, New York, 1955, pp. 367.
14. R.B. Roberts, The absolute scale of thermoelectricity, Phil. Mag. 36 (1977) 91–107.
15. A.L. Rockwood, Absolute half-cell entropy, Phys. Rev. A 36 (1987) 1525–1526.
16. C.A. Domenicali, Irreversible thermodynamics of thermoelectricity, Rev. Modern Phys. 26 (1954) 237–275.

CHAPTER 7

A SOLUTION OF THE BOLTZMANN EQUATIONS IN THE PRESENCE OF THREE COMPONENTS AND INELASTIC COLLISIONS

I.N. Shishkova, S.S. Sazhin

Keywords: Boltzmann equation, Inelastic collisions, n-Dodecane, p-Dipropylbenzene, Heat/mass transfer, Mixtures, Non-equilibrium state

INTRODUCTION

The limitations of widely used engineering approaches to modelling evaporation/condensation/ heat transfer processes based on the hydrodynamic approximation, even in the case when these processes take place at high pressures, have been discussed in a number of papers (see [1] for a review of the early models and [2] for a review of more recent developments). In a number of papers, including [3], [4],[5], [6] and [7], the evaporation of n-dodecane $C_{12}H_{26}$ (an approximation for Diesel fuel) has been studied and a new model has been developed based on a combination of the kinetic and hydrodynamic approaches. In the immediate vicinity of droplet surfaces (up to about one hundred molecular mean free paths), the vapor and ambient gas dynamics have been studied based on the Boltzmann equation (kinetic region), while at larger distances the analysis has been based on the hydrodynamic equations (hydrodynamic region). Mass, momentum and energy fluxes have been conserved at the interface between these regions.

The analysis of [3] was based on the assumption that n-dodecane evaporates into its own vapour (the contribution of air in the kinetic region was ignored). Although this assumption is rather far from most conditions in which fuel droplets evaporate, it allowed the authors of [3] to avoid the numerical solution of the Boltzmann equation and base their analysis on its approximate analytical solution for the mass flux. The authors of [4] and [5] considered a more realistic problem of n-dodecane evaporation into air and developed a new numerical algorithm for the solution of a system of two Boltzmann equations for n-dodecane and air, taking into account elastic collisions between: n-dodecane molecules; air molecules; and between n-dodecane and air molecules. One of the main limitations of the analysis presented in [4] and [5] was that the effects of heat transfer in the kinetic region were ignored. The latter was taken into account in a more general model presented in [6].

The approaches of [4], [5] and [6] were based on the assumption that all collisions between molecules are elastic. A new approach to taking into account the effects of inelastic collisions was developed in [7] and applied to the problem of n-dodecane droplet heating and evaporation in [8].

One of the important limitations of the approaches described in [3], [4], [5], [6], [7] and [8] is that they have been based on the assumption that Diesel fuel can be approximated by n-dodecane. A more detailed analysis of the composition of Diesel fuel showed that it includes hundreds of various hydrocarbon components [9] and [10]. These components can be subdivided into two main groups: alkanes and aromatics. The assumption that n-dodecane can approximate alkanes is a reasonable one (see [9] and [11]), while aromatics can be approximated by p-dipropylbenzene [11]. In this case a more accurate approximation of Diesel fuel, compared with the one based on its approximation by n-dodecane, could be its approximation by a mixture of n-dodecane and p-dipropylbenzene. Mass fractions of n-dodecane in this

---

Citation: I.N. Shishkova, S.S. Sazhin, A solution of the Boltzmann equations in the presence of three components and inelastic collisions, International Journal of Heat and Mass Transfer, Volume 71, April 2014, Pages 26-34, ISSN 0017-9310, http://dx.doi.org/10.1016/j.ijheatmasstransfer.2013.11.032.

mixture vary from 0.8 to 0.7[11] and [9]. To improve the accuracy of kinetic modelling of Diesel fuel droplet heating and evaporation, one would naturally expect that this modelling should take into account the contributions of both n-dodecane and p-dipropylbenzene. We cannot, however, perform this modelling before a kinetic algorithm of the solution of the system of three Boltzmann equations for n-dodecane, p-dipropylbenzene and air, taking into account the effects of inelastic collisions between respective molecules, has been developed. The results of the development of such an algorithm and its preliminary testing are presented in our paper.

The mathematical model, used in our analysis, is discussed in Section 2. In Section 3 this model is applied to the analysis of three test cases. The main results of the paper are summarised in Section 4 .

MATHEMATICAL MODELS

The evolution of the molecular velocity distribution functions of air $f_a \equiv f_a(\mathbf{r},t,\mathbf{v})$ (or any other neutral gas, including nitrogen), n-dodecane $f_n \equiv f_n(\mathbf{r},t,\mathbf{v})$, and p-dipropylbenzene $f_p \equiv f_p(\mathbf{r},t,\mathbf{v})$ is controlled by the corresponding Boltzmann equations:

$$\left.\begin{aligned} \frac{\partial f_a}{\partial t} + \mathbf{v}_a \frac{\partial f_a}{\partial \mathbf{r}} &= J_{aa} + J_{an} + J_{ap} \\ \frac{\partial f_n}{\partial t} + \mathbf{v}_n \frac{\partial f_n}{\partial \mathbf{r}} &= J_{na} + J_{nn} + J_{np} \\ \frac{\partial f_p}{\partial t} + \mathbf{v}_p \frac{\partial f_p}{\partial \mathbf{r}} &= J_{pa} + J_{pn} + J_{pp} \end{aligned}\right\}, \qquad (1)$$

where $J_{\alpha\beta}$ ($\alpha = a,n,p;\ \beta = a,n,p$) are collision integrals, taking into account the contribution of the collisions between molecules. Triple collisions are ignored.

Explicit expressions for the collision integrals $J_{\alpha\beta}$ can be presented as [4]:

$$J_{\alpha\beta} = \frac{\sigma^2_{\alpha\beta}}{2} \int_{-\infty}^{+\infty} d\mathbf{v}_1 \int_0^{\pi} \sin\theta d\theta \int_0^{2\pi} d\phi \left(f'_\alpha f'_{\beta 1} - f_\alpha f_{\beta 1}\right) |\mathbf{v}_\alpha - \mathbf{v}_{\beta 1}|, \qquad (2)$$

where $\sigma_{\alpha\beta} = (\sigma_\alpha + \sigma_\beta)/2$, σ_α and σ_β are the corresponding diameters of molecules of air 'a', n-dodecane 'n' and p-dipropylbenzene 'p', θ and ϕ are angular coordinates of molecules β relative to molecules α, superscript ′ indicates the velocities and the distribution functions after collisions, subscript $_1$ indicates that the function f_α is modified under the influence of collisions with molecules of the type $_\beta$. The first integral on the right hand side of (2) is calculated in the three dimensional velocity space. Expression (2) is the same as that used in [4], where the contribution of only 2 components in the kinetic region was taken into account. This is justified by the fact that triple collisions are ignored.

All collision integrals $J_{\alpha\beta}$ are calculated taking into account the contribution of internal degrees of freedom (inelastic collisions) as described in [7]. A degree of freedom is usually defined as a parameter corresponding to each independent variable necessary to describe the energy of a molecule [12]. A mono-atomic molecule has three degrees of freedom corresponding to its translational energies in x, y and z directions. Poly-atomic molecules have additional degrees of freedom corresponding to their rotational and vibrational motions (see [13] for the analysis of degrees of freedom of CO_2 molecules).

The main ideas of the approach used in [7] are summarized below.

Regardless of the nature of the collision between molecules, their center of mass is not affected by this collision. During the collisions, the energy of each molecule is redistributed between the degrees of freedom, but the total number of degrees of freedom remains the same. If we assume that none of these degrees of freedom is preferable to any of the others we can consider the redistribution of energy between these degrees of freedom during the collision process as random (this follows from the equipartition-of-energy theorem [12]). For each of these degrees of freedom we allocate one dimension in the N-dimensional space describing all degrees of freedom and consider a sphere in this space with its center at the origin (where energies of all degrees of freedom are equal to zero) and radius $r = \sqrt{\sum_{i=1}^{i=N} E_i}$, where E_i is the energy of the i th degree of freedom. Since r^2 gives the total energy of the system E_f, this equation can be considered as an equation for the conservation of energy. The location of the points on the surface of this sphere can be described by an N-dimensional vector $\mathbf{X} = (x_1, x_2, \ldots\ldots x_N)$ with the basis $(\mathbf{e}_1, \mathbf{e}_2, \ldots\ldots \mathbf{e}_N)$ and the norm $||\mathbf{X}|| = r = \sqrt{E_f}$. The redistribution of energy between the degrees of freedom during the collision process can be described in terms of the random rotation of vector **X** in the N-dimensional space.

The numerical algorithm used in the analysis is based on the method of direct numerical solution of the Boltzmann equation. Physical and velocity spaces are discretised along with time. Time and physical space are discretized as in conventional structured computational fluid dynamics (CFD) codes. The discretization of the velocity space is performed similarly to that of the physical space by replacing continuous values of **v** with a discrete set $\{\mathbf{v}^k\}^M$, where k indicates the position of a velocity cell and M is the total number of cells. The numerical solution of the discretized system of equations is performed in two steps. Firstly, molecular displacements are calculated ignoring the effect of collisions $(J^k_{\alpha\alpha} = J^k_{\alpha\beta} = 0)$. Secondly, the collisional relaxation is calculated under the assumption of spatial homogeneity.

APPLICATIONS

Evaporation and Condensation from Flat Walls

Let us consider two parallel walls kept at constant temperatures $T_{w1}=1$ and $T_{w2}=1$ as shown in Figure 1. The distance between the walls L is equal to 10 mean free paths for n-dodecane ℓ_n at temperature T_{w2}. Space between the walls is filled with:

(1) a mixture of n-dodecane, p-dipropylbenzene and nitrogen (three component mixture);

(2) a mixture of n-dodecane and nitrogen (two component mixture);

(3) pure n-dodecane or p-dipropylbenzene (single component).

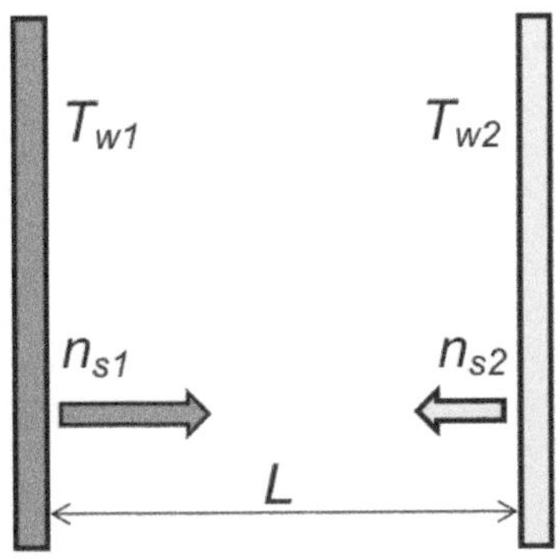

Figure 1. A schematic presentation of the setup used for the analysis of mass/heat transfer processes in up to three component mixtures. The distance between the walls L is assumed to be equal to 10 mean free molecular paths of n-dodecane molecules ℓ at the temperature of the second wall (T_{w2}).

Initially, molar fractions of all components are assumed to be equal to 1. It is also assumed that the distribution functions of n-dodecane and p-dipropylbenzene at both walls are half-Maxwellian with densities $n_{s1}=2$ and $n_{s2}=1$, describing evaporated species. N-dodecane and p-dipropylbenzene molecules hitting the walls are removed from the domain (they are condensed), while for nitrogen molecules it is assumed that the flux of incoming molecules is equal to the flux of molecules reflected from the walls.

Note that although, strictly speaking, the above mentioned combination of parameters might not be realistic for practical applications it turned out to be useful for testing the code.

The space grid in the numerical analysis has been taken equal to 0.5 ℓ_n, and12×12×12 velocity grid points have been used.

The predicted distributions of the number density of n-dodecane for three and two component mixtures and one component substance versus the distance from the first plate are shown in Figure 2 (the distance x is normalized by ℓ_n). Two cases have been considered: the contribution of internal degrees of freedom (corresponding to rotational and vibrational movements of the constituent molecules) is ignored (a); the numbers of these degrees for n-dodecane and p-dipropylbenzene are equal to 20, while nitrogen has 2 internal degrees of freedom (b). As follows from our previous analysis [7] and [8], only the first 20 internal degrees of freedom of such complex molecules as n-dodecane contribute to the results predicted by the kinetic calculations (the actual number of internal degrees of freedom of n-dodecane, taking into account translational, rotational and vibrational movements of the constituent molecules, is expected to be much more than 20). It is anticipated that the same conclusion should remain valid for p-dipropylbenzene. This is supported by our analysis presented later in this paper (see Figure 5).

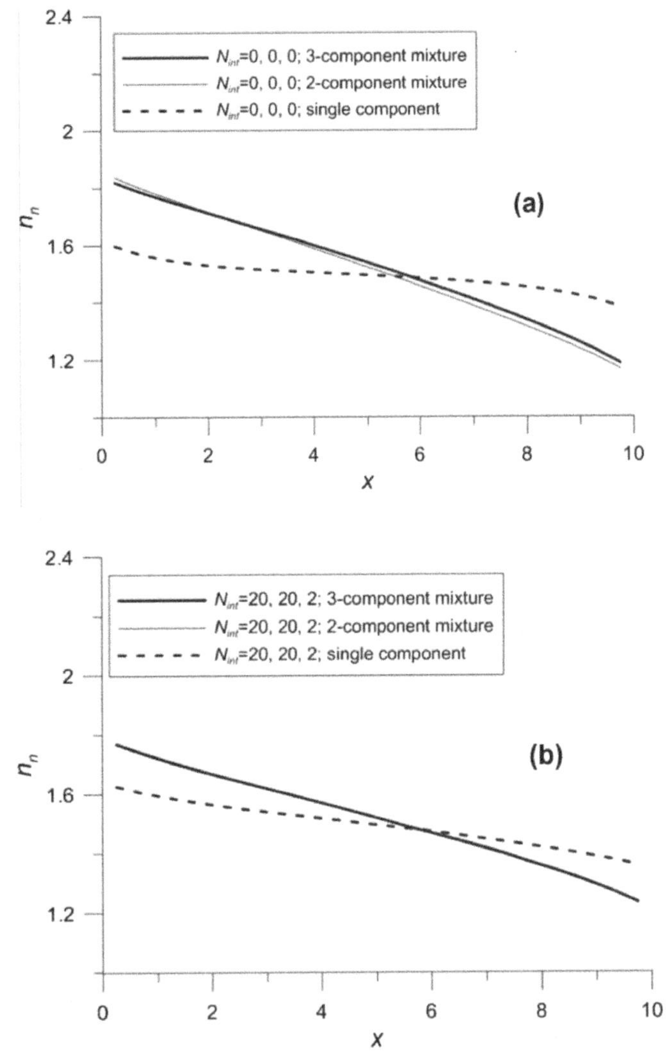

Figure 2. Plots of n-dodecane number density n_n versus the distance from the first plate x predicted by the model, taking into account the contributions of: three components, n-dodecane, p-dipropylbenzene and nitrogen (thick solid curves), two components: n-dodecane and nitrogen (thin solid curves), and a single component (n-dodecane) (dashed curves). The contribution of internal degrees of freedom is either ignored (a) or taken into account assuming that $N_{int}=20$ for n-dodecane and p- dipropylbenzene and $N_{int}=2$ for nitrogen (b).

As follows from Figure 2, the density distributions for three and two component mixtures are very close. They coincide within the accuracy of plotting when the contribution of internal degrees of freedom is taken into account (see Figure 2b). Both these distributions differ considerably from the distribution for the one component substance. In all cases the expected concentration jumps near the walls are clearly seen. These jumps are noticeably larger for the one component substance compared with three or two component mixtures. The physical nature of these density jumps has been widely discussed in the literature, starting with a pioneering monograph [1]. In [14] it was shown that the introduction of these jumps into a conventional CFD code (FLUENT) enables this code to be used for modelling rarefied gases.

The contribution of internal degrees of freedom has little effect on the distribution of number densities of n-dodecane. It leads to an increase in number density jumps for three and two component mixtures and a slight decrease in these jumps for the case of a single component.

The distributions of the number densities of p-dipropylbenzene for a three component mixture and a one component substance are shown in Figure 3. Comparing Figure 2 and Figure 3one can see that the distributions of n-dodecane and p-dipropylbenzene are very close which is related to the closeness of masses and sizes of the corresponding molecules (see Table 1).

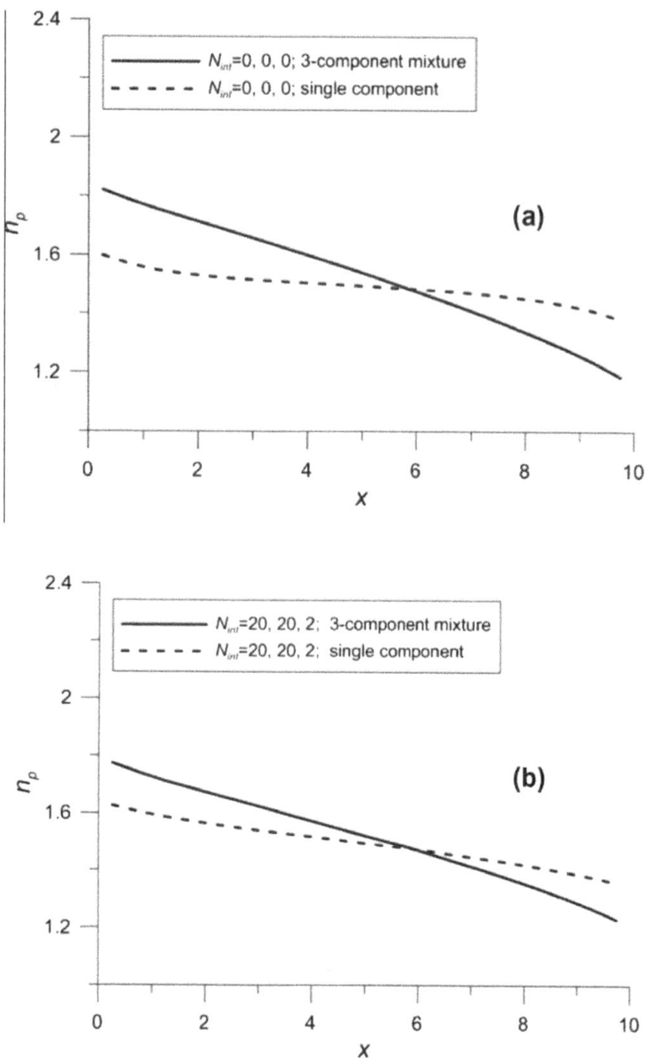

Figure 3. The same as Figure 2 but for p-dipropylbenzene, taking into account the contributions of three components: n-dodecane, p-dipropylbenzene and nitrogen (thick solid curves) and a single component (p-dipropylbenzene) (dashed curves).

Table 1. Chemical formulae, molar masses and molecular diameters of n-dodecane, p-dipropylbenzene and nitrogen are given in Table 1.

Component	Chemical formula	Molar mass (kg/kmol)	Molecular diameter
n-dodecane	$C_{12}H_{26}$	170.3	7.12 Å
p-dipropylbenzene	$C_{12}H_{18}$	162.27	6.73 Å
nitrogen	N_2	28.97	3.617 Å

The distributions of the number density of nitrogen for three and two component mixtures are shown in Figure 4. Comparing Figure 4 with Figure 2 and Figure 3 one can see that in contrast to n-dodecane and p-dipropylbenzene, the number density of nitrogen increases from the first plate to the second one. This means that nitrogen molecules are expected to move from the region of high concentration of n-dodecane and p-dipropylbenzene to the region of low concentration of these molecules, which is consistent with the corresponding thermodynamics equations for the mixtures. In contrast to the case shown in Figure 2, the distribution of nitrogen is rather different for three and two component mixtures. Since the number density of nitrogen at the walls is not specified we cannot infer any information about the number density jump in the vicinity of the walls for nitrogen. As in the cases shown in Figure 2 and Figure 3, the contribution of internal degrees of freedom produces only minor effects on the number density of nitrogen.

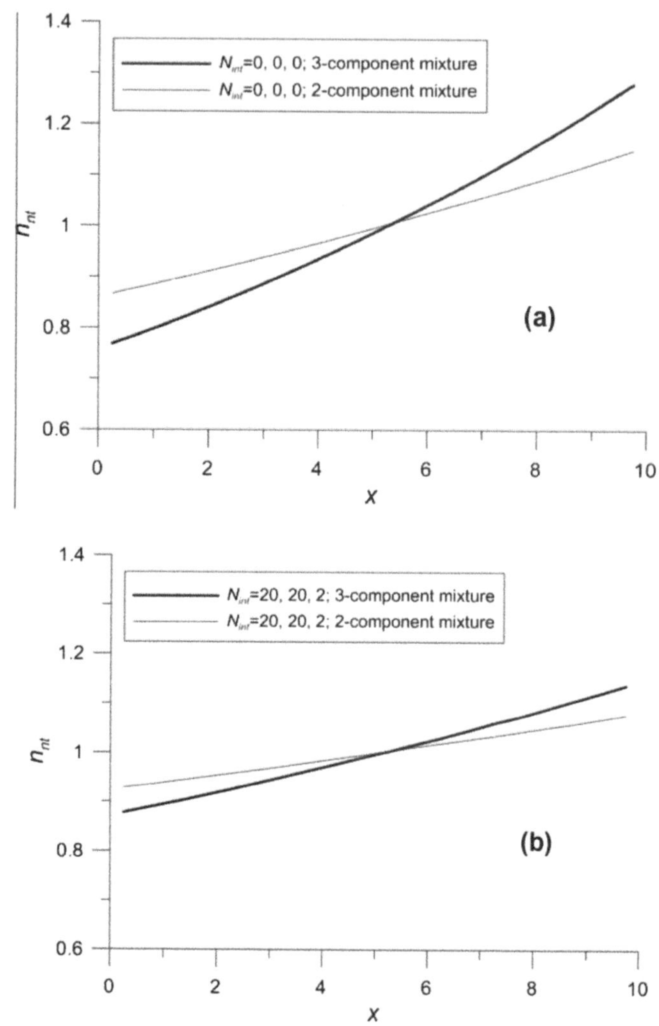

Figure 4. The same as Figure 2 and Figure 3 but for nitrogen, taking into account the contributions of: three components, n-dodecane, p-dipropylbenzene and nitrogen (thick solid curves); and two components, n-dodecane and nitrogen (thin solid curves).

The number fluxes of n-dodecane and p-dipropylbenzene for the three component mixture and single components versus the number of internal degrees of freedom N_{int} are shown in Figure 5. It is assumed that N_{int} for n-dodecane and p-dipropylbenzene are equal and N_{int} for nitrogen is equal to 2, except when N_{int} for n-dodecane and p-dipropylbenzene are equal to 0. In the latter case N_{int} for nitrogen has been taken equal to 0 as well. As one can see from this figure, although the number densities of n-dodecane and p-dipropylbenzene are weak functions of the number of internal degrees of freedom, the dependence of number fluxes of these components on this number is rather strong, at least at N_{int} 20. At larger N_{int} these fluxes do not depend on N_{int}, in agreement with our earlier results [7] and [8]. Mass fluxes of n-dodecane and p-dipropylbenzene are close for the three component mixture, but are noticeably different for the case of single components. Mass fluxes of n-dodecane and p-dipropylbenzene are both much larger for the single components than for the three component mixture as expected.

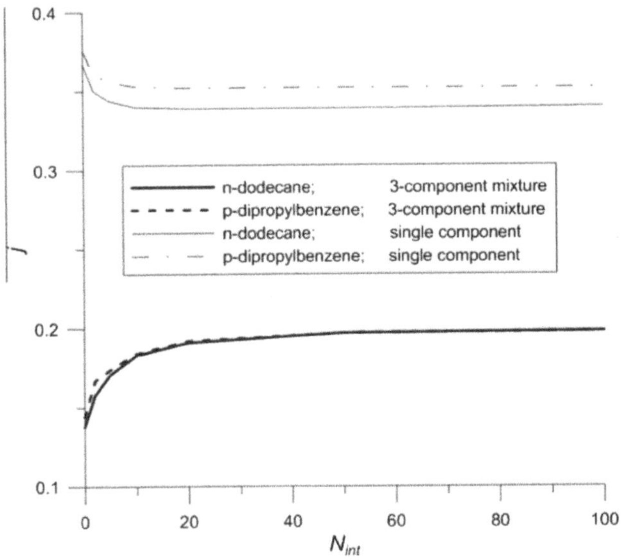

Figure 5. Plots of number fluxes of n-dodecane and p-dipropylbenzene versus the number of internal degrees of freedom N_{int}, predicted by the model, taking into account the contributions of: three components, n-dodecane, p-dipropylbenzene and nitrogen and the models of single components (n-dodecane or p-dipropylbenzene).

Heat Conduction between Two Flat Walls

Let us consider the same case as discussed in the previous section, but ignore the effects of evaporation and condensation and assume that T_{w1}=2 and T_{w2}=1 (see Figure 1). In this case the processes between the walls will lead to pure heat conduction between them.

The distributions of temperature of n-dodecane for three and two component mixtures and one component substance versus the distance from the first plate are shown in Figure 6. As in the case shown in Figure 2, two cases have been considered: the contribution of internal degrees of freedom is ignored (a); the numbers of these degrees for n-dodecane and p-dipropylbenzene are equal to 20, while nitrogen has 2 internal degrees of freedom (b).

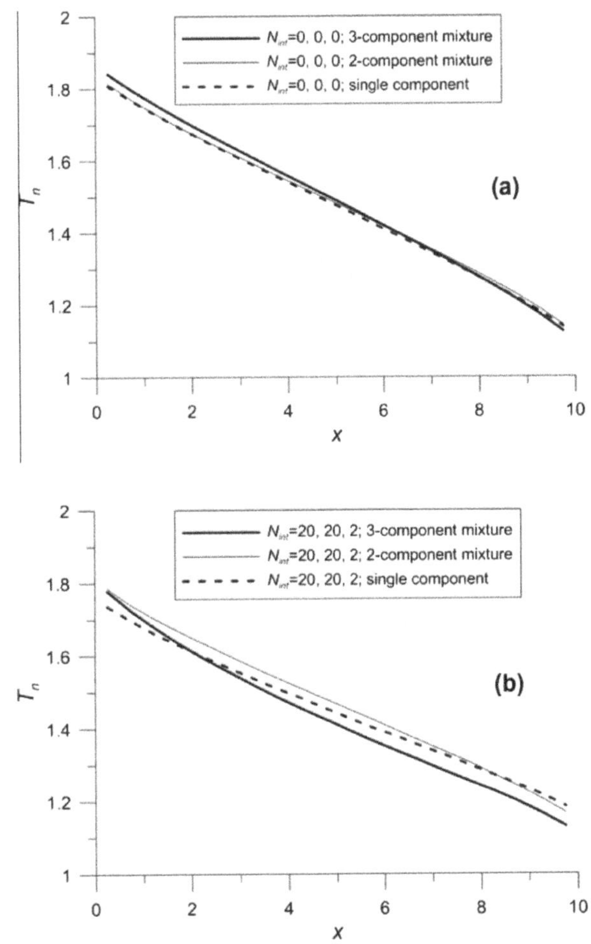

Figure 6. The same as Figure 2 but for n-dodecane temperature, taking into account only heat transfer between the walls.

As follows from Figure 6, the n-dodecane temperatures for three and two component mixtures and a single component are reasonably close, especially when the contribution of internal degrees of freedom is ignored (see Figure 6a). In all cases the expected temperature jumps near the walls are clearly seen. The physical nature of these temperature jumps is similar to that of the density jumps; see [1] and [14] for details. The contribution of internal degrees of freedom has little effect on the distribution of temperatures of n-dodecane, as in the case of number densities shown in Figure 2.

The distributions of the temperature of p-dipropylbenzene for the three component mixture and one component substance are shown in Figure 7. Comparing Figure 6 and Figure 7one can see that the distributions of temperature of n-dodecane and p-dipropylbenzene are very close which is related to the closeness of masses and sizes of the corresponding molecules (cf. the corresponding distributions of number densities shown in Figure 2 and Figure 3).

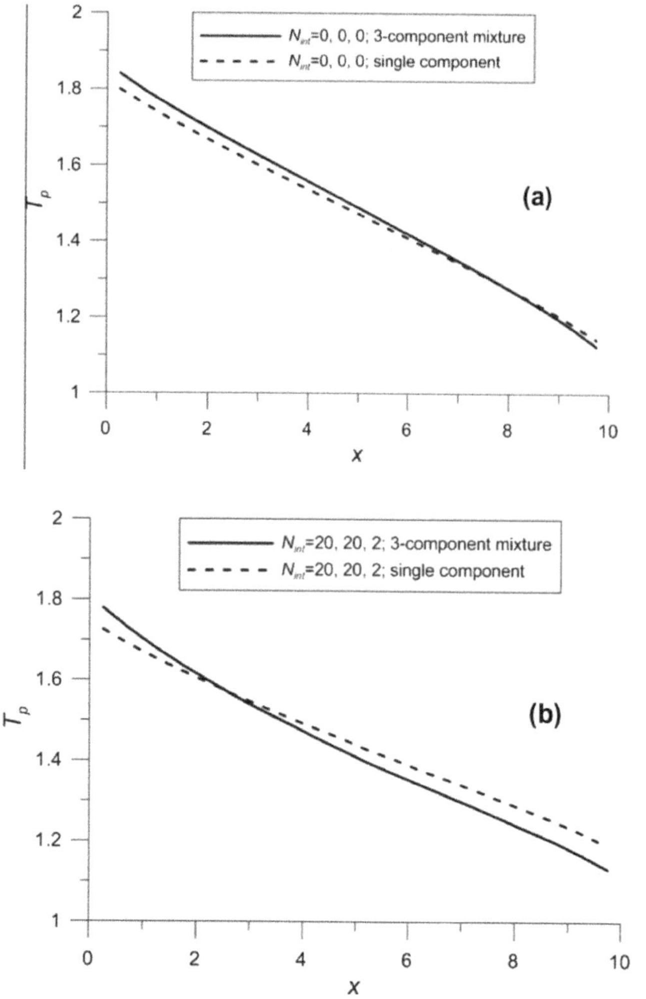

Figure 7. The same as Figure 3 but for p-dipropylbenzene temperature, taking into account only heat transfer between the walls.

The distributions of the temperature of nitrogen for three and two component mixtures and a single component are shown in Figure 8. Note that in contrast to the problem of evaporation and condensation, considered in the previous section, we considered the case when the space between the walls is filled with nitrogen alone. In contrast to the case of the distribution of mass densities, the distribution of temperatures of nitrogen is reasonably close to that of n-dodecane and p-dipropylbenzene. This shows that the system is close to local thermodynamic equilibrium as expected.

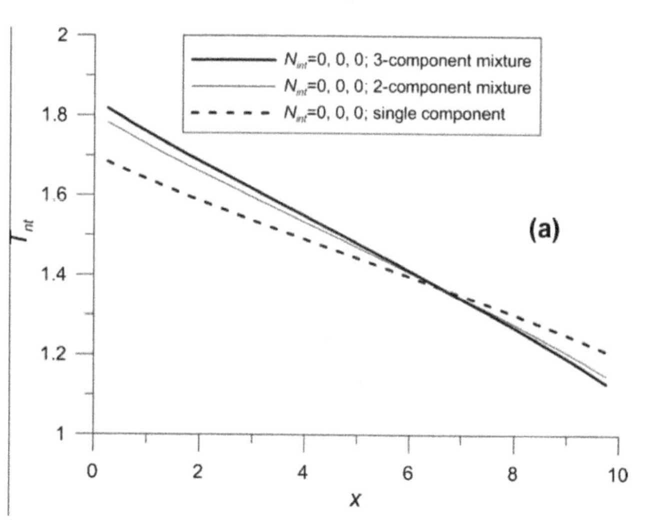

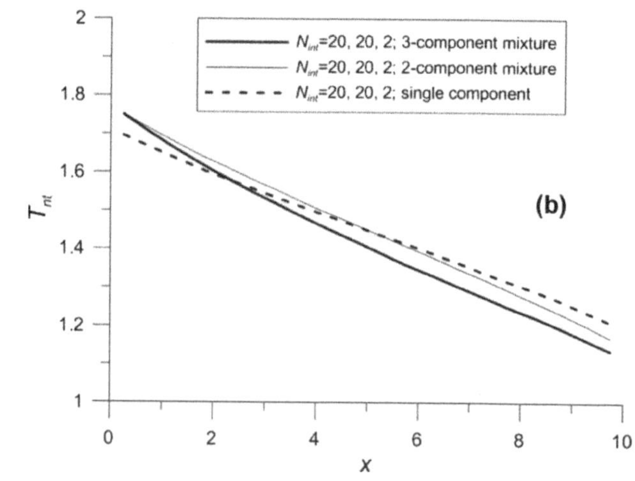

Figure 8. The same as Figure 6 but for nitrogen temperature.

The heat fluxes of n-dodecane, p-dipropylbenzene and nitrogen, predicted for the three component mixture, and single components, versus the number of internal degrees of freedom N_{int} are shown in Figure 9. As in the case shown in Figure 5, it is assumed that N_{int} for n-dodecane and p-dipropylbenzene are equal and N_{int} for nitrogen is equal to 2, except when N_{int} for n-dodecane and p-dipropylbenzene are equal to 0. In the latter case N_{int} for nitrogen has been taken equal to 0 as well. As one can see from this figure, the dependence of heat fluxes of components on N_{int} is weaker than in the case of number fluxes (cf. Figure 5). These heat fluxes remain practically unchanged for $N_{int}>10$. For the three component mixture the total heat flux between the walls is equal to the sum of the heat fluxes due to individual components. The largest heat flux for single components is predicted for nitrogen which reflects its largest thermal conductivity.

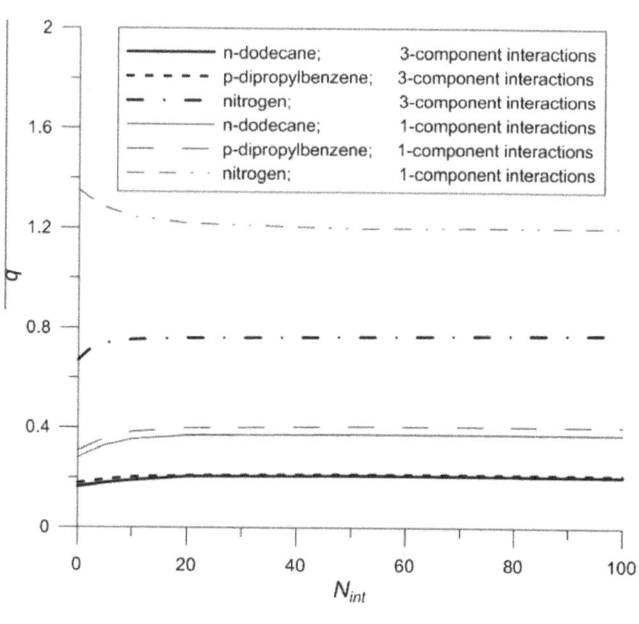

Figure 9. The same as Figure 5 but for heat fluxes due to various components, taking into account only heat transfer between the walls. The models, taking into account the contributions of three components (n-dodecane, p-dipropylbenzene and nitrogen) and single components (n-dodecane, p-dipropylbenzene or nitrogen) have been used.

Heat and Mass Transfer between Two Flat Walls

Let us consider the case when heat and mass transfer between walls take place simultaneously. As in the case of evaporation and condensation considered in Section3.1, it is assumed that the distribution functions of n-dodecane and p-dipropylbenzene at both walls are half-Maxwellian with densities $n_{s1}=2$ and $n_{s2}=1$, describing evaporated species. N-dodecane and p-dipropylbenzene molecules hitting the walls are removed from the domain (they are condensed), while for nitrogen molecules it is assumed that the flux of incoming molecules is equal to the flux of molecules reflected from the walls. Initially, molar fractions of all components are assumed to be equal to 1. In contrast to the case considered in Section 3.1, we took into account the difference between T_{w1} andT_{w2} and assumed that $T_{w1}=2$ and $T_{w2}=1$ (see Section 3.2). In all cases we have considered the three component mixture and assumed that $N_{int}=20$ for n-dodecane and p-dipropylbenzene, and $N_{int}=2$ for nitrogen.

The spatial distributions of the number densities of n-dodecane and nitrogen, predicted by the models taking into account both heat and mass transfer processes, only heat transfer process (see Section 3.2), and only mass transfer processes (see Section 3.1), are shown in Figure 10. The plot n_n versus x for the case when only mass transfer is taken into account is identical to the corresponding plot shown in Figure 2b (solid curve). The plot n *nt* versus x for the case when only mass transfer is taken into account is identical to the corresponding plot shown in Figure 4b (thick solid curve).

As follows from Figure 10a, the distributions of n-dodecane, predicted by the models taking into account both heat and mass transfer processes and only mass transfer processes are qualitatively similar. Both models predict a decrease in n-dodecane number density with x. This can be attributed to the fact that the rate of evaporation of n-dodecane from the hotter wall is stronger than that from the cooler wall. In the case when only heat transfer is taken into account, the number density of n-dodecane increases with x. This is related to an increase of n-dodecane number density with a decrease in temperature. The number density jumps near the walls, predicted by the model taking into account both heat and mass transfer, are larger than the ones predicted by the model taking into account only the mass transfer process. For all three models the number density of nitrogen increases with x. This is related to the fact that nitrogen does not evaporate and its number density decreases with increasing temperature.

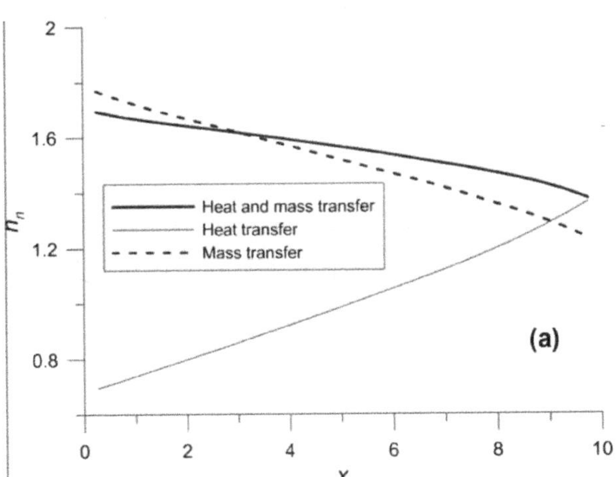

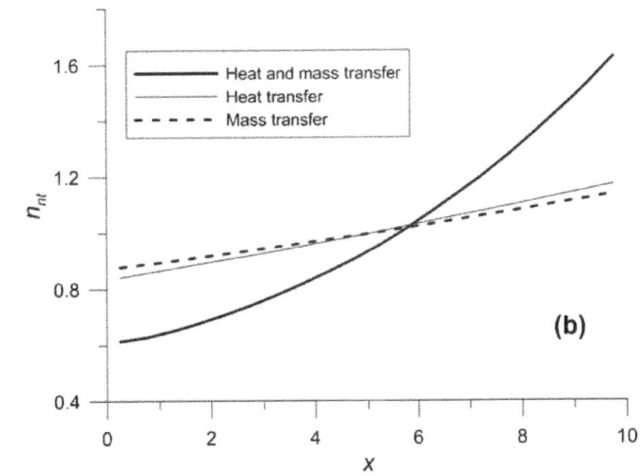

Figure 10. The same as Figure 2, but for n-dodecane (a) and nitrogen (b) number densities and the models taking into account both mass and heat transfer between the walls (thick solid curves), only heat transfer between them (thin solid curves), and only mass transfer between them (dashed curves). The model, taking into account the contributions of three components (n-dodecane, p-dipropylbenzene and nitrogen), and the contribution of internal degrees of freedom, assuming that $N_{int}=20$ for n-dodecane and p-dipropylbenzene and $N_{int}=2$ for nitrogen, has been used.

The spatial distributions of the temperature of n-dodecane, predicted by the models taking into account both heat and mass transfer processes, only heat transfer process (see Section 3.2), and only mass transfer processes (see Section 3.1), are shown in Figure 11. The plot T_n versus x for the case when only heat transfer is taken into account is identical to the corresponding plot shown in Figure 6b (thick solid curve).

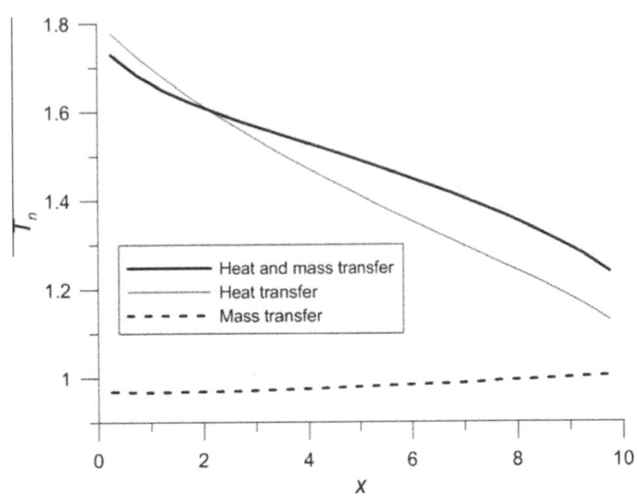

Figure 11. The same as Figure 10 but for n-dodecane temperature.

As follows from Figure 11, the distributions of n-dodecane temperature, predicted by the models taking into account both heat and mass transfer processes and only the heat transfer process are qualitatively similar. Both models predict a decrease in n-dodecane temperature with x, as expected. The temperature jumps near the walls, predicted by the model taking into account both heat and mass transfer, are larger than the one predicted by the model taking into account only the heat transfer process. For the model taking into account only mass transfer processes, the n-dodecane temperature slowly increases with x, which can be related to the corresponding decrease in the n-dodecane number density (see Figure 10a).

The velocity distribution functions of n-dodecane $f_n(v_x,\ v_y=0,\ v_z=0)$ at three positions x ($x=0.25,\ x=2.25$ and x=4.25), predicted by the three component model taking into account both heat and mass transfer processes, are shown in Figure 12. As can be seen from this figure, at all three positions these functions are close to Maxwellian. An increase in the peak value of this distribution function with x is related to the corresponding decrease in the n-dodecane temperature, as shown in Figure 11. The integrals of f_n over v_x, v_y and v_z would give us number densities of n-dodecane in the corresponding points x (see thick solid curve in Figure 10a).

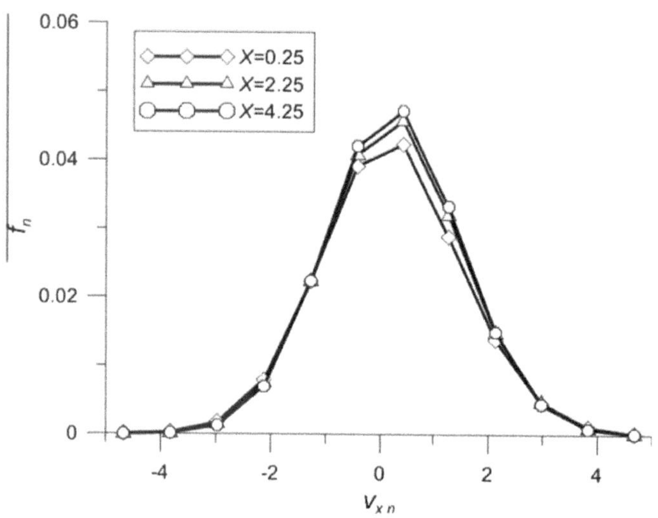

Figure 12. The velocity distribution functions of n-dodecane $f_n(v_x,\ v_y=0,\ v_z=0)$ at three positions x($x=0.25,\ x=2.25$ and x=4.25), predicted by the three component model taking into account both heat and mass transfer processes. As in the case of Figure 10 and Figure 11, the model, taking into account the contributions of three components (n-dodecane, p-dipropylbenzene and nitrogen), and the contribution of internal degrees of freedom, assuming that N_{int}=20 for n-dodecane and p-dipropylbenzene and N_{int}=2 for nitrogen, has been used.

Spatial distributions of the temperature of nitrogen, predicted by the models taking into account both heat and mass transfer processes, only heat transfer process (see Section3.2), and only mass transfer processes (see Section 3.1), are shown in Figure 13. The plot of T_{nt} versus x for the case when only heat transfer and only mass transfer are taken into account are qualitatively similar to the corresponding plots for n-dodecane temperatures shown in Figure 11 (thin solid and dashed curves). In the case when both heat and mass transfer processes are taken into account, however, the behavior of the curve T_{nt} versus x becomes more complex. At first T_{nt} increases with x until it reaches its maximal value. Then it decreases with x as in the case when only the heat transfer process is taken into account. The high nitrogen temperature jump near x=0 can be related to the low number density of nitrogen near x=0. Average temperature of the system decreases with increasing x as expected.

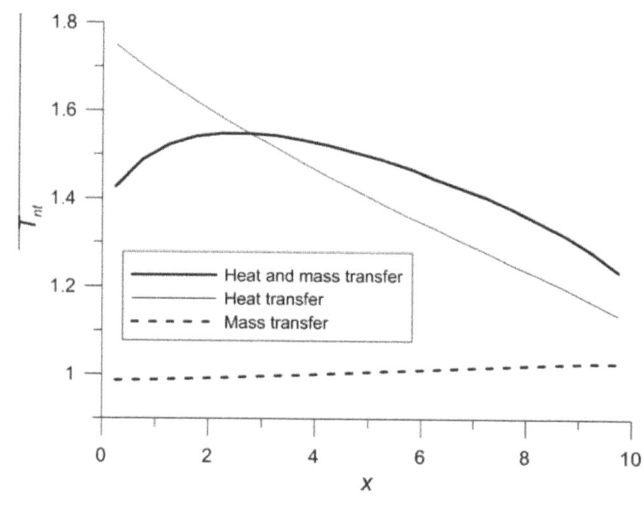

Figure 13. The same as Figure 11 but for the nitrogen temperature.

The velocity distribution functions of nitrogen $f_{nt}(v_x,\ v_y=0,\ v_z=0)$ at the same three positions x as in Figure 12, predicted by the three component model taking into account both heat and mass transfer processes, are shown in Figure 14. As can be seen from this figure, at all three positions these distributions are noticeably different from the Maxwellian ones. The range of velocities for nitrogen molecules shown in Figure 14 is wider than in the case of n-dodecane molecules as shown in Figure 12. This is attributed to the fact that nitrogen molecules are much lighter than n-dodecane molecules. Note that the dispersion of the distribution function, shown in Figure 14, at x=2.25 is noticeably larger than at other $x$which leads to increased temperature of nitrogen in the vicinity of this x as shown in Figure 13. The integrals of f_{nt} over v_x, v_y and v_z would give us number densities of nitrogen in the corresponding points x (see thick solid curve in Figure 10b).

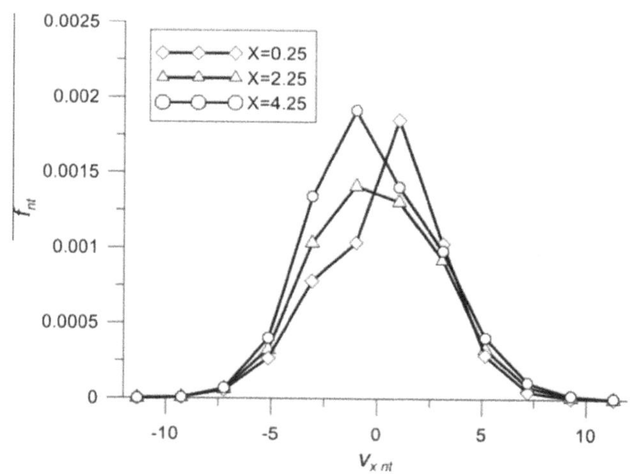

Figure 14. The same as Figure 12 but for the nitrogen velocity distribution functions.

CONCLUSIONS

A new algorithm for the solution of the system of three Boltzmann equations for three components in a mixture has been developed. Binary (but not triple) collisions between molecules have been taken into account; the collisions have been assumed to be elastic or inelastic and in the latter case they have been analyzed based on the model described in [7].

The functionality testing of the model has been performed using three simple test problems of heat and mass transfer in the mixture of n-dodecane, p-dipropylbenzene and nitrogen between two infinite parallel walls. The mixture of n-dodecane and p-dipropylbenzene has been chosen due to the fact that it can approximate realistic Diesel fuels reasonably well [11]. It has been shown that in the case of pure mass transfer the predicted distribution of n-dodecane number density between the walls is almost unaffected by the presence of p-dipropylbenzene, and the effects of internal degrees of freedom on this distribution have been shown to be relatively weak. The distribution of nitrogen between the walls, however, has been shown to be visibly affected by the presence of n-dodecane and p-dipropylbenzene.

In the case of pure mass transfer, the number fluxes of components between walls have been shown to change noticeably with increasing numbers of internal degrees of freedom N_{int} when $N_{int}<20$, but remain almost unchanged when $N_{int}>20$. This result is consistent with the one presented in [7] for similar problems but based on the analysis of two component mixtures. In the case of pure heat transfer between the walls the distributions of temperatures of various components between the walls have been shown to be qualitatively similar.

It has been pointed out that the velocity distribution functions of n-dodecane remain almost Maxwellian at various positions between the walls when both heat and mass transfer processes are taken into account. Noticeable deviations from the Maxwellian distributions have been observed for the corresponding distribution functions of nitrogen.

It is concluded that the predictions of the new kinetic algorithm are qualitatively consistent with the predictions of more basic models of the phenomena for which it has been tested. It is ready for the analysis of a wider range of heat mass transfer problems in which the contribution of three components in the mixture is essential.

ACKNOWLEDGEMENTS

The authors are grateful to Professor V. Gun'ko for discussions on the approximations of Diesel fuel composition, the EPSRC (UK) (Project EP/J006793/1), and the Russian Foundation for Basic Research (Grant 14-08-00467) for their financial support of this project.

REFERENCES

1. N.A. Fuchs, Evaporation and Droplet Growth in Gaseous Media, Pergamon Press, London, 1959.
2. S.S. Sazhin, Advanced models of fuel droplet heating and evaporation, Prog. Energy Combust. Sci. 32 (2006) 162–214.
3. A.P. Kryukov, V.Yu. Levashov, S.S. Sazhin, Evaporation of Diesel fuel droplets: kinetic versus hydrodynamic models, Int. J. Heat Mass Transfer 47 (2004) 2541–2549.
4. I.N. Shishkova, S.S. Sazhin, A numerical algorithm for kinetic modelling of evaporation processes, J. Comput. Phys. 218 (2006) 635–653.
5. S.S. Sazhin, I.N. Shishkova, A.P. Kryukov, V.Yu. Levashov, M.R. Heikal, Evaporation of droplets into a background gas: kinetic modelling, Int. J. Heat Mass Transfer 50 (2007) 2675–2691.
6. S.S. Sazhin, I.N. Shishkova, A kinetic algorithm for modelling the droplet evaporation process in the presence of heat flux and background gas, Atomization Sprays 19 (2009) 473–489.
7. I.N. Shishkova, S.S. Sazhin, J.-F. Xie, A solution of the Boltzmann equation in the presence of inelastic collisions, J. Comput. Phys. 232 (2012) 87–99.
8. S.S. Sazhin, J.-F. Xie, I.N. Shishkova, A.E. Elwardany, M.R. Heikal, A kinetic model of droplet heating and evaporation: effects of inelastic collisions and a non-unity evaporation coefficient, Int. J. Heat Mass Transfer 56 (2013) 525– 537.
9. C. Ledier, M. Orain, F. Grisch, J. Kashdan, G. Bruneaux, Vapour concentration measurements in biofuel sprays using innovative Planar Laser-Induced Fluorescence strategies, in: Proceedings of ILASS Europe 2011, 24th European Conference on Liquid Atomization and Spray Systems, Estoril, Portugal, September 2011 (Book of Abstracts ISBN 978-989-20-2409-7, page 100; full paper is on the conference CD).
10. V.M. Gun'ko, R. Nasiri, S.S. Sazhin, F. Lemoine, F. Grisch, A quantum chemical study of the processes during the evaporation of real-life Diesel fuel droplets, Fluid Phase Equilib. 356 (2013) 146–156.
11. Agency for Toxic Substances and Disease Registry (ATSDR), Toxicological profile for fuel oils, Atlanta, GA: U.S. Department of Health and Human Services, Public Health Service, 1995 (US Government Printing Office).
12. W.P. Crummet, A.B. Western, University Physics, Wm. C. Brown Publishers, Dubuque, USA, 1994.
13. S.S. Sazhin, P. Wild, C. Leys, D. Toebaert, E.M. Sazhina, The three temperature model for the fast-axial-flow CO2 laser, J. Phys. D: Appl. Phys. 26 (1993) 1872– 1883.
14. S.S. Sazhin, V.V. Serikov, Rarefied gas flows: hydrodynamic versus Monte Carlo modelling, Planet. Space Sci. 45 (1997) 361–368.

SECTION II: ADVANCED TOPICS IN TPSM

CHAPTER 8 QUANTUM LIQUIDS RESULTING FROM QUARK SYSTEMS WITH FOUR-QUARK INTERACTION

S. V. Molodtsov, G. M. Zinovjev

INTRODUCTION

The present paper is motivated by world-wide ambitious experimental programs to study the collisions of heavy ions at ultra-relativistic energies to create the quark–gluon plasma (QGP) and to investigate its properties. An obvious great challenge for the theoretical (and phenomenological) endeavors to perform this research is to be able to follow the evolution of a system created in such collisions from its initial state to a near-equilibrium plasma state and eventually to the final hadrons which are measured by detectors. Solving the problem of equilibrating and thermalizing in the framework of microscopic quark and gluon dynamics one needs to deal with the time-dependent strongly interacting systems not only at the asymptotically weak or strong coupling. Nowadays this task is obviously unrealizable and the current response on the requirements of progressing experiment is mainly based on the phenomenological consideration of relativistic heavy ion collisions as evolving through the certain stages.

As is believed, the initial heavy ion collisions at relativistic energies release the hadron constituents (partons) resulting in the highly non-equilibrium strongly interacting matter. The success of applying near-equilibrium hydrodynamics to describe the subsequent evolution of such a matter to describe the existing experimental data [1–5] suggests the system created is rather a quark–gluon liquid. Theoretically the transition between the nonequilibrium and equilibrium stages is still far from clear. But the data analysis foresees a very fast thermalization, i.e., sufficient degree of local equilibrium or, rather, isotropization, since the equations of hydrodynamics do not include the temperature of the produced matter with explicit collective properties. In recent years, there have appeared several scenarios of what could be the dynamics of a system transiting from the initial collision state to that when it becomes (almost) equilibrium [6]. However, this problem is not discussed in the present paper. Instead, we focus on another aspect of the problem, namely, the presence of the strong interaction in a system produced or, in other words, the small mean free path of its constituents, and on trying to understand the very nature of such interactions in a system whose dynamics is governed by the coupling constant, which is likely not too large (at the LHC energies the running coupling constant in QCD is $\alpha_s \sim 0.3$–0.4) avoiding AdS/CFT duality (holographic QCD) arguments very popular at the moment [7, 8].

The four-fermion (QCD-like) field theories still remain a most reliable source of quantitative information in the studies of the transport properties of strongly correlated systems and their thermodynamics, in particular, a chiral phase transition between massive hadrons and almost massless quarks. It is a thermodynamics that provides us with some general framework which lets one understand how the properties of macroscopic matter and, in particular, its collective behavior, emerge from the laws that govern microscopic dynamics. The results of this work allow us to suppose with a level of argumentation that is sufficiently high, in our view, that the picture based on the complex collective behavior of quarks (antiquarks, gluons), which is expressed in the presence of vacuum condensates even under normal conditions, can be set by the nontrivial thermodynamic properties of vacuum, which eventually determine the observable properties of strongly interacting matter. In our opinion, this possibility was not sufficiently widely discussed and, even more so, used already at the initial stage of studies of the quark–gluon matter due to purely accidental circumstances.

THERMODYNAMICS OF THE ENSEMBLE

In the present work we consider some aspects of the thermodynamical description of the quark ensemble with a

---

Citation: S. V. Molodtsov, G. M. Zinovjev, "Quantum liquids resulting from quark systems with four-quark interaction," DOI 10.1140/epjc/s10052-015-3268-1.

four-fermion interaction Hamiltonian density (generated, as it is believed, by a strong stochastic gluon field)

$$\mathcal{H} = -\bar{q}\,(i\boldsymbol{\gamma}\nabla + m)\,q - j^a_\mu \int \mathrm{d}\mathbf{y}\,\langle A^a_\mu A'^b_\nu\rangle\, j'^b_\nu, \tag{1}$$

where $j^a_\mu = \bar{q}t^a\gamma_\mu q$ is the quark current, with corresponding quark operators q, q¯, taken in the spatial point x (the variables with prime corresponds to the y point), mm is the current quark mass, $t^a = \lambda^a/2$ is the color gauge group SU(N_c) generators, μ,ν=0,1,2,3. We take the gluon field correlator $\langle A^a_\mu A'^b_\nu\rangle$ in the simple form of a color singlet, with contact (in time) interaction (without retarding)[1]

$$\langle A^a_\mu A'^b_\nu\rangle = G\,\delta^{ab}\,\delta_{\mu\nu}\,F(\mathbf{x}-\mathbf{y}) \tag{2}$$

(we do not include the corresponding delta-function on time in this formula). This simple correlation function is a fragment of the corresponding ordered exponent and besides the four-fermion interaction accompanied by an infinite number of multi-fermion vertices arises. But for our purposes here it would be quite enough restrict ourselves with this simple form. The effective interactions mentioned above appear in a natural way by the coarse-grained description of the system with handling the corresponding averaging procedure, and having in mind that the vacuum gluon field changed stochastically (for example, in the form of instanton liquid; see [9–11]). But this elaboration of the effective Hamiltonian resulting from the first principles of quantum chromodynamics (QCD) will be unessential for us, as will be demonstrated below. The choice of correlation function in the simplest form with instantaneous interaction does not generate any problem at transforming the final results from the Minkovski space to the Euclidean one and the form factor F(x) is interpreted in a simple way as an interaction 'potential' of point-like particles. The correlation function itself looks, formally, like a gauge non-invariant object.[2] Nevertheless, there exists an effective way to significantly compensate for this shortcoming, if all similar 'potentials' are looked through, in some sense, (to be elucidated below). For example, this set would be quite perceptible, if it becomes possible to confront two limits opposite in physics, for example, starting from the form factor with a delta-like function in the coordinate space (the Nambu–Jona–Lasinio (NJL) model [12], the correlation length is finite in this case) and extended to the form factor of a delta-like function in the momentum space (clearly, the correlation length tends to infinity in this case), analogous to what is well known in condensed matter physics as the Keldysh model (KKB) [13–16].

It is worth to remark here that we will need only one of its properties, although one of exceptional importance, which is related to the fact that, due to the special form factor behavior, all the momentum integrations in the problem get factorized and effectively the problem becomes one-dimensional (then only integrations over energy are in play).

In the KKB model the fermion behavior is considered in the stochastic random field with an infinite correlation length (the NJL model corresponds to the 'white noise' with zero correlation length). In this case one is lucky enough to be able to 'sum up' an entire diverging series and, therefore, to demonstrate that fermions are in general not on mass shell. Actually, it looks like one of the possible scenarios of quark confinement although there exist the standard statements that the four-fermion interaction models do not lead to the confinement of quarks and are non-renormalizable. However, recently it was shown that the KKB model (which is akin to the NJL one) possesses the remarkable property of exact integrability (in the sense by Thirring or Luttinger) [17] and in spite of the presence of divergences (as in the well-known practical field theories) in the intermediate calculations the final results turn out to be finite. Another peculiar fact of such a model is that the Dirac sea displays a finite relative depth with momentum cutoff increasing. Another interesting quark confinement mechanism is discussed in the model in which the quarks are 'living' off mass shell, similarly to the KKB model, given in Ref. [18–20].

From this point of view, other models with an arbitrary form factor (including the NJL model) could be represented as a superposition of elementary blocks obtained by using the KKB model. The utmost distributions mentioned above can be considered as a limiting case for the corresponding Gaussian correlators in the coordinate and momentum spaces, which, of course, look more realistic. The coupling constant scale GG, which will turn out to be interesting for applications, can be tuned by using the corresponding PDG meson observables. Comparing the results obtained (by continuity arguments) one can draw some conclusions about the behavior of the system with practically any interaction potential.

We consider it necessary to comment briefly on the case with a linear potential, which was always giving hope to discover an unusual feature in quark behavior, thereby shedding some light on the nature of confinement. Meanwhile, at present, however, it appears that such a singular 'potential' is even superfluous for our purposes, since the properties, we are interested in, are already revealed in the KKB model, which, in a sense, is like 'half way' from the NJL model to that with a linear potential. Secondly, the quasiparticles in the model with a linearly increasing potential cannot basically be distinguished from those in, for example, the NJL model, provided an integrable infrared singularity in the former is eliminated. As a result the same massive objects appear without the anomalies in the energy spectrum. Additionally, the analysis shows that the multi-fermion contributions present in the problem in the general case can be reduced to the four-fermion interaction in an acceptable way by inserting the respective vacuum expectation values. In other words, even the Hamiltonian of the form (1) seems to capture the essential features of quark interactions. Besides, already in the KKB model the imaginary parts of the appropriate effective scattering matrix in scalar, pseudoscalar, vector, and axial-vector channels are equal to zero, i.e. the free quarks do not appear in this scenario of dominating correlations. We give more comments on this point below in the chapter devoted to the polarization operator.

It is believed that at sufficiently large interaction the ground state of the system transforms from a trivial vacuum |0⟩ (the vacuum of free Hamiltonian) to the mixed state (with quark–antiquark pairs with the opposite momenta and vacuum quantum numbers) which is presented as the Bogolyubov trial function (in that way some separate reference frame is introduced and the chiral phase becomes fixed)

$$|\sigma\rangle = \mathcal{T}|0\rangle,\ \ \mathcal{T} = \prod_{p,s} \exp[\varphi_p\,(a^+_{p,s}b^+_{-p,s} + a_{p,s}b_{-p,s})].$$

Here a+, a, and b+, b are the quark creation and annihilation operators, a|0⟩=0, b|0⟩=0. The dressing transformation T transmutes the quark operators to the creation and annihilation operators of quasiparticles A=T a T†, B+=T b+T†.

The thermodynamic properties of a quark ensemble are known to be determined by solving the following problem. It is required to find a statistical operator

$$\xi = \frac{e^{-\beta\,\hat{H}_{\rm app}}}{Z_0},\ \ Z_0 = \mathrm{Tr}\,\{e^{-\beta\,\hat{H}_{\rm app}}\}, \tag{3}$$

that at fixed mean charge

$$\overline{Q}_0 = \mathrm{Tr}\{\xi\ Q_0\} = V\,\gamma \int d\widetilde{\boldsymbol{p}}\ (n-\bar{n}), \tag{4}$$

$d\widetilde{\mathbf{p}} = d\mathbf{p}/(2\pi)^3$, $(Q_0 = \bar{q}\gamma^0 q)$, and fixed mean entropy

$$\overline{S} = -\mathrm{Tr}\{\xi\ S\}$$
$$= -V\,\gamma \int d\widetilde{\boldsymbol{p}}\ [n\ln n + (1-n)\ln(1-n) +$$
$$+\bar{n}\ln\bar{n} + (1-\bar{n})\ln(1-n)], \tag{5}$$

$S = -\ln\xi$, provides a minimal value of mean energy of the quark ensemble

E=Tr{ξ H}

(H=∫dx H). In other words, we are interested in the minimum of the following functional:

$$\Omega = E - \mu\,\overline{Q}_0 - T\,\overline{S}, \tag{6}$$

where μ and T denote the Lagrangian multipliers for the chemical potential of the quark/baryon charge (which is usually taken to be three times larger than the baryon one in phenomenological considerations) and the temperature ($\beta=T^{-1}$), respectively. V is the volume the system is enclosed in, $\gamma=2N_c$ (in the case of several quark flavors $\gamma=2N_cN_f$, where N_f is the flavor number), $n=\mathrm{Tr}\{\xi A^+A\}$, $\bar{n}=\mathrm{Tr}\{\xi B+B\}$ are the components of the corresponding density matrix.

We restrict ourselves by considering the Bogolyubov–Hartree–Fock approximation in which the statistical operator is constructed on the basis of approximating the effective Hamiltonian H_{app}, quadratic in creation and annihilation operators for quasiparticles acting in the corresponding Fock space with a vacuum state |σ⟩. The average specific energy per quark w=E/(Vγ) results in [21–23]

$$w = \int d\widetilde{\boldsymbol{p}}\ p_0 - \int d\widetilde{\boldsymbol{p}}\ (1-n-\bar{n})\ p_0\ \cos\theta\ -$$
$$-\frac{1}{2}\int d\widetilde{\boldsymbol{p}}\ (1-n-\bar{n})\sin(\theta-\theta_m)\ M(\boldsymbol{p}), \tag{7}$$

where

$$M(\boldsymbol{p}) = 2G\int d\widetilde{\boldsymbol{q}}\ (1-n'-\bar{n}')\ \sin(\theta'-\theta'_m)\ F(\boldsymbol{p}+\boldsymbol{q}),$$

$\theta=2\varphi$, $p_0 = (p^2+m^2)^{1/2}$, the primed variables, correspond to the integration over momentum q. The auxiliary angle θ_m is determined from the relation $\sin\theta_m = m/p_0$. The first term in Eq. (7) is introduced in view of normalizing in such a way as to have the zero energy of ground state when the interaction is switched off. This constant is unessential for the following consideration and may be omitted, however, it should be kept in mind that it will appear as a regularizer in singular expressions further in the text.

The most stable extremals of the functional (7) are presented for comparison with the solid line for the NJL model and the dashed one for the KKB model under normal conditions (T=0, μ=0) in Figure 1. For the delta-like potential in coordinate space (the NJL model) the expression (7) diverges and to obtain the reasonable results the upper limit cutoff in the momentum integration Λ is introduced being one of the tuning model parameters along with the coupling constant GG and current quark mass mm. Below, we use one of the standard sets of the parameters for the NJL model [24]: Λ=631 MeV, $G\Lambda^2/(2\pi^2)\approx1.3$, m=5.5 MeV, whereas the KKB model parameters are chosen in such a way that for the same quark current masses the dynamical quark ones in both NJL and KKB models coincide at vanishing quark momentum. The momentum p_ϑ (parameter) corresponds to the maximal attraction between quark and antiquark. The value of this parameter inversed determines a characteristic size of the quasiparticle.

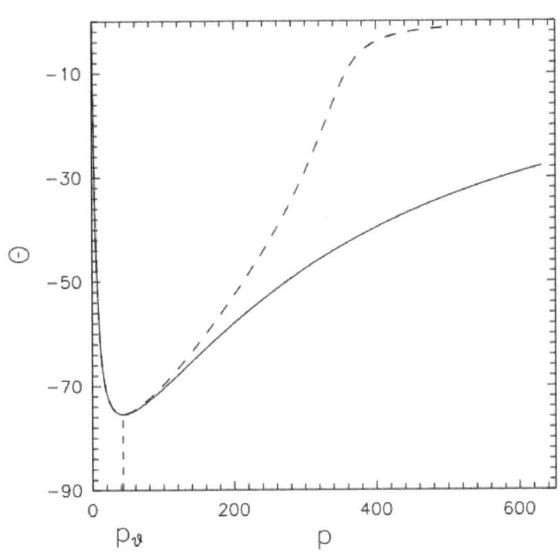

Figure 1: The most stable equilibrium angles θ (in degrees) as a function of momentum p in MeV. The *solid line* is for the NJL model, the *dashed one* corresponds to the KKB model.

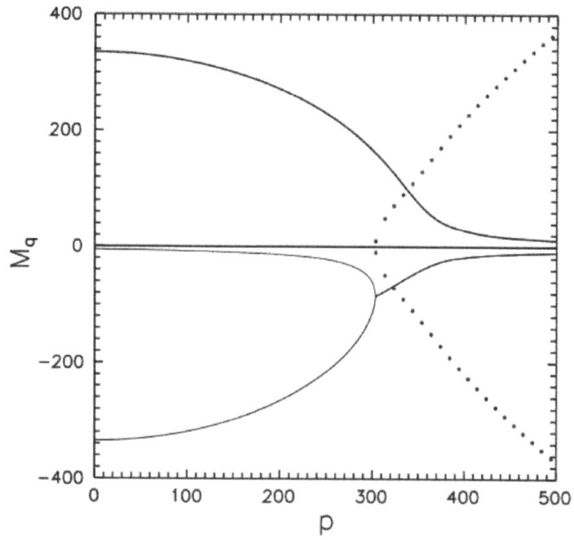

Figure 2: Three branches of solutions for the dynamical quark mass (in MeV) for the KKB model as a function of momentum (MeV). The imaginary parts of the solutions are shown by *dots*.

It is of order of $p_\theta \sim (mM_q)^{1/2}$, where M_q is a characteristic quark dynamical mass for the models considered, i.e. the quasiparticle size is comparable with the size the of π meson (Goldstone particle). It is a remarkable fact that the quasiparticle, as is seen from Figure 1, does not depend noticeably on the form factor profile or, in other words, on the scale, but rather depends on the coupling constant. Using the properties of extremals the functional expression (7) can be transformed to the form (see [21–23])

$$w = \int \mathrm{d}\widetilde{p}\; p_0 - \int \mathrm{d}\widetilde{p}\; (1 - n - \bar{n})\; P_0 + \\ + \frac{1}{4G} \int \mathrm{d}\widetilde{p}\mathrm{d}\widetilde{q}\;\; F(p+q)\; \widetilde{M}(p)\widetilde{M}(q), \tag{8}$$

where $P_0 = [p^2 + M_q^2(p)]^{1/2}$ is the energy of the quark quasiparticle with the dynamical quark mass

$$M_q(p) = m + M(p) = m + \int \mathrm{d}\widetilde{q}\; F(p+q)\; \widetilde{M}(q). \tag{9}$$

Below we omit often the arguments of the corresponding functions for the mass and quasiparticle energy. Varying the functional (8) with respect to the density of the induced quasiparticle mass $\widetilde{M}$ (in such a form it is convenient to take variational derivatives[3]) we obtain the equation for the dynamical quark mass as

$$M_q(p) = m + 2G \int \mathrm{d}\widetilde{q}\; (1 - n' - \bar{n}')\; \frac{M_q'}{P_0'}\; F(p+q), \tag{10}$$

which corresponds exactly to the mean field approximation. In particular, under normal conditions (T=0, μ=0) the dynamical quark mass in the NJL model is $M_q \sim 340$ MeV, whereas the dynamical quark mass of the KKB model is determined by the following equation:

$$M(p) = 2G\; \frac{M_q(p)}{P_0}. \tag{11}$$

In practice, it is convenient to use an inverse function $p(M_q)$. Then in the chiral limit $Mq=(4G^2-pp^2)^{1/2}$, at $|p|<2G$, and $M_q=0$ when $|p|>2G$. In this case the quark states with momenta $|p|<2G$ are degenerate in energy $P_0=2G$. Figure 2 demonstrates three branches of solutions of Eq. (11) for the dynamical quark mass. The dots show the imaginary part of the solutions which are generated at the point where two real solution branches merge.

MEAN ENERGY AS A FUNCTIONAL OF QUANTUM LIQUID THEORY

The goal that we pursued while passing from the expression for specific energy (7) to Eq. (8) was to derive such a form as would easily be recognized as the Landau energy functional of the Fermi-liquid theory [25–29]. Some aspects of this theory are interesting and useful apply for comparing the results obtained in the NJL and KKB models. We will also discuss the first order phase transition, which is apparently typical for interacting fermions (relativistic Fermi liquid).

Thus, the second term in (8) describes the contributions coming from quark and antiquark quasiparticles with occupation numbers n and $\bar{n}$ respectively. The unity in the expression $1 - n' - \bar{n}'$ corresponds to the vacuum fluctuations of quarks and antiquarks. The last term in (8) is due to the interaction of the quasiparticles. The presence of contributions coming from antiparticles and the relativistic form of the dynamics are the features which distinguish quark ensembles we study from the Fermi liquids considered in condensed matter physics. The first variation of the functional (8) with respect to the particle (antiparticle) density leads (as it should) to the energy of the quasiparticle:

$$\frac{\delta w}{\delta n} = P_0. \tag{12}$$

We consider, first, the situation of zero temperature and discuss some aspects of filling up the Fermi sphere by quarks. Let us assume that the momentum distribution of quarks (antiquarks) is determined by the following expressions taken at the β→0 limit:

$$n = [\mathrm{e}^{\beta(P_0-\mu)} + 1]^{-1},\;\; \bar{n} = [\mathrm{e}^{\beta(P_0+\mu)} + 1]^{-1}, \tag{13}$$

that is by the Fermi step function: n=1, at $P_0 \leq \mu$ and n=0 when $P_0 > \mu$. It is clear that for antiquarks $\bar{n} = 0$. The quark density is determined by using the Fermi momentum:

$$\rho = \frac{\gamma P_F^3}{6\pi^2},\;\; \rho = \frac{Q_0}{V}, \tag{14}$$

with the quark chemical potential that coincides with the quasiparticle energy on the Fermi surface, as follows from Eq. (12), i.e.

$$\mu = [P_F^2 + M_q^2(P_F)]^{1/2}. \tag{15}$$

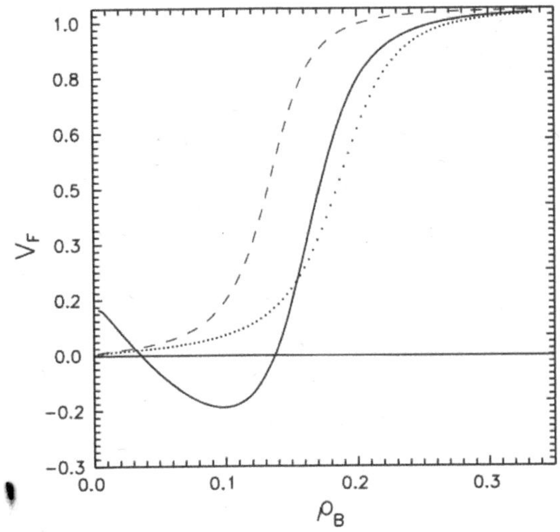

Figure 3: The group velocity of quasiparticles v_F on the Fermi surface. The *solid line* describes the NJL model, the *dashed one* corresponds to the KKB model, the *dots* show the data for the KKB model tuned to the $\pi\pi$-meson energy of Figs. 4, 5, 6, 7 and 8.

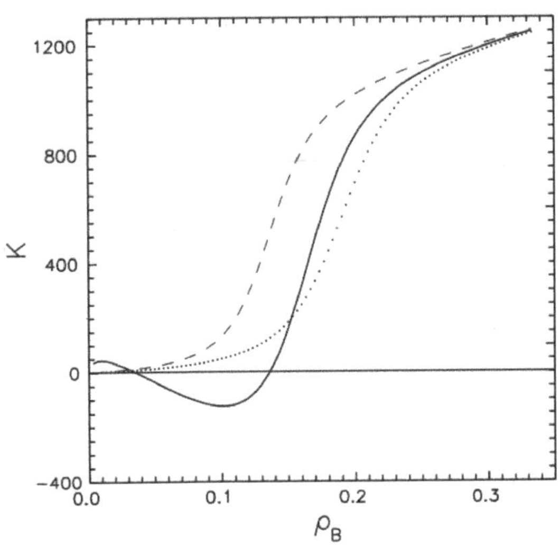

Figure 4: The compression module K in MeV.

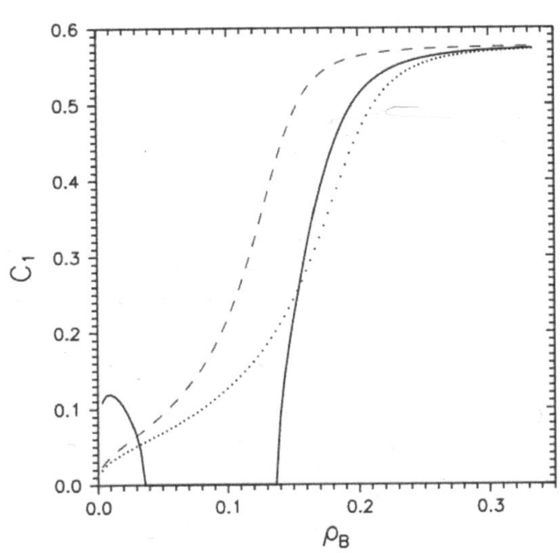

Figure 5: The first sound velocity C_1.

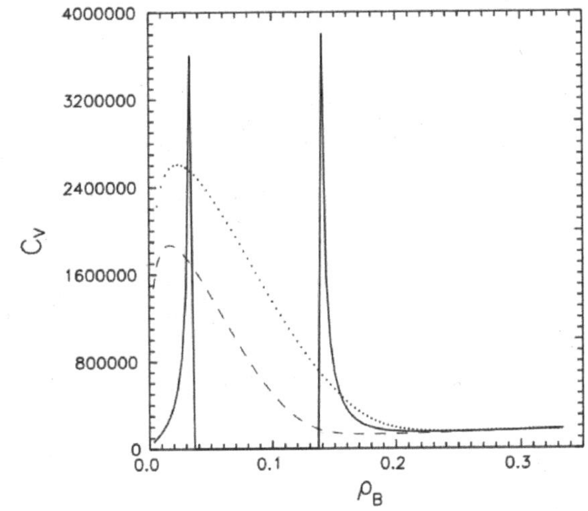

Figure 6: The slope factor at low temperatures $\frac{1}{3}\pi^2 N_F$ in the thermal conductivity expression at constant volume $C_V = \frac{1}{3}\pi^2 N_F T$.

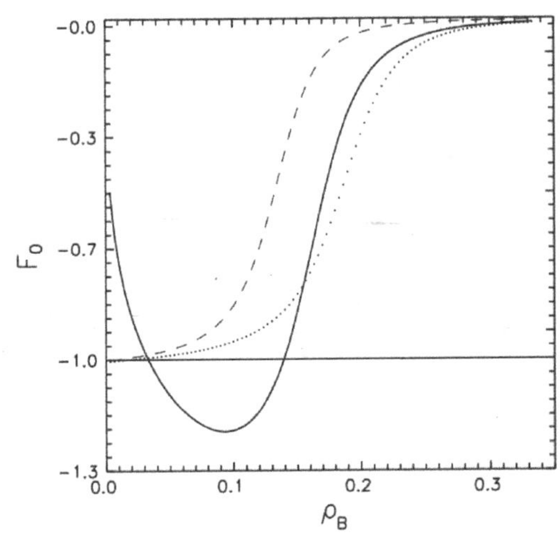

Figure 7: The factor F_0 of the Landau Fermi-liquid theory.

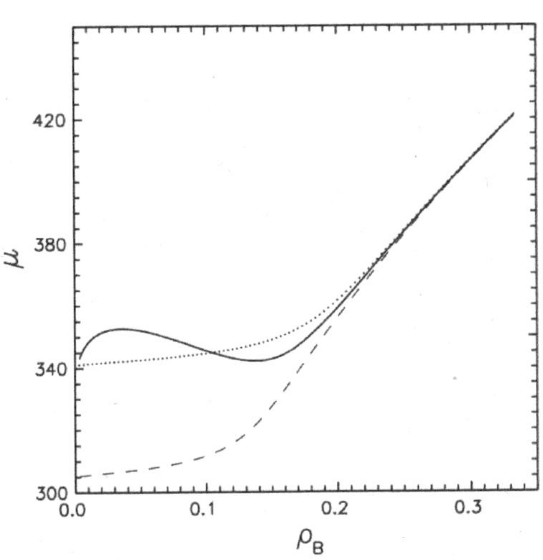

Figure 8: Chemical potential in MeV.

The group velocity of quasiparticles on the Fermi surface $v_f = \partial P_0/\partial \mathbf{p}|_{|\mathbf{p}|=P_F}$ is shown in Figure 3as a function of baryon (quark) density (by definition, the baryon density is three times smaller than the quark one $\rho_B=\rho/3$). A solid line describes the NJL model, while a dashed one corresponds to the KKB model. There are points for comparison that show a version for the KKB model when the parameters are tuned in such a way that the $\pi\pi$-meson masses coincide in the NJL and KKB models (a similar notation is used below in Figs. 4, 5, 6, 7, 8). The group velocity tending to unity in the region of normal nuclear densities corresponds to the chiral symmetry restoration when the induced quark mass tends to zero. The group velocity turns to zero for quarks with momenta |p|<2G in the chiral limit in the KKB model. The negative group velocities in the NJL model correspond to the regions of instability (see below). The points in which the group velocity vanishes give rise to the peaks in the density of states on the Fermi surface NF,

$$N_F = \gamma \int d\tilde{\mathbf{p}}\, \delta(P_0 - \mu) = \frac{\gamma}{2\pi^2} P_F P_F^0 \, (1 + F_0)^{-1},$$
$$F_0 = \frac{M_q}{P_F} \frac{dM_q}{dP_F}, \tag{16}$$

where $P_F^0 = P_0|_{|\mathbf{p}|=P_F}$, $N_F = d\rho/d\mu$. For more details on how to determine the parameter F_0, see below. The interaction term in the functional (8) vanishes in an ideal gas and causes the derivative of the quark dynamical mass in the Fermi momentum to turn to zero: $dM_q/dP_F=0$. Let us define the density of states of an ideal gas as

$$\widetilde{N}_F = \gamma/(2\pi^2) P_F P_F^0,$$

then the relation (16) can be written in the form

$$N_F = \widetilde{N}_F (1 + F_0)^{-1}.$$

Another important characteristic is the compression coefficient

$$K = 9\rho \frac{d\mu}{d\rho} = 3 \frac{P_F^2}{\mu} (1 + F_0). \tag{17}$$

Figure 4 demonstrates the data for the NJL and KKB models. They are consistent with the specific values obtained for a nuclear medium. One can also conclude that, in principle, these models admit a wide variety of equations of state including sufficiently restrictive ones. The negative values of the compression coefficient are not allowed and signal the region of instability. The first sound velocity, which is determined by the relation

$$c_1^2 = \frac{K}{9\,\mu} = \frac{v_F^2}{3} (1 + F_0), \tag{18}$$

is shown in Figure 5. When the baryon densities are somewhat higher than the density of normal nuclear matter, the sound velocity tends to its asymptotic value $c_1 = 1/\sqrt{3}$ which is a natural manifestation of the chiral symmetry restoration. If the sound velocity of an ideal Fermi gas $\widetilde{c}_1^2 = v_F^2/3$ is introduced in a way similar to the $\widetilde{N}_F$ definition, then the expressions (16), (18) can be endowed with the form whose physical meaning is an equality of flow coming through the Fermi sphere of quasiparticles of an (imaginary) ideal Fermi gas and interacting Fermi liquid (that is, there basically is a relativistic analog of the Luttinger theorem [30, 31]),

$$N_F\, c_1^2 = \widetilde{N}_F\, \widetilde{c}_1^2. \tag{19}$$

The thermal conductivity at a constant volume and a low temperature is given by the expression

$$C_V = \frac{1}{3} \pi^2 N_F\, T. \tag{20}$$

Figure 6 shows the slope (the factor $\frac{1}{3}\pi^2 N_F$ in Eq. (20), $N_F = d\rho/d\mu$), as a function of baryon/quark density, which demonstrates how informative it could be to measure the slope of a curve corresponding to the thermal conductivity. Yet another important characteristic of the Fermi liquid is defined by the second variational derivative, which for the functional (8) develops only a scalar component

$$f_0 = \frac{\delta^2 w}{\delta n^2} = \frac{M_q}{P_0} \frac{\delta M_q}{\delta n}. \tag{21}$$

For the Fermi liquid at zero temperature, in particular, we have

$$f_0 = \frac{2\pi^2}{\gamma P_F\, P_F^0} \frac{M_q}{P_F} \frac{dM_q}{dP_F}.$$

For example, in the NJL model

$$\frac{M_q}{P_F} \frac{dM_q}{dP_F} = -\frac{P_F}{P_F^0} \frac{1}{I + \pi^2 m/(G M_q^3)},$$

$$I = \ln \frac{\Lambda + P_\Lambda^0}{P_F + P_F^0} - \frac{\Lambda}{P_\Lambda^0} + \frac{P_F}{P_F^0}.$$

where $P_\Lambda^0 = P_0|_{|\mathbf{p}|=\Lambda}$. In the KKB model

$$\frac{M_q}{P_F} \frac{dM_q}{dP_F} = -\frac{M\, M_q^2}{M_q^3 + m P_F^3}.$$

In particular, in the chiral limit (when m=0) we have $(M_q/P_F)(dM_q/dP_F) = -1$. The collective oscillation modes of the Fermi liquid, corresponding to the so-called zero sound (the collisionless mode), are found by using the parameter

$$F_0 = \widetilde{N}_F\, f_0 = \frac{M_q}{P_F} \frac{dM_q}{dP_F},$$

which is shown in Figure 7. In particular, in the KKB model

$$F_0 = -\frac{M M_q^2}{M M_q^2 + (P_F^0)^2 m} \geq -1.$$

The zero sound oscillations are known to be determined by the solutions to the dispersion equation with a frequency parameter s (for details concerning this notation, see the section devoted to the polarization operator) of the form

$$F_0 = \frac{s}{2} \log \frac{s+1}{s-1} - 1. \tag{22}$$

When there is a repulsion in a system and the factor is positive $F_0>0$, the solutions to the dispersion equation $s=\lambda+i\eta$ describe continuous oscillations ($\eta=0$). In the case of weak attraction, when $-1<F_0<0$, the damped oscillations of zero sound are possible with a purely imaginary frequency ($\lambda=0$), which is given by the solutions to the following equation:

$$F_0 + 1 = \eta \ \arctan(1/\eta).$$

When the strong attraction is available and $F_0<-1$, the solutions reside on a second sheet of the complex plane s and describe the damped oscillations which are found from the solution to the equation

$$F_0 + 1 = \eta \ [-\pi + \arctan(1/\eta)].$$

It should, however, be recalled that these states of a Fermi liquid are unstable (as will be discussed below). It is hardly possible to apply directly the consideration of zero sound given above to the situation of interacting quarks and antiquarks under study, because here the contribution of vacuum fluctuations of antiquarks, which form along with quarks a chiral condensate, was completely ignored. On the other hand, zero sound oscillations are known to be interpreted as a bound state of a particle and hole in the vicinity of the Fermi sphere. Therefore, the excitations in a Fermi liquid should be described (in our case) by taking into account the interference between the bound states of a quark and antiquark, as well as of a quark and a hole of the Fermi sphere (the quantum numbers of that hole allows one to consider it as an anti-particle). We are doing that on calculating the polarization operator.

Turning now to the chemical potential of quasiparticles presented in Figure 8 let us emphasize that it is seen from the data for the NJL model that there is a region of occupied states almost degenerate with respect to the chemical potential with the vacuum chemical potential of a quasiparticle that quite naturally corresponds to the vanishing Fermi momentum.

In the pioneering papers of the NJL model there was a discontinuity in the chemical potential values [32] unlike the smooth curve in this plot.

Similarly, the chemical potential of occupied states in the KKB model differs from that in vacuum by a small quantity proportional to the quark current mass

$$\frac{d\mu}{d\rho} = \frac{\mu}{\rho}\frac{v_F^2}{3}\left(1 + \frac{M_q}{P_F}\frac{dM_q}{dP_F}\right) \sim m. \tag{23}$$

All the states with momentum $|p|<2G$ are degenerate with respect to the chemical potential in the chiral limit. We have $M_q=(4G^2-p^2)^{1/2}$, $P_0=2G$, when $P_F<|p|<2G$, $M_q=0$, $P_0=|p|$ if $|p|<P_F$, and $|p|>2G$. Such a behavior of the chemical potential is a consequence of a rapid decrease of the dynamical quark mass with increasing Fermi momentum [see also (15)]. It follows from Eq. (8) that the Fermi sphere is being filled as though from within. Those quarks with momenta smaller than the Fermi one $|p|<P_F$ do not take part in forming a condensate. As a result, the quark dynamical mass can only decrease with the Fermi momentum increasing. This dynamical mass is independent of the quark momentum in the NJL model because of the approximation assumed. This dependence should be taken into account in a more realistic case as an analysis of the KKB model shows.

It turns out that the pressure of some occupied states degenerate in the chemical potential almost coincides with that of vacuum (the pressure of a dilute Fermi gas; T=0)

$$P = -\frac{dE}{dV} = -\mathcal{E} + \mu\,\rho,$$

where E=E/V is the specific energy. Below we analyze respective data in more detail including the situation with nonzero temperature. The energy (and, hence, the pressure) of the ensemble is a discontinuous functional of the quark current mass (see [9–11]) in the KKB model. The integrands in (8) are then estimated as follows:

$$p_0 - P_0 + \frac{1}{4G}\,M^2 \sim -\frac{G\,m^2}{p^2},$$

and we find a linearly diverging integral for the specific energy of ensemble

$$w \sim -\int \frac{dp\;p^2}{2\pi^2}\;\frac{G\,m^2}{p^2},$$

despite the fact that the delta-like form factor in the momentum space is the strongest regularizer. It is paradoxical that any small value of the current mass m leads to the negative infinite energy of ensemble, while the expression $w|_{m=0}$ is well defined in the chiral limit. Even more so, a similar divergence occurs in the case of a delta-like form factor in the coordinate space. This fact is concealed by introducing the cutoff momentum Λ in the NJL model. Now it looks quite sensible to consider the relative pressure of quark ensemble in comparison with a (formally infinite) vacuum value because of the singular character (mentioned above) of the ensemble pressure in the KKB model. The pressure derivative in the ensemble density has the form: $dP/d\rho=\rho\ d\mu/d\rho$. Therefore, one can conclude by using an estimate given in (23) that the occupied states with momenta $|p|<2G$ are observed to be degenerate with respect to the pressure ($E=2G\rho$, $\mu=2G$) in the chiral limit in the KKB model. The deviations are proportional to the quark current mass beyond the chiral limit.

Now, we are able to analyze some thermodynamic properties of a system and to consider, first, the pressure of the quark ensemble in detail

$$P = -\frac{dE}{dV}.$$

By definition, the volume derivative should be calculated at the constant mean entropy, $d\bar{S}/dV = 0$. Implementing this constraint, one can, for example, extract the volume derivative of the chemical potential, $d\mu/dV$. However, this approach cannot be implied because the mean charge conservation might be broken. In fact, there is only one possibility to satisfy both conditions by introducing two independent chemical potentials for quarks and antiquarks separately. We use a symbol μ introduced earlier for the quark chemical potential, whereas the antiquark chemical potential is taken with an opposite charge and is denoted by $\bar{\mu}$. Then, we have

$$n = \frac{1}{e^{\beta\,(P_0-\mu)}+1}, \quad \bar{n} = \frac{1}{e^{\beta\,(P_0+\bar{\mu})}+1}$$

for the quark and antiquark densities, respectively. Some non-equilibrium states of quark ensemble could also be described on this way (formally with a loss of covariance, just similar to the electrodynamics as for the situation of electron–positron gas). However, we are here interested only in the special configuration when $\bar{\mu}$=μ. The partial derivative of the specific energy dw/dV can be presented in the following form:

$$\frac{dw}{dV} = \int d\tilde{\mathbf{p}} \left(\frac{dn}{d\mu}\frac{d\mu}{dV} + \frac{d\bar{n}}{d\bar{\mu}}\frac{d\bar{\mu}}{dV}\right)\Big[p_0 \cos\theta - 2G \times$$
$$\times \sin(\theta-\theta_m) \int d\tilde{\mathbf{q}}\ \sin(\theta'-\theta'_m)\,(n'+\bar{n}'-1)\,F\Big].$$

Dealing with the definition of an induced quark mass (9) and presenting the trigonometric factors via the quark dynamical mass we find the ensemble pressure as

$$P = -\frac{E}{V} - V\,2N_c \int d\tilde{\mathbf{p}} \left(\frac{dn}{d\mu}\frac{d\mu}{dV} + \frac{d\bar{n}}{d\bar{\mu}}\frac{d\bar{\mu}}{dV}\right) P_0. \tag{24}$$

The condition of mean charge conservation

$$\frac{d\bar{Q}_0}{dV} = \frac{\bar{Q}_0}{V} + V\,2N_c \int d\tilde{\mathbf{p}} \left(\frac{dn}{d\mu}\frac{d\mu}{dV} - \frac{d\bar{n}}{d\bar{\mu}}\frac{d\bar{\mu}}{dV}\right) = 0, \tag{25}$$

gives the first equation that interrelates the derivatives dμ/dV and d$\bar{\mu}$/dV. Here, a regularized expression for the mean charge of quarks and antiquarks is assumed modulo respective vacuum contribution.

Implementing the condition of constant mean entropy $d\bar{S}/dV = 0$ in a similar way one can obtain the second equation of the chemical potential derivatives system as follows:

$$\int d\tilde{\mathbf{p}} \frac{dn}{d\mu} \ln\frac{n}{1-n}\frac{d\mu}{dV} - \int d\tilde{\mathbf{p}} \frac{d\bar{n}}{d\bar{\mu}} \ln\frac{\bar{n}}{1-\bar{n}}\frac{d\bar{\mu}}{dV} = \frac{\bar{S}}{2N_c V^2}. \tag{26}$$

Substituting the expressions $T \ln\frac{n}{1-n} = \mu - P_0$ and $T \ln\frac{\bar{n}}{1-\bar{n}} = -\bar{\mu} - P_0$ into this equation and collecting similar terms we attain the following equation

$$\int d\tilde{\mathbf{p}} \left(\frac{dn}{d\mu}\frac{d\mu}{dV} + \frac{d\bar{n}}{d\bar{\mu}}\frac{d\bar{\mu}}{dV}\right) P_0 = -\frac{\bar{S}\,T}{2N_c\,V^2} - \frac{\bar{Q}_0\,\mu}{2N_c\,V^2}$$

if the condition $\bar{\mu} = \mu$ and Eq. (25) are satisfied. Finally, we have for the pressure

$$P = -\frac{E}{V} + \frac{\bar{S}\,T}{V} + \frac{\bar{Q}_0\,\mu}{V}. \tag{27}$$

Then the thermodynamic potential Ω should obey the following thermodynamic identity:

$$\Omega = -PV = E - \mu\,\overline{Q}_0 - T\,\overline{S}, \tag{28}$$

as it should.

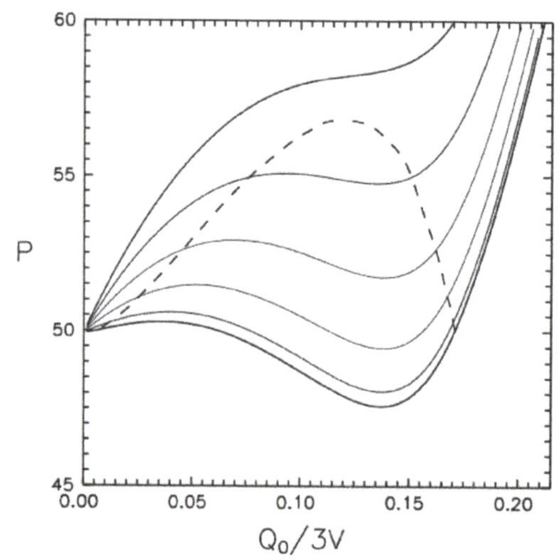

Figure 9: The ensemble pressure P (MeV/fm³) is shown as a function of charge density Q_0 at temperatures T=0 MeV, ..., T=50 MeV with spacing T=10 MeV. The *lowest curve* corresponds to zero temperature. The *dashed curve* shows the boundary of the liquid–gas phase transition; see the text.

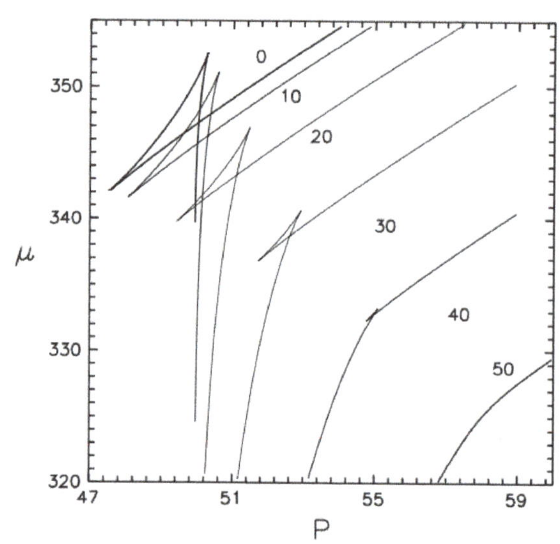

Figure 10: The fragments of the isotherms shown in Figure 9, see the text. The chemical potential μ(MeV) is plotted as a function of pressure P in MeV/fm³. The *top curve* corresponds to the zero isotherm and down with spacing 10 MeV till the isotherm 50 MeV (the *lowest curve*).

At low temperatures the antiquark contribution is small and the thermodynamic description can be approximately developed by using the chemical potential μ only. If the antiquark contribution becomes significant, the thermodynamic description is more sophisticated and should obviously include the chemical potential $\bar{\mu}$ with the additional condition $\bar{\mu}$ =μ. Figure 9shows the ensemble pressure P in MeV/fm³ as a function of the charge density Q_0/3V for various temperatures. The lowest curve is obtained at zero temperature. The next curves and the ones following upwards correspond to temperatures T=10 MeV, T=50 MeV (an upper curve) with a step T=10 MeV. Let us also remember that the pressure of the vacuum for the NJL model was estimated in [9–11] to be 40–50 MeV/fm^3, which is quite consistent with that obtained in the bag model. It was also demonstrated that there is a region of instability within a certain interval of the Fermi momenta generated by the anomalous behavior of pressure dP/dn<0 (see also [33–36]). Figure 10 displays fragments of isotherms shown in Figure 9 (but now in different coordinates) in the form of a chemical potential as a function of the ensemble pressure. A top curve is obtained at zero temperature. The isotherms below are shown in steps of 10 MeV. The lowest curve is obtained at a temperature 50 MeV. It is clearly seen from the figure that there are states on the isotherms which are in thermodynamic equilibrium. The pressure and chemical potential are the same for these states (see the characteristic Van der Waals triangle with intersecting curves). The equilibrium points obtained are shown in Figure 9 by a dashed curve. The points at which a dashed curve intersects with an isotherm give a boundary for a gas–liquid phase transition. The respective line P=const cuts off nonequilibrium and unstable fragments of an isotherm and describes a mixed phase. The critical temperature turns out to be equal to Tc≈46MeV with the critical charge density $\bar{Q}_0$ ≈0.12 charge/fm_3 for the above mentioned tuning parameters. Figure 11 shows the isobars. The pressure next to each curve is given in MeV/fm_3. The vacuum pressure corresponds to approximately ~50 MeV/fm^3. It is possible to extrapolate isobars into the region of small charge densities, however, it is not really necessary. The figure clearly demonstrates the presence of dilute (a gas) and dense (a liquid) phases in the vicinity of the vacuum isobar.

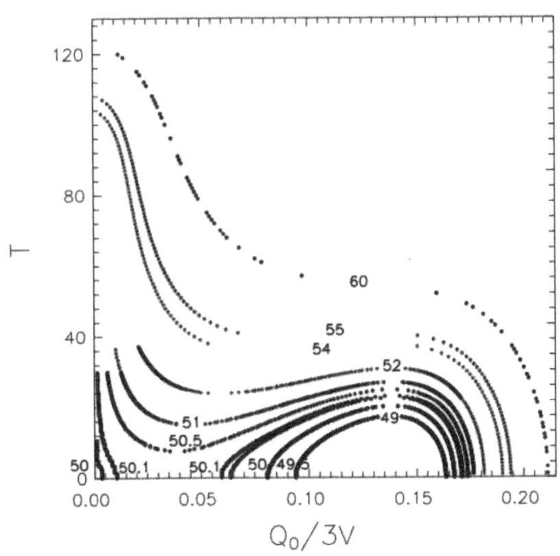

Figure 11: Isobars of NJL model. The pressure in MeV/fm^3 is indicated next to each *curve*. The vacuum pressure corresponds to approximately 50 MeV/fm^3.

Figure 12 shows the quark dynamical mass Mq (in MeV) as a function of the chemical potential μ (in MeV) for temperatures T=0 MeV, T=100 MeV in steps of T=10 MeV. The right-most curve corresponds to zero temperature. At low temperatures, below 50 MeV, the quark dynamical mass is a multi-ciphered function of the chemical potential (but the discontinuous solution has been presented in the pioneering papers).

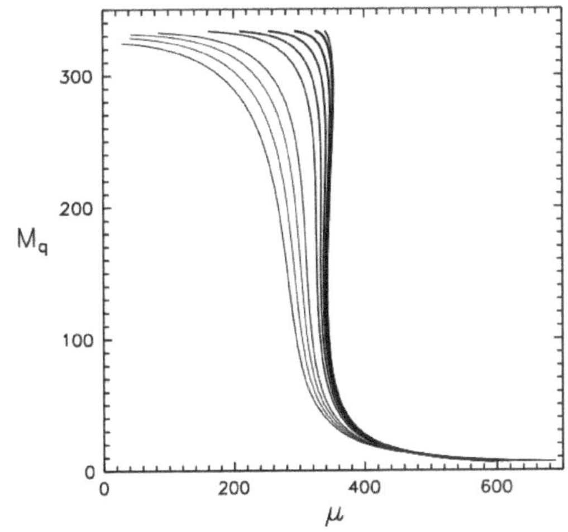

Figure 12: The dynamical quark mass |Mq| (MeV) as a function of chemical potential μ (MeV) at the temperatures T=0 MeV, ..., T=100 MeV with spacing T=10 MeV. The *right-most curve* corresponds to zero temperature.

Figure 13 shows the quark dynamical mass as a function of temperature at small charge density Q_0~0. This picture is easily recognizable in the context of the NJL model. It is the latter that is implied in a scenario of chiral invariance restoration under extreme temperatures higher than 100 MeV and with a highly diluted quark ensemble. We have already noted (see also [9–11]) that the momentum p_θ, which corresponds to the strongest quark–antiquark attraction dsinθ/dp=0, can be determined. For example, for the NJL model this parameter is equal to

$$p_\theta = (M_q\, m)^{1/2}. \qquad (29)$$

Its inverse value is given by the characteristic effective size of a quasiparticle $r_\theta = p_\theta^{-1}$. From the behavior of the quark dynamical mass as a function of temperature at small charge densities (see Figure 13) one can conclude that the quasiparticle size grows with the energy increasing.

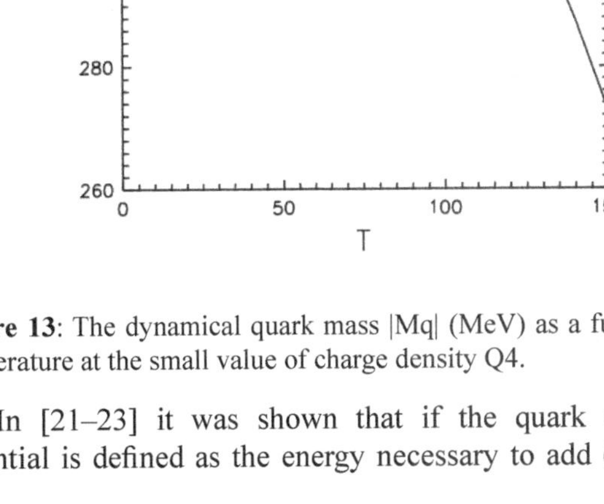

Figure 13: The dynamical quark mass |Mq| (MeV) as a function of temperature at the small value of charge density Q4.

In [21–23] it was shown that if the quark chemical potential is defined as the energy necessary to add (remove) one quasiparticle, μ=dE/dN, then the chemical potential in vacuum coincides with the quark dynamical mass (see also (12), (15)). Therefore, it seems to be reasonable to consider a QCD phase diagram by starting from this value of the chemical potential, though formally it can be taken smaller than the quark dynamical mass. In particular, we exactly reproduce the standard picture [24, 32] by taking the chemical potential equal zero. The results obtained allows one to conjecture that the phase transition of the (partial) restoration of the chiral invariance could already be realized in nature as a mixed phase of physical vacuum and baryonic matter. An indirect confirmation of this hypothesis can be seen in degenerate excited states of some baryons (see, for instance, [37]). It is, however, clear that the data presented (in particular, on the temperature and density of a critical point position) should be understood as just estimates. The critical temperature of the gas–liquid transition for nuclear matter extracted from experiment is estimated to be about 20 MeV.

(Let us mention here that the hadronization temperature as extracted from the RHIC and LHC data analysis in the framework of statistical hadronization models looks like 165 MeV.)

In addition, here (at T=0) a gas component possesses the nonzero density of order of 0.01 of the normal nuclear density, whereas the observed value should correspond to physical vacuum, i.e., to zero baryon density. It should be noted that although such an uncertainty is inherent in the other predictions of the chiral symmetry restoration phase transition which are widely discussed in many papers, they are somewhere around 2–6 normal nuclear matter densities.

POLARIZATION OPERATOR

Returning to the discussion of zero sound and excitations of a chiral condensate we would like also to recall that this knowledge is necessary for a more consistent analysis of the transition gas–liquid layer. To this end, we will need to know a polarization operator of the form

$$\Pi^{\Gamma}(p,q) = \int \frac{dk}{(2\pi)^4}\, i\,\pi^{\Gamma}(k+p,k-q), \tag{30}$$

where

$$\pi^{\Gamma}(k+p,k-q) = \mathrm{Tr}\{S(k+p)\Gamma S(k-q)\Gamma\},$$

is a respective density of the polarization operator in the channels Γ=1,iγ5,γμ,γ5γμ with the Green function of quark with the dynamical mass M_q

$$S(k) = \frac{1}{\hat{k}+\hat{\mu}-M_q(\boldsymbol{k})}, \tag{31}$$

$\hat{\mu}=\mu\gamma^0$, where p, q are the incoming and outgoing external momenta of the quark quasiparticles. It will be enough for our purposes to consider the quasiparticles with momenta pp=qq=QQ/2 in the center of mass frame. We analyze pseudoscalar and scalar channels, for which one can deduce

$$\Pi^{\pi,\sigma} = N_c \int d\tilde{\boldsymbol{k}}\, F(\boldsymbol{k}) \left[\frac{a+b\,\varepsilon}{\varepsilon^2-(E_p+E_q)^2} + \frac{c}{\varepsilon-E_p+E_q}\right],$$

$$a = \left(E_p+E_q\right)\left[2-n_p-n_q\right]\left[\frac{Q^2/4-k^2 \mp M_pM_q}{E_pE_q}-1\right],$$

$$b = \left[n_q-n_p\right]\left[\frac{Q^2/4-k^2 \mp M_pM_q}{E_pE_q}-1\right],$$

$$c = [n_p-n_q]\left[\frac{Q^2/4-k^2 \mp M_pM_q}{E_pE_q}+1\right], \tag{32}$$

where $\varepsilon=p_0-q_0$ is the transferred energy, $M_p=M_q(p)$, $E_p=[p^2+M_q^2(p)]^{1/2}$, the quantity F(k) is a form factor, and for the kinematics chosen p=k+Q/2 (here n_p is the occupation number for a quasiparticle with momentum p). In particular, at zero temperature we have the Fermi step: $n_p=n(E_p-\mu)$. A similar notation is also used for a quasiparticle with momentum q=k−Q/2.

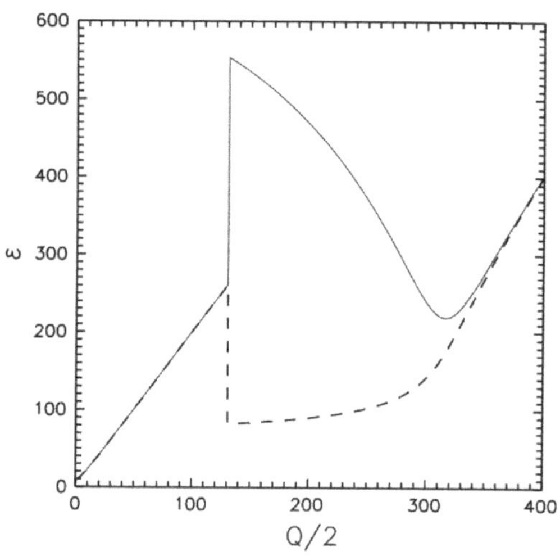

Figure 14: Energies (in MeV) of π (*dashed line*) and σ (*solid line*) mesons as a function of momentum Q/2 (in MeV) (T=0) for a gas of low baryon density such that P_F~130 MeV.

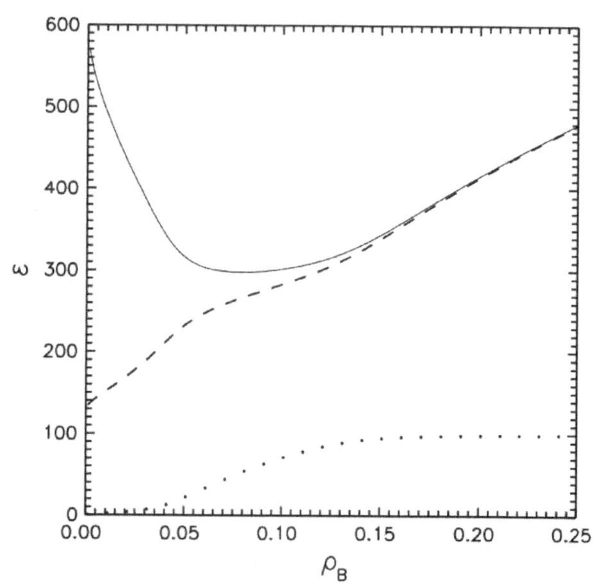

Figure 15: Energies (in MeV) of π (*dashed line*) and σ (*solid line*) mesons as a function of baryon density (T=0). The branch corresponding to a bound quark–hole state is shown by *dots*; Qq=−Qh=50 MeV.

The first term in Eq. (32) corresponds to the quark and antiquark contributions, whereas the second one comes from the quark and hole configuration. It is easy to see that at F(kk)=δ(kk) (in the KKB model) we have cubic dispersion relations to determine the bound states: 1−2G Ππ,σ(ε,Q)= 0. As an example, Figure 14 shows the energies (in units of MeV) calculated of π (dashed line) and σ (solid line) mesons as functions of the momentum Q/2 (in units of MeV) at zero temperature for the gas with small baryon density corresponding to the Fermi momentum PF~130MeV. The region of degeneracy, seen in Figure 14 at low quark momenta, is a consequence of the above discussed fact that the Fermi sphere is filled from the inside, and quarks with momenta smaller than PFPF do not participate in forming the quark dynamical mass. Such a behavior is not observed in the NJL model because of the approximation adopted (the quark mass is independent of the momentum). Figure 15 shows the energies (in units of MeV) of π (dashed line) and σ (solid line) mesons as the functions of the baryon density (T=0). The dots indicate a branch corresponding to the quark-hole bound state which appears to be degenerate for π and σ mesons. Just these branches correspond to the third additional root of the dispersion equation mentioned above, albeit so that there are only two roots (see the discussion of the NJL model). To be specific, the quark momentum is assumed to be 50 MeV larger than the Fermi one but the hole momentum is 50 MeV smaller than the latter in this example. As follows from Eq. (32), the polarization operator in the NJL model is defined by integrating over the running quark momentum k and is represented as a superposition of branches of the KKB model, which has already been mentioned in the introduction. The most significant contributions (for the kinematics we chose) are those coming from the terms denoted as a and c in Eq. (32). Integrating over the angle (it is more convenient to express the final formula by going to a nonsymmetric integration point; the corrections become negligibly small) one can obtain (at T=0) the following results:

$$\Pi^{\pi,\sigma} = A^{\pi,\sigma} + B^{\pi,\sigma},$$

$$A^{\pi,\sigma} = \int^{\Lambda P_F} \frac{dk\, k}{2\pi^2 Q}\left[(E_+ - E_-)\left(1+\frac{E_+ + E_-}{2E_k}\right) - \frac{Q^2-\varepsilon^2+2(M_q^2 \mp M_q^2)}{2E_k}\ln\left(\frac{\varepsilon+E_k+E_+}{\varepsilon+E_k+E_-}\frac{\varepsilon-E_k-E_+}{\varepsilon-E_k-E_-}\right)\right],$$

$$B^{\pi,\sigma} = \int_0^{P_F} \frac{dk\, k}{2\pi^2 Q}\left[(E_+ - E_-)\left(1-\frac{E_+ + E_-}{2E_k}\right) + \frac{Q^2-\varepsilon^2+2(M_q^2 \mp M_q^2)}{2E_k}\ln\left(\frac{\varepsilon-E_k+E_+}{\varepsilon-E_k+E_-}\frac{\varepsilon+E_k-E_+}{\varepsilon+E_k-E_-}\right)\right],$$

where $E_\pm = [(k \pm Q)^2 + M_q^2]^{1/2}$ and $E_k = [k^2 + M_q^2]^{1/2}$

At small momentum Q the component Bσ is transformed into Eq. (22) with the parameter s=EFε/(kFQ). The first component $A^{\pi,\sigma}$ results from the contribution coming from a quark–antiquark pair, whereas in the case of the second one $B^{\pi,\sigma}$ arises due to the coupling of quark and hole residing in the vicinity of the Fermi sphere. It should be noted that for a quark ensemble we consider the medium properties which are mainly governed by the term $A^{\pi,\sigma}$ responsible for the quark–antiquark condensate, contrary to what we have in condensed matter physics, where the dominant contribution, as is well known, is given by $B^{\pi,\sigma}$. Therefore, the results obtained exclusively by using an analogy with the condensed matter physics should be taken with a grain of salt. In particular, in the present paper we have analyzed in detail a situation with the zero sound description taken as an example illustrating just this point. The zero sound would represent in itself the highly damped oscillations described by the only scalar parameter F_0, while with no an antiquark present. A more accurate analysis shows that, for example, there is a stable branch of quark and hole excitations in the Fermi sphere in addition to a paired quark–antiquark state in the KKB model. We observe a regular mass convergence for π and σ mesons when the baryon density increases by performing a numerical integration in the NJL model. This effect is clearly related to the restoration of chiral symmetry. The influence of a bound quark-hole state in the Fermi sphere turns out to be insignificant. For instance, for the densities of the order of normal nuclear matter the dispersion law changes by a few MeV when the quark and hole momentum differs more than 200MeV, but there are no damped oscillations as in the KKB model.

One of the drawbacks of the models studied so far is the lack of quark confinement that is understood here simply as the impossibility to observe a single particle state with a regular (real) dispersion law. We see formally that one quasiparticle can freely propagate, indeed. But adding just another quasiparticle can dramatically change the picture due to the existence of a bound channel. For example, in the KKB model the bound states in scalar, pseudoscalar, vector, and axial-vector channels appear at any quasiparticle momenta (details can be found in [38–40]). In particular, the bound state energy, obtained by using the dispersion equation 1−2G Π=0, has the form

$$\varepsilon^2_{\pi,\sigma} = (E_p + E_q)^2 - 2G\frac{E_p + E_q}{E_p E_q}(E_p E_q \pm M_p M_q - \mathbf{pq}),$$

in the π and σ channels (an upper sign corresponds to the pseudoscalar channel). The first term in this expression is the energy of free particle motion. The second one is strictly positive at any momenta p and q and plays a role of binding energy in π and σ channels (only in the configuration of q=p the binding energy vanishes for a scalar channel). Similarly, one can show that a quark and antiquark are always coupled in vector and axial-vector channels, i.e. the scattering matrix is always singular except for a tensor channel where it is trivial because of the initial interaction Hamiltonian, which is taken as a product of two color currents. Similar bound states exist in the diquark channel. As a consequence, the states with any number of quark quasiparticles turn out to be the bound states in the channels we have just mentioned. The same behavior is observed in the NJL model where the bound states appear for the quarks with momenta somewhat lower than the cutoff momentum, i.e. the scattering matrix is also singular within this momentum interval as in the KKB model. It seems the bound states appear rather due to the fermion correlations than the physical influence of a field that is familiar in the quantum electrodynamics. Then, in order to understand what may take place beyond the cutoff momentum one apparently has to study the appropriate nonlocal models.

TRANSITION LAYER BETWEEN GAS AND LIQUID

The concept of a mixed phase of physical vacuum and baryonic matter would receive a substantial confirmation if we are able to demonstrate the existence of the boundary (transition) layer where a transformation of the quark ensemble from one aggregate state to another takes place. As was argued above the indicative characteristic to explore a homogeneous phase (at finite temperature) is the mean charge (density) of the ensemble. All the other characteristics, for example, a chiral condensate, dynamical quark mass, etc. can be reconstructed if one knows the ensemble mean charge. So, here we analyze a specific case of the surface (transition) layer at zero temperature.

We assume that the quark ensemble parameters in a gaseous phase are approximately the same as those at zero charge $\rho_g=0$, i.e. as in vacuum (minor differences in pressure, chemical potential, and quark condensate are neglected). The dynamical quark mass develops here the maximal value, and it is M=335 MeV for the parameter choice standard for the NJL model. Then, as the Van der Waals diagram shows, a liquid phase, being in equilibrium with a gas phase, gains the density $\rho_l=3\times0.185$ charge/fm^3 (by a reason which becomes clear below we correct it to take the value $\rho_l=3\times0.157$ charge/fm^3). The detached factor 3 here links again the magnitudes of quark and baryon matter densities. The quark mass is approximately $\overset{*}{M}\approx 70$ MeV in this phase. Hereafter we focus on describing two adjoining semi-infinite layers (i.e. assuming a plane symmetry of the corresponding one-dimensional problem).

The previous experience teaches that an adequate description of heterogeneous states can be reached with the mean field approximation [41, 42]. In our particular case it means making use of the corresponding effective quark–meson Lagrangian [43–45] (functional of the Ginzburg–Landau type)

$$\mathcal{L} = -\bar{q}\,(\hat{\partial} + M)\,q - \frac{1}{2}\,(\partial_\mu\sigma)^2 - U(\sigma) - \frac{1}{4}\,F_{\mu\nu}F_{\mu\nu}$$
$$-\frac{m_v^2}{2}\,V_\mu V_\mu - g_\sigma\,\bar{q}q\,\sigma + ig_v\,\bar{q}\,\gamma_\mu\,q\,V_\mu, \qquad (33)$$

where

$$F_{\mu\nu} = \partial_\mu V_\nu - \partial_\nu V_\mu, \quad U(\sigma) = \frac{m_\sigma^2}{2}\,\sigma^2 + \frac{b}{3}\,\sigma^3 + \frac{c}{4}\,\sigma^4,$$

and σ is the scalar field, Vμ is the field of vector mesons, mσ, mv are the masses of scalar and vector mesons and gσ, gv are the coupling constants of the quark–meson interaction. The U(σ) potential includes the nonlinear σ field interaction terms up to the fourth order, for example. For the sake of simplicity we do not include the contributions coming from the pseudoscalar and axial-vector mesons.

The meson component of such a Lagrangian should be self-consistently treated by considering the corresponding quark loops. Here we do not see any reason to go beyond the well elaborated and reliable one loop approximation (33) [43–45], although recently considerable progress has been reached in scrutinizing the non-homogeneous quark condensates by applying the powerful methods of exact integration [46–51]. Here we believe it is more practical to adjust phenomenologically the effective Lagrangian parameters based on the transparent physical picture. It is easy to see that handling (33) in one loop approximation we arrive, in actual fact, at the Walecka model [52–54] but adopted for the quarks. In what follows we are working with the designations of that model and we hope that it does not lead to misunderstandings.

In the context of our paper we propose to interpret Eq. (33) in the following way. Each phase might be considered, in a sense, with regard to another phase as an excited state which requires the additional (apart from a charge density) set of parameters (for example, the meson fields) for its complete description, and those are characterizing the measure of deviation from the equilibrium state. Then the crucial question becomes whether it is possible to adjust the parameters of the effective Lagrangian (33) to obtain the solutions in which the quark field interpolates between the quasiparticles in the gas (vacuum) phase and the quasiparticles of the filled-up states. For all that, the density of the filled-up state ensemble should asymptotically approach the equilibrium value of ρlρl and should turn to the zero value in the gas phase (vacuum).

The scale inherent in this problem may be assigned by one of the masses referred to in the Lagrangian (33). In particular, we bear in mind the dynamical quark mass in the vacuum MM. Besides, there are another four independent parameters in the problem and in order to compare them with the results of a study of the nuclear matter we employ the form characteristic for the (nuclear) Walecka model,

$$C_s = g_\sigma\,\frac{M}{m_\sigma}, \quad C_v = g_v\,\frac{M}{m_v}, \quad \bar{b} = \frac{b}{g_\sigma^3\,M}, \quad \bar{c} = \frac{c}{g_\sigma^4}.$$

Parameterizing the potential $U(\sigma)$ as $b_\sigma = 1.5\,m_\sigma^2\,(g_\sigma/M)$ $c_\sigma = 0.5\,m_\sigma^2\,(g_\sigma/M)^2$ and we arrive at the sigma model whereas the choice b=0, c=0 results in the Walecka model. As to standard nuclear matter, application of the parameters bb and cc demonstrates the vital model dependent character and they are quite different from the parameter values of the sigma model. Truly, in that case their values are also regulated by additional requirement of an accurate description of the saturation property. On the other hand, for the quark Lagrangian (33) we could intuitively anticipate some resemblance with the sigma model and, hence, could introduce two dimensionless parameters η and ζ in the form of $b = \eta\,b_\sigma, c = \zeta^2\,c_\sigma$, which characterizes some fluctuations of the effective potential. Then the scalar field potential is presented as follows:

$$U(\sigma) = \frac{m_\sigma^2}{8}\,\frac{g_\sigma^2}{M^2}\left(4\,\frac{M^2}{g_\sigma^2} + 4\,\frac{M}{g_\sigma}\,\eta\,\sigma + \zeta^2\sigma^2\right)\sigma^2.$$

The meson and quark fields are determined by solving the following system of stationary equations:

$$\Delta\,\sigma - m_\sigma^2\,\sigma = b\,\sigma^2 + c\,\sigma^3 + g_\sigma\,\rho_s,$$
$$\Delta\,V - m_v^2\,V = -g_v\,\rho,$$
$$(\hat{\nabla} + \overset{*}{M})\,q = (E - g_v\,V)\,q, \qquad (34)$$

where $\overset{*}{M}$=M+$g_\sigma\sigma$ is the running value of the dynamical quark mass, E stands for the quark energy and V=–iV_4. The density matrix describing the quark ensemble at T=0 has the form

$$\xi(x) = \int^{P_F} d\tilde{\mathbf{p}}\, q_{\mathbf{p}}(x)\, \bar{q}_{\mathbf{p}}(x),$$

in which p is the quasiparticle momentum and the Fermi momentum PF is defined by the corresponding chemical potential. The densities ρs and ρ at the right hand sides of Eq. (34) are by definition

$$\rho_s(x) = \mathrm{Tr}\{\xi(x), 1\}, \quad \rho(x) = \mathrm{Tr}\{\xi(x), \gamma_4\}.$$

Here we confine ourselves to the Thomas–Fermi approximation while describing the quark ensemble. Then the densities which we are interested in are given with some local Fermi momentum PF(x) as

$$\rho = \gamma \int^{P_F} d\tilde{\mathbf{p}} = \frac{\gamma}{6\pi^2}\, P_F^3,$$

$$\rho_s = \gamma \int^{P_F} d\tilde{\mathbf{p}}\, \frac{\overset{*}{M}}{E} =$$

$$= \frac{\gamma}{4\pi^2}\, \overset{*}{M}\, P_F^2 \left\{ \left(1+\lambda^2\right)^{1/2} - \frac{\lambda^2}{2} \ln\left[\frac{(1+\lambda^2)^{1/2}+1}{(1+\lambda^2)^{1/2}-1}\right] \right\}, \tag{35}$$

where γ is a quark gamma-factor which for one flavor is $\gamma = 2N_c$, $E = (\mathbf{p}^2 + \overset{*}{M}{}^2)^{1/2}$ and $\lambda = \overset{*}{M}/P_F$. Under the assumption adopted the ensemble chemical potential is constant and, therefore, the local value of the Fermi momentum is defined by the running value of dynamical quark mass and vector field as

$$\mu = M = g_v\, V + (P_F^2 + \overset{*}{M}{}^2)^{1/2}. \tag{36}$$

Now we should tune the Lagrangian parameters in Eq. (33). For asymptotically large distances (in a homogeneous phase) we may neglect the gradients of scalar and vector fields, and the equation for the scalar field of the system (34) leads to the first equation that relates the parameters Cs, Cv, $\bar{b}, \bar{c}$ by

$$\frac{M^2(\overset{*}{M} - M)}{C_s^2} + \bar{b}\, M\, (\overset{*}{M} - M)^2 + \bar{c}(\overset{*}{M} - M)^3 = -\rho_s. \tag{37}$$

The vector field asymptotically is given by the ensemble density $V = C_v^2\, \rho/(g_v M^2)$. The second equation derived from the relation (36) for the chemical potential looks like

$$M = \frac{C_v^2 \rho}{M^2} + (P_F^2 + \overset{*}{M}{}^2)^{1/2}. \tag{38}$$

If we know the liquid density we obtain the Fermi momentum (PF=346 MeV) from (35). Applying the identities (37), (38) we have for the particular case b=0, c=0 the results $C_s^2 = 25.3$, $C_v^2 = -0.471$, i.e. the vector component C2v is small (compared to C_s^2) and acquires a negative value that is unacceptable. Apparently, it looks necessary to abandon the contribution coming from the vector field or to reduce the dynamical quark mass $\overset{*}{M}$ up to the value which retains the identity (38) valid with positive C_v^2 or even zero value. In the gaseous phase the dynamical quark mass can also be corrected to a value larger than the vacuum value. It is clear that in the situation of the liquid with the density ρl=3×0.185ch/fm³ the dynamical quark mass should coincide (or exceed) M=346 MeV in the gaseous phase. However, here we correct the liquid density (as was argued above) to decrease its value up to ρl=3×0.157ch/fm³, which is quite acceptable in the nucleation capacity. In fact, this possibility can be simply justified by another choice of the NJL model parameters. Thus, we obtain at $\overset{*}{M}$=70M MeV and b=0, c=0 the result that $C_s^2 = 28.4$, $C_v^2 = 0.015$, i.e. we have a small but positive value for the vector field coefficient. At the same time, aiming here to estimate the surface tension effects only we do not strive for the precise fit of the parameters. In the Walecka model these coefficients are $C_s^2 = 266.9$, $C_v^2 = 145.7$ (b=0, c=0). Moreover, there is another parameter set with $C_s^2 = 64$, $C_v^2 \approx 0$ [55–59] but it is rooted in the essential nonlinearity of the sigma-field due to the nontrivial values of the coefficients b and c. The option (formally unstable) with negative c (b) has also been discussed.

The coupling constant of scalar field is fixed by the standard (for the NJL model) relation between the quark mass and the π-meson decay constant g_σ=M/f_π (we put f_π=100 MeV) although there is not any objection to treat this coupling constant as an independent parameter. As a result of the whole agreement we have for the σ-meson mass $m_\sigma = g_\sigma\, M/C_s$. In principle, we could even fix the σ-meson mass and coupling constant g_σ, but all relations above mentioned lead eventually to quite suitable values of the σ-meson mass as will be demonstrated below. The vector field plays, as we see, a secondary role because of the small magnitude of the constant C_v. Then taking the vector meson mass as mv≈740 MeV (a slightly smaller value than the mass of the ω meson because of simple technical reasons only) we calculate the coupling constant of the vector field from the relation similar to the scalar field m_v=g_v M/C_v. Amazingly, its value turns out to be steadily small being compared to the value characteristic for the NJL model, $g_v = \sqrt{6} g_\sigma$. However, at the values of constant C_v which we are interested in it is very difficult to maintain the reasonable balance, and to be specific in this paper we prefer to choose the massive vector field. Actually, it is unessential because we need this parameter (as we remember) to estimate the vector field strength only.

The key point of our interest here is the surface tension coefficient [55–59] which can be defined as

$$u_s = 4\pi\, r_0^2 \int_{-\infty}^{\infty} dx \left[\mathcal{E}(x) - \frac{\mathcal{E}_l}{\rho_l}\, \rho(x)\right]. \tag{39}$$

The parameter r_o will be discussed in the next section on considering the features of the quark liquid droplet, and for the present we would like to notice only that for the parameters considered its magnitude for $N_f=1$ is around $r_o=0.79$ fm. Recalling the factor $3^{1/3}$ which connects the baryon and quark numbers, we find the magnitude ($\tilde{r}_0 = 3^{1/3}0.79 \approx 1.14$ fm) in full agreement with the magnitude standard for the nuclear matter calculations (in the Walecka model) $\tilde{r}_0=1.1$–1.3 fm.

In order to proceed we calculate E(x) in the Thomas–Fermi approximation as

$$\mathcal{E}(x) = \gamma \int^{P_F(x)} \widetilde{d\mathbf{p}}\, [\mathbf{p}^2 + \overset{*}{M}(x)]^{1/2} + \frac{1}{2} g_V\, \rho(x)\, V(x) - \frac{1}{2} g_\sigma\, \rho_s(x)\, \sigma(x).$$

To give some idea for the 'setup' prepared we present here the characteristic parameter values for some fixed b and c with $\rho l=3\times0.157$ ch/fm³. In the liquid phase they are M∗=70MeV (PF=327 MeV) and el=310. 5 MeV (the index l stands for a liquid phase and e(x)=E(x)/ρ(x) defines the density of the specific energy). Both relations (37) and (38) are obeyed by this state. There exists a solution with larger value of quark mass $\overset{*}{M}=306$ MeV, (PF=135MeV) (we have faced a similar situation in the first section dealing with the gas of quark quasiparticles) and e=338 MeV ~eg (eg is the specific energy in the gas phase) that satisfies both equations as well. The specific energy of this solution appears to be larger than the specific energy of the previous solution. It is also worthwhile to mention the existence of an intermediate state corresponding to the saturation point with the mass $\overset{*}{M}=95$ MeV, (PF=291MeV) and e=306 MeV. Obviously, it is the most favorable state with the smallest value of specific energy (and with the zero pressure of quark ensemble), and the system can fall into this state only in the presence of a significant vector field. This state (already discussed in the first section) corresponds to the minimal value of chemical potential (T=0) and can be reached at the densities typical for the normal nuclear matter. However, Eq. (38) is not valid for this state.

Two other parameters, η, ζ, are fixed by looking through all the configurations in which the solution of the equation system (34) with a stable kink of the scalar field does exist and describes the transition of quarks from the gaseous phase to the liquid one. First, it is reasonable to scan the η, c (ζ=c η)-plane, in order to identify the domain in which the increase of specific energy E−El ρ/ρl≤0is revealed at running through all possible states which provide the necessary transition (without taking into account the field gradients). In practice one needs to follow a simple heuristic rule. The state with $P_F\sim1$ MeV (i.e. e and the corresponding ρ) and the state of the characteristic liquid energy E_l (together with ρ_l) should be compared while scanning the Lagrangian parameters η and c. Just the domain in which they are commensurable could provide us with the solutions which we are interested in, and Figure 5 shows its boundary. The curve could be continued beyond the value η=2.5, but the values of corresponding parameter η are unrealistic and not shown in the plot.

We calculate the solution of equation system (34) numerically by the Runge–Kutta method with the initial conditions $\sigma(L) \approx 0, \sigma'(L) \approx 0$ imposed at the large distance L≫t, where t is a characteristic thickness of transition layer (about 2 fm). Such a simple algorithm appears to be quite suitable if the vector field contribution is considered as a small correction (this just takes place in the situation under consideration) and is presented as

$$V(x) = \frac{1}{2m_V} \int_{-L}^{L} dz\, e^{-m_V|x-z|}\, g_V\, \rho(z),$$

where the charge (density) ρ is directly defined by the scalar field. We considered the solutions including the contribution of the vector field as well and the corresponding results confirm the estimates obtained.

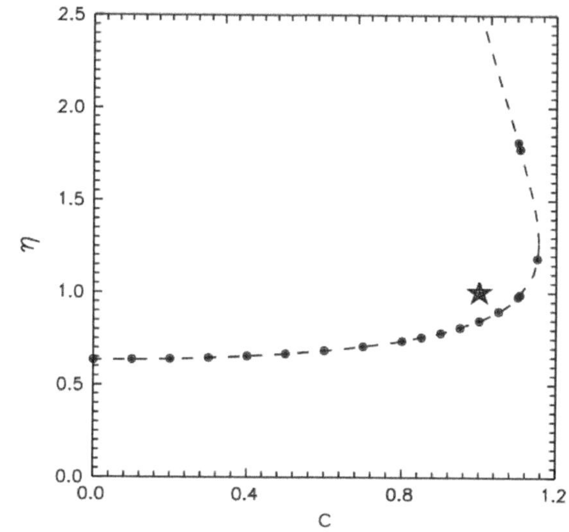

Figure 16: The domain of the η, c (ζ=c η) plot in which an increase of specific energy occurs; see the text. The *dots* represent a stable kink. The *star* shows the position of canonical (chiral) kink; see the text.

A rather simple analysis shows that interesting solutions are located along the boundary of discussed domain. Some of those are depicted in Figure 16 as the dots. Figure 17 shows the stable kinks of the σ field with the parameter c=1.1 for two solutions with η≈0.977 (mσ≈468MeV) (solid line) and η≈1.813 (mσ≈690 MeV) (dashed line). For the sake of clarity we consider the gas (vacuum) phase is on the right. Then the asymptotic value of σ field on the left hand side (σ≈80 MeV) corresponds to M∗=70 MeV. The thickness of the transition layer for the solution with η≈0.977is t≈2 fm, whereas for the second solution we have t≈1 fm.

Characterizing the whole spectrum of the solutions obtained we should mention that there exist other more rigid (chiral) kinks, which correspond to the transition into the state with the dynamical quark mass changing its sign, i.e. M→−M. In particular, for the kink with the canonical parameter values η=1, c=1 is clearly seen (marked by the star in Figure 16) and its surface tension coefficient is about 2mπ (mπ is the π-meson mass). The most populated class of solutions consists of those having metastable character. The system comes back to the starting point (after an evolution) pretty rapidly, and usually the σ field does not evolve to such an extent as to reach the asymptotic value (which corresponds to the dynamical quark mass in the liquid phase M*=70 MeV). Switching on the vector field changes the solutions insignificantly (for our situation with small C_v it does not exceed 2 MeV in the maximum).

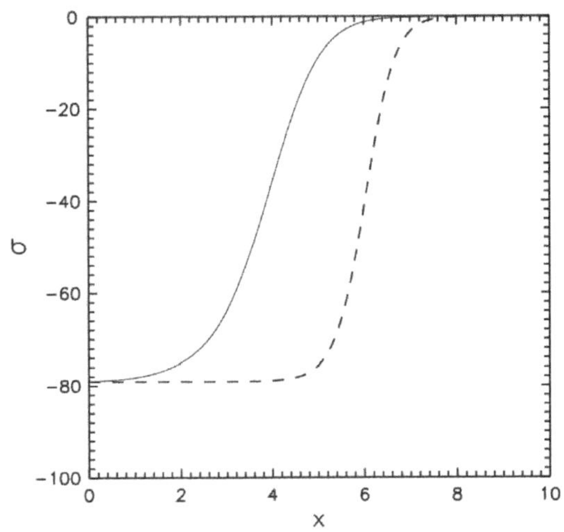

Figure 17: The stable kink solutions with c=1.1, the *solid line* corresponds to η≈0.977 (mσ≈468MeV) and the *dashed line* corresponds to η≈1.813 (mσ≈690 MeV), x is given in the units of fm and σ is given in MeV.

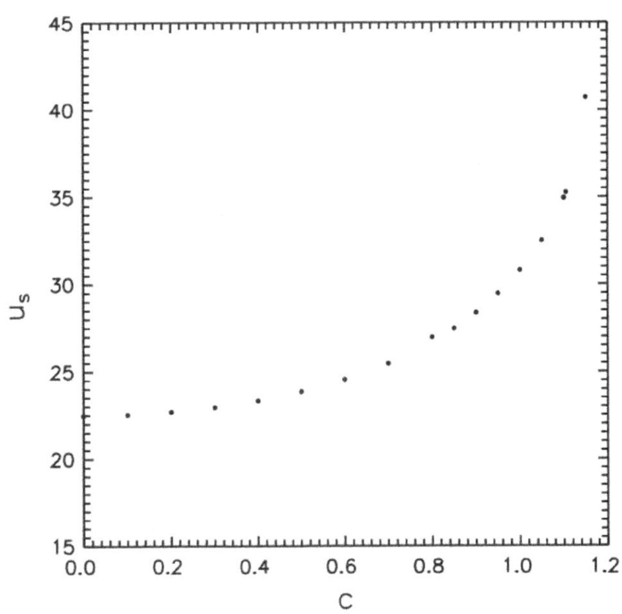

Figure 18: The surface tension coefficient us in MeV as a function of the parameter c (ζ=c η) for the curve of stable kinks (with η≤1.2).

The surface tension coefficient us in MeV for the curve of stable kinks with parameter η≤1.2 as a function of the parameter c (ζ=c η) is depicted in Figure 18. The σ-meson mass at c≈0 is mσ≈420 MeV and changes smoothly up to the value mσ≈500 MeV at c≈1.16 (the maximal value of the coefficient cc beyond which the stable kink solutions are not observed). In particular, mσ≈450 MeV at c=1. Two kink solutions with c=1.1 for η≈0.977and for η≈1.813 (shown in Figure 17, and the second one is not shown in Figure 18) have the tension coefficient values us≈35 MeV and us≈65 MeV, respectively. The maximal value of the tension coefficient for the normal nuclear matter does not exceed us=50 MeV. The nuclear Walecka model claims the value us≈19 MeV [55–59] as acceptable and calculable. The reason to have this higher value of the surface tension coefficient for the quarks is rooted in the different values of the mass deficit. Indeed, for nuclear matter it does not exceed $\overset{*}{M}\approx 0.5$Malbeit so that more realistic values are considered around $\overset{*}{M}\approx 0.7$M and for the quark ensemble the mass deficit amounts to $\overset{*}{M}\approx 0.3$M. We are also able to estimate the compression coefficient of the quark matter K, which occurs significantly larger than the nuclear one. Actually, we see a quite smooth analogy between the results obtained and the results of the bag soliton model [60–63]. The thermodynamic treatment developed in the present paper allows us to formulate the adequate boundary conditions for the bag in physical vacuum and to diminish considerably the uncertainties in searching the true soliton Lagrangian. We believe, and it was also shown here, that to single out one soliton solution among others (including even those obtained by the exact integration method [46–51]), which describes the transitional layer between two media, is not an easy problem if the boundary conditions above formulated are not properly imposed.

DROPLET OF QUARK LIQUID

The results of the previous sections have led us to put forward the challenging question about the creation and properties of finite quark systems or the droplets of quark liquid which are in equilibrium with the vacuum state. Thus, as a droplet we imply the spherically symmetric solution of the equation system (34) for σ(r) and V(r) with the obvious boundary conditions σ′(0)=0 and V′(0)=0 in the origin (the primed variables denote the first derivatives in r) and rapidly decreasing at large distances, σ→0, V→0, when r→∞.

A quantitative analysis of similar nuclear physics models which includes the detailed tuning of parameters is usually based on the comprehensive fitting of the available experimental data. This way is obviously irrelevant in studying the quark liquid droplets. This global difficulty dictates the specific tactics of analyzing. We propose to start, first of all, with selecting the parameters which could be worthwhile to play the role of physical observables. Naturally, the total baryon number which phenomenologically (via factor 33) related to the number of valence quarks in an ensemble, is a reasonable candidate for this role. Besides, the density of the quark ensemble ρ(r), the mean size of the droplet R_0, and the thickness of the surface layer t looks suitable for such an analysis.

It is argued above that the vector field contribution is negligible because of the small value of the coefficient Cv compared to the Cs magnitude, and we follow this conclusion (or assumption) albeit so that we understand it is scarcely justified in the context of a finite quark system. Thus, we put gv=0, V=0 in what follows and it simplifies all the calculations enormously.

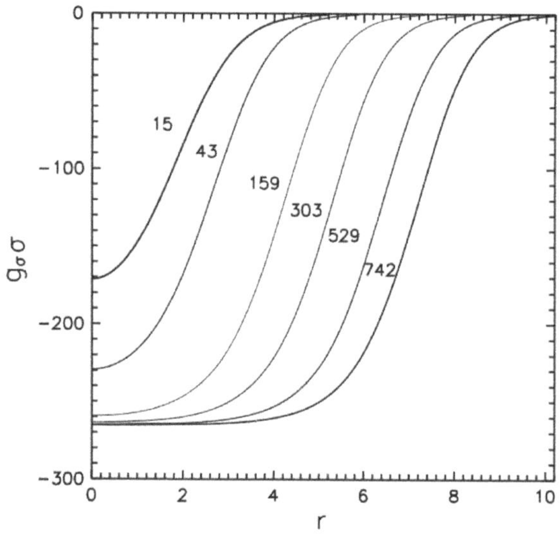

Figure 19: σ field (MeV) as a function of the distance r (fm) for several solutions of the equation system (34) which are characterized by the net quark number Nq written to the *left of each curve.*

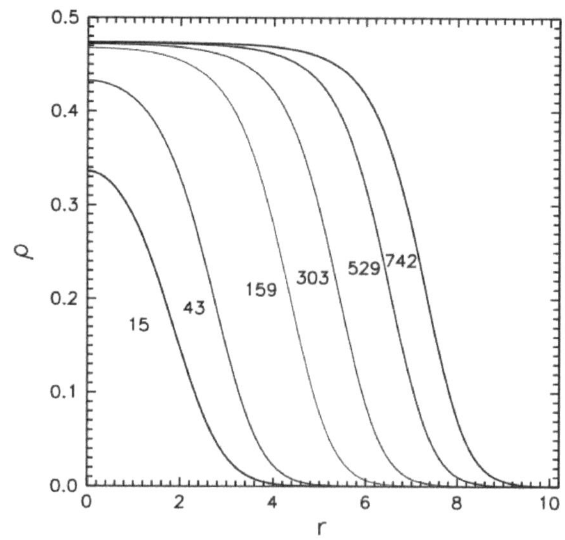

Figure 20: Distribution of the quark density ρ (ch/fm³) for the corresponding solutions presented in Figure 20

Figure 19 shows the number of solutions (σ field in MeV) to the system (34) at N_f=1 and Figure20 presents the corresponding distributions of ensemble density ρ (ch/fm³). The parameters C_s, C_v, b, and c are derived by the same algorithm as in the previous section, i.e. the chemical potential of the quark ensemble M=335 MeV (and σ→0) is fixed at spatial infinity. The filled-up states (of a liquid) are characterized by the parameters $\overset{*}{M}$ =70MeV, $\rho_0=\rho_1=3\times0.157$ ch/fm³. The σ-meson mass and the coupling constant gσ are derived at fixed coefficients η and ζ, and they just define the behavior of the solutions σ(r), ρ(r), etc. The magnitudes of the functions σ(r) and ρ(r) at the origin are not strongly correlated with the values characteristic for the filled-up states and are practically determined by solving the boundary value problem for system (34). In particular, the solutions presented in Figure 19 have been obtained with the running coefficient η at ζ=η. The most relevant parameter (instead of η) from the physical view point is the total number of quarks in the droplet N_q (as discussed above) and it is depicted to the left of each curve. (The variation of $\overset{*}{M}$, ρ_0 and fπ could be considered as well instead of the two mentioned parameters η and ζ.)

Analyzing the full spectrum of solutions obtained by scanning, one can reveal a recurrent picture (at a certain scale) of kink droplets which are easily parameterized by the total number of quarks N_q in a droplet and by the density ρ_0. These characteristics are obviously fixed at completing the calculations. The property which allows us to single out these solutions is related to the value of droplet specific energy (see below).

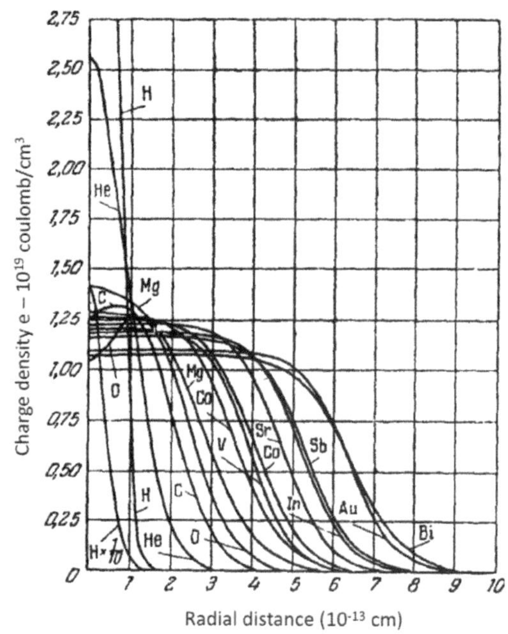

Figure 21: A summary of the approximate charge density distributions found for various nuclei by studying the electron scattering (R. Hofstadter).

Table 1 exhibits the results of fitting the density ρ(r) with the Fermi distribution

$$\rho_F(r) = \frac{\widetilde{\rho_0}}{1 + e^{(R_0 - r)/b}}, \tag{40}$$

where $\widetilde{\rho_0}$ is the density in the origin, R_0 is the mean size of droplet, and the parameter b defines the thickness of the surface layer t=4ln(3)b. Besides, the coefficient r_0, which is absorbed in the surface tension coefficient (39), the σ-meson mass, $R_0=r_0N^{1/3}$, and the coefficient η at which all other values have been obtained are also presented in Table 1.

Table 1: The results of fitting the Fermi distribution with Nf=1, $\widetilde{\rho}_0$ (ch/fm^3), R_0, t, r_0 (fm), b (fm^{-1}), m_σ (MeV)

NqNq	ρ˜0ρ~0	R0R0	bb	tt	r0r0	mσmσ	ηη
15	0.34	1.84	0.51	2.24	0.74	351	0.65
43	0.43	2.19	0.52	2.28	0.75	384	0.73
159	0.46	4.19	0.52	2.29	0.77	409	0.78
303	0.47	5.23	0.52	2.29	0.78	417	0.795
529	0.47	6.37	0.52	2.27	0.79	423	0.805
742	0.47	7.15	0.52	2.27	0.79	426	0.81

The curves plotted in Figure 19 and results of Table 1 justify us to conclude that the density distributions at Nq≥50 are in full agreement with the corresponding data typical for the nuclear matter. The thicknesses of the transition layers in both cases are also similar and the coefficient r0 with the factor $3^{1/3}$ included is in full correspondence with r˜0. The values of the σ-meson mass in Table 1 look quite reasonable as well.

Obviously, we are unable to tune our model, unlike the Walecka one, by drawing the experimental data on the quark drops. However, the subject is getting increasingly interesting because of possible astrophysical applications in studying the anomalous quark stars [55–59]. However, there is the possibility to trace some qualitative objective observations, in particular, the central density of drops is getting smaller with the quark quasiparticles number in the drop decreasing. Meanwhile, we know from the experiments that a nuclear matter develops an opposite tendency of a certain increase of nuclear density that becomes quite noticeable (factor 2.5) for helium and is much larger (factor 10) than the standard nuclear density for hydrogen. Figure 21 gives a summary of the approximate charge density distributions studied by R. Hofstadter in electron scattering off various nuclei.

Obviously, we understand that the Thomas–Fermi approximation which is used for evaluating becomes hardly justified at small number of quarks, and we should deal with the solutions of the complete equation system (34). However, one very encouraging hint comes from the chiral soliton model of nucleon [64], where it has been demonstrated that solving this system (34) the good description of nucleon and Δ can be obtained.

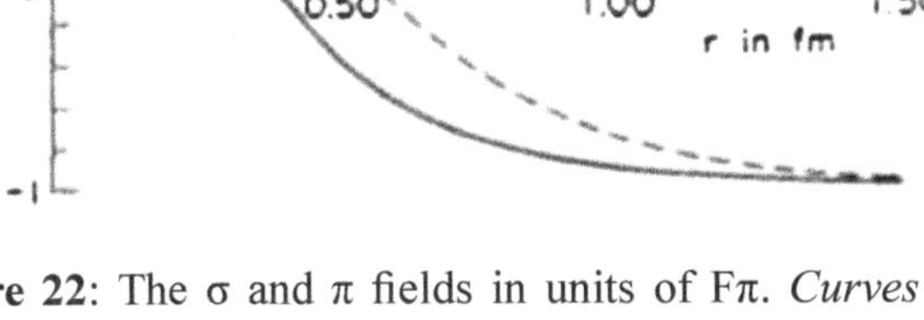

Figure 22: The σ and π fields in units of Fπ. *Curves 1 and 2* show σ, *3 and 4* show π, *5 and 6* show σ2+π2, (*curves 1 and 3* with vector and axial-vector fields contributions included; see Ref. [64]).

In a sense we consider analyzing just a three quark system as a central result of our paper. Looking at Figure 1 of Ref. [64] (we represent it here as Figure 22) we see that the curve describing the behavior of the scalar field at large distances reaches its minimal value (according to the sign choice done in [64] it corresponds to the largest quark mass of order 300300 MeV). It looks like it that by approaching the center of baryon, the chiral symmetry is partially restored and the scalar field in the region of ~0.5 fm disappears. One of the possible scenarios for solving the system of equations (34) could be a variant in which the scalar field reaches maximal (zero) value (with a zero value of the derivative over coordinate) at this point (or the center of the baryon). Then a scalar field can, in principle, smoothly approach its minimal value coming to the center of the baryon. It allows us to conclude that we could deal with an "ordinary" quark with positive (zero) mass for the solutions of such a type. However, the baryon is getting a large width (size) in this scenario. There is another type of solutions, in which the "speed" of passing by the point 0.50.5 fm is not getting slower. In fact such a situation could be realized by doing a chiral rotation where a quark inside a baryon falls in the metastable region of negative quark masses. Such a solution develops an already quite suitable width of order ~1 fm due to the presence of a massive (11 GeV) scalar field. Clearly, the problem of the existence of σ meson so heavy (strengthening the chiral effect) is crucial to collect the necessary information on the phase diagram of strongly interacting matter. Such solutions develop a surface tension coefficient which is larger by a factor 2 than the corresponding coefficient of a single kink and, as we believe, signals some instability of the single kink solution.

Figure 23 displays the specific binding energy of the ensemble. It is defined by the expression similar to Eq. (39) in that the integration over the quark droplet volume is performed. The specific energy is normalized (compared) to the ensemble energy at spatial infinity, i.e. in vacuum. Actually, Figure 23 shows several curves in the upper part of the plot which corresponds to the calculations with N_f=1. The solid line is obtained by scanning over the parameter η and corresponds to the data presented in Table 1. The dashed curve is calculated at fixed η=0.4 but by scanning over the parameter M∗M∗. It is clearly seen that if the specific energy data are presented as a function of quark number N_q, then the solutions, which we are interested in, rally in the local vicinity of the curve where the maximal binding energy–|Eb| is reached. A similar solution scanning can be performed over the central density parameter ρ0 in origin. The corresponding data are dotted for a certain fixed M∗ and ρ0. It is interesting to notice that on scanning over any variable discussed a saturation property is observed and it looks like the minimum in eb at N_q~200–250. The results for the specific binding energy as a function of particle number are in qualitative agreement with the corresponding experimental data. One may even observe quantitative agreement if the factor 33 (the energy necessary to remove one baryon) is taken into account. In fact, the equation system (34) represents an equation of balance for the current quarks circulating between liquid and gas phases.

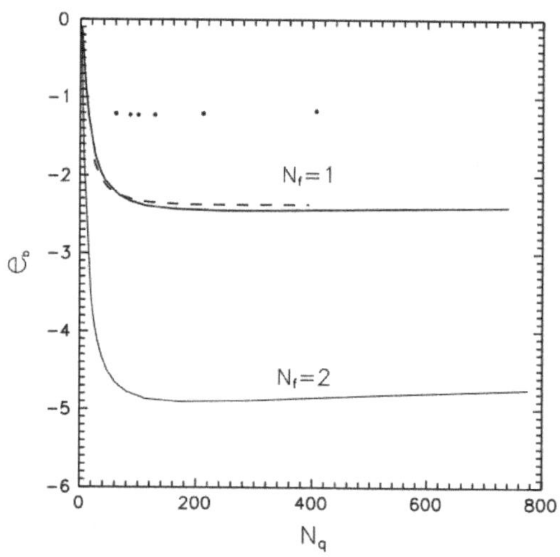

Figure 23: The specific binding energy at N_f=1and N_f=2 in MeV as a function of quark number N_q.

Summarizing we would like to emphasize that in the present paper we have demonstrated how a phase transition of the liquid–gas kind (with reasonable values of parameters) emerges in the NJL-type models. The constructed quark ensemble displays some interesting features of the nuclear ground state (for example, the existence of the state degenerate with the vacuum one), and the results of our study are suggestive of speculating that the quark droplets could coexist in equilibrium with the vacuum under the normal conditions.

It seems that if the quark drops really exist, then their features could be quite similar to the nuclear matter, and the scenario of partial chiral symmetry restoration has been realized as a chiral soliton. In the mean field (Ginsburg–Landau) approximation we are able to show the strengthening of chiral effects in the baryon that appears as the existence of a massive 1 GeV scalar meson, and a realistic phase diagram of strongly interacting matter may be obtained by studying the mechanism of chiral effects strengthening.

CONCLUSION

In the present paper we described quantum liquids (Landau Fermi liquids) resulting from the quark models with a four-fermion interaction. This consideration is based on the identity of results obtained in [21–23] by using a dressing Bogolyubov transformation and mean field approximation. We demonstrated that the mean energy of the ensemble serves as an energy functional of the Landau theory. It was shown that in a wide range of potentials interesting for applications one can expect the quantum liquids to behave in essentially the same way. For some of their properties a band of estimates was obtained. A comparison of NJL and KKB models, substantially different in many aspects, demonstrates that the properties of quantum liquids do not actually depend on the shape of the form factor (a natural interaction length); rather, they are mainly determined by the coupling constant of the interaction. It was shown that a common distinctive feature of ensembles is the presence of occupied states degenerate with respect to the vacuum in the chemical potential and pressure. Based on this observation, the inhomogeneous states, which allowed describing a transition layer, estimating the surface tension, as well as studying some properties of quark liquid droplets, were considered. It is noted that in the case of a small number of quarks in a droplet instability associated with lowering of the energy barrier, separating chiral phases, apparently manifests itself. This instability is seen in two kinks merging into one chiral soliton. The idea of a dynamical equilibrium of a mixed phase consisting of baryon matter and vacuum is discussed as a possible scenario for explaining the stability of nuclear matter.

ACKNOWLEDGMENTS

The authors are indebted D. V. Anchishkin, K. A. Bugaev, E.-M. Ilgenfritz, V. V. Skalozub, A. M. Snigirev, V. D. Tonnev and many other colleagues for numerous fruitful discussions. S.V.M. is deeply grateful to Professor L. V. Keldysh for interesting discussions. The work was partially supported by the State Fund for Fundamental Research of Ukraine, Grant No Ph58/04.

REFERENCES

1. P.F. Kolb, P. Huovinen, U. Heinz, H. Heiselberg, Phys. Lett. B 500, 232 (2001)
2. P.F. Kolb, P. Huovinen, U. Heinz, H. Heiselberg, Phys. Lett. B 503, 58 (2001)
3. P.F. Kolb, P. Huovinen, U. Heinz, H. Heiselberg, T. Hirano, Phys. Rev. Lett. 86, 2754 (2001)
4. P.F. Kolb, P. Huovinen, U. Heinz, H. Heiselberg, Phys. Rev. C 65, 011901 (2002)
5. P.F. Kolb, P. Huovinen, U. Heinz, H. Heiselberg, J. Phys. G 36, 064031 (2009)
6. J. Berges, J.-P. Blaizot, F. Gelis, arXiv:1203.2042 [hep-ph]
7. C.P. Herzog, arXiv:0904.1975 [hep-th]
8. D. Mateos, arXiv:1106.3295 [hep-th]
9. G.M. Zinovjev, S.V. Molodtsov, Theor. Math. Phys. 160, 444 (2009)
10. S.V. Molodtsov, G.M. Zinovjev, Phys. Rev. D 80, 076001 (2009)
11. S.V. Molodtsov, A.N. Sissakian, G.M. Zinovjev, Europhys. Lett. 87, 61001 (2009)
12. Y. Nambu, G. Jona-Lasinio, Phys. Rev. 122, 345 (1961)
13. L.V. Keldysh, Doctor thesis. FIAN, Singapore (1965)
14. E.V. Kane, Phys. Rev. 131, 79 (1963)
15. V.L. Bonch-Bruevich, Physics of Solid States (VINITI, Moscow, 1965)
16. M.V. Sadovskii, *Diagrammatics* (World Scientific, Singapore, 2006)
17. S.V. Molodtsov, G.M. Zinovjev, arXiv:1406.5561v2

(International School on Nuclear Physics 36-Course "Nuclei in the laboratory and in the cosmos", EriceSicily, 2014, Italy)
18. G.V. Efimov, S.N. Nedelko, Eur. Phys. J. C1, 343 (1998)
19. A.C. Kolloniatis, S.N. Nedelko, Phys. Rev. D 64, 114025 (2001)
20. S.N. Nedelko, V.E. Voronin, arXiv:1403.0415
21. S.V. Molodtsov, G.M. Zinovjev, Europhys. Lett. 93, 11001 (2011)
22. S.V. Molodtsov, G.M. Zinovjev, Phys. Rev. D 84, 036011 (2011)
23. G.M. Zinovjev, S.V. Molodtsov, Yad. Phys. 75, 262 (2012)
24. T. Hatsuda, T. Kunihiro, Phys. Rep. 247, 221 (1994)
25. L.D. Landau, JETP 30, 1058 (1956)
26. L.D. Landau, JETP 32, 59 (1957)
27. D. Pines, Ph Nozieres, *Theory of Quantum Liquids* (W. A. Benjamin, New York, 1966)
28. Ph Nozieres, *Theory of Interacting Fermi Systems* (Westview Press, Boulder, 1997)
29. A.J. Leggett, *Quantum Liquids* (Oxford University Press, Oxford, 2006)
30. J.M. Luttinger, J.C. Ward, Phys. Rev. 118, 1417 (1960)
31. J.M. Luttinger, Phys. Rev. 119, 1153 (1960)
32. M. Asakawa, K. Yazaki, Nucl. Phys. A504, 668 (1989)
33. H. Tezuka, Phys. Rev. C 22, 2585 (1980)
34. H. Tezuka, Phys. Rev. C 24, 288 (1981)
35. G. Baym, S.A. Chin, Nucl. Phys. A262, 537 (1976)
36. T. Matsui, Nucl. Phys. A370, 369 (1981)
37. L. Ya, Glozman. Phys. Rep. 444, 1 (2007)
38. G.M. Zinovjev, S.V. Molodtsov, Yad. Phys. 75, 1387 (2012)
39. G.M. Zinovjev, M.K. Volkov, S.V. Molodtsov, Theor. Math. Phys. 161, 408 (2010)
40. S.V. Molodtsov, M.K. Volkov, G.M. Zinovjev, arXiv:0812.2666
41. A.I. Larkin, YuN Ovchinnikov, JETP 47, 1136 (1964)
42. P.G. De Gennes, *Superconductivity of Metals and Alloys* (W. A. Benjamin, New York, 1966)
43. T. Eguchi, H. Sugawara, Phys. Rev. D 10, 4257 (1974)
44. K. Kikkawa, Prog. Theor. Phys. 56, 947 (1976)
45. M.K. Volkov, Fiz. Elem. Chast. Atom. Yadra 17, 433 (1986)
46. S. Carignano, D. Nickel, M. Buballa, Phys. Rev. D 82, 054009 (2010)
47. D. Nickel, Phys. Rev. Lett. 103, 072301 (2009)
48. G. Basar, G.V. Dunne, Phys. Rev. Lett. 100, 200404 (2008)
49. T. Kunihiro, Y. Minami, Z. Zhang, arXiv:1009.4534 [nucl-th]
50. M. Buballa, S. Carignano, arXiv:1210.7155 [hep-ph], arXiv:1406.1367 [hep- ph]
51. S. Carignano, M. Buballa, B.-J. Schafer, arXiv:1404.0057 [hep-ph]
52. J.D. Walecka, Ann. Phys. 83, 491 (1974)
53. F.E. Serr, J.D. Walecka, Phys. Lett. B79, 10 (1978)
54. N.V. Jai, L.N. Savushkin, Fiz. Elem. Chast. Atom. Yadra 23, 847 (1992)
55. J. Boguta, A. Bodmer, Nucl. Phys. A292, 413 (1977)
56. M.B. Pinto, V. Koch, J. Randrup, Phys. Rev. C 86, 025203 (2012). arXiv:1207.5186[hep-ph]
57. J.J. Bjerrum-Bohr, I.N. Mishustin, T. Dossing arXiv:1112.2514 [nucl-th]
58. N. Itoh, Prog. Theor. Phys. 44, 291 (1970)
59. P.W. Anderson, N. Itoh, Nature 256, 25 (1975)
60. K. Huang, D.R. Stump, Phys. Rev. D 14, 223 (1976)
61. R. Friedberg, T.D. Lee, Phys. Rev. D 18, 2623 (1978)
62. R. Goldflam, L. Wilets, Phys. Rev. D 25, 1951 (1982)
63. E.G. Lübeck, M.C. Birse, E.M. Henley, L. Wilets, Phys. Rev. D 33, 234 (1986)
64. W. Broniowski, M.K. Banerjee, Phys. Lett. B158, 335 (1978)

CHAPTER 9

BOSE–EINSTEIN CONDENSATION IN THE RINDLER SPACE

Shingo Takeuchi

INTRODUCTION

The Bose–Einstein condensation is getting a lot of attention recently as a quantum fluid for the test of the analogy between sound waves in quantum fluids and scalar field fluctuation in curved space–times [1]. Thanks to this analogy, we can expect new progress and insight which are difficult only in the gravitational analysis in the quantum gravity and cosmology. Further, the experiments of the quantum gravity and cosmology are difficult due to the required energy level and the scale of phenomena. However, the experiments of the condensed matters will be free from such problems to some extent. It is thought for these reasons that pseudo experiments of the gravity are possible in the quantum fluids through the analogy.

In order to understand the quantum phenomena in the gravity and cosmology such as the Hawking radiation [2] and particle creation [3] and [4], the Unruh effect [5], [6], [7] and [8] is important in terms of the role that the event-horizons play.

The Unruh effect is a prediction that one moving in the Minkowski space–time with a linear constant acceleration experiences the space–time as a thermal-bath with the Unruh temperature, $T_U = \hbar a/(2\pi c k_B) \approx 4 \times 10^{-23}\, a/(\mathrm{cm/s^2})\,[\mathrm{K}]$, where a is the acceleration.

Now, various experimental attempts in the condensed matters to observe the gravitational phenomena are being invented (see Ref. [9] for example). Particularly as for the experiments to detect the Unruh effect, there are attempts in Bose–Einstein condensates [11], graphenes [12] and Berry phases [13]. For other attempts see Ref.[14], for example, and related references. We also address the issue of the Unruh effect in the Bose–Einstein condensation. Whether the Bose–Einstein condensation occurs or not is determined by temperature. We assume in this paper that the Unruh temperature exists in the constantly accelerating system according to the Unruh effect mentioned above. At this time we can think that the Unruh effect affects the Bose–Einstein condensation. Although an enormous number of studies have been done on the Unruh effect and the Bose–Einstein condensation so far, these are performed separately and little is known about the Unruh effect in the Bose–Einstein condensation at this moment. Since both the Unruh effect and the Bose–Einstein condensation are very important in the fundamental physics, new understandings could be expected by combining the Unruh effect and the Bose–Einstein condensation. In this paper, from such a background, we calculate the critical acceleration for the Bose–Einstein condensation based on the thermal excitation brought into by the Unruh effect.

Let us here explain the Bose–Einstein condensation briefly (for more details, see Ref.[10] for example). The Bose–Einstein condensation state is the situation that all the particle stay in the least energy state uniformly owing to the Bose–Einstein statistics, and it appears as a phase transition that all the particles uniformly drop to the least energy state as the entropically-favored state due to the Bose–Einstein statistics at some time. (In the Bose–Einstein condensation state, since all the particles are in a state such as the least energy state, de Broglie wave of each particle becomes longer and even, and eventually the system itself becomes a de Broglie wave.) The situation that the effect of the Bose–Einstein statistics is dominant can be considered as the low temperature region. Actually it is considered at the low temperature region that the particle›s thermal motion energy is extremely small so that it does not excite the state of the particles from the least energy state. Hence, the Bose–Einstein condensation state emerges in the system composed of bosonic particle at extremely low temperature.

Citation: Shingo Takeuchi, Bose–Einstein condensation in the Rindler space, http://dx.doi.org/10.1016/j.physletb.2015.09.013

We calculate in this paper the critical acceleration at which the Bose–Einstein condensation begins to appear in the accelerating system by the Unruh effect when lowering the acceleration. We mention our critical acceleration more precisely. We first consider an ideal gas in the Minkowski space–time at zero-temperature, where the gas is composed of particles described by a free complex scalar field at finite density. As the gas is now in zero-temperature, it can be considered to be in the Bose–Einstein condensation state. We then start to accelerate such an ideal gas uniformly. At this time, according to the Unruh effect, since each constantly accelerating particle composing the gas will experience the temperature $T_U = \hbar a/(2\pi c k_B)$ (a is the acceleration) in the accelerating frame, the ideal gas performing uniformly accelerating motion will experience the temperature T_U as a whole in the accelerating frame. Here, what the gas experiences thermal means that the space which the gas observes is filled with some medium that are performing thermal fluctuation. In this paper, we consider that there is no interaction between the medium and the particles composing the gas without the thermal excitation. Hence it is considered that, as growing the acceleration gradually from lower acceleration, in the accelerating frame, the dissolution of the Bose–Einstein condensation is observed eventually, and finally the Bose–Einstein condensation disappears entirely at some acceleration. The critical acceleration we calculate is the acceleration at that time.

We here mention the emergence mechanism of the Bose–Einstein condensation in our model and its critical moment. It is composed of three steps. We first calculate the effective potential of the field meaning the particles composing the gas, and then obtain the particle density by performing a derivative to it with regard to the particle›s chemical potential. Then we find that, if we decrease the acceleration with fixing the particle density to constant, either the particle›s chemical potential or the absolute value of the zero-mode of the field has to grow. Here, the acceleration plays the role of the temperature in the accelerating frame as mentioned in the above paragraph, and the zero-mode of the field can be considered to correspond to the least energy state. Hence the absolute value of the zero-mode is considered as the expectation value of the Bose–Einstein condensation state, and whether it is zero or not corresponds to its disappearance or appearance, respectively. We consider to start with the high acceleration situation where there is no Bose–Einstein condensation in the accelerating frame, and decrease the acceleration gradually. At this time, we keep the particle density to constant. Namely, the number of particle is always constant in our model. Then, corresponding to no Bose–Einstein condensation, the absolute value of the zero-mode should be zero while we decrease acceleration. Thus the chemical potential must grow to keep the particle density to constant (step 1). However we find that there is an upper bound for the value of the chemical potential, which is need to avoid a divergence of the probability amplitude. Hence, when the chemical potential reaches the upper bound (step 2), there is no way but the absolute value of the zero-mode starts to grow (step 3), if we further decrease acceleration with keeping the particle density to constant. Thus, the moment that the chemical potential reaches the upper bound is the critical moment that the Bose–Einstein condensation appears, and this is the Bose–Einstein condensation in our model.

The situation where our analysis is actually performed is just before the Bose–Einstein condensation state starts to occur. To be more specific, our calculation to obtain the critical acceleration is performed in the situation that the absolute value of the zero-mode and the chemical potential are put to zero and the upper bound value respectively in the equation of the particle density as the critical moment mentioned above. Then the acceleration obtained from that equation is the critical acceleration.

In our analysis, due to some technical difficulty mentioned later, we consider the situation $m/a_c \ll 1$ (m and a_c are the mass of the particle and the critical acceleration, respectively), and take the leading contribution of this in our calculation, where m is the mass of the complex scalar particle considered in this paper and a_c is the critical acceleration. As the acceleration is proportional to the temperature in the relation given by the Unruh effect, this can be read as $k_B T_c/2 \gg mc^2$ (T_c is the critical temperature fixed with a_c), which is the relativistic situation that the particle's thermal motion energy is much higher than its static energy at the critical moment.

The result in this paper is hence applicable to the system of the complex scalar field with the critical temperature for the Bose–Einstein condensation that is much higher than the mass. Since the kind of the field considered in this paper is a complex scalar field composing an ideal gas, the actual particle that can correspond to this paper is some ideal complex scalar particle system whose critical temperature of the Bose–Einstein condensation locates in the region $k_B T/2 \gg mc^2$.

As stated above, in this study we perform the calculation for the critical acceleration in the Bose–Einstein condensation. So far several kinds of the critical accelerations for the spontaneous symmetry breaking induced by the Unruh effect have been carried out.

The critical acceleration for the chiral symmetry restoration was studied in Nambu–Jona–Lasinio model at zero and finite chemical potentials for quarks in Refs. [15] and [16] respectively. In Ref. [17] the critical acceleration for the restoration of the spontaneous symmetry breaking of the Z_2 symmetry in the real scalar field theory was studied.

There are also papers concluding that the Unruh effect does not contribute to the restoration of the spontaneous symmetry breaking [18]. The system in these papers is not the finite density, and their discussion is based on the effective potential. On the other hand, our Bose–Einstein condensation occurs from the relation between the chemical potential and the acceleration playing the role of the temperature. This is a different point between us and him. Because of this, their conclusion would not be applied to our study readily. We should take this issue to a future work.

Lastly, we introduce interesting papers that will have some connection to this study. Ref.[19] concludes that a larger acceleration should enhance a condensate as compared to those in a non-accelerated vacuum. In Ref. [20], although the background space–time is not the Rindler space, whether the Bose–Einstein condensation can occur or not is shown in an ideal boson gas model with a point-like impurity at finite temperature in some uniform gravitational force in each of D=1,2 and 3. Next, it is shown in Ref. [21] that free massless scalar particles are detected by a constantly accelerating detector with not the Bose–Einstein distribution but the

Fermi–Dirac distribution in odd dimensions, and this problem is solved in Ref. [22].

THE MODEL

The model in this paper is a free complex scalar model at the finite density corresponding to an ideal gas in the constantly accelerating system with the acceleration a. The Lagrangian density is given as

$$\mathcal{L} = \hbar^2 g^{\mu\nu}\partial_\mu\phi^*\partial_\nu\phi - c^2m^2\phi^*\phi \tag{1}$$

with $\phi \equiv \frac{1}{\sqrt{2}}(\phi_1 + i\phi_2)$ that means the field of the particle composing a gas, and μ, ν are the coordinate in the Rindler space explained below.

In the Minkowski space–time as an inertial system, the background space–time of a constantly accelerating system is given by the Rindler space. The Rindler coordinate in our notation is given as follows:

$$(\eta, \rho, y, z) \equiv (\eta, \rho, x_\perp) \quad \text{with} \quad x_\perp \equiv (y, z). \tag{2}$$

This relates with the Minkowski coordinate $(t,x,x_\perp)$ as

$$(t, x) = \frac{c}{a}\left(\sinh\frac{at}{c},\, c\cosh\frac{at}{c}\right) \equiv \rho\,(\sinh\eta,\, c\cosh\eta)\,, \tag{3}$$

where the accelerating direction has been thought to be in the x-direction. The Rindler metric in our notation is

$$ds^2 = (c\rho)^2 d\eta^2 - d(c\rho)^2 - dx_\perp^2. \tag{4}$$

In what follows, we use the unit system: $c=\hbar=k_B=1$.

The constantly accelerating one in the Minkowski space–time corresponds to the one moving along a line on a constant ρ in the Rindler space. The relation between ρ and a are ρ=1/a as can be seen from eq. (3). Since constantly accelerating one experiences the system as a thermal-bath with the Unruh temperature T_U (the acceleration are in the relation: $T_U=a/2\pi$) by the Unruh effect, one moving along a line on a constant ρ in the Rindler space gets the temperature

$$T_U=1/2\pi\rho. \tag{5}$$

Hence, the gas in our study is considered to be in such a thermal-bath with Unruh temperature T_U in the accelerating frame. Here, although our gas is considered to be in a thermal-bath with some medium performing thermal fluctuation giving the Unruh temperature T_U in the accelerating frame, we assume that there is no interaction between our gas and the medium other than the thermal excitation.

We can see from eq. (5) that varying ρ means varying the temperature. For this reason, how to interpret the results would be unclear if the four-dimensional space–time integration including the ρ-integration was performed.

Let us here turn to how this problem is handled in other papers on the critical acceleration given by the Unruh effect. In Refs. [15], [16] and [17], the action is given with the four-dimensional space–time integration including the ρ-integration. This point is the problem. However, their calculation is once performed using the Green›s function that has dependence on the ρ-direction, the coordinate ρ is treated as a constant in the calculation of the effective potential. As a result, the ρ-integration becomes just a volume factor in the calculation of the effective potential.

The difficulty in the treatment of the coordinate ρ is also mentioned at the chapter of conclusions and discussions in Ref. [23] in the context of the study of the Larmor radiation with the correction rooted in the Unruh effect.

In our analysis, as well as Refs. [15], [16] and [17], we obtain the Green›s function that has dependence on the ρ-direction. However, the space–time integration in our action is given without the ρ-integration. As a result, ρ is a parameter in our model.

We can see at this time that the action is needed to be multiplied by a quantity with the dimension of length so that the action becomes dimensionless. To this purpose, we put dρin our action. This means that the process to obtain the path-integral representation given in eq. (6) for the probability amplitude from the operator formalism representation has been performed with fixing ρ. If the analysis in Refs. [15], [16] and [17] is performed again by our way, there is no difference in the final result.

There are four regions separated by the event-horizons in the Rindler space. The region treated in this paper is the right wedge only.

THE EFFECTIVE POTENTIAL

Performance of the Path-Integral

We start with the probability amplitude:

$$Z = \int \mathcal{D}\pi_\eta \mathcal{D}\phi \times \exp\left[i\int d^3x\, d\rho\, \gamma\left(\pi^\eta\partial_\eta\phi + \pi^{\eta*}\partial_\eta\phi^* - (\mathcal{H} - \mu q)\right)\right], \tag{6}$$

where $\gamma \equiv \sqrt{-\det g_{\mu\nu}}$ and $d^3x \equiv d\eta\, d^2x_\perp$, and μ and q are the chemical potential and the density of particles, respectively. For the reason mentioned in Section 2, to fix the acceleration and Unruh temperature, the integration in the ρ-direction is not included in the space–time integration in the action. As a result, ρ is a parameter in our analysis, and the Unruh temperature that the gas experiences is given according to the relation in eq. (5). At this time, we can see that the action is needed to be multiplied by a quantity with the dimension of length so that it becomes dimensionless. We put dρ to this purpose. However the space–time integration is the one without the ρ -direction. Explanation for π^η and H is in what follows.

$(\pi^\mu, \pi^{\mu*}) \equiv \left(\frac{\partial\mathcal{L}}{\partial(\partial_\mu\phi^*)}, \frac{\partial\mathcal{L}}{\partial(\partial_\mu\phi)}\right)$ are the momenta given as

$$(\pi^\eta, \pi^{\eta*}) = (g^{\eta\eta}\partial_\eta\phi^*,\, g^{\eta\eta}\partial_\eta\phi). \tag{7}$$

H is the Hamiltonian density given as

$$\begin{aligned}\mathcal{H} &= \pi^\eta\,\partial_\eta\phi + \pi^{\eta*}\,\partial_\eta\phi - \mathcal{L} \\ &= \pi^{\eta*}\pi_\eta - g^{ij}\partial_i\phi^*\partial_j\phi + m^2\phi^*\phi,\end{aligned} \tag{8}$$

where π^μ is defined as $\pi^\mu \equiv \frac{1}{\sqrt{2}}(\pi_1^\mu + i\pi_2^\mu)$. Correspondingly, the functional integration measure of π_η changes as $D\pi_\eta = D\pi_{1\eta} D\pi_{2\eta}$. From eq. (7), we can see

$$\pi_{1,2}^\eta = g^{\eta\eta}\partial_\eta \phi_{1,2}. \tag{9}$$

It turns out that the conserved current associated with the U(1) global symmetry in our model is obtained as

$$J^\mu = -i\, g^{\mu\nu}(\phi\, \partial_\nu \phi^* - \phi^*\, \partial_\nu \phi). \tag{10}$$

Using this J^μ, the integral of conserved charge density can be written as

$$\int d^3x d\rho\, \gamma\, q = \int d^3x d\rho\, \gamma\, J^\eta = \int d^3x d\rho\, \gamma\, (-\pi_2^\eta \phi_1 + \pi_1^\eta \phi_1). \tag{11}$$

Substituting eqs. (8) and (11) into eq. (6), we can obtain the following Z:

$$\begin{aligned} Z = \int \mathcal{D}\pi_{1\eta}\mathcal{D}\pi_{2\eta}\mathcal{D}\phi \exp\Big[\frac{i}{2}\int d^3x d\rho\, \gamma\, \Big\{ & -\frac{1}{2}(\pi_1^\eta \pi_{1\eta} - 2(\partial_\eta\phi_1 + \mu\,\phi_2)\pi_1^\eta) \\ & -\frac{1}{2}(\pi_2^\eta \pi_{2\eta} - 2(\partial_\eta\phi_2 + \mu\,\phi_1)\pi_2^\eta) \\ & +\frac{1}{2}((\partial_i\phi_1)^2 + (\partial_i\phi_2)^2 - m^2(\phi_1^2+\phi_2^2))\Big\}\Big], \end{aligned} \tag{12}$$

where $i,j=x_\perp$. Here, in the above, we perform the following rewritings:

$$\begin{aligned} &\pi_1^\eta \pi_{1\eta} - 2(\partial_\eta\phi_1 + \mu\,\phi_2)\pi_1^\eta \\ &= g^{\eta\eta}\Big\{(\pi_{1\eta} - (\partial_\eta\phi_1 + \mu\,\phi_2))^2 - (\partial_\eta\phi_1 + \mu\,\phi_2)^2\Big\}, \end{aligned} \tag{13a}$$

$$\begin{aligned} &\pi_2^\eta \pi_{2\eta} - 2(\partial_\eta\phi_2 + \mu\,\phi_1)\pi_2^\eta \\ &= g^{\eta\eta}\Big\{(\pi_{2\eta} - (\partial_\eta\phi_2 + \mu\,\phi_1))^2 - (\partial_\eta\phi_2 + \mu\,\phi_1)^2\Big\}. \end{aligned} \tag{13b}$$

Furthermore, we redefine the fields as

$$\pi_{1\eta} - (\partial_\eta\phi_1 + \mu\,\phi_2) \to \pi_{1\eta}, \tag{14a}$$

$$\pi_{2\eta} - (\partial_\eta\phi_2 + \mu\,\phi_1) \to \pi_{2\eta}. \tag{14b}$$

As a result, we can write Z as

$$\begin{aligned} Z = C\int \mathcal{D}\phi \exp\Big[\frac{i}{2}\int d^3x d\rho\, \gamma\, \Big(& g^{\eta\eta}(\partial_\eta\phi_1 + \mu\,\phi_2)^2 + (\partial_i\phi_1)^2 \\ & + g^{\eta\eta}(\partial_\eta\phi_2 + \mu\,\phi_1)^2 + (\partial_i\phi_2)^2 - m^2({\phi_1}^2 + {\phi_2}^2)\Big)\Big]. \end{aligned} \tag{15}$$

with

$$C \equiv \int \mathcal{D}\pi_{1\eta}\mathcal{D}\pi_{2\eta} \exp\Big[i\int d^3x d\rho\, \gamma\, g^{\eta\eta}((\pi_{1\eta})^2 + (\pi_{2\eta})^2)\Big]$$

. Performing the path-integral of $\pi_{1\eta}$ and $\pi_{2\eta}$ formally, we think that C become some factor. We ignore C in what follows.

As a result, with some straight forward calculation, we can write Z in the following form,

$$\begin{aligned} Z = \int \mathcal{D}\phi \exp\Big[-\frac{i}{2}\int d^3x d\rho\, \gamma\, \Big(&\phi_1 G \phi_1 + \phi_2 G \phi_2 \\ &+ 2g^{\eta\eta}\mu\,(\phi_2\partial_\eta\phi_1 - \phi_1\partial_\eta\phi_2)\Big)\Big], \end{aligned} \tag{16}$$

where $G \equiv \partial_\eta^2 + \gamma^{-1} g^{ij}\partial_i(\gamma\,\partial_j) + m^2 - g^{\eta\eta}\mu^2$.

Let us now rewrite the real and imaginary parts of the field into a convenient expression for the analysis of the Bose–Einstein condensation. As we have written in the introduction, the Bose–Einstein condensation can be considered as the situation that all particles get to be the least energy state. The least energy state can be considered as the zero-mode of the field in our model, and the situation that all the particles are in the least energy state can be considered as the condensation of the zero-mode. Hence, a convenient expression for the analysis of the Bose–Einstein condensation in our analysis is the one in which the zero-mode is separated as

$$\phi_1 \equiv \sqrt{2}\alpha\, \cos\theta + \hat{\phi}_1, \tag{17a}$$

$$\phi_2 \equiv \sqrt{2}\alpha\, \sin\theta + \hat{\phi}_2, \tag{17b}$$

where α plays the role of the expectation value of the condensation of the zero-mode, θ is a phase in the zero-mode and $\hat{\phi}_1$, and $\hat{\phi}_2$ are the non-zero modes. Depending on just before or after the condensation starts, α behaves as follows:

At this time, $\phi_1 G \phi_1$ and $\phi_2 G \phi_2$ can be written as

$$\phi_1 G\,\phi_1 = 2\alpha^2(m^2 - g^{\eta\eta}\mu^2)\cos^2\theta + \hat{\phi}_1 G\,\hat{\phi}_1, \tag{18}$$

$$\phi_2 G\,\phi_2 = 2\alpha^2(m^2 - g^{\eta\eta}\mu^2)\sin^2\theta + \hat{\phi}_2 G\,\hat{\phi}_2. \tag{19}$$

As a result, we can rewrite Z into

$$\begin{aligned} Z = \exp\Big[-i\alpha^2\int d^3x d\rho\, \gamma\, (m^2 - g^{\eta\eta}\mu^2)\Big]\int \mathcal{D}\hat{\phi} \exp\Big[& -\frac{i}{2}\int d^3x d\rho\, \gamma\, (\hat{\phi}_1 \;\; \hat{\phi}_2)\begin{pmatrix} G & -2g^{\eta\eta}\mu\,\partial_\eta \\ 2g^{\eta\eta}\mu\,\partial_\eta & G \end{pmatrix}\begin{pmatrix}\hat{\phi}_1 \\ \hat{\phi}_2\end{pmatrix} \\ & + \sqrt{2}\alpha\cos\theta\Big(G\,(\phi_1+\phi_2) + (\phi_1+\phi_2)\,G\Big)\Big]. \end{aligned} \tag{20}$$

Our analysis will be performed just before the condensation as mentioned later. For this reason, we put α=0 in what follows. As a result, we can write Z as

$$Z = \int \mathcal{D}\hat{\phi} \exp\Big[-\frac{i}{2}\int d^3x d\rho\, \gamma\, (\hat{\phi}_1 \;\; \hat{\phi}_2) \times \begin{pmatrix} G & -2g^{\eta\eta}\mu\,\partial_\eta \\ 2g^{\eta\eta}\mu\,\partial_\eta & G \end{pmatrix}\begin{pmatrix}\hat{\phi}_1 \\ \hat{\phi}_2\end{pmatrix}\Big]. \tag{21}$$

Then, we can see from the form of G given below eq. (16) that there should be the condition:

$m^2-g^{\eta\eta}\mu^2\geq 0.$ (22)

Otherwise the path-integrals of Z diverges at the configuration that all the momenta are zero. Further, the above relation gives the upper limit of the chemical potential for given a mass and an Unruh temperature.

Performing the diagonalization as

$$Z=\int\mathcal{D}\hat{\phi}\exp\left[-\frac{i}{2}\int d^3xd\rho\,\gamma\,(\hat{\phi}_1\;\;\hat{\phi}_2)\,UU^{-1}\times\begin{pmatrix}G & -2g^{\eta\eta}\mu\,\partial_\eta\\ 2g^{\eta\eta}\mu\,\partial_\eta & G\end{pmatrix}UU^{-1}\begin{pmatrix}\hat{\phi}_1\\ \hat{\phi}_2\end{pmatrix}\right].$$
$$=\int\mathcal{D}'\hat{\phi}'\exp\left[-\frac{i}{2}\int d^3xd\rho\,\gamma\,(\hat{\phi}'_1\;\;\hat{\phi}'_2)\times\begin{pmatrix}G+2g^{\eta\eta}\mu\,\partial_\eta & 0\\ 0 & G-2g^{\eta\eta}\mu\,\partial_\eta\end{pmatrix}\begin{pmatrix}\hat{\phi}'_1\\ \hat{\phi}'_2\end{pmatrix}\right],\tag{23}$$

where $U\equiv\frac{1}{\sqrt{2}}\begin{pmatrix}i & -i\\ 1 & 1\end{pmatrix}$ is a unitary matrix defined to perform the above diagonalization, and correspondingly $\begin{pmatrix}\hat{\phi}'_1\\ \hat{\phi}'_2\end{pmatrix}\equiv U^{-1}\begin{pmatrix}\hat{\phi}_1\\ \hat{\phi}_2\end{pmatrix}$. At this transformation, the functional measure is also transformed, which we have described as $\mathcal{D}\hat{\phi}\to\mathcal{D}'\hat{\phi}'$. However, since U is a constant unitary matrix, the difference between $\mathcal{D}\hat{\phi}$ and $\mathcal{D}'\hat{\phi}'$ contribute only to some constant factor in the path-integral, and we ignore it in what follows.

Performing the path-integral, we can obtain

$$Z=\mathrm{Det}\Big((G+2g^{\eta\eta}\mu\,\partial_\eta)(G-2g^{\eta\eta}\mu\,\partial_\eta)\Big)^{-1/2}.\tag{24}$$

In the above, with regard to the treatment of dρ, we consider that the integration of this has been performed by assigning a value at one point. Hence, the effective action W defined as Z=expiW can be written as

$$\begin{aligned}W&=\frac{i}{2}\mathrm{Log\,Det}\Big((G+2g^{\eta\eta}\mu\,\partial_\eta)(G-2g^{\eta\eta}\mu\,\partial_\eta)\Big)\\&=\frac{i}{2}\mathrm{Tr\,Log}\Big((G+2g^{\eta\eta}\mu\,\partial_\eta)(G-2g^{\eta\eta}\mu\,\partial_\eta)\Big)\\&=\frac{i}{2}V\int\frac{dk^3}{(2\pi)^3}\Big(\mathrm{Log}(\partial_\eta^2+\gamma^{-1}g^{ij}\partial_i(\gamma\,\partial_j)+M^2+2ig^{\eta\eta}\mu\,\partial_\eta)\\&\quad+\mathrm{Log}(\partial_\eta^2+\gamma^{-1}g^{ij}\partial_i(\gamma\,\partial_j)+M^2-2ig^{\eta\eta}\mu\,\partial_\eta)\Big),\end{aligned}\tag{25}$$

where $M^2\equiv m^2-g^{\eta\eta}\mu^2$ and $V\equiv\int d^3x\gamma$ is the volume for the $(\eta,x_\perp)$ space–time, which appears from the rewriting of a functional trace into an integration:

$$\mathrm{Tr}\to\left(\frac{L}{2\pi}\right)^3\left(\frac{2\pi}{L}\right)^3\sum_k\equiv V\int\frac{dk^3}{(2\pi)^3}\tag{26}$$

with $\left(\frac{2\pi}{L}\right)^3\sum_k=\int dk^3$ and $V\equiv L^3=\int d^3x\gamma$, where L means just length in each space for the η, $x_\perp$-directions, and $\sqrt{-\det g_{ab}}=\sqrt{-g_{\rho\rho}\det g_{ab}}=\gamma\;(a,b=\eta,x_\perp$ (a,b=η,$x_\perp$ except for ρ, and $g_{\rho\rho}=1$). $d^3k\equiv d\omega d^2k_\perp$ with $k_\perp\equiv(k_y,k_z)$, which are momenta corresponding to the coordinate $(\eta,x_\perp)$ as in eq. (32).

Defining $\delta_\pm\equiv g^{\eta\eta}\partial_\eta^2+\gamma^{-1}g^{ij}\partial_i(\gamma\,\partial_j)\pm 2ig^{\eta\eta}\mu\,\partial_\eta$, we further rewrite W as

$$\begin{aligned}W&=\frac{i}{2}V\int\frac{dk^3}{(2\pi)^3}\Big(\mathrm{Log}(\delta_++M^2)+\mathrm{Log}(\delta_-+M^2)\Big)\\&=\frac{i}{2}V\int\frac{dk^3}{(2\pi)^3}\Big(\int_0^{M^2}d\Delta^2(\delta_++\Delta^2)^{-1}+\mathrm{Log}\,\delta_+\\&\quad+\int_0^{M^2}d\Delta^2(\delta_-+\Delta^2)^{-1}+\mathrm{Log}\,\delta_-\Big).\end{aligned}\tag{27}$$

Here, we can ignore $\int\frac{dk^3}{(2\pi)^3}\mathrm{Log}\,\delta_\pm$ for the reason in what follows: First, since we will perform the derivative with regard to the chemical potential to obtain the particle density at the end, we look at only the part concerning the chemical potential as

$$\begin{aligned}&\int\frac{dk^3}{(2\pi)^3}\mathrm{Log}\,\delta_\pm\\&=\int\frac{dk^3}{(2\pi)^3}\mathrm{Log}\,\big(\partial_\eta^2+\gamma^{-1}g^{ij}\partial_i(\gamma\,\partial_j)\big)\\&\quad+\int dk_\eta\frac{dk_\perp^2}{(2\pi)^3}\mathrm{Log}\left(1\pm\frac{2ig^{\eta\eta}\mu\,\partial_\eta}{\partial_\eta^2+\gamma^{-1}g^{ij}\partial_i(\gamma\,\partial_j)}\right)\\&\sim\int dk_\eta\frac{dk_\perp^2}{(2\pi)^3}\mathrm{Log}\left(1\pm\frac{2ig^{\eta\eta}\mu\,\partial_\eta}{\partial_\eta^2+\gamma^{-1}g^{ij}\partial_i(\gamma\,\partial_j)}\right).\end{aligned}\tag{28}$$

In the above, we have described the integral $\int dk^3$ separately as $\int dk_\eta\int dk_\perp^2$. Then, replacing ∂_i with ik_i, from the fact that $\int dk\,k\log(1+\frac{c_1k}{c_2+k^2})=0$ ($c_{1,2}$ are some constants), we can see that the μ -dependent part vanishes in the $k_\perp$-integrals.

Finally, we can write the effective action W in eq. (27) as

$$W=\frac{i}{2}V\gamma\int\frac{dk^3}{(2\pi)^3}\int_0^{M^2}d\Delta^2\,(\widetilde{D}_++\widetilde{D}_-),\tag{29}$$

where $\widetilde{D}_\pm$ is

$$\gamma^{-1}\left(\delta_\pm+\Delta^2\right)^{-1}\equiv\widetilde{D}_\pm=\widetilde{D}_\pm(k_\eta,\rho,k_\perp).\tag{30}$$

In the above, we wrote the arguments concerning the momenta as k_η and $k_\perp$ despite that these are given by the differential operators in the actual expression, which may be allowed.

Calculation of $\widetilde{D}_\pm$

From the definition in eq. (30), we can have the following equation:

$$\delta^4(x-x') = \gamma\left(\delta_\pm + \Delta^2\right) D_\pm(x-x')$$
$$= \gamma\left(g^{\eta\eta}\partial_\eta^2 + \gamma^{-1}g^{ij}\partial_i(\gamma\partial_j) \pm 2i\, g^{\eta\eta}\mu\,\partial_\eta + \Delta^2\right)$$
$$\times D_\pm(x-x') \tag{31}$$

with $D_\pm(x-x')$ defined as

$$D_\pm(x-x') = \int \frac{d^3k}{(2\pi)^3}\,\widetilde{D}_\pm\, e^{i\,(\omega\,(\eta-\eta')-k_\perp(x_\perp-x'_\perp))}. \tag{32}$$

We can rewrite eq. (31) as

$$\delta(\rho-\rho')\int \frac{d^3k}{(2\pi)^3}\, e^{i\,(\omega\,(\eta-\eta')-k_\perp(x_\perp-x'_\perp))}$$
$$= \int \frac{d^3k}{(2\pi)^3}\,\gamma\left(g^{\eta\eta}\partial_\eta^2 + \gamma^{-1}g^{ij}\partial_i(\gamma\partial_j) \pm 2i\, g^{\eta\eta}\mu\,\partial_\eta + \Delta^2\right)$$
$$\times \widetilde{D}_\pm\, e^{i\,(\omega\,(\eta-\eta')-k_\perp(x_\perp-x'_\perp))}$$
$$= \int \frac{d^3k}{(2\pi)^3}\,\gamma\left(\rho^{-2}\partial_\eta^2 - \rho^{-1}\partial_\rho(\rho\partial_\rho) - (\partial_x^2+\partial_y^2)\right.$$
$$\left.\pm 2i\, g^{\eta\eta}\mu\,\partial_\eta + \Delta^2\right)\widetilde{D}_\pm\, e^{i\,(\omega\,(\eta-\eta')-k_\perp(x_\perp-x'_\perp))}, \tag{33}$$

where $d^3k \equiv d\omega\, dk_\perp^2$. As a result, we can obtain the equation which $\widetilde{D}_\pm$ should satisfy as

$$\gamma^{-1}\delta(\rho-\rho') = (-\rho^{-2}\omega^2 - \rho^{-1}\partial_\rho - \partial_\rho^2 + k_y^2$$
$$+ k_z^2 \pm 2\rho^{-2}\mu\,\omega + \Delta^2)\widetilde{D}_\pm. \tag{34}$$

Finally, we can obtain the equation that determines $\widetilde{D}_\pm$ as

$$(\mathcal{F} + \Omega_\pm^2)\,\widetilde{D}_\pm = -\rho\,\delta(\rho-\rho'), \tag{35}$$

where $\mathcal{F} \equiv \rho^2\partial_\rho^2 + \rho\,\partial_\rho - \rho^2\kappa^2$ with $\kappa^2 \equiv k^2+\Delta^2$ $(k^2 \equiv k_y^2 + k_z^2)$, and $\Omega_\pm^2 \equiv \omega(\omega \mp 2\mu)$.

Let us now determine $\widetilde{D}_\pm$ from eq. (35). First, we can see that F can satisfy the following differential equation:

$$\mathcal{F}\,\Theta_\lambda(\rho,k) = (i\lambda)^2\,\Theta_\lambda(\rho,k), \tag{36}$$

with

$$\Theta_\lambda(\rho,k) = C_\lambda K_{i\lambda}(\kappa\rho), \quad C_\lambda = \frac{1}{\pi}\sqrt{2\lambda\sinh(\pi\lambda)}, \tag{37}$$

where the values of λ are positive real numbers, $K_{i\lambda}(\kappa\rho)$ is the second kind modified Bessel function, and C_λ was obtained from the equation that the second kind modified Bessel functions satisfy,

$$\int_0^\infty \frac{d\rho}{\rho}\,\Theta_{\lambda'}(\rho,k)\,\Theta_\lambda(\rho,k) = \delta(\lambda'-\lambda). \tag{38}$$

We then assume that $\widetilde{D}_\pm$ can be written by taking $\Theta_\lambda(\rho,k)$ as the bases for the orthogonal directions labeled by λ as

$$\widetilde{D}_\pm = \int_0^\infty d\lambda\, f_{\lambda,\pm}(\omega,\rho')\,\Theta_\lambda(\rho,k), \tag{39}$$

where the above means an orthogonal base expansion of $\Theta_\lambda(\rho,k)$ labeled by λ , and $f_{\lambda,\pm}(\omega,\rho')$ are the coefficients of each direction. If we can find $f_{\lambda,\pm}(\omega,\rho')$ that satisfies eq. (35), this assumption is right. We now obtain such $f_{\lambda,\pm}(\omega,\rho')$.

We can see that now we can have the following two independent equations as

$$\int \frac{d\rho}{\rho}\,\widetilde{D}_\pm\,\mathcal{F}\,\Theta_\lambda(\rho,k) = -\lambda^2\int \frac{d\rho}{\rho}\,\widetilde{D}_\pm\,\Theta_\lambda(\rho,k), \tag{40a}$$

$$\int \frac{d\rho}{\rho}\,\Theta_\lambda(\rho,k)\,\mathcal{F}\widetilde{D}_\pm = -\Omega_\pm^2\int \frac{d\rho}{\rho}\,\Theta_\lambda(\rho,k)\,\widetilde{D}_\pm - \Theta_\lambda(\rho',k). \tag{40b}$$

Here, in obtaining eqs. (40a) and (40b), we have used eqs. (36) and (35), respectively.

Eq. (40a) and eq. (40b) are equivalent to each other. Actually, this equivalence can be seen easily by what eq. (40a) and eq. (40b) can be represented as $\langle\widetilde{D}_\pm|\mathcal{F}|\Theta_\lambda(\rho,k)\rangle$ and $\langle\Theta_\lambda(\rho,k)|\mathcal{F}|\widetilde{D}_\pm\rangle$, respectively. Then, by subtracting these two equations each other,

$$0 = (-\lambda^2 + \Omega_\pm^2)\int \frac{d\rho}{\rho}\widetilde{D}_\pm\,\Theta_\lambda(\rho,k) + \Theta_\lambda(\rho',k)$$
$$= (-\lambda^2 + \Omega_\pm^2)\int \frac{d\rho}{\rho}\cdot\int d\lambda'\, f_{\lambda',\pm}(\omega,\rho')\Theta_{\lambda'}(\rho,k)\cdot\Theta_\lambda(\rho,k)$$
$$+ \Theta_\lambda(\rho',k)$$
$$= (-\lambda^2 + \Omega_\pm^2)\int d\lambda'\cdot\int \frac{d\rho}{\rho}\,\Theta_{\lambda'}(\rho,k)\Theta_\lambda(\rho,k)\cdot f_{\lambda',\pm}(\omega,\rho')$$
$$+ \Theta_\lambda(\rho',k)$$
$$= (-\lambda^2 + \Omega_\pm^2)\int d\lambda'\cdot\delta(\lambda'-\lambda)\cdot f_{\lambda',\pm}(\omega,\rho') + \Theta_\lambda(\rho',k)$$
$$= (-\lambda^2 + \Omega_\pm^2)\,f_{\lambda,\pm}(\omega,\rho') + \Theta_\lambda(\rho',k). \tag{41}$$

In the above, we have used eqs. (38) and (39). From the above, $f_{\lambda,\pm}(\omega,\rho')$ can be determined as

$$f_{\lambda,\pm}(\omega,\rho') = \frac{\Theta_\lambda(\rho',k)}{\lambda^2 - \Omega_\pm^2} \equiv \frac{\Theta_\lambda(\rho',k)}{\lambda^2 - \Omega^2}. \tag{42}$$

In the above, we have regarded Ω_+^2 and Ω_-^2 as the same each other, and written as $\Omega^2 \equiv \Omega_+^2 = \Omega_-^2$. Because the integrations of $f_{\lambda,+}(\omega,\rho')$ and $f_{\lambda,-}(\omega,\rho')$ with regard to ω in eq. (32) do not produce any difference under the transformation ω→−ω. Namely,

$$D_+(x-x')$$
$$= \int \frac{d\omega\, d^2k_\perp}{(2\pi)^3}\cdot\int d\lambda\,\frac{\Theta_\lambda(\rho',k)\,\Theta_\lambda(\rho,k)}{\lambda^2 - \omega(\omega-2\mu)}\,e^{i\,(\omega\,(\eta-\eta')-k_\perp(x_\perp-x'_\perp))}$$
$$= \int \frac{d\omega\, d^2k_\perp}{(2\pi)^3}\cdot\int d\lambda\,\frac{\Theta_\lambda(\rho',k)\,\Theta_\lambda(\rho,k)}{\lambda^2 - \omega(\omega+2\mu)}\,e^{i\,(\omega\,(\eta-\eta')-k_\perp(x_\perp-x'_\perp))}$$
$$= D_-(x-x'). \tag{43}$$

As a result, we can put $\Omega_\pm^2 \equiv \Omega^2$. Correspondingly, there is no difference between $\widetilde{D}_+$ and $\widetilde{D}_-$. Hence, we put $\widetilde{D}_\pm \equiv \widetilde{D}$ and $D_\pm \equiv D$.

Now, substituting eq. (42) into eq. (39), we can write $\widetilde{D}$ as

$$\widetilde{D} = \int d\lambda\, \frac{\Theta_\lambda(\rho',k)\Theta_\lambda(\rho,k)}{\lambda^2-\Omega^2}. \tag{44}$$

Finally, the effective action W can be written as

$$W = i\,V\gamma \int \frac{dk^3}{(2\pi)^3} \int_0^{M^2} d\Delta^2\, \widetilde{D}. \tag{45}$$

Calculation with the Euclidianization

Now that $\widetilde{D}_\pm = \widetilde{D}$ have been obtained, backing to eq. (29), we perform the integrations. We start with performing the Wick rotation toward the η-direction. At this time, the metric given in eq. (4) changes to

$$ds_E^2 = \rho^2 d\eta^2 + d\rho^2 + dx_\perp^2. \tag{46}$$

We can see from this form that the η -direction is S^1-compactified with the period$\beta=2\pi$.

Then, the following replacement arises:

$$V \equiv \int d^3x\gamma \to -i\beta \int d^2x_\perp\, \gamma_E \equiv -i\beta\, V_E \tag{47}$$

with

$$\gamma \equiv \sqrt{-g} \to \gamma_E \equiv \sqrt{g} = \rho, \tag{48}$$

$$\int d\omega \to \frac{2\pi i}{\beta}\sum_n \quad \text{and} \quad \omega \to \omega_n = n, \tag{49}$$

where V is defined first under eq. (26), and the reason for $\omega_n=n$ is that, generally, ω_n in the S^1-compactification with a period R is given as $\omega_n=2\pi n/R$, and in our situation $R=2\pi$. Correspondingly, the values Ω takes are discretized, and we change the notation of Ω as $\Omega\to\Omega_n$.

As a result, the effective potential Γ defined as $W_E \equiv \beta V_E \Gamma$ with $\exp(iW)\equiv\exp(-W_E)$can be obtained as

$$\Gamma = \frac{\rho}{\pi^3}\sum_{n=-\infty}^{\infty}\int \frac{d^2k_\perp}{(2\pi)^2}\int_0^{M^2} d\Delta^2 \int d\lambda\,\lambda\,\sinh(\pi\lambda)\,\frac{K_{i\lambda}^2(\kappa\rho)}{\lambda^2-\Omega_n^2} \tag{50}$$

In the above, $d^2k_\perp=2\pi k dk$ $(0\leq k\leq\infty)$,

$$\Psi \equiv \int_0^{M^2} d\Delta^2\, K_{i\lambda}^2(\kappa\rho), \tag{51}$$

and we have evaluated the summation of n as

$$\sum_{n=-\infty}^{\infty}\frac{1}{\lambda^2-\Omega_n^2} = \frac{\pi}{\sqrt{\lambda^2+\mu^2}}\Big(\coth\big(\pi(\mu-\sqrt{\lambda^2+\mu^2})\big) - \coth\big(\pi(\mu+\sqrt{\lambda^2+\mu^2})\big)\Big) \equiv \pi\,\Phi. \tag{52}$$

Next, since we finally calculate the critical acceleration of the Bose–Einstein condensation, we focus on the critical moment. The value of the chemical potential at the critical moment is given by m/a_c, where a_c is the critical acceleration. We explain this in what follows.

The Chemical Potential at the Critical Moment

Using the effective potential Γ given in eq. (50), the particle density $d=-\partial\Gamma/\partial\mu$ can be written as

$$d = -\frac{\rho}{2\pi^3}\int d\lambda\,\lambda\,\sinh(\pi\lambda)\Big(\frac{\partial\Phi}{\partial\mu}\int dkk\,\Psi + \Phi\int dkk\,\frac{\partial\Psi}{\partial\mu}\Big). \tag{53}$$

Here, to perform $\frac{\partial\Psi}{\partial\mu}$, we show the μ -dependence of Ψ by expanded it around μ=0 as

$$\Psi = \Psi\Big|_{\mu=0} + \frac{\partial\Psi}{\partial\mu}\Big|_{\mu=0}\mu + \frac{1}{2}\frac{\partial^2\Psi}{\partial\mu^2}\Big|_{\mu=0}\mu^2 + \cdots = \int_0^{m^2} d\Delta\, K_{i\lambda}^2(\kappa\rho) - a^2 K_{i\lambda}^2(\kappa\rho)\Big|_{\Delta^2=m^2}\mu^2 + \cdots \tag{54}$$

with

$$\frac{\partial\Psi}{\partial\mu}\Big|_{\mu=0} = \frac{\partial M^2}{\partial\mu}\frac{\partial\Psi}{\partial M^2}\Big|_{\mu=0} = 0, \tag{55a}$$

$$\frac{1}{2}\frac{\partial^2\Psi}{\partial\mu^2}\Big|_{\mu=0} = \frac{1}{2}\frac{\partial}{\partial\mu}\Big(\frac{\partial M^2}{\partial\mu}\frac{\partial\Psi}{\partial M^2}\Big)\Big|_{\mu=0} = -a^2\frac{\partial\Psi}{\partial M^2}\Big|_{\Delta^2=m^2}. \tag{55b}$$

Using the above expansion, we can write the particle density as

$$d = -\frac{\rho}{2\pi^3}\int d\lambda\,\lambda\,\sinh(\pi\lambda) \times\Big(\frac{\partial\Phi}{\partial\mu}\int dkk\,\Psi - 2a^2\mu\,\Phi K_{i\lambda}^2(\kappa\rho)\Big|_{\Delta^2=m^2}\Big). \tag{56}$$

We can always confirm numerically that $\frac{\partial\Phi}{\partial\mu}<0$ and $\Phi>0$. Further, we can see $\Psi>0$, since the integrand in Ψ is given by the square and the integral direction is positive, as can be seen in eq. (51).

Hence, looking at eq. (56), we can say the following two things: When the chemical potential is increased with fixing all the other parameters, the particle density always grows up. On the other hand, when the acceleration is decreased with fixing all the other parameters, the particle density always decreased.

Hence, when decreasing the acceleration from higher accelerations, where there is no condensation, say α=0, with keeping the particle density to constant, the chemical potential should grow up to keep the particle density to constant. However, as can be seen from eq. (22), there is the upper bound for the value the chemical potential can take, which is m/a. Hence, when the chemical potential reaches the upper bound in such a situation, α should start to grow up, where α plays the role of the condensation as written under eq. (17a) and (17b).

Hence, we can see that the value of the chemical potential at the critical moment is m/a_c.

Calculation in the Relativistic Situation

As mentioned under eq. (52), our analysis is the one at the critical moment. Hence, for the reason mentioned above, m/a_c is assigned to the value of the chemical potential withα=0 and $\rho = a_c^{-1}$.

We put an assumption that m/a_c is small. This means that the chemical potential in this study is also small. The situation that m/a_c is small corresponds to the relativistic situation as mentioned in the introduction.

In what follows, we use two symbols: μ , μ'. μ means the true chemical potential, where "true" means that it has been given at eq. (6) and the derivative with regard to μ can act on. On the other hand, μ' means just a value m/a_c. μ' is not the chemical potential and the derivative cannot act on.

Then, it turns out that we can expand Φ and Ψ with regard to μ and μ' as

$$\Phi = \Phi_0 + \Phi_2\,\mu^2 + \cdots, \tag{57a}$$

$$\Psi = \Psi_0\,\mu'^2 + \Psi_2\,\mu^2 + \cdots, \tag{57b}$$

where

$$\Phi_0 \equiv \frac{2}{\lambda}\coth(\pi\lambda), \tag{58a}$$

$$\Phi_2 \equiv -\frac{1}{\lambda^3}\Big(\coth(\pi\lambda) + \frac{\pi\lambda}{\sinh^2(\pi\lambda)}\big(1 - 2\pi\lambda\coth(\pi\lambda)\big)\Big), \tag{58b}$$

$$\Psi_0 \equiv -\Psi_2 \equiv a_c^2\, K_{i\lambda}(k/a_c). \tag{58c}$$

At this time, substituting (57a) and (57b) into eq. (50), Γ at the critical moment in the relativistic situation is given as

$$\Gamma = \frac{1}{2a_c\pi^3}\int d\lambda\,\lambda\sinh(\pi\lambda)\int dk\,k\left(\Phi_0\Psi_0\,\mu'^2 + \Phi_0\Psi_2\,\mu^2\right). \tag{59}$$

THE PARTICLE DENSITY AND CRITICAL ACCELERATION

We now obtain the particle density d=−∂Γ/∂μ using eq. (59) as

$$\begin{aligned} d &= -\frac{\mu}{a_c\pi^3}\int_0^\infty d\lambda\lambda\sinh(\pi\lambda)\int dkk\,\Phi_0\Psi_2 \\ &= \frac{2a_c\mu}{\pi^3}\int_0^\infty d\lambda\,\cosh(\pi\lambda)\int dkk\,K_{i\lambda}^2(k/a_c) \\ &= \frac{a_c^3\mu}{\pi^2}\int d\lambda\,\lambda\,\coth(\pi\lambda), \end{aligned} \tag{60}$$

where, in the above calculation, we have used $\sinh(\pi\lambda)\,\Phi_0 = \frac{2}{\lambda}\cosh(\pi\lambda)$, and performed the integration with regard to k as $\int dkk\,K_{i\lambda}^2(k/a) = \frac{\pi\lambda}{2}a_c^2\,\mathrm{csch}(\pi\lambda)$.

Writing $\coth(\pi\lambda) = \frac{1}{2} + \frac{1}{e^{2\pi\lambda}-1}$ in eq. (60), we can see that the contribution from the 1/2 diverges. A similar divergence appeared in a similar scalar field model with ours [17], which was canceled by a mass renormalization. There also appeared the divergence in the Nambu–Jona–Lasinio model in the accelerating frame [15] and [16], which was contained by a cutoff. We may be also able to show that the 1/2 can be canceled by some regularization. However there are some unclear points in its physical meaning. The source of the divergence can be thought clearly as the ultra divergence in the one-loop order. Because our analysis can be regraded as an one-loop calculation from eq. (24), and λ can be regarded as some momentum from eq. (36). However, if we perform a mass renormalization by replacing m^2 with $m_0^2 + \delta m^2$ toward the m in the replacement ofμ=m/a_c to be performed right after this as the critical moment, it seems that the counter term δm^2 becomes independent of the acceleration. In this case, the counter term determined in the accelerating frame can survive even in the inertial frame. As a result, the consistency in the link between the accelerating and inertial frames in a model is unclear. On the other hand, if we perform the regularization by using a cutoff, how to determine the cutoff concretely is unclear for us at this moment. We do not examine this divergence problem in this paper, and handle it by simply ignoring the 1/2, considering that it is canceled by some regularization.

As a result, by putting μ=m/a_c as the critical moment, we can obtain the particle density as $d = a_c^2 m/24\pi^2$. Hence we can obtain the critical acceleration as

$$a_c = 2\pi\sqrt{\frac{6d}{m}}. \tag{61}$$

For the actual observation, we show the result in the MKS units as

$$a_c = 2\pi\sqrt{c^3\hbar\frac{6d}{m}} \approx 32.763\sqrt{2\pi}\sqrt{\frac{d}{m}}\,[\mathrm{cm/s^2}], \tag{62}$$

where d and m in the last equation are the number of particle in a centimeter unit cube and the weight of particle measured by kilogram, respectively.

REMARK

In this paper, based on the thermal excitation brought into by the Unruh effect, we have calculated the critical acceleration at which the Bose–Einstein condensation starts to occur, as in eq. (61).

The system we have considered is an ideal gas composed of a complex scalar particle in the uniformly accelerating Minkowski space–time at zero-temperature. To this purpose, we have considered a free complex scalar field at finite density given in eq. (1) in the Rindler space given in eq. (4), and adopted an expression as in eqs. (17a) and (17b) to describe the Bose–Einstein condensation.

In our analysis, we have given the space–time integration in the action as the one without the ρ-integration to fix the acceleration and the Unruh temperature. This is different point from Refs. [15], [16] and [17]. If performing the analysis like them, there will be no difference in the final result. However, it is thought that we should describe explicitly that the space–time integration in the action is the one without the ρ-integration to make clear the physical meaning.

To carry out the calculation, we have picked up the leading terms in the integrals(57a) and (57b). As a result, we could obtain a result as in eq. (61). This manipulation means to assume that the complex scalar field considered in this paper has the critical temperature for the Bose–Einstein condensation in the temperature region $k_B T/2 \gg mc^2$, which is the relativistic situation that the particle's thermal motion energy is much higher than its static energy at the critical moment.

We can consider the analysis in another extremal region such as the non-relativistic situation in which the particle's thermal motion energy at the critical moment is much less than the particle's static energy as $k_B T/2 \ll mc^2$, where this T is around T_c. However, the middle calculations appearing in the expansion in this limit are very complicated, and the analysis in this limit is technically difficult. We take this to a future work.

Furthermore, the problem on the divergence in eq. (60) has not been examined in this paper. We also take this to a future work.

However, as mentioned in the introduction, there are papers concluding that the Unruh effect does not contribute to the restoration of the spontaneous symmetry breaking [18]. The system in these papers is not the finite density unlike us, and their conclusion would not be applied to our study readily.

However, we can see intuitively that the situation which our study and Refs. [15],[16] and [17] can lead is strange. Because, for a situation, it can be observed in different ways depending on the frame in which observer stays. For example, for an observer in an inertial frame, it can be observed as a Bose–Einstein condensation state. But for an accelerating observer, it can be observed as an normal gas. Therefore they might be right after all.

If we perform this study more, before that, it is thought that we should make clear whether their conclusion can hold or not in the finite density models like ours.

In Ref. [24], a critical temperature has been obtained in the same model with ours but the background space–time is taken to the Euclid space at finite temperature. Their result is $\sqrt{3d/m}$ as written there, and we can see that there is difference from our result in the factor. We think that this discrepancy originates in the difference between the Rindler space that has the event-horizons and the Euclid space and what our calculation has been performed only in the right wedge of the Rindler space.

We can see from our result that the critical acceleration rooted in the thermal excitation brought into by the Unruh effect has different behavior from the critical temperature obtained by the true thermal excitation in the flat space. We may expect from this that the observational data›s behaviors of the critical temperatures measured in the accelerating and the flat static frames are different each other. It would be interesting, if we can find this discrepancy in the experiments.

However, on the other hand, we think at last that the significance of this study would be what we could recognize a puzzle mentioned above.

ACKNOWLEDGEMENTS

The author thanks Seckson Sukhasena for various things, Sujiphat Janaun who had discussions on the manuscript and checked it and Robert Soldati who informed me about very interesting papers. Lastly, the author would like to thank the Institute for Fundamental Study and Naresuan University.

REFERENCES

1. W.G. Unruh, Experimental black hole evaporation, Phys. Rev. Lett. 46 (1981) 1351.
2. S.W. Hawking, Particle creation by black holes, Commun. Math. Phys. 43 (1975) 199; Erratum: S.W. Hawking, Commun. Math. Phys. 46 (1976) 206.
3. M. Novello, M. Visser, G. Volovik (Eds.), Artificial Black Holes, World Scientific, 2002.
4. C. Barcelo, S. Liberati, M. Visser, Analogue gravity, Living Rev. Relativ. 8 (2005) 12, Living Rev. Relativ. 14 (2011) 3, arXiv:gr-qc/0505065.
5. P.C.W. Davies, Scalar particle production in Schwarzschild and Rindler metrics, J. Phys. A 8 (1975) 609.
6. W.G. Unruh, Notes on black hole evaporation, Phys. Rev. D 14 (1976) 870.
7. W.G. Unruh, N. Weiss, Acceleration radiation in interacting field theories, Phys. Rev. D 29 (1984) 1656.
8. L.C.B. Crispino, A. Higuchi, G.E.A. Matsas, The Unruh effect and its applications, Rev. Mod. Phys. 80 (2008) 787, arXiv:0710.5373 [gr-qc].
9. The sixth paragraph of the first page in Y. Kurita, M. Kobayashi, T. Morinari, M. Tsubota, H. Ishihara, Spacetime analogue of Bose–Einstein condensates: Bogoliubov–de Gennes formulation, Phys. Rev. A 79 (2009) 043616, arXiv:0810.3088 [cond-mat.other].
10. M. Ueda, Fundamentals and New Frontiers of Bose–Einstein Condensation, World Scientific Pub. Co. Inc., 2010.
11. A. Retzker, J.I. Cirac, M.B. Plenio, B. Reznik, Detection of acceleration radiation in a Bose–Einstein condensate, Phys. Rev. Lett. 101 (2008) 110402.
12. A. Iorio, G. Lambiase, The Hawking–Unruh phenomenon on graphene, Phys. Lett. B 716 (2012) 334, arXiv:1108.2340 [cond-mat.mtrl-sci].

13. E. Martin-Martinez, I. Fuentes, R.B. Mann, Using Berry's phase to detect the Unruh effect at lower accelerations, Phys. Rev. Lett. 107 (2011) 131301, arXiv:1012.2208 [quant-ph].

14. The second paragraph of the first page in [11].

15. T. Ohsaku, Dynamical chiral symmetry breaking and its restoration for an accelerated observer, Phys. Lett. B 599 (2004) 102, arXiv:hep-th/0407067.

16. D. Ebert, V.C. Zhukovsky, Restoration of dynamically broken chiral and color symmetries for an accelerated observer, Phys. Lett. B 645 (2007) 267, arXiv:hep-th/0612009.

17. P. Castorina, M. Finocchiaro, Symmetry restoration by acceleration, Physics 3 (2012) 1703, arXiv:1207.3677 [hep-th].

18. C.T. Hill, Can the Hawking effect thaw a broken symmetry?, Phys. Lett. B 155 (1985) 343; C.T. Hill, One loop operator matrix elements in the Unruh vacuum, Nucl. Phys. B 277 (1986) 547.

19. S. Benic, K. Fukushima, Unruh effect and condensate in and out of an accelerated vacuum, arXiv:1503.05790 [hep-th].

20. R.M. Cavalcanti, P. Giacconi, G. Pupillo, R. Soldati, Bose–Einstein condensation in the presence of a uniform field and a point-like impurity, Phys. Rev. A 65 (2002) 053606, arXiv:cond-mat/0111490.

21. S. Takagi, On the response of a Rindler particle detector. 2, Prog. Theor. Phys. 74 (1985) 142.

22. H. Ooguri, Spectrum of Hawking radiation and Huygens' principle, Phys. Rev. D 33 (1986) 3573.

23. S. Iso, Y. Yamamoto, S. Zhang, Stochastic analysis of an accelerated charged particle – transverse fluctuations, Phys. Rev. D 84 (2011) 025005, arXiv:1011.4191 [hep-th].

24. J.I. Kapusta, C. Gale, Finite-Temperature Field Theory: Principles and Applications, Univ. Press, Cambridge, UK, 2006.

CHAPTER 10

BURG'S MODIFIED ENTROPY IN STATISTICAL MECHANICS

Amritansu Ray and S. K. Majumder

Keywords: Entropy optimization, probability distribution, Shannon's Entropy, Burg's Entropy

INTRODUCTION

The concept of entropy [1] figured strongly in the physical sciences during the 19th century, especially in thermodynamics and statistical mechanics [2], as a measure of equilibrium and evolution of thermodynamic systems. Two main views were developed which were the macroscopic view formulated originally by Clausius and Carnot and the microscopic approach associated with Boltzmann and Maxwell. Since then, both the approaches have made introspection in natural thermodynamic and microscopically probabilistic systems possible. Entropy is defined as the measure of a system's thermal energy per unit temperature that is unavailable for doing useful work. Because work is obtained from ordered molecular motion, the amount of entropy is also a measure of molecular disorder or randomness of a system. The concept of entropy provides deep insight into the direction of spontaneous change for many day-to-day phenomena. Now, how entropy was developed by Rudolf Clausius [3] is discussed below.

Clausius's Entropy

To provide a quantitative measure for the direction of spontaneous change, Clausius introduced the concept of entropy as a precise way of expressing the second law of thermodynamics. The Clausius form of the second law states that spontaneous change for an irreversible process [4] in an isolated system (that is, one that does not exchange heat or work with its surroundings) always proceeds in the direction of increasing entropy. By the Clausius definition, if an amount of heat Q flows into a large heat reservoir at temperature T above absolute zero, then $\Delta s=Q/T$. This equation effectively gives an alternate definition of temperature that agrees with the usual definition. Assume that there are two heat reservoirs R_1 and R_2 at temperatures T_1 and T_2. (such as the stove and the block of ice). If an amount of heat Q flows from R_1 to R_2, then the net entropy change for the two reservoirs is $\Delta s = Q\left(\frac{1}{T_2} - \frac{1}{T_1}\right)$ which is positive, provided that $T_1>T_2$. Thus, the observation that heat never flows spontaneously from cold to hot is equivalent to requiring the net entropy change to be positive for a spontaneous flow of heat. When the system is in thermodynamic equilibrium, then dS=0, *i.e.*, if $T_1=T_2$, then the reservoirs are in equilibrium, no heat flows, and $\Delta s=0$. If the gas absorbs an incremental amount of heat dQ from a heat reservoir at temperature T and expands reversibly against the maximum possible restraining pressure P, then it does the maximum work and dW=PdV. The internal energy of the gas might also change by an amount dU as it expands. Then, by conservation of energy, dQ=dU+PdV. Because the net entropy change for the system plus reservoir is zero when maximum work is done and the entropy of the reservoir decreases by an amount $dS_{reservoir} = -\frac{dQ}{T}$, this must be counterbalanced by an entropy increase of $dS_{system} = \frac{dU + PdV}{T} = \frac{dQ}{T}$ for the working gas so that $dS_{system} + dS_{reservoir} = 0$. For any real process, less than the maximum work would be done (because of friction, for example), and so the actual amount of heat dQ′ absorbed from the heat reservoir would be less than the maximum amount dQ. For example, the gas could be allowed to expand freely into a vacuum and do no work at all. Therefore, it can be stated that $dS_{system} = \frac{dU + PdV}{T} \geqslant \frac{dQ'}{T}$ with, dQ′=dQ in the case of maximum work corresponding to a reversible process. This equation defines S_{system} as a thermodynamic state variable, meaning that its value is completely determined by the current state of the system and not by how the system reached that state. Entropy is a comprehensive property

Citation: Ray, A.; Majumder, S.K. Burg's Modified Entropy in Statistical Mechanics. Mathematics 2016, 4, 10. doi:10.3390/math4010010.

in that its magnitude depends on the amount of material in the system. In one statistical interpretation of entropy, it is found that for a very large system in thermodynamic equilibrium, entropy S is proportional to the natural logarithm of a quantity Ω corresponding to S and can be realized; that is, S=KlnΩ, in which K is related to molecular energy. On the other hand, entropy generation analysis [5,6,7,8,9,10,11] is used to optimize the thermal engineering devices for higher energy efficiency; it has attracted wide attention to its applications and rates in recent years. In order to access the best thermal design of systems, by minimizing the irreversibility, the second law of thermodynamics could be employed. Entropy generation is a criterion for the destruction of a systematized work.The development of the theory followed two conceptually different lines of thought. Nevertheless, they are symbiotically related, in particular through the work of Boltzmann.

Boltzmann's Entropy

In addition to thermodynamic (or heat-change) entropy, physicists also study entropy statistically [12,13]. The statistical or probabilistic study of entropy is presented in Boltzmann's law. Boltzmann's equation is somewhat different from the original Clausius (thermodynamic) formulation of entropy. Firstly, the Boltzmann formulation is structured in terms of probabilities, while the thermodynamic formulation does not consist in the calculation of probabilities. The thermodynamic formulation can be characterized as a mathematical formulation, while the Boltzmann formulation is statistical. Secondly, the Boltzmann equation yields a value of entropy S while the thermodynamic formulation yields only a value for the change in entropy (dS). Thirdly, there is a shift in content, as the Boltzmann equation was developed for research on gas molecules rather than thermodynamics. Fourthly, by incorporating probabilities, the Boltzmann equation focuses on microstates, and thus explicitly introduces the question of the relationship between macrostates and microstates. Boltzmann investigated such microstates and defined entropy in a new way such that the macroscopic maximum entropy state corresponded to a thermodynamic configuration which could be formulated by the maximum number of different microstates. He noticed that the entropy of a system can be considered as a measure of the disorder in the system and that in a system having many degrees of freedom, the number measuring the degree of disorder also measured the uncertainty in a probabilistic sense about the particular microstates.

The value W was originally intended to be proportional to the Wahrscheinlichkeit (means probability) of a macrostate for some probability distribution of a possible microstate, in which the thermodynamic state of a system can be realized by assigning different ξ and ρ of different molecules. The Boltzmann formula is the most general formula for thermodynamic entropy; however, his hypothesis was for an ideal gas of N identical particles, of which N_i(i=1,2,.... ,l) are the ith microscopic condition of position and momentum of a given distribution D_i=(N_1,N_2,....,N_l). Here, D_1=(N,0,....,0), D_2=(N−1,1,....,0),...*etc.* For this state, the probability of each microstate system is equal, so it was equivalent to calculating the number of microstates associated with a macrostate. Then the statistical disorder is given by [14]

$$W = \frac{N!}{N_1!N_2!\ldots.N_l!}, [N_1 + N_2 + \ldots.. + N_l = N].$$

Therefore, the entropy given by Boltzmann is: SB=KlnWWhere

$$\ln W = \ln\frac{N!}{N_1!N_2!\ldots.N_l!} = \ln N! - \sum_{i=1}^{l} \ln(N_i!)$$

Let us now take an approximate value of W for a large N!. Using Stirling's approximation lnN!≅NlnN−N, we have:

$$\ln W \cong N\ln N - N - \left[\sum_{i=1}^{l} N_i \ln(N_i) - \sum_{i=1}^{l}(N_i)\right]$$

$$= -\sum_{i=1}^{l} N_i \ln N_i = -N\sum_{i=1}^{l} p_i \ln p_i$$

$S_B = -K\sum_{i=1}^{n} p_i \ln p_i$ where p_i=N_i/N is the probability of the occurrence of i th microstates. Boltzmann was the first to emphasize the probabilistic meaning of entropy and the probabilistic nature of thermodynamics.

Information Theory and Shannon's Entropy

Unlike the first two entropy approaches (thermodynamic entropy by Clausius and Boltzmann's entropy), the third major form of entropy did not fall within the field of physics, but was developed instead in a new field known as information theory [15,16,17] (also known as communication theory). A fundamental step in using entropy in new contexts unrelated to thermodynamics was provided by Shannon [18], who came to conclude that entropy could be used to measure types of disorder other than that of thermodynamic microstates. Shannon was interested in information theory [19,20], particularly in the ways in which information can be conveyed via a message. This led him to examine probability distributions in a very general sense and he worked to find a way of measuring the level of uncertainty in different distributions.

For example, suppose the probability distribution for the outcome of a coin toss experiment is *P(H)* = 0.999 and *P(T)* = 0.001. One is likely to notice that there is much more "certainty" than "uncertainty" about the outcome of this experiment and, consequently, the probability distribution. If, on the other hand, the probability distribution governing that same experiment were *P(H)* = 0.5 and *P(T)* = 0.5, then there is much less "certainty" and much more "uncertainty" when compared to the previous distribution. However, how can these uncertainties can be quantified? Is there some algebraic function which measures the amount of uncertainty in any probabilistic distribution in terms of the individual probabilities? From these types of simple examples and others, Shannon was able to devise a set of criteria which any measure of uncertainty may satisfy. He then tried to find an algebraic form which would satisfy his criteria and discovered that there was only one formula which fit. Let the probabilities of n possible outcomes E_1, E_2,.......,E_n of an experiment be p_1, p_2,.....,p_n, giving rise to the probability distribution P=(p_1,p_2,...,p_n; $\sum_{i=1}^{n} p_i = 1; p_i's \geqslant 0$. There is an uncertainty as to the outcome when the experiment is performed. Shannon suggested the measure $-\sum_{i=1}^{n} p_i \ln p_i$, which is identical to the previous entropy relation if the constant of probability is taken as the Boltzmann constant K. Thus, Shannon showed that entropy, which measures the amount of disorder in a thermodynamic system, also measures the amount of uncertainty in any probability distribution. Let us now give the formal definition of Shannon's entropy as follows: Consider a random experiment $P=(p_1,p_2,\ldots,p_n)$ whose possible outcomes have probabilities p_i,i=1,2,...,n that are known. Can we guess in advance which outcome we shall obtain? Can we measure the amount of uncertainty? We shall denote such an uncertainty measure by $H(P)=H_n(p_1,p_2,\ldots,p_n)$. The most common as well as the most useful measure of uncertainty is Shannon's informational entropy (which should satisfy some basic requirements), which is defined as follows:

Definition I:

Let $(p_1,p_2,\ldots,p_n)$ be the probability of the occurrence of the events $E_1,E_2,\ldots,E_n$ associated with a random experiment. The Shannon's entropy probability distribution $(p_1,p_2,\ldots,p_n)$ of the random experiment system P is defined by $H(P) = H_n(p_1,p_2,\ldots,p_n) = -\sum_{i=1}^{n} p_i \ln p_i$ where, $0\ln 0 \cong 0$. The above definition is generalized straightforwardly as the definition of entropy of a random variable.

Definition II: Let $X \in R$ be a discrete random variable which takes the value xi(i=1,2,3,....,n) with the probability pi,i=1,2,...,n; then the entropy H(X) of X is defined by the expression $H(X) = -\sum_{i=1}^{n} p_i \ln p_i$ Examination of *H* orHn(P) reveals why Shannon's measure is the most satisfactory measure of entropy because of the following:

i. $H_n(P)$ is a continuous function of $p_1,p_2,\ldots,p_n$.

ii. $H_n(P)$ is a symmetric function of its arguments.

iii. $H_{n+1}(p_1,p_2,\ldots,p_n,0)=H_n(p_1,p_2,\ldots,p_n)$, *i.e.*, it should not change if there is an impossible outcome to the probability.

iv. Its minimum is 0 when there is no uncertainty about the outcome. Thus, it should vanish when one of the outcomes is certain to happen so that

$$H_n(p_1,p_2,\ldots,p_n) = 0; p_i = 1, p_j = 0, i \neq j, i = 1,2,\ldots,n$$

v. It is the maximum when there is maximum uncertainty, which arises when the outcomes are equally likely so that $H_n(P)$ is the maximum when $p_1 = p_2 = \ldots\ldots = p_n = \frac{1}{n}$.

vi. The maximum value of $H_n(P)$ increases with n.

vii. For two independent probability distributions $P(p_1,p_2,\ldots,p_n)$ and $Q(q_1,q_2,\ldots,q_m)$, the uncertainty of the joint scheme P∪Q should be the sum of their uncertainties: $H_{n+m}(P\cup Q)=H_n(P)+H_m(Q)$

Shannon's entropy has various applications in the field of portfolio analysis, the measurement of economic analysis, transportation, and urban and regional planning as well as in the fields of statistics, thermodynamics, queuing theory, parametric estimation, *etc.* It has been used in the field noncommensurable and conflicting criteria [21] and in the nonlinear complexity of random sequences [22] as well.

DISCUSSION

Jaynes' Maximum Entropy (MaxEnt) Principle

Let the random variable of an experiment be X, and assume the probability mass associated with the value xi is pi, *i.e.*, $P_x(i)\equiv P(X)=p_i$,i=1,2,....,n. The set $(X,P) = \{(x_1,p_2),(x_2,p_2),\ldots,(x_n,p_n)\}, \sum_{i=1}^{n} p_i = 1$ is called the source ensemble as described by Karmeshu [23]. In general, we may find expected values of the functions $g_1(X)$, $g_2(X)$,.....,$g_z(X)$ to get $\sum_{i=1}^{n} p_i g_{ri} = a_r, r = 1,2,\ldots\ldots,z, p_i \geqslant 0$ and with natural constraint $\sum_{i=1}^{n} p_i = 1$,given a number of constraints. Thus, we have z+1 relations between $(p_1,p_2,\ldots\ldots,p_n)$. There may be infinite probability distributions $(p_1, p_2,\ldots\ldots,p_n)$ satisfying the above equation. If we know only $g_1(x)$, $g_2(x)$,.....,$g_z(x)$, then we get a family of max entropy distributions. If, in addition, we know the values of $a_1,a_2,\ldots\ldots,a_z$, we get a specific member of this family and we call it the max entropy probability distribution. According to a great article by Jaynes [24,25,26], we choose the probability distribution out of all these which maximizes the measure of entropy as shown by Shannon's equation, $S(P) = -\sum_{i=1}^{n} p_i \ln p_i$.

Any distribution of the form $p_i = \exp[-\lambda_0 - \lambda_1 g_1(x_i) - \lambda_2 g^2_{(x_i)} - \ldots\ldots - \lambda_z g_z(x_i)]$, i = 1, 2, ... , n may be regarded as the maximum entropy distribution where $\lambda_0,\lambda_1,\ldots\ldots,\lambda_z$ are determined as functions of $a_1,a_2,\ldots\ldots,a_z$; then the maximum entropy S_{max} is given by $S_{max}=\lambda_0+\lambda_1 a_1+\lambda_2 a_2+\ldots\ldots+\lambda_{zaz}$.

Kapur [27,28] showed that there is always a concave function of a_1,a2,......,a_m. We also note that all the probabilities given by pi are always positive. We naturally want to know whether there is another measure of entropy other than Shannon's entropy which, when maximized, subject to, $\sum_{i=1}^{n} p_i = 1, \sum_{i=1}^{n} p_i g_{ri} = a_r, r = 1,2,\ldots\ldots,z, p_i \geqslant 0$, gives positive probabilities and for which S_{max} is possibly a concave function [29] of parameters. Kapur [30] studied that Burg's [31] measure of entropy, which has been very successfully used in spectral analysis, does always give positive probabilities. The maximum entropy principal of Jaynes has been used frequently to derive the distribution of statistical mechanics by maximizing the entropy of the system subject to some given constraints. The Maxwell-Boltzman distribution is obtained when there is only one constraint on a system which prescribes the expected energy per particle of the system by Bose-Einstein (B.E.) distribution, Fermi-Dirac (F.D.) distribution and intermediate statistics (I.S.) distributions; these are obtained by maximizing the entropy subject to two constraints by Kapur and Kesavan, and Kullback [32,33] and also by the present authors [34].

Formulation of MEPD in Statistical Mechanics Using Shannon's Measure of Entropy

Let $p_1,p_2,\ldots\ldots,p_n$ be the probabilities of a particle having energy levels $\varepsilon_1,\varepsilon_2,\ldots,\varepsilon_n$, respectively, and let the expected value of energy be prescribed as $\bar{\varepsilon}$; then, to get MEPD, we maximize the Shannon's measure of entropy:

$$S(P) = -\sum_{i=1}^{n} p_i \ln p_i \tag{1}$$

Subject to

$$\sum_{i=1}^{n} p_i = 1, \sum_{i=1}^{n} p_i \varepsilon_i = \bar{\varepsilon} \tag{2}$$

Let the Lagrangian be

$$L = -\sum_{i=1}^{n} p_i \ln p_i - (\lambda - 1)\left(\sum_{i=1}^{n} p_i - 1\right) - \mu\left(\sum_{i=1}^{n} p_i \varepsilon_i - \bar{\varepsilon}\right) \tag{3}$$

Differentiating with respect to p_i's, we get:

$\ln p_i + \lambda + \mu\varepsilon_i = 0 \Rightarrow p_i = \exp(-\lambda - \mu\varepsilon_i)$ where, λ,μ are to be determined by using Equation (2) so that

$$p_i = \frac{\exp(-\mu\varepsilon_i)}{\sum_{i=1}^{n} \exp(-\mu\varepsilon_i)}, i = 1,2,\ldots,n, \tag{4}$$

Where

$$\frac{\sum_{i=1}^{n} \varepsilon_i \exp(-\mu\varepsilon_i)}{\sum_{i=1}^{n} \exp(-\mu\varepsilon_i)} = \bar{\varepsilon} \tag{5}$$

Equation (4) is the well-known Maxwell-Boltzmann distribution from statistical mechanics which is used in many areas [35,36,37].

Burg's Entropy Measure and MEPD

When $S(P) = -\sum_{i=1}^{n} p_i \ln p_i$ was replaced by Burg's measure of entropy $B(P) = \sum_i \ln p_i$, it gave interesting results as shown by Kapur. Burg's measure of entropy is always negative, but this does not matter in entropy maximization, where it has been found that a probability distribution with maximum entropy satisfies the same constraint and it does not matter if all the entropies are negative. So, in Equation (1) when we use

$$B(P) = \sum_i \ln p_i, \text{ with the constraints } \sum_{i=1}^{n} p_i = 1, \sum_{i=1}^{n} i p_i = m \tag{6}$$

we get

$$p_i = \frac{1}{\lambda + \mu i}, i = 1,2,3,\ldots\ldots,n \tag{7}$$

where λ&μ are obtained by solving the equations

$$\sum_i \frac{1}{\lambda + \mu i} = 1, \sum_i \frac{i}{\lambda + \mu i} = m \tag{8}$$

Multiplying first and second Equation (8) by λ&μ respectively then adding

We get

$$\sum_i \frac{\lambda + \mu i}{\lambda + \mu i} = \lambda + \mu m \text{ or, } n = \lambda + \mu m \tag{9}$$

so that from Equation (8),

$$\sum_{i=1}^{n} \frac{1}{n - \mu(m - i)} = 1 \tag{10}$$

Then, μ=0 is an obvious solution but that will give us λ=n and this will satisfy the second equation of (8)

if $m = \frac{n+1}{2}$. Now, Equation (10) is the nth degree polynomial in μ, and one of its roots is zero. Its non-zero solutions will be obtained by solving an equation of (n−1)th degree in μ. *Lemma* has been proved by Kapur as the following:

Lemma:

All the roots of $\sum_{r=1}^{n} \frac{1}{n - \mu(m - i)} = 1$ are real; in other words, none of the roots can be complex.

Proof:

Let $\mu = \alpha + i\beta, i = \sqrt{-1}$ be a pair of complex conjugate roots of Equation (10). Then,

$\sum_{r=1}^{n} \frac{1}{n - (\alpha + i\beta)(m - i)} = 1$ and $\sum_{r=1}^{n} \frac{1}{n - (\alpha - i\beta)(m - i)} = 1$; subtracting the second from the first, we get

$\sum_{r=1}^{n} \frac{2i\beta(m - r)}{\{n - \alpha(m - r)\}^2 + \beta^2} = 0$, which gives $\beta = 0$, or $\sum_{r=1}^{n} \frac{(m - r)}{\{n - \alpha(m - r)\}^2 + \beta^2} = 0$. The second possibility can easily be ruled out. To find the actual location of the n real roots, let us assume

$f(\mu) \equiv \sum_r \frac{1}{n - \mu(m - r)} - 1$ and this function is discontinuous at the following points:

$$-\frac{n}{n - m}, -\frac{n}{n - (m - 1)}, \ldots\ldots, -\frac{n}{1 - l}, \frac{n}{l}, \ldots\ldots, \frac{n}{m - 2}, \frac{n}{m - 1}$$

, where m=k+l, k=[m] and l is a "+" fraction. More precisely, f(μ)→+∞ when μ→ points from one side and f(μ)→−∞ when μ→ points from other side. Again,

$$f(\mu) \equiv \sum_r \frac{1}{n - \mu(m - r)} - 1 = \frac{1}{n}\left[\sum_r \left\{1 - \mu\left(\frac{m - r}{n}\right)\right\}^{-1}\right] - 1$$

$$= \frac{1}{n}\left[\sum_r \left\{1 + \mu\left(\frac{m - r}{n}\right)\right\} + O\left(\mu^2\right)\right] - 1 = \frac{1}{n}\left[n + \frac{\mu}{n}\left\{mn - \frac{n(n+1)}{2}\right\}\right] - 1 + O\left(\mu^2\right)$$

$$= \frac{\mu}{n}\left\{m - \frac{(n+1)}{2}\right\} + O\left(\mu^2\right)$$

d Burg's Modifie Entropy (MBE) Measure and MEPD

Monotonic Character of MBE

We propose use of Burg's modified entropy instead of Burg's entropy. Maximizing the Burg's modified measure of entropy:

$$B_{\text{mod}}(P) = \sum_{i=1}^{n} \ln(1 + ap_i) - \ln(1 + a), a > 0 \tag{11}$$

$$\frac{d}{da}(B_{\text{mod}}(P)) = \sum_{i=1}^{n} \frac{p_i}{1 + ap_i} - \frac{1}{1 + a} = \sum_{i=1}^{n} p_i \left[\frac{1}{1 + ap_i} - \frac{1}{1 + a}\right] = \sum_{i=1}^{n} \frac{ap_i(1 - p_i)}{(1 + a)(1 + ap_i)} > 0 \tag{12}$$

Therefore, $B_{mod}(P)$ is the monotonic increasing function of a. For the probability distribution

(p,1−p)

$$\frac{d}{da}(B_{\text{mod}}(P)) = \sum_{i=1}^{n} \frac{p_i}{1 + ap_i} - \frac{1}{1 + a} = \sum_{i=1}^{n} p_i \left[\frac{1}{1 + ap_i} - \frac{1}{1 + a}\right] = \sum_{i=1}^{n} \frac{ap_i(1 - p_i)}{(1 + a)(1 + ap_i)} > 0 \tag{13}$$

it is showed (see: Table 1.) that $\underset{P}{Max}(B_{\text{mod}}(P)) = \ln\left(1 + \frac{a^2}{4(1 + a)}\right) = \ln\left(\frac{(a + 2)^2}{4(1 + a)}\right)$.

Table 1. The a values & maximum values of Burg's Modified Entropy

a	$\underset{P}{Max}(B_{\textbf{mod}}(P))$
0.5	0.04082
1.0	0.11778
1.5	0.20294
2.0	0.28768
5.0	0.71376
10.0	1.1856
20.0	1.7512
30.0	2.1111
40.0	2.3754
50.0	2.5843
∞	∞

The measure of entropy $\underset{P}{Max}(B_{\text{mod}}(P))$ is the Burg's modified entropy. This is a better measure than the standard Burg's measure since it is always positive and there is no computational problem when p_i is very small. In the above case, the maximum value increases with the number of possible outcomes n.

MBE and Its Relation with Burg's Entropy

$$B_{\text{mod}}(P) = \sum_{i=1}^{n} \ln(1 + ap_i) - \ln(1 + a) = \sum_{i=1}^{n}\left[\ln a + \ln\left(p_i + \frac{1}{a}\right)\right] - \ln a - \ln\left(1 + \frac{1}{a}\right)$$
$$= (n - 1)\ln a + \sum_{i=1}^{n} \ln\left(p_i + \frac{1}{a}\right) - \ln\left(1 + \frac{1}{a}\right)$$

So, when a→∞, maximizing (B_{mod}(P)) and (B(P)) will give the same result in both cases; again, if (B_{mod}(P)) is maximized under the constraints

$$\sum_{i=1}^{n} p_i = 1, \sum_{i=1}^{n} p_i g_{ri} = a_r, r = 1, 2, \ldots\ldots, m, p_i \geqslant 0 \tag{14}$$

we get $\frac{1}{1 + ap_i} = [\lambda_0 + \lambda_1 g_1(x_i) + \lambda_2 g^2_{(x_i)} + \ldots\ldots + \lambda_m g_m(x_i)]$, i=1,2,...........,n. Letting, we have

$$\frac{1}{p_i} = \lambda'_0 + \lambda'_1 g_1(x_i) + \lambda'_2 g^2_{(x_i)} + \ldots\ldots\ldots + \lambda'_m g_m(x_i) \tag{15}$$

The λ'i's are determined by using Constraints (14) and (15) and this gives the MEPD when Burg's entropy is maximized as subject to Equation (14).

Therefore, when a→∞, the MEPD of BME → Burg's MEPD; in fact:

$$(B_{\text{mod}}(P)) - (B(P)) = (n-1)\ln a + \sum_{i=1}^{n} \ln \frac{\left(p_i + \frac{1}{a}\right)}{p_i} - \ln\left(1 + \frac{1}{a}\right) \to \infty \text{ as}$$

$$a \to \infty.$$

MBE and Its Concavity of Smax under Prescribed Mean

$$\text{Maximize } B_{\text{mod}}(P) = \sum_{i=1}^{n} \ln(1 + ap_i) - \ln(1+a) \tag{16}$$

subjectto $\sum_{i=1}^{n} p_i = 1, \sum_{i=1}^{n} ip_i = m.$

We obtain this using Lagranges multiplier mechanics:

$$\left(p_i + \frac{1}{a}\right) = \frac{1}{\lambda + \mu i} \text{ Where, } \sum_i \frac{1}{\lambda + \mu i} = \sum_i \left(p_i + \frac{1}{a}\right) = 1 + \frac{n}{a} \tag{17}$$

From the above equation:

$$\sum_i \frac{i.1}{\lambda + \mu i} = \sum_i \left(p_i + \frac{1}{a}\right) = \sum_{i=1}^{n} \left(ip_i + \frac{i}{a}\right) \tag{18}$$

$$\sum_i \frac{i}{\lambda + \mu i} = m + \frac{n(n+1)}{2a}, \text{ i.e., } \frac{n(n+1)}{2a} + m = \sum_i \frac{i}{\lambda + \mu i} \text{ and}$$

$$\frac{\mu n(n+1)}{2a} + \mu m = \sum_i \frac{\mu i}{\lambda + \mu i}, \text{ therefore } \sum_i \frac{\lambda}{\lambda + \mu i} = \sum_i \left(p_i + \frac{1}{a}\right) = \lambda\left(1 + \frac{n}{a}\right) \tag{19}$$

So

$$\sum_i \frac{\lambda}{\lambda + \mu i} + \sum_i \frac{\mu i}{\lambda + \mu i} = \lambda\left(1 + \frac{n}{a}\right) + \frac{\mu n(n+1)}{2a} + \mu m = \sum_i \frac{\lambda + \mu i}{\lambda + \mu i} = \sum_i^{n} 1 = n \tag{20}$$

$$\lambda\left(1 + \frac{n}{a}\right) + \frac{\mu n(n+1)}{2a} + \mu m = n \Rightarrow \lambda = \frac{n - \mu\left(m + \frac{n(n+1)}{2a}\right)}{1 + \frac{n}{a}} \tag{21}$$

Therefore,

$$\lambda + \mu i = \frac{n + \mu\left[\left(1 + \frac{n}{a}\right)i - m - \frac{n(n+1)}{2a}\right]}{1 + \frac{n}{a}} \tag{22}$$

and

$$p_i = \frac{a+n}{an+\mu\left[(n+a)i - am - \frac{n(n+1)}{2}\right]} - \frac{1}{a}, i = 1,2,\ldots,n, \tag{23}$$

where μ is determined as a function of m and that is

$$1+\frac{n}{a} = \sum_i^n \frac{a+n}{an+\mu\left[(n+a)i - am - \frac{n(n+1)}{2}\right]} \;\&$$

$$\frac{1}{a} = \sum_i^n \frac{1}{an+\mu\left[(n+a)i - am - \frac{n(n+1)}{2}\right]},$$

Therefore,

$$S_{\max} = \sum_i \ln(1+ap_i) - \ln(1+a)$$
$$= \sum_i^n \ln\frac{a}{\lambda+\mu i} - \ln(1+a) = \ln\frac{a^n}{1+a} - \sum_i \ln(\lambda+\mu i)$$
$$= \ln\frac{a^n}{1+a} - \sum_i^n \ln\left[n+\mu\left(\left(1+\frac{n}{a}\right)i - m - \frac{n(n+1)}{2}\right)\right] + n\ln\left(1+\frac{n}{a}\right)$$

$$\frac{dS_{\max}}{dm} = -\sum_{i=1}^n \frac{\left[\frac{d\mu}{dm}\left\{\left(1+\frac{n}{a}\right)i - m - \frac{n(n+1)}{2a}\right\} - \mu\right]}{n+\mu\left[\left(1+\frac{n}{a}\right)i - m - \frac{n(n+1)}{2a}\right]}$$
$$= -\sum \frac{\left(p_i+\frac{1}{a}\right)}{a+n} a\left[\frac{d\mu}{dm}\left[\left(1+\frac{n}{a}\right)i - m - \frac{n(n+1)}{2a}\right] - \mu\right]$$
$$= \frac{d\mu}{dm}\left[m + \frac{n(n+1)}{2} - m - \frac{n(n+1)}{2}\right] + \frac{\mu a}{a+n}\left(1+\frac{n}{a}\right)$$

So that

$$\frac{dS_{\max}}{dm} = \frac{d\mu}{dm} \;\&\; \frac{d^2S_{\max}}{dm^2} = \frac{d\mu}{dm} = \frac{\mu\sum_i^n\left(p_i+\frac{1}{a}\right)^2}{\sum_i^n\left(p_i+\frac{1}{a}\right)^2\left[\left(1+\frac{n}{a}\right)i - m - \frac{n(n+1)}{2a}\right]} \tag{24}$$

$S_{\max}$ will be a concave function of m if dμ/dm<0, that is, if either μ>0,<0 when the denominator <0,>0, respectively.

In the above case, when a=1 we get $S_{\max}$ and the derivative of $S_{\max}$ as follows:

$$\frac{dS_{\max}}{dm} = -\sum_i \frac{1}{\lambda+\mu i}\left[\frac{d}{dm}(\lambda+\mu i)\right] = \sum_i\left[\frac{n}{2}\frac{d\mu}{dm} + \frac{1}{n+1}\left(m\frac{d\mu}{dm}+\mu\right) - \frac{d\mu}{dm}i\right] \tag{25}$$

$$= \frac{d\mu}{dm}\sum_i(1+p_i)\left(\frac{n}{2}+\frac{m}{n+1}-i\right) + \frac{\mu}{n+1}\sum_i(1+p_i)$$
$$= \frac{d\mu}{dm}\sum_i\left(\frac{n}{2}+\frac{m}{n+1}-i\right) + \frac{d\mu}{dm}\left(\frac{n}{2}+\frac{m}{n+1}\right) - \frac{d\mu}{dm}\sum_i ip_i + \frac{\mu n}{n+1} + \frac{\mu}{n+1}$$
$$= \frac{d\mu}{dm}\sum_i\left(\frac{n}{2}.n + \frac{m}{n+1}.n - \frac{n(n+1)}{2}\right) + \frac{d\mu}{dm}\left(\frac{n}{2}+\frac{m}{n+1}\right) - \frac{d\mu}{dm}m + \frac{\mu n}{n+1} + \frac{\mu}{n+1}$$
$$= \frac{d\mu}{dm}\left(\frac{mn}{n+1} - \frac{m}{n+1} - m\right) + \frac{\mu n}{n+1} + \frac{\mu}{n+1} = \mu\left(\frac{n}{n+1}+\frac{1}{n+1}\right) = \mu \tag{26}$$

So that

$$\frac{d^2S_{\max}}{dm^2} = \frac{d\mu}{dm} \tag{27}$$

Additionally, we have

$$0 = -\sum_i (1+p_i)^2 \left[\frac{n}{2}\frac{d\mu}{dm} + \frac{1}{n+1}\left(m\frac{d\mu}{dm} + \mu\right) - \frac{d\mu}{dm}i\right] \tag{28}$$

So that

$$\frac{d\mu}{dm}\sum_i (1+p_i)^2 \left(\frac{n}{2} + \frac{m}{n+1} - i\right) + \frac{\mu}{n+1}\sum_i (i+p_i)^2 = 0 \tag{29}$$

Therefore,

$$\frac{d\mu}{dm} = \frac{\frac{\mu}{n+1}\sum_i (1+p_i)^2}{\sum_i (1+p_i)^2 \left(i - \frac{n}{2} - \frac{m}{n+1}\right)} \tag{30}$$

So,

$$\frac{d^2S_{\max}}{dm^2} < 0 \Rightarrow \frac{d\mu}{dm} < 0 \tag{31}$$

Therefore, from Equations (27) and (30),

$S_{\max}$ will be a concave function of m if either

$$\text{(i) } \mu > 0, \sum_i (1+p_i)^2 \left(i - \frac{n}{2} - \frac{m}{n+1}\right) < 0 \tag{32}$$

Or

$$\text{(ii) } \mu < 0, \sum_i (1+p_i)^2 \left(i - \frac{n}{2} - \frac{m}{n+1}\right) > 0 \tag{33}$$

Additionally, when μ=0, $\frac{d^2S_{\max}}{dm^2} = 0$, all the probabilities are equal, and from Equation (24) we have:

We have,

$$m + \frac{n(n+1)}{2} = \frac{1}{\lambda}\frac{n(n+1)}{2} \text{ or, } m = \frac{n(n+1)}{2}\left[\frac{1}{\lambda} - 1\right] \Rightarrow m = \frac{n+1}{2}. \tag{34}$$

Now, if we proceed algebraically as done Section 2.3, we get:

$$1 + p_i = \frac{1}{\lambda + \mu i} = \frac{1}{\frac{1}{n+1}(n - \mu m) - \mu\left(\frac{n}{2} - i\right)}, i = 1,2,\ldots,n,$$

$\sum_{i=1}^{n} \frac{1}{\frac{1}{n+1}(n - \mu m) - \mu\left(\frac{n}{2} - i\right)} = \sum_{1}^{n}(1+p_i) = n+1$. The obvious solution of the above problem is μ=0 which will give $\lambda = \frac{1}{p_i} = n$, *i.e.*, uniform distribution, and thus we get, $m = \frac{n+1}{2}$.

AN ILLUSTRATIVE EXAMPLE IN STATISTICAL MECHANICS

Example

Let $p_1, p_2, \ldots, p_{10}$ be the probabilities of a particle having energy levels $\varepsilon_1, \varepsilon_2, \ldots, \varepsilon_{10}$; respectively, and let the expected value of energy be prescribed as m; then, we get the maximum entropy probability distribution (MEPD) with MBE as follows:

Maximize $\sum_{i=1}^{10} \ln(1+p_i) - \ln 2$ such that, $\sum_{i=1}^{10} p_i = 1$ and $\sum_{i=1}^{10} i p_i = m$ for $m = 1\,(0.25)\,10$

Solution:

Maximizing the measure of entropy subject to the given constraints, we have $(1+p_i) = \dfrac{1}{\lambda + \mu i}$.

When $\sum_i \dfrac{1}{\lambda + \mu i} = 10 + 1 = 11, \dfrac{10(10+1)}{2} + m = \sum_i \dfrac{i}{\lambda + \mu i}$ *i.e.,* $\sum_i \dfrac{i}{\lambda + \mu i} = 55 + m$, we get the probability distribution in the form of a table and also get the values of S_{max} as described by Kapur and Kesavan.

There may be two cases,

Case (i) when $m < \dfrac{n+1}{2} = 5.5$.

In this case, when m lies between 1 and 5.5, μ>0 implies $\dfrac{dS_{max}}{dm} > 0$ and then S_{max} is increasing.

Case (ii) when $m > \dfrac{n+1}{2} = 5.5$.

In this case, when m lies between 5.5 and 10, μ<0 implies $\dfrac{dS_{max}}{dm} < 0$ and then S_{max} is decreasing.

When $m = \dfrac{n+1}{2} = 5.5, \mu = 0, \dfrac{dS_{max}}{dm} = 0$.

it can be shown that S_{max} will be concave if we prescribe E(g(x)) instead of E(x), where g(x) is a monotonic increasing function of x; then we will apply the necessary changes. Again, since the concavity of S_{max} has already been proven, this will enable us to handle the inequality constraint of the type $m_1 \leqslant \sum_{i=1}^{n} i p_i \leqslant m_2$.

Simulated Results

Following Table (Table 2) Found Using LINGO Software 2011 where Different Max-Entropy Values Are Given for Different m Values:

Table 2. Comparative Maximum Entropy Values

m	Maximum Entropy Value of $B_{mod}(P)$	Maximum Entropy Value of B(P)	Maximum Entropy Value of S(P)
1.00	0	−158.225	0
1.25	0.0896122	−45.13505	0.6255029
1.50	0.1345336	−38.98478	0.9547543
1.75	0.1633092	−35.43113	1.194875
2.00	0.1836655	−32.94620	1.385892
2.25	0.1989097	−31.05219	1.542705
2.50	0.2107459	−29.53735	1.675885
2.75	0.2201939	−28.28982	1.790029
3.00	0.2279397	−27.24394	1.888477
3.25	0.2344146	−26.35851	1.973503
3.50	0.2399041	−25.60649	2.046725
3.75	0.24461	−24.96958	2.109324
4.00	0.248682	−24.43512	2.162186
4.25	0.252127	−23.99419	2.205980
4.50	0.2549453	−23.64047	2.241209

4.75	0.257137	−23.36942	2.268253
5.00	0.2587024	−23.17789	2.287386
5.25	0.2596416	−23.06376	2.298794
5.50	**0.2599546**	**−23.02585**	**2.302585**
5.75	0.2596416	−23.06376	2.298794
6.00	0.2587024	−23.17789	2.287386
6.25	0.257137	−23.36942	2.268253
6.50	0.2549453	−23.64047	2.241209
6.75	0.252127	−23.99419	2.205980
7.00	0.248682	−24.43512	2.162186
7.25	0.24461	−24.96958	2.109324
7.50	0.2399041	−25.60649	2.046725
7.75	0.2344146	−26.35851	1.973503
8.00	0.2279397	−27.24394	1.888477
8.25	0.2201939	−28.28982	1.790029
8.50	0.2107459	−29.53735	1.675885
8.75	0.1989097	−31.05219	1.542705
9.00	0.1836655	−32.94620	1.385892
9.25	0.1633092	−35.43113	1.194875
9.50	0.1345336	−38.98478	0.9547543
9.75	0.0896122	−45.13505	0.6255029
10.0	0	−141.25285	0

Graphs obtained from the above table are given on the next page (Figure 1, Figure 2 and Figure 3):

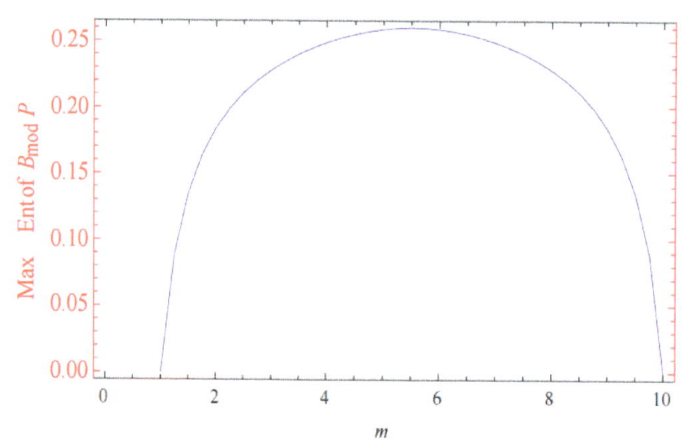

Figure 1. Maximum Burg's modified entropy.

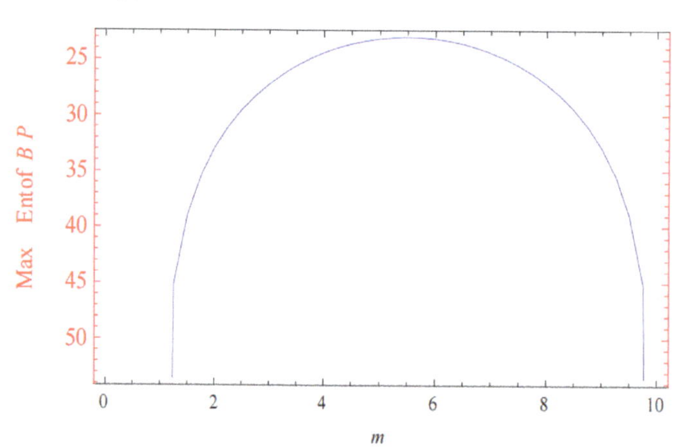

Figure 2. Maximum Burg's entropy.

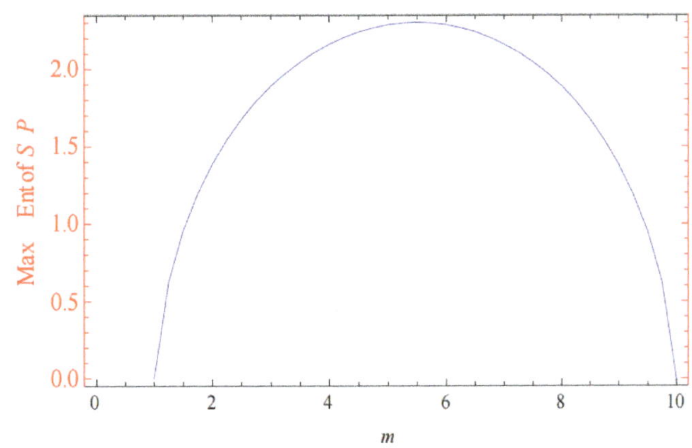

Figure 3. Maximum Shannon's entropy.

If p_i is plotted against i, we get rectangular hyperbolic types of curves (Table 3 and Figure 4).

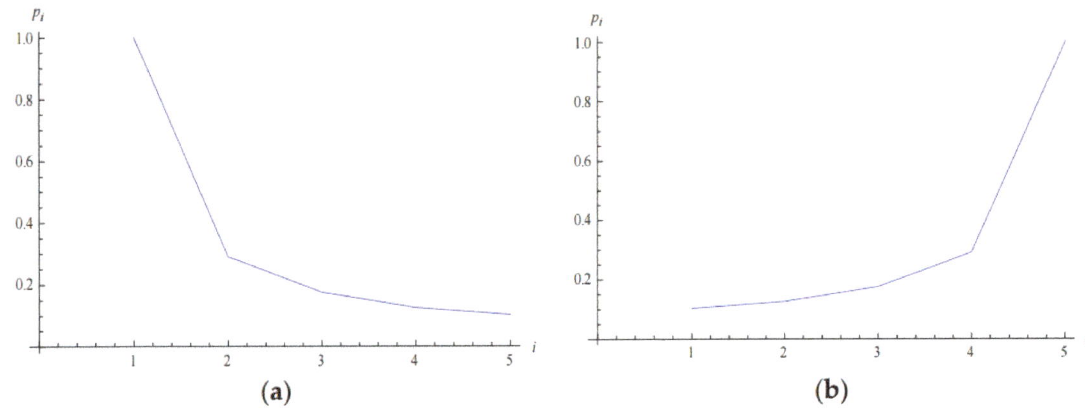

Figure 4. Rectangular hyperbolic types of graphs. (**a**) when $m < \frac{n+1}{2}, \mu > 0$; (**b**) when $m > \frac{n+1}{2}, \mu < 0$.

Table 3. p_i and i graph

i	p_i
1	1.000000
2	0.2899575
3	0.1747893
4	0.1253783
5	0.1027619
6	0.1027620
7	0.1253782
8	0.1747891
9	0.2899574
10	1.000000

CONCLUSIONS

In the present paper we have presented different MEPDs and respective entropy measures with their properties. It has been found that MBE is a better measure than Burg's entropy when the maximized subject to the mean is prescribed, and it also has been shown that unlike Burg's entropy, the maximum value of MBE increases with n. The main problem here will consist of solving z+1 simultaneous transcendental equations for the Lagranges multipliers. An application in statistical mechanics with simulated data has been studied with the help of Lingo11 software and corresponding graphs are provided. Now, one question arises: Will this result continue to hold for other moment constraints also? When we take generalized moment expectation of g(x) instead of expectation of x, then g(x) must be a monotonic increasing function of x, and if pibecomes negative for some values of the moments, then we have to set those probabilities to zero and reformulate the problem for the remaining probabilities over the remaining range and solve it.

ACKNOWLEDGMENTS

Authors would like to express their sincere gratitude and appreciation to the reviewers for their valuable comments and good suggestions to make this paper enriched and impactful, and a special thank you to the Assistant Editor.

AUTHOR CONTRIBUTIONS

A. Ray and S. K. Majumder conceived and designed the experiments; A. Ray performed the experiments, analyzed the data and both A. Ray and S. K. Majumder contributed in analysis tools; A. Ray wrote the paper.

REFERENCES

1. Andreas, G.; Keller, G.; Warnecke, G. *Entropy*; Princeton University Press: Princeton, NJ, USA, 2003.
2. Phil, A. *Thermodynamics and Statistical Mechanics: Equilibrium by Entropy Maximisation*; Academic Press: Cambridge, MA, USA, 2002.
3. Rudolf, C. Ueber verschiedene für die Anwendung bequeme Formen der Hauptgleichungen der mechanischen Wärmetheorie. *Annalen der Physik* **1865**. (In German).
4. Wu, J. Three Factors Causing the Thermal Efficiency of a Heat Engine to Be Less than Unity and Their Relevance to Daily Life. *Eur. J. Phys.* **2015**, *36*, 015008.
5. Rashidi, M.M.; Shamekhi, L. Entropy Generation Analysis of the Revised Cheng-Minkowycz Problem for Natural Convective Boundary Layer Flow of Nanofluid in a Porous Medium. *J. Thermal Sci.* **2015**, *19*.
6. Rashidi, M.M.; Mohammadi, F. Entropy Generation Analysis for Stagnation Point Flow in a Porous Medium over a Permeable Stretching Surface. *J. Appl. Fluid Mech.* **2014**, *8*, 753–763.
7. Rashidi, M.M.; Mahmud, S. Analysis of Entropy Generation in an MHD Flow over a Rotating Porous Disk with Variable Physical Properties. *Int. J. Energy* **2014**.
8. Abolbashari, M.H.; Freidoonimehr, N. Analytical Modeling of Entropy Generation for Casson Nano-Fluid Flow Induced by a Stretching Surface. *Adv. Powder Technol.* **2015**, *231*.
9. Baag, S.S.R.; Dash, M.G.C.; Acharya, M.R. Entropy Generation Analysis for Viscoelastic MHD Flow over a Stretching Sheet Embedded in a Porous Medium. *Ain Shams Eng. J.* **2016**, *23*.
10. Shi, Z.; Tao, D. Entropy Generation and Optimization of Laminar Convective Heat Transfer and Fluid Flow in a Microchannel with Staggered Arrays of Pin Fin Structure with Tip Clearance. *Energy Convers. Manag.* **2015**, *94*, 493–504.
11. Hossein, A.M.; Ahmadi, M.A.; Mellit, A.; Pourfayaz, F.; Feidt, M. Thermodynamic Analysis and Multi Objective Optimization of Performance of Solar Dish Stirling Engine by the Centrality of Entransy and Entropy Generation. *Int. J. Electr. Power Energy Syst.* **2016**, *78*, 88–95.
12. Giovanni, G. *Statistical Mechanics: A Short Treatise*; Springer Science & Business Media: New York, NY, USA, 2013.
13. Reif, F. *Fundamentals of Statistical and Thermal Physics*; Waveland Press: Long Grove, IL, USA, 2009.
14. Rudolf, C.; Shimony, A. *Two Essays on Entropy*; University of California Press: Berkeley, CA, USA, 1977.
15. Shu-Cherng, F.; Rajasekera, J.R.; Tsao, H.S.J. *Entropy Optimization and Mathematical Programming*; Springer Science & Business Media: New York, NY, USA, 2012.
16. Robert, M.G. *Entropy and Information Theory*; Springer Science & Business Media: New York, NY, USA, 2013.
17. Silviu, G. *Information Theory with New Applications*; MacGraw-Hill Books Company: New York, NY, USA, 1977.
18. Shannon, C.E. A Mathematical Theory of Communication. *SigmobileMob. Comput. Commun. Rev.* **2001**, *5*, 3–55.
19. Thomas, M.C.; Thomas, J.A. *Elements of Information Theory*; John Wiley & Sons: Hoboken, NJ, USA, 2012.
20. Christodoulos, A.F.; Pardalos, P.M. *Encyclopedia of Optimization*; Springer Science & Business Media: New York, NY, USA, 2008.
21. Arash, M.; Azarnivand, A. Application of Integrated Shannon's Entropy and VIKOR Techniques in Prioritization of Flood Risk in the Shemshak Watershed, Iran. *Water Resour. Manag.* **2015**, *30*, 409–425.
22. Liu, L.; Miao, S.; Liu, B. On Nonlinear Complexity and Shannon's Entropy of Finite Length Random Sequences. *Entropy* **2015**, *17*, 1936–1945.
23. Karmeshu. *Entropy Measures, Maximum Entropy Principle and Emerging Applications*; Springer: Berlin, Germany; Heidelberg, Germany, 2012.
24. Jaynes, E.T. Information Theory and Statistical Mechanics. II. *Phys. Rev.* **1957**, *108*, 171–190.
25. Jaynes, E.T. On the Rationale of Maximum-Entropy Methods. *Proc. IEEE* **1982**, *70*, 939–952.
26. Jaynes, E.T. Prior Probabilities. *IEEE Trans. Syst. Sci. Cybernet.* **1968**, *4*, 227–241.
27. Kapur, J.N. *Maximum-Entropy Models in Science and Engineering*; John Wiley & Sons: New York, NY, USA, 1989.
28. Kapur, J.N. Maximum-Entropy Probability Distribution for a Continuous Random Variate over a Finite Interval. *J. Math. Phys. Sci.* **1982**, *16*, 693–714.
29. Ray, A.; Majumder, S.K. Concavity of maximum entropy through modified Burg's entropy subject to its prescribed mean. *Int. J. Math. Oper. Res.* **2016**, *8*. to appear.
30. Kapur, J.N. *Measures of Information and Their Applications*; Wiley: New York, NY, USA, 1994.
31. Burg, J. The Relationship between Maximum Entropy Spectra and Maximum Likelihood Spectra. *Geophysics* **1972**, *37*, 375–376.
32. Narain, K.J.; Kesavan, H.K. *Entropy Optimization Principles with Applications*; Academic Press: Cambridge, MA, USA, 1992.
33. Solomon, K. *Information Theory and Statistics*; Courier Corporation: North Chelmsford, MA, USA, 2012.

34. Amritansu, R.; Majumder, S.K. Derivation of some new distributions in statistical mechanics using maximum entropy approach. *Yugoslav J. Oper. Res.* **2013**, *24*, 145–155.

35. Ulrych, T.J.; Bishop, T.N. Maximum Entropy Spectral Analysis and Autoregressive Decomposition. *Rev. Geophys.* **1975**,*13*, 183–200.

36. Michele, P.; Ferrante, A. On the Geometry of Maximum Entropy Problems. **2011**. arXiv:1112.5529.

37. Ke, J.-C.; Lin, C.-H. Maximum Entropy Approach to Machine Repair Problem. *Int. J. Serv. Oper. Inform.* **2010**, *5*, 197–208.

CHAPTER 11 ENTANGLEMENT ENTROPIES OF NON-EQUILIBRIUM FINITE-SPIN SYSTEMS

Koichi Nakagawa

Recently, a new approach to understanding quantum entanglement using thermofield dynamics (TFD) [1–3] has been proposed in Ref. [4]. In this new treatment of quantum entanglement with TFD, extended density matrices have been formulated on the double Hilbert space (ordinary and tilde Hilbert spaces), and the entanglement states show a quantum-mechanically complicated behavior. The new TFD-based method allows the entanglement states to be easily understood, because the states in the TFD tilde space play the role of tracers of the initial states. In the new analysis, a general formulation of the extended density matrices has been constructed and applied to some simple cases. Consequently, it has been found that intrinsic quantum entanglement can be distinguished from the thermal fluctuations included in the definition of ordinary quantum entanglement at finite temperatures. Based on the analysis presented in Ref. [4], it has been argued that the new TFD-based method is applicable not only to equilibrium states but also to non-equilibrium states. However, analysis of the entanglement entropies of non-equilibrium systems was not conducted in Ref. [4] and, therefore, examination of the entanglement entropies of non-equilibrium systems with the use of TFD is of current interest. In the present communication, therefore, the "extended" entanglement entropies of non-equilibrium spin systems are intensively investigated in both the dissipative and non-dissipative cases, based upon a TFD algorithm.

Let us consider the S=1/2S=1/2 spin system described by the Hamiltonian

$$\mathcal{H} = -J\boldsymbol{S}_{\mathrm{A}} \cdot \boldsymbol{S}_{\mathrm{B}}, \tag{1}$$

incorporating the spin operators $\boldsymbol{S}_{\mathrm{A}} = (S_{\mathrm{A}}^{x}, S_{\mathrm{A}}^{y}, S_{\mathrm{A}}^{z})$ and $\boldsymbol{S}_{\mathrm{B}} = (S_{\mathrm{B}}^{x}, S_{\mathrm{B}}^{y}, S_{\mathrm{B}}^{z})$ of the subsystems A and B, respectively. The state, $|s\rangle$, of the total system is then denoted by the direct product, $|s\rangle=|s_A,s_B\rangle=|s_A\rangle|s_B\rangle$. Using the base $\{|{++}\rangle,|{+-}\rangle,|{-+}\rangle,|{--}\rangle\}$ the matrix form of the Hamiltonian (1) is then expressed as

$$\begin{aligned}\mathcal{H} &= \sum_{s_{\mathrm{A}},s_{\mathrm{B}},s'_{\mathrm{A}},s'_{\mathrm{B}}} h_{s_{\mathrm{A}},s_{\mathrm{B}},s'_{\mathrm{A}},s'_{\mathrm{B}}} \left|s_{\mathrm{A}}, s_{\mathrm{B}}\right\rangle \left\langle s'_{\mathrm{A}}, s'_{\mathrm{B}}\right| \\ &= -\frac{J}{4}\left(|++\rangle\langle++| + |--\rangle\langle--|\right) + \frac{J}{4}\left(|+-\rangle\langle+-| + |-+\rangle\langle-+|\right) \\ &\quad - \frac{J}{2}\left(|+-\rangle\langle-+| + |-+\rangle\langle+-|\right).\end{aligned} \tag{2}$$

For the equilibrium states in terms of the Hamiltonian expressed in Eq. (2), the ordinary density matrix, ρ_{eq}, of this system can be obtained as

$$\begin{aligned}\rho_{\mathrm{eq}} := \frac{e^{-\beta\mathcal{H}}}{Z(\beta)} = \frac{e^{-\beta J/4}}{Z(\beta)} \Big(& e^{\beta J/2}\left(|++\rangle\langle++| + |--\rangle\langle--|\right) \\ & + \cosh\frac{\beta J}{2}\left(|+-\rangle\langle+-| + |-+\rangle\langle-+|\right) \\ & + \sinh\frac{\beta J}{2}\left(|-+\rangle\langle+-| + |+-\rangle\langle-+|\right)\Big),\end{aligned} \tag{3}$$

where β is the inverse temperature and the partition function, Z(β), is defined as

$$Z(\beta) := \mathrm{Tr} e^{-\beta\mathcal{H}} = 2e^{-\beta J/4}\left(e^{\beta J/2} + \cosh\frac{\beta J}{2}\right). \tag{4}$$

Let us turn our attention to a non-equilibrium system with dissipation, which is described by the Hamiltonian of Eq. (2). The time dependence of the ordinary density matrix, ρ(t), of this system is given by the dissipative von Neumann equation [5–7], where

Citation: Koichi Nakagawa, "Entanglement entropies of non-equilibrium finite-spin systems," DOI: 10.1093/ptep/ptv005.

$$i\hbar\frac{\partial}{\partial t}\rho(t) = \left[\mathcal{H}, \rho(t)\right] - \epsilon\left(\rho(t) - \rho_{\rm eq}\right), \tag{5}$$

with □ being a dissipation parameter. The solution of Eq. (5) is expressed as

$$\rho(t) = e^{-\epsilon t}U^{\dagger}(t)\rho_0 U(t) + \left(1 - e^{-\epsilon t}\right)\rho_{\rm eq}, \tag{6}$$

for any initial density matrix, ρ_0, where the unitary operator, U(t), denotes

$$\begin{aligned} U(t) := e^{i\mathcal{H}t/\hbar} = e^{i\omega t/4}\Big(& e^{-i\omega t/2}\left(|++\rangle\langle++| + |--\rangle\langle--|\right) \\ &+ \cos\frac{\omega t}{2}\left(|+-\rangle\langle+-| + |-+\rangle\langle-+|\right) \\ &- i\sin\frac{\omega t}{2}\left(|-+\rangle\langle+-| + |+-\rangle\langle-+|\right)\Big), \end{aligned} \tag{7}$$

and $\omega:=J/\hbar$. Because the explicit expression of $\rho(t)$ in Eq. (6) is complicated for any initial condition, hereafter, let us confine ourselves to the initial condition $\rho_0=|+-\rangle\langle+-|$. Inserting Eqs. (3) and (7), along with the initial condition, into Eq. (6), we then obtain

$$\begin{aligned} \rho(t) = \frac{e^{-\epsilon t}}{2}\Bigg(& \frac{2\left(e^{\epsilon t}-1\right)}{3+e^{-\beta\omega\hbar}}\left(|++\rangle\langle++| + |--\rangle\langle--|\right) \\ &+ \frac{\cos\omega t + e^{\epsilon t} + e^{\beta\omega\hbar}\left(3\cos\omega t + e^{\epsilon t} + 2\right)}{1+3e^{\beta\omega\hbar}}|+-\rangle\langle+-| \\ &+ \frac{-\cos\omega t + e^{\epsilon t} + e^{\beta\omega\hbar}\left(-3\cos\omega t + e^{\epsilon t} + 2\right)}{1+3e^{\beta\omega\hbar}}|-+\rangle\langle-+| \\ &+ \left(\frac{\left(e^{\epsilon t}-1\right)\left(-1+e^{\beta\omega\hbar}\right)}{1+3e^{\beta\omega\hbar}} - i\sin\omega t\right)|+-\rangle\langle-+| \\ &+ \left(\frac{\left(e^{\epsilon t}-1\right)\left(-1+e^{\beta\omega\hbar}\right)}{1+3e^{\beta\omega\hbar}} + i\sin\omega t\right)|-+\rangle\langle+-|\Bigg). \end{aligned} \tag{8}$$

The ordinary entanglement entropy, S, is defined by

$$S := -k_{\rm B}{\rm Tr}_{\rm A}\left[\rho_{\rm A}\log\rho_{\rm A}\right], \tag{9}$$

with $\rho A := {\rm Tr}B\rho(t)$ where TrA and TrB represent traces over the variables of subsystems A and B, respectively. The insertion of Eq. (8) into Eq. (9) yields

$$S = -k_{\rm B}\left(\frac{1+e^{-\epsilon t}\cos\omega t}{2}\log\frac{1+e^{-\epsilon t}\cos\omega t}{2} + \frac{1-e^{-\epsilon t}\cos\omega t}{2}\log\frac{1-e^{-\epsilon t}\cos\omega t}{2}\right). \tag{10}$$

It is also possible to argue that SS in Eq. (10) is directly proportional to an entanglement, E(C), which is a function of the "concurrence", $C := \sqrt{1-e^{-2\epsilon t}\cos^2\omega t}$ [8]. The time dependence of S and C is displayed in Fig. 1 (in units of $k_B=1$). In the dissipative system, S and C converge to the constants $k_B\log 2$ and 1, respectively, at $t\rightarrow\infty$, so it is reasonable to think that S and C include not only the contribution of the quantum fluctuation, but also the contribution of the classical and thermal fluctuations. However, this fact is not manifest in the above expressions of S and C.

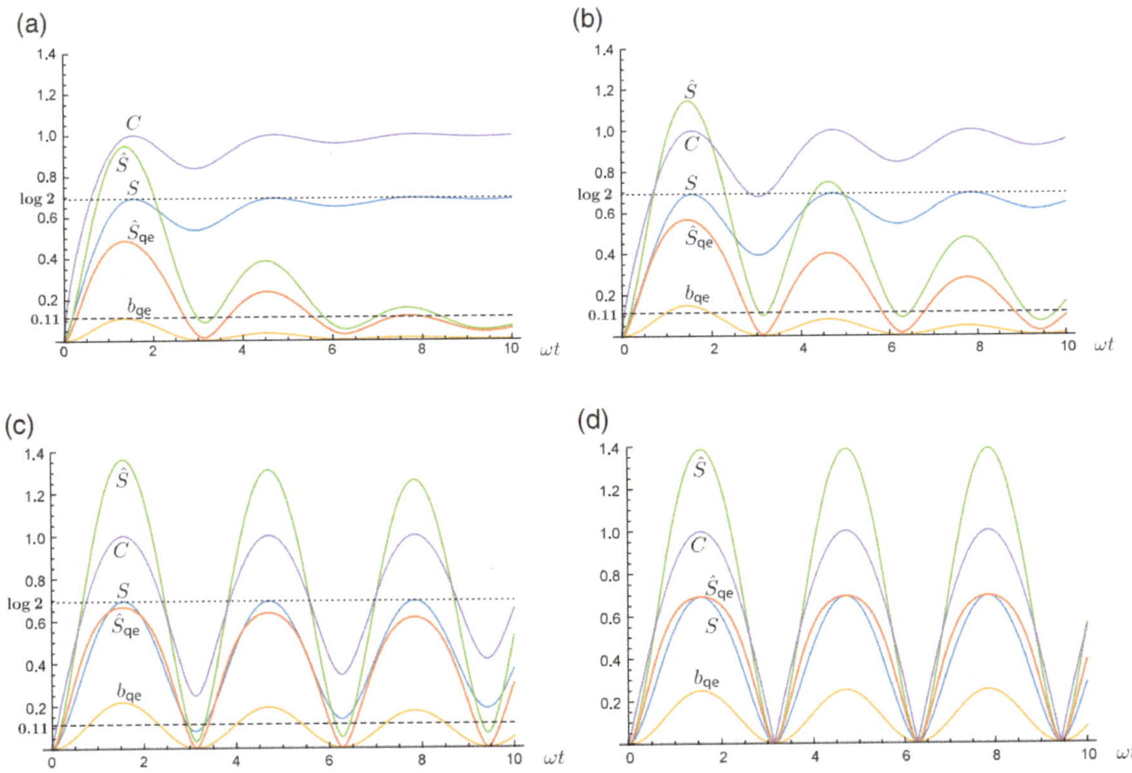

Figure 1. Time dependence of entropies C,S, $\hat{S}$, and $\hat{S}_{qe}$, along with parameter b_{qe}, in dissipative and non-dissipative systems with scaled temperature, T/J=0.7. Parts (a), (b), and (c) show cases with a scaled dissipation rate of ϵ/ω=0.2,0.1, and 0.01, respectively. The dotted and dashed lines in parts (a), (b), and (c) represent the asymptotes of the S and $\hat{S}$ curves, respectively. Part (d) is the non-dissipation case (ϵ=0).

We are now in a position to investigate the extended density matrix, $\hat{\rho}$, in the TFD double Hilbert space. Note that ρ^ρ^ has been defined in Ref. [4] as follows:

$$\hat{\rho} := |\Psi\rangle\langle\Psi|,\ |\Psi\rangle := \rho(t)^{1/2}\sum_{s}|s,\tilde{s}\rangle = \rho(t)^{1/2}\sum_{s}|s\rangle|\tilde{s}\rangle, \tag{11}$$

using the ordinary density matrix, ρ(t), in Eq. (6), where {|s⟩} is the orthogonal complete set in the original Hilbert space and {|s~⟩} is the same set in the tilde Hilbert space of the TFD [9, 10]. If entanglement subsystems A and B are being examined, each of the |s⟩ and |s~⟩ states are represented as the direct products |sA,sB⟩=|sA⟩|sB⟩ and |s~A,s~B⟩=|s~A⟩|s~B⟩, respectively. We are then led to the renormalized extended density matrix, $\hat{\rho}_A$, as

$$\begin{aligned}\hat{\rho}_A &:= \mathrm{Tr}_B\hat{\rho} := \sum_{s_B,\tilde{s}'_B}\langle s_B,\tilde{s}'_B|\hat{\rho}|s_B,\tilde{s}'_B\rangle\\ &= b_{d1}|+\rangle\langle+||\tilde{+}\rangle\langle\tilde{+}| + b_{d2}|-\rangle\langle-||\tilde{-}\rangle\langle\tilde{-}| + b_{cf}\left(|+\rangle\langle-||\tilde{+}\rangle\langle\tilde{-}| + |-\rangle\langle+||\tilde{-}\rangle\langle\tilde{+}|\right)\\ &\quad + b_{qe}\left(|+\rangle\langle+||\tilde{-}\rangle\langle\tilde{-}| + |-\rangle\langle-||\tilde{+}\rangle\langle\tilde{+}|\right),\end{aligned} \tag{12}$$

where the matrix elements b_{d1}, b_{d2}, b_{cf}, and b_{qe} are respectively obtained as analytic functions of t, β, ϵ, and ω, and correspond to the two diagonal components (d1 and d2), the classical fluctuations (c_f), and the quantum entanglements (qe) of $\hat{\rho}_A$, respectively. The parameter b_{qe} in Eq. (12) expresses the quantum entanglement effect. This quantum fluctuation plays a crucial role in various quantum systems, and it has been used as an order parameter of 2D quantum systems [11, 12]. The time dependences of the parameter b_{qe} in several cases are shown in Fig. 1. As can be seen from Eq. (12), only the intrinsic quantum entanglement is extracted clearly in the TFD formulation. In particular, it can be understood that the entangled state of the system emerges through a single product, such as |+⟩⟨+||−~⟩ ⟨−~|, in $\hat{\rho}_A$.

The "extended" entanglement entropy is defined as

$$\hat{S} := -k_B\mathrm{Tr}_A\left[\hat{\rho}_A\log\hat{\rho}_A\right], \tag{13}$$

using the renormalized $\hat{\rho}_A$ in Eq. (12) [4]. The insertion of Eq. (12) into Eq. (13) and subsequent simplification eventually yield

$$\hat{S} = \hat{S}_{\mathrm{cl}} + \hat{S}_{\mathrm{qe}}, \tag{14}$$

where

$$\hat{S}_{\mathrm{cl}} := -k_{\mathrm{B}}\left(\sqrt{4b_{\mathrm{cf}}^2 + (b_{\mathrm{d}1} - b_{\mathrm{d}2})^2}\ \operatorname{arccoth}\frac{b_{\mathrm{d}1} + b_{\mathrm{d}2}}{\sqrt{4b_{\mathrm{cf}}^2 + (b_{\mathrm{d}1} - b_{\mathrm{d}2})^2}} + \frac{b_{\mathrm{d}1} + b_{\mathrm{d}2}}{2}\log\left(b_{\mathrm{d}1}b_{\mathrm{d}2} - b_{\mathrm{cf}}^2\right)\right), \tag{15}$$

and

$$\hat{S}_{\mathrm{qe}} := -2k_{\mathrm{B}}b_{\mathrm{qe}}\log b_{\mathrm{qe}}, \tag{16}$$

respectively. In Eqs. (14), (15), and (16), the expressions of S^, the classical and thermal fluctuation parts, $\hat{S}_{\mathrm{cl}}$, and the quantum entanglement part, $\hat{S}_{\mathrm{qe}}$, also incorporate analytic functions of t, β, ϵ, and ω, respectively; however, the full calculation is quite tedious. So, we show the numerical behavior of C,S, $\hat{S}$, and $\hat{S}_{\mathrm{qe}}$, and b_{qe} for a few cases in Figs. 1(a)–(c) (in units of k_B=1). As can be seen from these figures, at t→∞, $\hat{S}$ converges to the value 0.11351···, and both $\hat{S}_{\mathrm{qe}}$ and b_{qe} vanish, respectively. As a consequence, the traditional entanglement entropy, S, becomes larger than the extended entanglement entropies, $\hat{S}$ and $\hat{S}_{\mathrm{qe}}$, at t→∞. $\hat{S}_{\mathrm{qe}}$ is then smaller than S when ϵ is relatively larger. As ϵ becomes smaller, $\hat{S}_{\mathrm{qe}}$ becomes compatible with S and, at ϵ=0, $\hat{S}_{\mathrm{qe}} \lesssim S$. These results suggest that the quantum entanglement is enhanced as the dissipation becomes weaker.

For non-dissipative systems, $\hat{S}$ in Eq. (14) and $\hat{S}_{\mathrm{qe}}$ in Eq. (16) reduce to

$$\hat{S} = -k_{\mathrm{B}}\left(\cos^4\frac{\omega t}{2}\cdot\log\left(\cos^4\frac{\omega t}{2}\right) + \sin^4\frac{\omega t}{2}\cdot\log\left(\sin^4\frac{\omega t}{2}\right) + \frac{1}{2}\sin^2\omega t\cdot\log\left(\frac{\sin^2\omega t}{4}\right)\right) \tag{17}$$

and

$$\hat{S}_{\mathrm{qe}} = -\frac{k_{\mathrm{B}}}{2}\sin^2\omega t\cdot\log\left(\frac{\sin^2\omega t}{4}\right), \tag{18}$$

respectively, at ϵ=0. The time dependence of $\hat{S}$ and $\hat{S}_{\mathrm{qe}}$ at ϵ=0 is shown in Fig. 1(d) (in units of k_B=1). It is apparent in this figure that all the curves (C, S, $\hat{S}$, $\hat{S}_{\mathrm{qe}}$, and b_{qe}) showing the entanglement have the same phase; however, their amplitudes differ. Specifically, $\hat{S}$ is larger than S and $\hat{S}_{\mathrm{qe}}$≈S at ϵ=0, a result that differs from that of Ref. [4] and that can be seen in Eqs. (17) and (18). It appears that a mistake was made in Ref. [4] in counting the non-zero eigenvalues of $\hat{\rho}_{\mathrm{A}}$.

In this communication, we have examined the extended entanglement entropies of non-equilibrium spin systems in both the dissipative and non-dissipative cases, based upon the TFD formulation. In the dissipative case in particular, the extended entanglement entropy is derived using the extended density matrix and is proven to separate into the classical and thermal fluctuation parts and the quantum entanglement part. These quantities are compared to the traditional entanglement entropy, the concurrence, and b_{qe} in $\hat{\rho}_{\mathrm{A}}$. These results are summarized in Fig. 1 and show that the conditions yielding the maximum entangled state can be obtained using these five quantities.

We have clearly indicated that, in the TFD formulation, the extended quantum entanglement entropy part and the parameter b_{qe} are recognized as effective quantities for measurement of the quantum entanglement. It is apparent that the new TFD-based method enables us to clearly distinguish between the various states of quantum systems.

REFERENCES

1. U. Fano, Mod. Phys. A, 42, 74 (1957).
2. I. Prigogine, C. George, F. Henin, and L. Rosenfeld, Chem. Scripta, 4, 5 (1973).
3. Y. Takahashi and H. Umezawa, Collect. Phenom., 2, 55 (1975).
4. Y. Hashizume and M. Suzuki, Physica A, 392, 3518 (2013).
5. M. Suzuki, In Quantum bio-informatics II: from quantum information to bio-informatics, Tokyo University of Science, Japan, 12–16 March 2008 (QP-PQ: Quantum probability and white noise analysis; v. 24), eds. L. Accardi, W. Freundberg, and M. Ohya (World Scientific, Singapore, 2008).
6. M. Suzuki, Physica A, 390, 1904 (2011).
7. M. Suzuki, Physica A, 391, 1074 (2012).
8. W. K. Wootters, Phys. Rev. Lett., 80, 2245 (1998).
9. M. Suzuki, J. Phys. Soc. Jpn., 54, 4483 (1985).
10. M. Suzuki, Statistical Mechanics (Iwanami, Tokyo, 2000).
11. J.-M. Stephan, S. Furukawa, G. Misguich, and V. Pasquier, Phys. Rev. B, 80, 184421 (2009).
12. S. Tanaka, R. Tamura, and H. Katsura, Phys. Rev. A, 86, 032326 (2012).

CHAPTER 12

INTERFACIAL MASS TRANSFER DURING GAS–LIQUID PHASE CHANGE IN DEFORMABLE POROUS MEDIA WITH HEAT TRANSFER

Wolfgang Ehlers, Kai Häberle

Keywords: Phase transition, Thermoelasticity, Theory of Porous Media, Singular surface, Interfacial areas

INTRODUCTION

Transitions between gas, liquid, and solid phases of a certain substance are important physical processes. Such processes, for example drying or freezing, do not occur only in well-investigated "open systems" but also in porous media. However, phase changes in the latter are only scarcely investigated but of great importance, for example in geomechanics (CO_2 sequestration, steam injection for enhanced oil recovery, or soil remediation), in food industries (drying and baking processes), or in other areas. In the present contribution, the focus is laid on "gas-into-liquid" or "liquid-into-gas" transitions of a single substance. This has to be clearly distinguished from phase changes between mixtures of different substances such as water evaporating into air, which are not part of this work.

The treatment of phase change processes in a continuum description by introducing a singular front (interface) has already been tackled by different groups, compare, for example, the work by Jamet (2014), Juric and Tryggvason (1998), Morland and Gray (1995), Morland and Sellers (2001), Tanguy et al. (2007), and Wang and Oberlack (2011). Most of these contributions use the level-set method to describe the moving interface inside the continuum.

Regarding phase transitions in the pore space of a porous aggregate, the earliest work to the authors' knowledge goes back to Lykov (1974). Further articles in this direction have been presented by the group around Bénet (Lozano et al. 2009; Ruiz and Bénet 2001; Chammari et al. 2005; Lozano et al. 2008), by Hassanizadeh and Gray (Gray 1983; Hassanizadeh and Gray 1990; Niessner and Hassanizadeh 2009a, b), or by Bedeaux and Kjelstrup (1999). Examples of applying these models to actual physical problems are the simulation of drying processes by Kowalski (2000) or the bread-baking problem by Huang et al. (2006).

The previously mentioned articles consider phase-change processes in porous media, where the solid matrix is usually idealised as a rigid body. Including solid deformations into the model, we proceed from the well-founded concept of the Theory of Porous Media (TPM). This approach provides an ideal framework for multiphasic and multicomponent continua including arbitrary solid deformations based on elasticity, viscoelasticity, or elasto-plasticity, as well as an arbitrary pore content of either miscible or immiscible fluids, liquids and gases. The reader who is interested in the basics of the TPM is referred, for example, to the publications of de Boer (2000), de Boer and Ehlers (1986), Bowen (1980, 1982), Ehlers (1991, 1989, 2002), Ehlers et al. (2004), Wieners et al. (2005), Ehlers and Graf (2007), Ehlers (2009) or Schrefler and Zhan (1993), Schrefler and Scotta (2001), and citations therein.

The development of a thermodynamically consistent description of phase-change processes in porous media based on the necessity of satisfying the requirements of the entropy inequality[1] of the TPM starts with contributions by de Boer (1995), de Boer and Bluhm (1999), de Boer and Kowalski (1995) and continues with an article by Ehlers and Graf (2007). These articles have in common that they provide a derivation of the mass production term, which describes the mass transfer from one phase to another, but do not present any numerical example exhibiting the impact of this term during the phase-transition process. Furthermore, no comments are

Citation: Wolfgang Ehlers and Kai Häberle, "Interfacial Mass Transfer During Gas–Liquid Phase Change in Deformable Porous Media with Heat Transfer," DOI 10.1007/s11242-016-0674-2.

made on how to determine the mass-transition coefficients. The model considered in this contribution is a partially saturated solid as was discussed as a triphasic material by Ehlers (2009). In the present article, the model consists of a materially incompressible, thermoelastic porous solid together with liquid and gaseous phases of a certain fluid, such as water (H_2O) or carbon dioxide (CO_2). The fluid phases are considered compressible and are described by an appropriate equation of state, where, in the present article, use is made of the *van der Waals* equation (1873) together with the *Antoine* equation (1888) for the vaporization–condensation line. Extensions of this model towards the simultaneous existence of two pore fluids, such as H_2O and CO_2, including mixing processes are possible and are planned for a follow-up publication.

Phase changes introduce a discontinuity in the density and mass transfer between the two coexisting phases. This induces so-called production terms in the mass, momentum, and energy balance relations. To determine these terms, the interface between the gaseous and the liquid phases is described by a separating, immaterial, smooth surface together with certain jump conditions for the description of the phase transition as a discontinuity in the fluid density. Therewith, the phase transition is considered on the microscale by a jump over the immaterial discontinuous surface, which is then averaged over the volume element to find a constitutive relation for the mass production term in the mass balance equations. This procedure includes the concept of interfacial areas, compare, for example, Joekar-Niasar et al. (2008) and Sahimi (2011) and others.

BASIC SETTING AND GOVERNING EQUATIONS

It is the goal of the present contribution to describe the transition between the gaseous and the liquid phases of a single substance in the pore space of a deformable porous solid, demonstrated by the condensation/evaporation problem of a pore fluid component due to cooling/heating, cf. Figure 1. At the microstructure of the pore scale, locally taken as Representative Elementary Volume (REV), this process is characterized by a mass transfer over local interfaces Γ separating the gas from the liquid, while the solid remains continuous over Γ, since it is not affected by the fluid mass exchange. For the description of this process, the article concerns a bicomponent, triphasic aggregate of a thermoelastic porous solid and a fluid component such as CO_2, where the three phases are given by the porous solid φ^S together with the liquid phase φ^L and the gaseous phase φ^G of the overall fluid matter φ^{FM}.

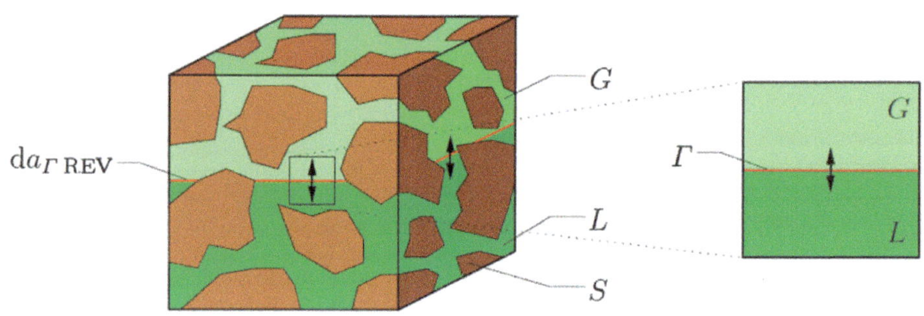

Figure 1: Sketch of a porous microstructure filled with gaseous and liquid phases of a certain substance. Mass can be transferred from liquid to gas or vice versa across the interface ΓΓ, locally denoted by $da_{\Gamma REV}$.

Consider a triphasic aggregate $\mathcal{B} = \bigcup_\alpha \mathcal{B}^\alpha$ with boundary surface S, where α={S,L,G}. By introducing a separating and immaterial smooth and local surface indicating the interface Γ, a body B can locally be separated into two parts given by B^+ and B^-, which represent the porous solid saturated by either the pore gas (B^+) or by the pore liquid (B−), respectively, cf. Figure 2. Locally, the body itself and its total surface are then given by B=B+∪B− and S=S+∪S−, whereas S±∪Γ yields the entire surface of the body parts B±. Mass can be transferred across Γ by $\hat{\varrho}^G_\Gamma$ from B^+ to B^- or, vice versa, by $\hat{\varrho}^L_\Gamma$ from B− to B+.

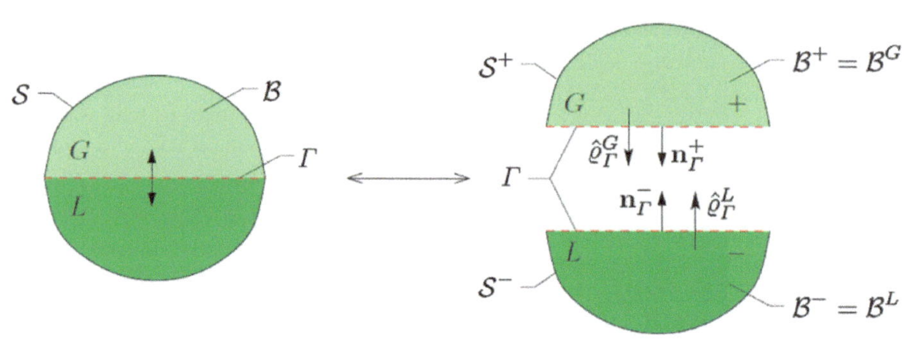

Figure 2: Local microstructural interface Γ dividing the gas-saturated and the liquid-saturated parts into the partitions B+ and B−.

Consider a scalar-valued function Ψ(x,t), which is continuous in B^+ and B^-, and where the jump of Ψ over the interface Γ is defined as the difference between its values in B^+ and B^-, viz.,

$$[\![\Psi]\!] := \Psi^+ - \Psi^- . \tag{1}$$

The orientation of Γ at B^+ and B^- is given by the outward-oriented surface normals n^+_Γ and n^-_Γ yielding

$$\mathbf{n}^+_\Gamma = -\mathbf{n}^-_\Gamma , \quad \text{where} \quad \mathbf{n}^+_\Gamma =: \mathbf{n}_\Gamma , \quad \mathbf{n}^-_\Gamma = -\mathbf{n}_\Gamma . \tag{2}$$

Basic Assumptions of the Theory of Porous Media

Deformable porous media, such as liquid- and/or gas-saturated porous solid materials, can be described by the well-founded continuum-mechanical approach of the Theory of Porous Media (TPM), compare, for example, Ehlers (2002, 2009). The TPM proceeds from a formal or a virtual homogenization of the microstructure of the components under consideration such that one obtains a set of superimposed continua with mutual interactions, which are introduced by production terms incorporated in the balance equations. In contrast to the Theory of Mixtures (TM), cf. Bowen (1976), the TPM makes use of the concept of volume fractions firstly applied by Biot (1941) in order to measure the volumetric portions of the individual materials composing the overall aggregate. This concept yields

$$V = \sum_\alpha V^\alpha \quad \text{with} \quad V^\alpha = \int_{\mathcal{B}} \mathrm{d}v^\alpha =: \int_{\mathcal{B}} n^\alpha(\mathbf{x}, t)\, \mathrm{d}v ,$$

$$n^\alpha = \frac{\mathrm{d}v^\alpha}{\mathrm{d}v} \quad \text{and} \quad \sum_\alpha n^\alpha = 1, \tag{3}$$

where n^α is the volume fraction of φ^α at the actual position x and time *t*, and dv and dvα are the bulk and the partial volume element.

Following this, n^S is the solid volume fraction and $n^F=n^L+n^G$ is the overall fluid volume fraction or the porosity, respectively. In case that the pore space is filled with immiscible fluids, such as wetting and non-wetting phases, one introduces additional saturations s^β with β={L,G}, which are defined as the volume fractions of φ^β with respect to the pore space. Thus,

$$s^\beta = \frac{n^\beta}{n^F}, \quad \text{where} \quad \sum_\beta s^\beta = 1. \tag{4}$$

Assuming immiscible phases occupying separate volumes within the overall medium, two different densities can be defined by relating the local mass dm^α of φ^α either to its partial volume dvαdvα or to the bulk volume dv:

$$\rho^{\alpha R} = \frac{dm^\alpha}{dv^\alpha}, \quad \rho^\alpha = \frac{dm^\alpha}{dv}, \quad \rho^\alpha = n^\alpha \rho^{\alpha R} \quad \text{and} \quad \rho := \sum_\alpha \rho^\alpha . \tag{5}$$

Therein, $\rho^{\alpha R}$ is the effective (or realistic) density representing the real material density of φ^α at its actual position x, while the partial density ρα relates the local mass to the bulk volume of the overall porous medium. The so-called mixture density ρ is the sum of the partial densities taken over all constituents, cf. (5)44.

Kinematical Relations

Following the concept of the TPM with superimposed and interacting continua, each constituent φαφα is assigned its own unique motion function χα such that

$$\mathbf{x} = \chi_\alpha(\mathbf{X}_\alpha, t). \tag{6}$$

In the setting of superimposed continua, this implies that each spatial point x is simultaneously occupied by material points P^α of all constituents. Since χ^α must not only be unique but also uniquely invertible, it is concluded that each P^α stems from a unique reference position X^α at time t0. With (6), one easily finds the velocity and acceleration fields of φ^α as

$$\overset{\prime}{\mathbf{x}}_\alpha = \mathbf{v}_\alpha = \frac{\partial \chi_\alpha(\mathbf{X}_\alpha, t)}{\partial t}, \quad \overset{\prime\prime}{\mathbf{x}}_\alpha = \frac{\partial^2 \chi_\alpha(\mathbf{X}_\alpha, t)}{\partial t^2}. \tag{7}$$

In a solid–fluid aggregate, the porous solid is described in a *Lagrange*an framework by its displacement vector u_S, while the pore fluids φ^β are specified by a modified *Euler*ian setting through their seepage velocities w^β, viz.,

$$\mathbf{u}_S = \mathbf{x} - \mathbf{X}_S, \quad \mathbf{w}_\beta = \mathbf{v}_\beta - \mathbf{v}_S. \tag{8}$$

With (6), one also concludes to the material solid deformation gradient and its inverse

$$\mathbf{F}_S = \mathrm{Grad}_S\, \mathbf{x}, \quad \mathbf{F}_S^{-1} = \mathrm{grad}\, \mathbf{X}_S \tag{9}$$

as well as to the material and the spatial solid velocity gradients:

$$(\mathbf{F}_S)'_S = \mathrm{Grad}_S \mathbf{v}_S \quad \text{and} \quad \mathbf{L}_S = (\mathbf{F}_S)'_S \mathbf{F}_S^{-1} = \mathrm{grad}\, \mathbf{v}_S. \tag{10}$$

In the above equations, the material and the spatial gradient operators are defined by GradS(·)=d(·)/dXS and grad(·)=d(·)/dx.

In analogy to (10), the fluid velocity gradients read

$$\mathbf{L}_\beta = \mathrm{grad}\, \mathbf{v}_\beta . \tag{11}$$

Regarding the immaterial character of the singular surface Γ with its outward-oriented unit surface normal n_Γ shown in Figure 2, Γ is allowed to propagate through B by its own velocity v_Γ. This also leads to the definition of relative velocities $w_{\beta\Gamma}$ of the fluid phases φ^β with respect to Γ. Thus,

$$\mathbf{v}_\Gamma := \overset{\prime}{\mathbf{x}}_\Gamma, \quad \mathbf{w}_{\beta\Gamma} = \mathbf{v}_\beta - \mathbf{v}_\Gamma . \tag{12}$$

Note in passing that the surface normal and the velocity of Γ are jump-free:

$$[\![\mathbf{n}_\Gamma]\!] = \mathbf{0}, \quad [\![\mathbf{v}_\Gamma]\!] = \mathbf{0}. \tag{13}$$

Balance Relations

The general balance relations of the TPM follow the arguments of Truesdell (1984) and are formulated according to Ehlers (1996, 2002):

$$\frac{d_\alpha}{dt}\int_{\mathcal{B}} \Psi^\alpha \, dv = \int_{\mathcal{S}} \Phi^\alpha \cdot \mathbf{n}\, da + \int_{\mathcal{B}} \sigma^\alpha \, dv + \int_{\mathcal{B}} \hat{\Psi}^\alpha \, dv,$$

$$\frac{d_\alpha}{dt}\int_{\mathcal{B}} \mathbf{\Psi}^\alpha \, dv = \int_{\mathcal{S}} \mathbf{\Phi}^\alpha \, \mathbf{n}\, da + \int_{\mathcal{B}} \boldsymbol{\sigma}^\alpha \, dv + \int_{\mathcal{B}} \hat{\mathbf{\Psi}}^\alpha \, dv. \tag{14}$$

By use of the *Gauss*ian integral theorem and the transport theorem for dv, (14) is easily transferred towards

$$\int_{\mathcal{B}} \left[(\Psi^\alpha)'_\alpha + \Psi^\alpha \operatorname{div} \mathbf{v}_\alpha \right] dv = \int_{\mathcal{B}} (\operatorname{div} \boldsymbol{\phi}^\alpha + \sigma^\alpha + \hat{\Psi}^\alpha)\, dv,$$

$$\int_{\mathcal{B}} \left[(\mathbf{\Psi}^\alpha)'_\alpha + \mathbf{\Psi}^\alpha \operatorname{div} \mathbf{v}_\alpha \right] dv = \int_{\mathcal{B}} (\operatorname{div} \mathbf{\Phi}^\alpha + \boldsymbol{\sigma}^\alpha + \hat{\mathbf{\Psi}}^\alpha)\, dv. \tag{15}$$

In the above equations, Ψ^α and $\mathbf{\Psi}^\alpha$ are the scalar and vectorial volume-specific densities of the physical quantities in BB that have to be balanced, Φ^α and $\mathbf{\Phi}^\alpha$ are the effluxes of the physical quantities through the external surface S (action at the vicinity), σ^α and $\boldsymbol{\sigma}^\alpha$ are the supplies of the physical quantities (action from a distance), and $\hat{\Psi}^\alpha$ and $\hat{\mathbf{\Psi}}^\alpha$ represent the total productions of the physical quantities due to the mutual interaction between the constituents φ^α. Furthermore, $d_\alpha(\cdot)/dt$ is the material time derivative of $(\cdot)$ with the convective part following the motion of φ^α. Finally, $(\cdot)'_\alpha$ abbreviates $d_\alpha(\cdot)/dt$, and $\operatorname{div}(\cdot)$ is the divergence operator corresponding to $\operatorname{grad}(\cdot)$.

By use of standard arguments, the global balance equations (14) are transferred to local balance equations reading

$$(\Psi^\alpha)'_\alpha + \Psi^\alpha \operatorname{div} \mathbf{v}_\alpha = \operatorname{div} \boldsymbol{\phi}^\alpha + \sigma^\alpha + \hat{\Psi}^\alpha,$$

$$(\mathbf{\Psi}^\alpha)'_\alpha + \mathbf{\Psi}^\alpha \operatorname{div} \mathbf{v}_\alpha = \operatorname{div} \mathbf{\Phi}^\alpha + \boldsymbol{\sigma}^\alpha + \hat{\mathbf{\Psi}}^\alpha. \tag{16}$$

For the generation of the individual forms of the balances for mass, momentum, and energy, the physical quantities and their fluxes, supplies, and productions can be taken from Table 1. Therein, $\mathbf{T}^\alpha$ are the *Cauchy* stresses of φ^α, $\mathbf{b}^\alpha$ the body forces, ε^α the internal energies, $\mathbf{q}^\alpha$ the heat influxes, r^α the heat supplies, and $\hat{\rho}^\alpha$, $\hat{\mathbf{s}}^\alpha$ and $\hat{e}^\alpha$ the total productions of mass, momentum, and energy.

Table 1: Physical quantities of the individual balances equations

Balance	$\Psi^\alpha, \mathbf{\Psi}^\alpha$	$\boldsymbol{\phi}^\alpha, \mathbf{\Phi}^\alpha$	$\sigma^\alpha, \boldsymbol{\sigma}^\alpha$	$\hat{\Psi}^\alpha, \hat{\mathbf{\Psi}}^\alpha$
Mass	ρ^α	$\mathbf{0}$	0	$\hat{\rho}^\alpha$
Momentum	$\rho^\alpha \mathbf{v}_\alpha$	$\mathbf{T}^\alpha$	$\rho^\alpha \mathbf{b}^\alpha$	$\hat{\mathbf{s}}^\alpha$
Energy	$\rho^\alpha (\varepsilon^\alpha + \frac{1}{2}\mathbf{v}^\alpha \cdot \mathbf{v}^\alpha)$	$(\mathbf{T}^\alpha)^T \mathbf{v}_\alpha - \mathbf{q}^\alpha$	$\rho^\alpha (\mathbf{b}^\alpha \cdot \mathbf{v}_\alpha + r^\alpha)$	$\hat{e}^\alpha$

In case that B is intersected by a singular surface Γ, one has to derive balance equations for B+ and B− with external surfaces S(B+)=S+∪Γ and S(B−)=S−∪Γ. Summarizing these equations, one obtains instead of (15), also compare, for example, Alts and Hutter (1988):

$$\int_{\mathcal{B}\setminus\Gamma} \left[(\Psi^\alpha)'_\alpha + \Psi^\alpha \operatorname{div} \mathbf{v}_\alpha \right] dv + \int_\Gamma [\![\Psi^\alpha \mathbf{w}_{\alpha\Gamma}]\!] \cdot \mathbf{n}_\Gamma \, da$$

$$= \int_{\mathcal{B}\setminus\Gamma} (\operatorname{div} \boldsymbol{\phi}^\alpha + \sigma^\alpha + \hat{\Psi}^\alpha)\, dv + \int_\Gamma [\![\boldsymbol{\phi}^\alpha]\!] \cdot \mathbf{n}_\Gamma \, da,$$

$$\int_{\mathcal{B}\setminus\Gamma} \left[(\mathbf{\Psi}^\alpha)'_\alpha + \mathbf{\Psi}^\alpha \operatorname{div} \mathbf{v}_\alpha \right] dv + \int_\Gamma [\![\mathbf{\Psi}^\alpha \otimes \mathbf{w}_{\alpha\Gamma}]\!] \, \mathbf{n}_\Gamma \, da$$

$$= \int_{\mathcal{B}\setminus\Gamma} (\operatorname{div} \mathbf{\Phi}^\alpha + \boldsymbol{\sigma}^\alpha + \hat{\mathbf{\Psi}}^\alpha)\, dv + \int_\Gamma [\![\mathbf{\Phi}^\alpha]\!] \, \mathbf{n}_\Gamma)\, da. \tag{17}$$

In comparison with (15), the additional terms $[\![\cdot]\!]$ denote the jump of physical quantities and their fluxes across Γ. Furthermore, since the local versions of (17) have to hold at the same time as (15), one obtains

$$\left.\begin{aligned} (\Psi^\alpha)'_\alpha + \Psi^\alpha \operatorname{div} \mathbf{v}_\alpha &= \operatorname{div} \boldsymbol{\phi}^\alpha + \sigma^\alpha + \hat{\Psi}^\alpha, \\ (\mathbf{\Psi}^\alpha)'_\alpha + \mathbf{\Psi}^\alpha \operatorname{div} \mathbf{v}_\alpha &= \operatorname{div} \mathbf{\Phi}^\alpha + \boldsymbol{\sigma}^\alpha + \hat{\mathbf{\Psi}}^\alpha \end{aligned}\right\} \quad \forall\, \mathbf{x} \in \mathcal{B}\setminus\Gamma, \tag{18}$$

and

$$\left.\begin{aligned} [\![\Psi^\alpha \mathbf{w}_{\alpha\Gamma} - \boldsymbol{\phi}^\alpha]\!] \cdot \mathbf{n}_\Gamma &= 0, \\ [\![\mathbf{\Psi}^\alpha \otimes \mathbf{w}_{\alpha\Gamma} - \mathbf{\Phi}^\alpha]\!] \, \mathbf{n}_\Gamma &= \mathbf{0} \end{aligned}\right\} \quad \forall\, \mathbf{x} = \mathbf{x}_\Gamma \in \Gamma \tag{19}$$

substituting (16). Note that the local balances (18) are unchanged in comparison with (16) but accompanied by jump conditions (19) summarizing the jump of the physical quantities across Γ.

Combining (18) and (19) with the physical quantities of Table 1, one obtains the local balance equations and jump conditions for mass, momentum, and energy:

- mass:

$$(\rho^\alpha)'_\alpha + \rho^\alpha \operatorname{div} \mathbf{v}_\alpha = \hat{\rho}^\alpha \quad \forall\, \mathbf{x} \in \mathcal{B}\backslash\Gamma,$$
$$[\![\rho^\alpha \mathbf{w}_{\alpha\Gamma}]\!] \cdot \mathbf{n}_\Gamma = 0 \quad \forall\, \mathbf{x} = \mathbf{x}_\Gamma \in \Gamma, \tag{20}$$

- momentum:

$$\rho^\alpha (\mathbf{v}_\alpha)'_\alpha = \operatorname{div} \mathbf{T}^\alpha + \rho^\alpha \mathbf{b} + \hat{\mathbf{p}}^\alpha \quad \forall\, \mathbf{x} \in \mathcal{B}\backslash\Gamma,$$
$$[\![\rho^\alpha \mathbf{v}_\alpha \otimes \mathbf{w}_{\alpha\Gamma} - \mathbf{T}^\alpha]\!]\, \mathbf{n}_\Gamma = \mathbf{0} \quad \forall\, \mathbf{x} = \mathbf{x}_\Gamma \in \Gamma, \tag{21}$$

- energy:

$$\rho^\alpha (\varepsilon^\alpha)'_\alpha = \mathbf{T}^\alpha \cdot \mathbf{L}_\alpha - \operatorname{div} \mathbf{q}^\alpha + \rho^\alpha r^\alpha + \hat{\varepsilon}^\alpha \quad \forall\, \mathbf{x} \in \mathcal{B}\backslash\Gamma,$$
$$\left[\!\!\left[\rho^\alpha \left(\varepsilon^\alpha + \tfrac{1}{2}\, \mathbf{v}_\alpha \cdot \mathbf{v}_\alpha\right) \mathbf{w}_{\alpha\Gamma} - (\mathbf{T}^\alpha)^T \mathbf{v}_\alpha + \mathbf{q}^\alpha\right]\!\!\right] \cdot \mathbf{n}_\Gamma = 0 \quad \forall\, \mathbf{x} = \mathbf{x}_\Gamma \in \Gamma. \tag{22}$$

In addition to the total mass production $\hat{\rho}^\alpha$, the above equations also contain the direct momentum and energy productions $\hat{\mathbf{p}}^\alpha$ and $\hat{\varepsilon}^\alpha$. These terms are related to their total counterparts via

$$\hat{\mathbf{s}}^\alpha = \hat{\mathbf{p}}^\alpha + \hat{\rho}^\alpha\, \mathbf{v}_\alpha,$$
$$\hat{e}^\alpha = \hat{\varepsilon}^\alpha + \hat{\mathbf{p}}^\alpha \cdot \mathbf{v}_\alpha + \hat{\rho}^\alpha \left(\varepsilon^\alpha + \tfrac{1}{2}\, \mathbf{v}_\alpha \cdot \mathbf{v}_\alpha\right). \tag{23}$$

Following the basic TPM assumptions, cf. Ehlers (1996, 2002), the total production terms are constrained by

$$\sum_\alpha \hat{\rho}^\alpha = 0, \quad \sum_\alpha \hat{\mathbf{s}}^\alpha = \mathbf{0}, \quad \sum_\alpha \hat{e}^\alpha = 0. \tag{24}$$

Considering the individual production terms and their physical meaning, the mass production $\hat{\rho}^\alpha$ can be understood as the mass transferred to φ^α either due to chemical reactions or due to phase-change processes. The direct momentum production $\hat{\mathbf{p}}^\alpha$ is interpreted as the local volume average of the internal contact forces acting on φ^α, while $\hat{\varepsilon}^\alpha$ represents the local heat exchange between φ^α and the other constituents in the overall aggregate.

CONSTITUTIVE RELATIONS

The problem under discussion consists of a bicomponent, triphasic aggregate of an inert, materially incompressible solid φ^S and two immiscible and compressible pore-fluid phases φ^L and φ^G of the same component in a non-isothermal environment. The problem is governed by the following balance relations taken from (20)11, (21)11 and (22)11 under the assumption of quasi-static conditions ($\overset{''}{\mathbf{x}}_\alpha = \mathbf{0}$) and constant gravitational forces ($\mathbf{b}^\alpha$=$\mathbf{g}$):

$$
\begin{aligned}
(\rho^S)'_S + \rho^S \operatorname{div} \mathbf{v}_S &= 0, && \\
(\rho^\beta)'_S + \rho^\beta \operatorname{div} \mathbf{v}_S + \operatorname{div}(\rho^\beta \mathbf{w}_\beta) &= \hat{\rho}^\beta, && \beta = \{G,\ L\}, \\
\mathbf{0} &= \operatorname{div} \mathbf{T}^\alpha + \rho^\alpha \mathbf{g} + \hat{\mathbf{p}}^\alpha, && \alpha = \{S,\ L,\ G\}, \\
\rho^\alpha (\varepsilon^\alpha)'_\alpha &= \mathbf{T}^\alpha \cdot \mathbf{L}_\alpha - \operatorname{div} \mathbf{q}^\alpha + \rho^\alpha r^\alpha + \hat{\varepsilon}^\alpha, && \alpha = \{S,\ L,\ G\}.
\end{aligned} \tag{25}
$$

The gas and liquid mass balances (25)22 have been rearranged in comparison with (20)11 such that the material time derivatives of the gas and liquid phases can be expressed by the solid time derivative and additional terms according to the modified *Euler*ian setting of the fluid components. Note in passing that the mass balance of a materially incompressible porous solid reduces under isothermal conditions to a volume balance of the form $(n^S)'_S + n^S \operatorname{div} \mathbf{v}_S = 0$ when ρ^{SR}= const. Furthermore, if one considers a purely continuum mechanical problem with prescribed values for the complete motion and temperature states, it is easily concluded that (25) is insufficient to determine the open fields consisting of the stress tensors, internal energies, heat fluxes, and production terms. These terms must be found by constitutive equations solving the so-called closure problem. Obviously, the constitutive equations have to fulfil the entropy inequality of the whole aggregate in order to represent a thermodynamically admissible constitutive environment. The complete procedure of generating sound constitutive equations for multiphasic–multicomponent models usually leads to a lengthy formalism, which is not included in this article. Note in passing that the assumption of quasi-static conditions is always justified when creeping-flow conditions are concerned. In this case, it is sufficient to proceed with the assumption of single temperature processes as far as chemical reactions with sudden temperature variations are excluded, which is the case in the present article. Based on this and following the line of Ehlers (2002, 2009), the exploitation of the entropy principle yields (θ^α=θ) the following results:

- partial *Cauchy* continua:

$$\mathbf{T}^\alpha = (\mathbf{T}^\alpha)^T.$$

- concept of extra stresses:

$$\begin{cases} \mathbf{T}^S = -\, n^S\, p^{FR}\, \mathbf{I} + \mathbf{T}^S_E, \\ \mathbf{T}^\beta = -\, n^\beta\, p^{\beta R}\, \mathbf{I} + \mathbf{T}^\beta_E. \end{cases}$$

- negligible fluid extra (frictional) stresses:

$$\mathbf{T}^\beta_E \approx \mathbf{0}.$$

- *Dalton*'s law:

$$p^{FR} = s^L p^{LR} + s^G p^{GR}, \tag{26}$$

- direct momentum productions:

$$\begin{cases} \hat{\mathbf{p}}^{L} = p^{LR}\,\text{grad}\,n^{L} + p^{C}(s^{G}\,\text{grad}\,n^{L} - s^{L}\,\text{grad}\,n^{G}) + \hat{\mathbf{p}}^{L}_{E}, \\ \hat{\mathbf{p}}^{G} = p^{GR}\,\text{grad}\,n^{G} + \hat{\mathbf{p}}^{G}_{E}. \end{cases}$$

- concept of phase separation:

$$\psi^{S} = \psi^{S}(\mathbf{F}_{S}, \theta),\ \psi^{L} = \psi^{L}(\rho^{LR}, \theta, s^{L}),\ \psi^{G} = \psi^{G}(\rho^{GR}, \theta).$$

In the above setting, $\mathbf{T}^{S}_{E}$ and $\mathbf{T}^{\beta}_{E}$ are the so-called extra-stress tensors, $p^{\beta R}$ are the effective pressures of the fluid constituents, p^{FR} is the overall pore-fluid pressure, and II is the second-order identity tensor. Furthermore, $\hat{\mathbf{p}}^{\beta}_{E}$ are the extra terms of the direct momentum productions, and ψ^{α} are the *Helmholtz* free energy functions of the constituents φ^{α}.

Given the above results, the exploitation of the entropy inequality at equilibrium furthermore yields

- entropy free energy relations:

$$\eta^{\alpha} = -\frac{\partial \psi^{\alpha}}{\partial \theta}.$$

- solid extra (effective) stress:

$$\mathbf{T}^{S}_{E} = \rho^{S}\frac{\partial \psi^{S}}{\partial \mathbf{F}_{S}}(\mathbf{F}_{S})^{T}. \tag{27}$$

- effective fluid pressures:

$$p^{\beta R} = (\rho^{\beta R})^{2}\frac{\partial \psi^{\beta}}{\partial \rho^{\beta R}}.$$

- capillary pressure:

$$p^{C} = -s^{L}\rho^{LR}\frac{\partial \psi^{L}}{\partial s^{L}}.$$

Thermoelastic Porous Solid

The thermomechanical behavior of the solid constituent basically depends on a multiplicative split of the solid deformation gradient FS into purely mechanical and purely thermal parts:

$$\mathbf{F}_{S} = \mathbf{F}_{Sm}\mathbf{F}_{S\theta}. \tag{28}$$

While F_S only depends on the solid displacement u_S through the displacement gradient $H_S = \text{Grad}_S u_S$, the thermal part $F_{S\theta}$ is described by a constitutive assumption following Lu and Pister (1975), viz.,

$$\begin{aligned} \mathbf{F}_{S} &= \mathbf{I} + \mathbf{H}_{S}, \\ \mathbf{F}_{S\theta} &= (\det \mathbf{F}_{S\theta})^{1/3}\,\mathbf{I} \quad \text{with} \quad \det \mathbf{F}_{S\theta} = \exp(3\alpha^{S}\Delta\theta), \\ \mathbf{F}_{Sm} &= \mathbf{F}_{S}\mathbf{F}^{-1}_{S\theta}. \end{aligned} \tag{29}$$

Therein, α^{S} is the coefficient of thermal expansion, and $\Delta\theta = \theta - \theta_0$ is the temperature variation compared to the initial temperature θ_0.

As in elasto-plasticity, cf., for example, Ehlers (1991), the multiplicative split of the deformation gradient is associated with the existence of an intermediate configuration and an additive split of strain tensors in the solid reference, intermediate and actual configuration. For example, one obtains the following decomposition of the *Green–Lagrange*an strain in the solid reference configuration:

$$\begin{aligned} \mathbf{E}_{S} &= \tfrac{1}{2}\left(\mathbf{F}^{T}_{S}\mathbf{F}_{S} - \mathbf{I}\right) = \mathbf{E}_{Sm} + \mathbf{E}_{S\theta}, \\ \mathbf{E}_{S\theta} &= \tfrac{1}{2}\left(\mathbf{F}^{T}_{S\theta}\mathbf{F}_{S\theta} - \mathbf{I}\right), \\ \mathbf{E}_{Sm} &= \mathbf{E} - \mathbf{E}_{S\theta}. \end{aligned} \tag{30}$$

If only small strains are expected, a formal linearization of the above strain measures around the natural state given by F_S and $F_{S\theta}$ equal to II yields

$$\begin{aligned} \boldsymbol{\varepsilon}_{S} &:= \text{lin}\,\mathbf{E}_{S} = \tfrac{1}{2}\left(\mathbf{F}_{S} + \mathbf{F}^{T}_{S}\right) - \mathbf{I} = \tfrac{1}{2}\left(\mathbf{H}_{S} + \mathbf{H}^{T}_{S}\right), \\ \boldsymbol{\varepsilon}_{S\theta} &:= \text{lin}\,\mathbf{E}_{S\theta} = \tfrac{1}{2}\left(\mathbf{F}_{S\theta} + \mathbf{F}^{T}_{S\theta}\right) - \mathbf{I}, \\ \boldsymbol{\varepsilon}_{Sm} &= \boldsymbol{\varepsilon}_{S} - \boldsymbol{\varepsilon}_{S\theta}. \end{aligned} \tag{31}$$

If the temperature variation is such that $\det F_{S\theta}$ is approximately equal to lin $(\det F_{S\theta})$, a formal linearization of $(\det F_{S\theta})^{1/3}$ around $\Delta\theta = 0$ furthermore yields

$$\text{lin}\,(\det \mathbf{F}_{S\theta})^{1/3} = 1 + \alpha^{S}\Delta\theta. \tag{32}$$

Given this result, the thermal part of the deformation gradient, cf. $(29)_2$, and the thermal strain of $(31)_2$ read

$$\begin{aligned} \text{lin}\,(\mathbf{F}_{S\theta}) &= (1 + \alpha^{S}\Delta\theta)\,\mathbf{I}, \\ \boldsymbol{\varepsilon}_{S\theta} &= \alpha^{S}\Delta\theta\,\mathbf{I}. \end{aligned} \tag{33}$$

By integration of the solid mass balance $(25)_1$, one obtains

$$\rho^{S} = \rho^{S}_{0S}(\det \mathbf{F}_{S})^{-1}, \tag{34}$$

where ρ^{S}_{0S} is the partial solid density in the solid reference configuration at $t = t_0$. Splitting the partial density in effective density and volume fraction yields by use of (28)

$$n^{S}\rho^{SR} = n^{S}_{0S}\,\rho^{SR}_{0S}\,(\det \mathbf{F}_{Sm})^{-1}(\det \mathbf{F}_{S\theta})^{-1} \tag{35}$$

with n^{S}_{0S} and ρ^{SR}_{0S} as the solid volume fraction and effective solid density at $t = t_0$.

As was explained before, $\rho^{SR} = \rho^{SR}_{0S}$ is constant at constant temperatures for materially incompressible solids. Consequently, variations in ρ^{SR} can only be initiated by temperature variations. Thus, it is obvious that (35) can be split as follows, where (29)33 is used:

$$\begin{aligned} \rho^{SR} &= \rho^{SR}_{0S}\,(\det \mathbf{F}_{S\theta})^{-1} = \rho^{S}_{0S}\exp(-3\alpha^{S}\Delta\theta), \\ n^{S} &= n^{S}_{0S}\,(\det \mathbf{F}_{Sm})^{-1} = n^{S}_{0S}\,(\det \mathbf{F}_{S})^{-1}\exp(3\alpha^{S}\Delta\theta). \end{aligned} \tag{36}$$

Linearizing $(\det \mathbf{F}_{S})^{-1}$, $(\det \mathbf{F}_{S\theta})^{-1}$ and $(\det \mathbf{F}_{Sm})^{-1}$ via

$$\begin{aligned} \text{lin}\,(\det \mathbf{F}_{S})^{-1} &= 1 - \text{Div}_{S}\,\mathbf{u}_{S}, \\ \text{lin}\,(\det \mathbf{F}_{Sm})^{-1} &= 1 - \text{Div}_{S}\,\mathbf{u}_{S} + 3\alpha^{S}\Delta\theta, \\ \text{lin}\,(\det \mathbf{F}_{S\theta})^{-1} &= 1 - 3\alpha^{S}\Delta\theta, \end{aligned} \tag{37}$$

where $\mathrm{Div}_S(\cdot)$ is the divergence operator corresponding $\mathrm{Grad}_S(\cdot)$, one obtains the following relations for the solid densities and volume fractions:

$$\begin{aligned} \rho^S &= \rho^S_{0S}\,(1 - \mathrm{Div}_S\,\mathbf{u}_S), \\ n^S &= n^S_{0S}\,(1 - \mathrm{Div}_S\,\mathbf{u}_S + 3\,\alpha^S\,\Delta\theta), \\ \rho^{SR} &= \rho^{SR}_{0S}\,(1 - 3\,\alpha^S\,\Delta\theta). \end{aligned} \tag{38}$$

In the geometrically linearized setting, the solid extra stress and the solid entropy can be obtained from the solid *Helmholtz* energy, which, for an isotropic, thermoelastic solid can be given as the sum of a purely mechanical and a purely thermal part:

$$\rho^S_{0S}\,\psi^S\,(\boldsymbol{\varepsilon}_{Sm},\,\theta) = \rho^S_{0S}\,\psi^S_m\,(\boldsymbol{\varepsilon}_{Sm}) + \rho^S_{0S}\,\psi^S_\theta\,(\theta). \tag{39}$$

Therein, the mechanical part is given by

$$\rho^S_{0S}\,\psi^S_m\,(\boldsymbol{\varepsilon}_{Sm}) = \mu^S\,\boldsymbol{\varepsilon}_{Sm}\cdot\boldsymbol{\varepsilon}_{Sm} + \tfrac{1}{2}\,\lambda^S\,(\boldsymbol{\varepsilon}_{Sm}\cdot\mathbf{I})^2, \tag{40}$$

where μ^S and λ^S are the *Lamé* constants, while the thermal part has to be found from the condition

$$c^S_V = -\theta\,\frac{\partial^2\psi^S_\theta}{\partial\theta^2}, \tag{41}$$

where c^S_V is the specific heat at constant volume. Substituting ε_{Sm} by $\boldsymbol{\varepsilon}_S - \boldsymbol{\varepsilon}_{S\theta}$ from (31)33 and (33)22 and integrating (41) together with the side conditions $\psi^S_\theta(\theta_0) = 0$ and $\partial\psi^S_\theta/\partial\theta(\theta_0) = 0$, one obtains

$$\begin{aligned} \rho^S_{0S}\psi^S_m &= \mu^S(\boldsymbol{\varepsilon}_S\cdot\boldsymbol{\varepsilon}_S) + \tfrac{1}{2}\lambda^S(\boldsymbol{\varepsilon}_S\cdot\mathbf{I})^2 - 3\,k^S\alpha^S\Delta\theta\,(\boldsymbol{\varepsilon}_S\cdot\mathbf{I}) + \tfrac{1}{2}k^S(3\,\alpha^S\Delta\theta)^2, \\ \rho^S_{0S}\psi^S_\theta &= -\tfrac{1}{2}k^S(3\,\alpha^S\Delta\theta)^2 - \rho^S_{0S}c^S_V\left(\theta\ln\tfrac{\theta}{\theta_0} - \Delta\theta\right), \end{aligned} \tag{42}$$

where $k^S = \frac{2}{3}\,\mu^S + \lambda^S$ is the compression modulus. Addition of the mechanical and the thermal parts of (42) yields the free energy of a linear thermoelastic solid skeleton:

$$\begin{aligned} \rho^S_{0S}\psi^S &= \mu^S(\boldsymbol{\varepsilon}_S\cdot\boldsymbol{\varepsilon}_S) + \tfrac{1}{2}\lambda^S(\boldsymbol{\varepsilon}_S\cdot\mathbf{I})^2 + m^S\Delta\theta\,(\boldsymbol{\varepsilon}_S\cdot\mathbf{I}) \\ &\quad - \rho^S_{0S}c^S_V\left(\theta\ln\frac{\theta}{\theta_0} - \Delta\theta\right). \end{aligned} \tag{43}$$

Therein, $m^S = -3\,k^S\alpha^S$ is the stress–temperature modulus.

Proceeding from the basic constitutive relations, the *Cauchy* stress T^S_E from $(27)_2$ can be related to the second *Piola–Kirchhoff* stress S^S_E yielding

$$\mathbf{S}^S_E = \rho^S_{0S}\,\frac{\partial\psi^S}{\partial\mathbf{E}_S} = \det\mathbf{F}_S\,\mathbf{F}_S^{-1}\,\mathbf{T}^S_E\,\mathbf{F}_S^{T-1}. \tag{44}$$

Under small-strain conditions, where $\mathbf{T}^S_E \approx \mathbf{S}^S_E \approx \boldsymbol{\sigma}^S_E$ and $\mathbf{E}^S \approx \boldsymbol{\varepsilon}_S$, the first and the second partial derivatives of $\rho^S_{0S}\,\psi^S$ with respect to ε^S yield the solid extra stress and the mechanical tangent. Thus,

$$\begin{aligned} \boldsymbol{\sigma}^S_E &= 2\,\mu^S\,\boldsymbol{\varepsilon}_S + \lambda^S\,(\boldsymbol{\varepsilon}_S\cdot\mathbf{I})\,\mathbf{I} + m^S\Delta\theta\,\mathbf{I}, \\ \overset{4}{\mathbf{B}}_{0S} &= 2\,\mu^S\,(\mathbf{I}\otimes\mathbf{I})^{\overset{23}{T}} + \lambda^S\,(\mathbf{I}\otimes\mathbf{I}), \end{aligned} \tag{45}$$

where $\overset{4}{(\cdot)}$ indicates a tensor of fourth order and $(\cdot)^{\overset{23}{T}}$ denotes its transposition with respect to its second and third basis vectors. Finally, note in passing that (45) can also be found directly from (40), when (40) is differentiated with respect to ε_{Sm} which is then substituted by $\varepsilon_S - \varepsilon_{S\theta}$.

Given the above, the solid entropy is obtained on the basis of $(27)_1$ and (42). Thus,

$$\eta^{S} = -\frac{1}{\rho_{0S}^{S}} m^{S}(\boldsymbol{\varepsilon}^{S} \cdot \mathbf{I}) + c_{V}^{S} \ln\frac{\theta}{\theta_0}. \tag{46}$$

The final step of the constitutive setting of the solid constituent is the determination of the internal energy by use of the *Legendre* transformation

$$\varepsilon^{S} = \psi^{S} + \theta\eta^{S}. \tag{47}$$

Considering the initial condition $\varepsilon^{S}(\boldsymbol{\varepsilon}_S = \mathbf{0}, \theta_0) = 0$, the solid internal energy reads:

$$\rho_{0S}^{S}\varepsilon^{S} = \mu^{S}(\boldsymbol{\varepsilon}_S \cdot \boldsymbol{\varepsilon}_S) + \tfrac{1}{2}\lambda^{S}(\boldsymbol{\varepsilon}_S \cdot \mathbf{I})^2 + \rho_{0S}^{S} c_{V}^{S}\,\Delta\theta. \tag{48}$$

Pore Fluids

While the volume fraction of the overall pore fluid can be determined from (3), $(36)_2$, and $(37)_2$ through

$$n^{F} = 1 - n^{S} = 1 - n_{0S}^{S}\left(1 - \mathrm{Div}_S \mathbf{u}_S + 3\alpha^{S}\Delta\theta\right), \tag{49}$$

further considerations have to be made for the determination of n^L and n^G or s^L and s^G, respectively. Here, we proceed from the Brooks and Corey law (1964) given by

$$s_{\text{eff}}^{L} = \left(\frac{p^D}{p^C}\right)^{\lambda}. \tag{50}$$

Therein, s_{eff}^{L} is the effective saturation given by van Genuchten (1980) as

$$s_{\text{eff}}^{L} = \frac{s^L - s_{\text{res}}^{L}}{1 - s_{\text{res}}^{L} - s_{\text{res}}^{G}}, \tag{51}$$

where s_{res}^{L} and s_{res}^{G} are constants describing the residual saturations remaining in the fully liquid- or gas-saturated domains. Furthermore,

$$p^C = p^{GR} - p^{LR} \tag{52}$$

is the capillary pressure defined as the difference between the effective pressures of the non-wetting and the wetting fluid (Brooks and Corey 1964), p^D is the so-called bubbling or entry pressure, and λ is an adaptation parameter, which *Brooks* and *Corey* called pore-size distribution index.

Inverting (50), one obtains

$$p^C = p^D\left(s_{\text{eff}}^{L}\right)^{-1/\lambda}. \tag{53}$$

Substituting the entry pressure by $p^D = \rho^{LR} g h^D$, where g is the value of the local gravitational force g and h^D is the macroscopic capillary pressure head, an integration of (53) with respect to $(27)_4$ yields

$$\psi^L = g h^D \lambda \left(s_{\text{eff}}^{L}\right)^{-1/\lambda} + f(\rho^{LR}, \theta). \tag{54}$$

For the determination of the effective fluid and gas pressures, one usually proceeds from a so-called equation of state (EOS), describing a certain fluid substance in the liquid, gaseous, and supercritical states. Here, we proceed from the classical van der Waals equation (1873) given by

$$p^{\beta R} = \bar{R}^{\beta}\theta \frac{\rho^{\beta R}}{1 - b\,\rho^{\beta R}} - a\,(\rho^{\beta R})^2, \tag{55}$$

where $\bar{R}^{\beta}$ is the specific gas constant of φ^{β}, and a and b are constants describing the cohesion pressure (a) and the co-volume (b) as functions of the critical temperature $\theta_{\text{crit}}^{\beta}$ and the critical pressure $p_{\text{crit}}^{\beta R}$:

$$a = \frac{27\left(\bar{R}^{\beta}\theta_{\text{crit}}^{\beta}\right)^2}{64\,p_{\text{crit}}^{\beta R}}, \qquad b = \frac{\bar{R}^{\beta}\theta_{\text{crit}}^{\beta}}{8\,p_{\text{crit}}^{\beta R}}. \tag{56}$$

Since (55) provides three possible solutions for the density in the two-phase region for a given pair of temperature and pressure, an additional criterion is needed to choose the correct value. To cope with this problem, the saturation pressure p_{sat} at the given temperature can be estimated on the basis of the Antoine equation (1888)

$$\log_{10}(p_{\text{sat}}) = A - \frac{B}{\theta + C - 273.15}, \tag{57}$$

where A, B, and C are empirical parameters of the considered fluid material. Comparing the given pressure $p^{\beta R}$ with p_{sat}, one can distinguish whether the current conditions are gaseous or liquid.

Given the *van der Waals* EOS, an integration of (55) with respect to $(27)_3$ yields

$$\begin{aligned} \psi_{vdW}^{\beta} &= \bar{R}^{\beta}\theta \ln\frac{\rho^{\beta R}}{1 - b\,\rho^{\beta R}} - a\,\rho^{\beta R} + k(\theta), \\ k(\theta) &= -c_{V}^{\beta R}\theta\,(\ln\theta - 1), \end{aligned} \tag{58}$$

where the determination of $k(\theta)$ follows the same procedure as was used for the solid constituent, when the temperature-dependent part of the free energy ψ^S had to be identified.

Based on $(27)_{2,3}$, (54), and (58), it is concluded that $\psi_{vdW}^{G} = \psi^G$ and $\psi_{vdW}^{L} = f(\rho^{LR}, \theta)$, such that

$$\begin{aligned} \psi^G &= \bar{R}^{\beta}\theta \ln\frac{\rho^{GR}}{1 - b\,\rho^{GR}} - a\,\rho^{GR} - c_{V}^{GR}\theta\,(\ln\theta - 1), \\ \psi^L &= \bar{R}^{\beta}\theta \ln\frac{\rho^{LR}}{1 - b\,\rho^{LR}} - a\,\rho^{LR} - c_{V}^{LR}\theta\,(\ln\theta - 1) + g h^D \lambda \left(s_{\text{eff}}^{L}\right)^{-1/\lambda}. \end{aligned} \tag{59}$$

Note in passing that the constants a and b included in ψ^G and ψ^G are the same, since both describe the same matter; however, in different phase states, only c_{V}^{LR} and c_{V}^{GR} are different.

Based on (59), the entropies and internal energies of the fluid constituents can be determined. As a result, one obtains

$$\begin{aligned} \eta^{\beta} &= -\bar{R}^{\beta}\ln\frac{\rho^{\beta R}}{1 - b\,\rho^{\beta R}} + c_{V}^{\beta R}\ln\theta, \\ \varepsilon^{G} &= -a\,\rho^{GR} + c_{V}^{GR}\theta, \\ \varepsilon^{L} &= -a\,\rho^{LR} + c_{V}^{LR}\theta + g h^D \lambda \left(s_{\text{eff}}^{L}\right)^{-1/\lambda}. \end{aligned} \tag{60}$$

where $(27)_1$ and (47) referring to the fluid constituents have been used.

Once the fluid pressures and energies are properly defined, the dissipation inequality as the non-reversible part of the entropy inequality of the overall aggregate reveals the extra terms of the direct fluid momentum productions as

$$\hat{\mathbf{p}}_E^{\beta} = -(n^{\beta})^2 \rho^{\beta R} g\, (\mathbf{K}_r^{\beta})^{-1} \mathbf{w}_{\beta}, \tag{61}$$

where K_r^{β} is the tensor of relative permeabilities. Inserting (61) into the fluid momentum balances $(25)_2$ yields under the assumption of creeping-flow or quasi-static conditions, respectively, and negligible fluid extra stresses the following *Darcy*-like equations for the filter velocities $n^{\beta}\mathbf{w}_{\beta}$ of the fluid phases, cf. Ehlers (2009):

$$\begin{aligned} n^G \mathbf{w}_G &= -\frac{\mathbf{K}_r^G}{\rho^{GR} g} [\,\mathrm{grad}\, p^{GR} - \rho^{GR}\, \mathbf{g}\,], \\ n^L \mathbf{w}_L &= -\frac{\mathbf{K}_r^L}{\rho^{LR} g} \left[\mathrm{grad}\, p^{LR} - \rho^{LR}\, \mathbf{g} - \frac{p^C}{n^L} (s^G \mathrm{grad}\, n^L - s^L \mathrm{grad}\, n^G) \right]. \end{aligned} \tag{62}$$

Note in passing that the above *Darcy*-like equations have not been introduced as constitutive equations but as a result of an exploitation of the dissipative part of the entropy inequality of the overall aggregate in combination with the assumptions stated above. The tensor of relative permeabilities can be related to the tensor K^{β} of hydraulic conductivities and to the intrinsic permeability tensor K^S of the deformed solid skeleton through

$$\mathbf{K}_r^{\beta} = k_r^{\beta}\, \mathbf{K}^{\beta} \quad \text{and} \quad \mathbf{K}^{\beta} = \frac{\rho^{\beta R} g}{\mu^{\beta R}} \mathbf{K}^S. \tag{63}$$

While $\mu^{\beta R}$ indicates the effective shear viscosity of the pore fluids, k_r^{β} are the so-called relative permeability factors, which, following Brooks and Corey (1964), read

$$\begin{aligned} k_r^G &= (1 - s_{\mathrm{eff}}^L)^2 \left(1 - (s_{\mathrm{eff}}^L)^{\frac{2+\lambda}{\lambda}} \right), \\ k_r^L &= (s_{\mathrm{eff}}^L)^{\frac{2+3\lambda}{\lambda}}. \end{aligned} \tag{64}$$

Considering isotropic permeabilities, where the entries of K^S reduce to the single value K^S, one obtains after Markert (2007) the following relation for the deformation-dependent permeability K^S with respect to its initial value K_{0S}^S of the solid reference configuration:

$$K^S = \left(\frac{1 - n^S}{1 - n_{0S}^S} \det \mathbf{F}_{Sm} \right)^{\pi} K_{0S}^S. \tag{65}$$

Therein, $\pi > 0$ is an additional parameter governing the deformation dependency. Furthermore, note that $\det F_{Sm}$ can be substituted in geometrically linear approaches by

$$\mathrm{lin}\,(\det \mathbf{F}_{Sm}) = 1 + \mathrm{Div}\, \mathbf{u}_S - 3\,\alpha^S \Delta\theta. \tag{66}$$

Heat Transfer

On the basis of the dissipation inequality of the model under consideration, one easily concludes to the applicability of the *Fourier*'s law for the heat influx vectors q^{α}. Thus,

$$\mathbf{q}^{\alpha} = -\mathbf{H}^{\alpha}\, \mathrm{grad}\, \theta, \quad \text{where } \mathbf{H}^{\alpha} = n^{\alpha} \mathbf{H}^{\alpha R} \tag{67}$$

with $H^{\alpha R}$ as the effective constituent-specific heat-conduction tensor. In case of isotropic heat conduction, HαHα reduces to

$$\mathbf{H}^{\alpha} = H^{\alpha}\, \mathbf{I}, \quad \text{where } H^{\alpha} = n^{\alpha} H^{\alpha R}. \tag{68}$$

According to $(22)_1$, the direct energy production $\hat{\varepsilon}^{\alpha}$ is part of the energy balance of the individual constituents. In case that all components share the same temperature, as it is assumed in the present considerations, one has to sum up the energy balances of the components towards the total energy balance for the computation of the temperature change. In this case, $\hat{\varepsilon}^{\alpha}$ is substituted by $\hat{e}^{\alpha}$, where the sum of which vanishes over all constituents according to $(24)_3$. Thus, one obtains

$$\sum_{\alpha} \rho^{\alpha} (\varepsilon^{\alpha})'_{\alpha} = \sum_{\alpha} \left[\mathbf{T}^{\alpha} \cdot \mathbf{L}_{\alpha} - \mathrm{div}\, \mathbf{q}^{\alpha} + \rho^{\alpha} r^{\alpha} - \hat{\mathbf{p}}^{\alpha} \cdot \mathbf{v}_{\alpha} - \hat{\rho}^{\alpha} \left(\varepsilon^{\alpha} + \tfrac{1}{2} \mathbf{v}_{\alpha} \cdot \mathbf{v}_{\alpha} \right) \right], \tag{69}$$

where no constitutive equation is needed for $\hat{\varepsilon}^{\alpha}$, which is usually taken for the description of the heat exchange between the constituents as a result of different temperatures, cf. Ghadiani (2005) for details. However, frictional effects induced by $\hat{\mathbf{p}}^{\alpha} \cdot \mathbf{v}_{\alpha}$ and energetic quantities induced by phase transitions through $\hat{\rho}^{\alpha}$ also lead to mechanical and non-mechanical energy exchanges and heat transfers between the constituents.

Mass Transitions

Up to now, the model is completed apart from a proper formulation of the mass transition terms appearing in the balance relations as macroscopic density productions $\hat{\rho}^{\alpha}$, cf. $(20)_1$ and (23). In order to find a constitutive relation for these terms, a closer look is taken at the microscopic behavior at the interface between the liquid and the gaseous phases of a substance, which is mathematically represented by a singular surface, cf. Sect. 2.3.

Wherever phase transitions occur at the microscale or the REV scale, respectively, there are local jumps across singular surfaces, such that the jumps can be described by $(20)_2$. If we recall that the solid is not affected by the jump across Γ, we only have to consider the phase jump and its consequences of the fluid matter (component) φ^{FM} under consideration. Furthermore, it has been assumed that φ^{FM} exists in the gaseous phase only in B^+ and in the liquid phase only in B^-, cf. Figure 2. Following this, one has to proceed with jump conditions for φ^{FM}.

Applying the mass jump Eq. (20)22 to φ^{FM} yields

$$[\![\rho^{FM} \mathbf{w}_{FM\Gamma}]\!] \cdot \mathbf{n}_{\Gamma} = (\rho^{FM+} \mathbf{w}_{FM\Gamma}^{+} - \rho^{FM-} \mathbf{w}_{FM\Gamma}^{-}) \cdot \mathbf{n}_{\Gamma} = 0. \tag{70}$$

With $\rho^{FM+} \mathbf{w}_{FM\Gamma}^{+} = \rho^G \mathbf{w}_{G\Gamma}$ and $\rho^{FM-} \mathbf{w}_{FM\Gamma}^{-} = \rho^L \mathbf{w}_{L\Gamma}$, (70) becomes

$$(\rho^G \mathbf{w}_{G\Gamma} - \rho^L \mathbf{w}_{L\Gamma}) \cdot \mathbf{n}_{\Gamma} = 0. \tag{71}$$

Following the work of Whitaker (1977), we specify the interfacial mass transfer of φ^{β} through

$$\hat{\varrho}_\Gamma^\beta := \rho^\beta \, \mathbf{w}_{\beta\Gamma} \cdot \mathbf{n}_\Gamma^\beta, \tag{72}$$

such that

$$\begin{aligned}\hat{\varrho}_\Gamma^{\mathrm{G}} &= \rho^{\mathrm{FM+}} \mathbf{w}_{\mathrm{FM}\Gamma}^{+} \cdot \mathbf{n}_\Gamma^{\mathrm{FM+}} = \rho^{\mathrm{G}} \mathbf{w}_{\mathrm{G}\Gamma} \cdot \mathbf{n}_\Gamma, \\ \hat{\varrho}_\Gamma^{\mathrm{L}} &= \rho^{\mathrm{FM-}} \mathbf{w}_{\mathrm{FM}\Gamma}^{-} \cdot \mathbf{n}_\Gamma^{\mathrm{FM-}} = -\rho^{\mathrm{L}} \mathbf{w}_{\mathrm{L}\Gamma} \cdot \mathbf{n}_\Gamma \\ &\text{and } \hat{\varrho}_\Gamma^{\mathrm{G}} + \hat{\varrho}_\Gamma^{\mathrm{L}} = 0,\end{aligned} \tag{73}$$

where $\mathbf{n}_\Gamma^{\mathrm{FM+}} = \mathbf{n}_\Gamma$ and $\mathbf{n}_\Gamma^{\mathrm{FM-}} = -\mathbf{n}_\Gamma$ have been used. From (73)33, it is clearly seen that, for example, during evaporation, mass is removed from the liquid phase and added to the gaseous phase through $\hat{\varrho}_\Gamma^{\mathrm{G}} = -\hat{\varrho}_\Gamma^{\mathrm{L}}$. Since the interfacial mass production $\hat{\varrho}_\Gamma^\beta$ and the continuum mass production $\hat{\rho}^\beta$ have to be equivalent in the sense that $\hat{\varrho}_\Gamma^{\mathrm{G}}$ leaves B+ and generates the density production $\hat{\rho}^{\mathrm{L}}$ in B− and vice versa, $\hat{\varrho}_\Gamma^\beta$ and $\hat{\rho}^\beta$ can be related to each other by

$$\hat{\rho}^{\mathrm{G}} \, \mathrm{d}v \mathrel{\hat{=}} \hat{\varrho}_\Gamma^{\mathrm{L}} \, \mathrm{d}a_\Gamma \quad \text{and} \quad \hat{\rho}^{\mathrm{L}} \, \mathrm{d}v \mathrel{\hat{=}} \hat{\varrho}_\Gamma^{\mathrm{G}} \, \mathrm{d}a_\Gamma, \tag{74}$$

where dv is the unit volume of the REV and d_{ar} is the unit area at the interface in the REV given by

$$\mathrm{d}a_\Gamma := \int_{A_{\mathrm{REV}}} \mathrm{d}a_{\Gamma\,\mathrm{REV}}. \tag{75}$$

Based on (74) and an idea of Niessner and Hassanizadeh (2008), we introduce the so-called interfacial area a_Γ as the density of internal phase-change surfaces measured with respect to the unit volume of the REV:

$$a_\Gamma := \frac{\mathrm{d}a_\Gamma}{\mathrm{d}v}. \tag{76}$$

Given this equation, the mass-transition-depending density production $\hat{\rho}^\beta$ and the interfacial mass transfer $\hat{\varrho}_\Gamma^\beta$ are related to each other through

$$\hat{\rho}^{\mathrm{G}} = a_\Gamma \hat{\varrho}_\Gamma^{\mathrm{L}} \quad \text{and} \quad \hat{\rho}^{\mathrm{L}} = a_\Gamma \hat{\varrho}_\Gamma^{\mathrm{G}}. \tag{77}$$

The interfacial area a_Γ comprises all menisci separating the liquid and the gaseous phases in the pore space of the REV. In turn, the menisci depend on the surface tensions of the involved phases and the pore structure, or in other words, on the capillary pressure, which is given in (53) as a function of the effective saturation. This justifies that a_Γ can be assumed as $a_\Gamma = a_\Gamma(s_{\mathrm{eff}}^{\mathrm{L}})$ or as $a_\Gamma = a_\Gamma(s^{\mathrm{L}})$, respectively. However, it should be noted that the concept of interfacial areas as it was introduced by Niessner and Hassanizadeh (2008) concerns the hysteresis of imbibition and drainage curves. This effect is not included in this study, since cooling or, alternatively, heating of a pure substance in a deformable porous solid either yields imbibition or drainage and does not switch between these two effects. Finally, it has to be mentioned that the influence of the common lines on the phase-change process, i.e. the influence of the contact lines of the interface with the solid material, is also neglected.

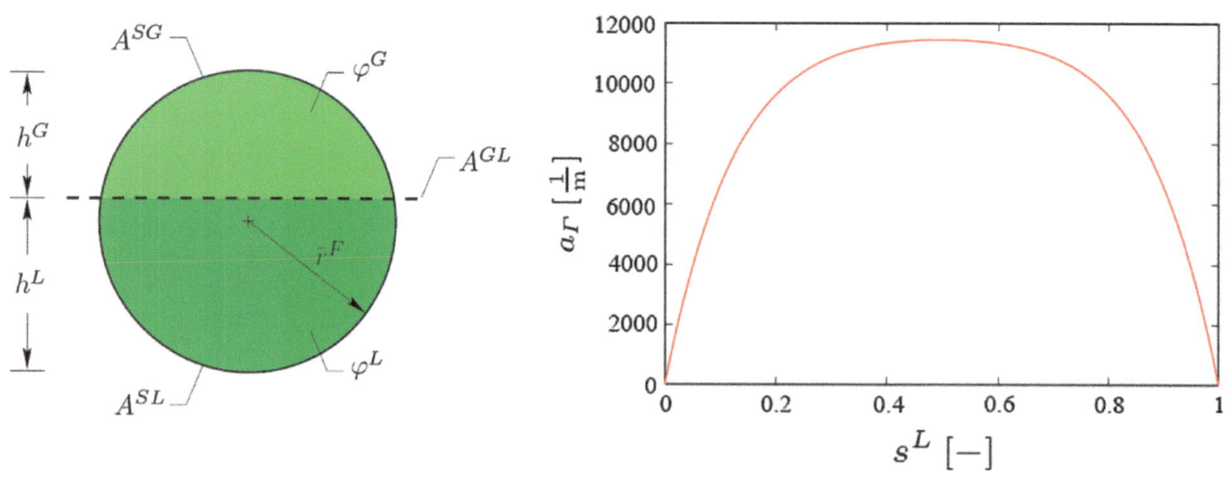

Figure 3: *Left* Volume-equivalent sphere of the pore space with the hydraulic/equivalent radius $\bar{r}^{\mathrm{F}}$, the contact areas between the solid phase and the fluid phases, A^{SL} and A^{SG}, the gas–liquid contact area A^{GL} and the filling heights h^L and h^G. *Right* Interfacial area $a_\Gamma(s^L)$ given by (84) and presented for d_{50}=0.06mmand n^S=0.9.

Niessner and Hassanizadeh (2008) also presented an empirical derivation of the interfacial area a_Γ based on data obtained by Joekar-Niasar et al. (2008) combined with a pore-network model as it has been introduced by Sahimi (2011) and others. In the present article, use is made of an approximation by Graf (2008), who described the interfacial area between the fluid phases as a function of the liquid saturation. The basic idea is to approximate the pore space by introducing a sphere with the pore-space-equivalent volume VF composed of the fluid volumes V^L and V^G given as a function of the filling heights h^L and h^G, cf. Figure 3 (left):

$$\begin{aligned}V^{\mathrm{F}} &= V^{\mathrm{L}} + V^{\mathrm{G}} = \tfrac{4}{3}\pi\,(\bar{r}^{\mathrm{F}})^3, \\ V^\beta &= \tfrac{1}{3}\pi\,(h^\beta)^2\,(3\,\bar{r}^{\mathrm{F}} - h^\beta), \quad \beta = \{\mathrm{L},\,\mathrm{G}\}.\end{aligned} \tag{78}$$

Furthermore, the surface area between the liquid and gaseous volumes reads

$$A^{\mathrm{GL}} = \pi\, h^\beta\, (2\,\bar{r}^{\mathrm{F}} - h^\beta), \tag{79}$$

where h^β in (78)22 and (79) has to be taken as the larger value out of h^L and h^G such that $h^\beta \geq \bar{r}^{\mathrm{F}}$.

Based on (4), the saturation s^β is defined by the local ratio of the V^β over V^F. Thus,

$$s^\beta = \frac{n^\beta}{n^{\mathrm{F}}} = \frac{V^\beta}{V^{\mathrm{F}}} = \frac{(h^\beta)^2\,(3\,\bar{r}^{\mathrm{F}} - h^\beta)}{4\,(\bar{r}^{\mathrm{F}})^3}. \tag{80}$$

Given (80), the filling height h^β can be determined as a function of the saturation s^β and the equivalent pore-fluid radius $\bar{r}^{\mathrm{F}}$:

$$\begin{aligned} h^\beta &= \Big[\frac{36864}{8910}\,(s^\beta)^3 - \frac{18432}{2970}\,(s^\beta)^2 + \frac{12084}{2970}\,s^\beta\Big]\,\bar{r}^{\mathrm{F}} \\ &\approx [4.137\,(s^\beta)^3 - 6.206\,(s^\beta)^2 + 4.069\,s^\beta]\,\bar{r}^{\mathrm{F}}. \end{aligned} \tag{81}$$

Since the distribution, sizes, and forms of the solid particles as well as the tortuosity and connectivity of the pores are unknown, it is not possible to calculate the exact pore volume V^F, such that an approximation is needed. Comparing a spherical pore with radius $\bar{r}^{\mathrm{F}}$ with a characteristic spherical solid particle with radius $\bar{r}^{\mathrm{S}}$ yields

$$\left.\begin{aligned} V^{\mathrm{S}} &= n^{\mathrm{S}}\,V \\ V^{\mathrm{F}} &= n^{\mathrm{F}}\,V \end{aligned}\right\} \text{ such that } \frac{n^{\mathrm{F}}}{n^{\mathrm{S}}} = \frac{V^{\mathrm{F}}}{V^{\mathrm{S}}} = \frac{(\bar{r}^{\mathrm{F}})^3}{(\bar{r}^{\mathrm{S}})^3} \text{ and thus, } \bar{r}^{\mathrm{F}} = \left(\frac{n^{\mathrm{F}}}{n^{\mathrm{S}}}\right)^{1/3} \bar{r}^{\mathrm{S}}. \tag{82}$$

Proceeding from d_{50} as the medial grain diameter of a granular soil, one ends up with

$$\bar{r}^{\mathrm{F}} = \tfrac{1}{2}\left(\frac{n^{\mathrm{F}}}{n^{\mathrm{S}}}\right)^{1/3} d_{50}. \tag{83}$$

Given the above results, the interfacial area a_Γ can be specified. Based on (76), $(78)_1$, and (79) with $A_\Gamma = A^{GL}$, a_Γ can be obtained as a function of s^β via

$$\begin{aligned} a_\Gamma(s^\beta) &= \frac{\mathrm{d}a_\Gamma}{\mathrm{d}V} = \frac{A_\Gamma}{V} = n^{\mathrm{F}}\frac{A_\Gamma}{V^{\mathrm{F}}} \\ &= \frac{3\,n^{\mathrm{F}}\,h^\beta\,(2\,\bar{r}^{\mathrm{F}} - h^\beta)}{4\,(\bar{r}^{\mathrm{F}})^3}. \end{aligned} \tag{84}$$

While $\bar{r}^{\mathrm{F}}$, at a certain state of the solid deformation, is a function of d_{50}, h^β depends on s^L, which is given as a function of the capillary pressure p^C. As a result, the a_Γ–s^L curve, cf. Figure 3 (right), is comparable to the curve found by Joekar-Niasar et al. (2008), when their curve is cut at a certain value of p^C. Once a_Γ is known, it is still necessary to determine the interfacial mass production $\hat{e}^\beta_\Gamma$ such that $\hat{\rho}^\beta$ can be fixed, cf. (77). For this purpose, we proceed from the energy jump across the singular surface Γ, cf. $(22)_2$,

$$\Big[\!\!\Big[\rho^\alpha\,\big(\varepsilon^\alpha + \tfrac{1}{2}\,\mathbf{v}_\alpha\cdot\mathbf{v}_\alpha\big)\,\mathbf{w}_{\alpha\Gamma} - \mathbf{T}^\alpha\,\mathbf{v}_\alpha + \mathbf{q}^\alpha\Big]\!\!\Big]\cdot\mathbf{n}_\Gamma = 0, \tag{85}$$

where $T^\alpha = (T^\alpha)T$ has been used according to $(26)_1$. Applying (85) to the solid constituent φ^S, one obtains with the aid of $(26)_1$:

$$\Big[\!\!\Big[\rho^{\mathrm{S}}\,\big(\varepsilon^{\mathrm{S}} + \tfrac{1}{2}\,\mathbf{v}_{\mathrm{S}}\cdot\mathbf{v}_{\mathrm{S}}\big)\,\mathbf{w}_{\mathrm{S}\Gamma} - \big(\mathbf{T}^{\mathrm{S}}_{\mathrm{E}} - n^{\mathrm{S}}\,p^{\mathrm{FR}}\mathbf{I}\big)\,\mathbf{v}_{\mathrm{S}} + \mathbf{q}^{\mathrm{S}}\Big]\!\!\Big]\cdot\mathbf{n}_\Gamma = 0. \tag{86}$$

Since the solid material is inert and not involved in the phase-change process, all terms related to the solid material itself are considered continuous over the singular surface. Thus, it remains that

$$n^{\mathrm{S}}\,[\![p^{\mathrm{FR}}]\!]\,\mathbf{v}_{\mathrm{S}}\cdot\mathbf{n}_\Gamma = 0. \tag{87}$$

However, since the solid velocity v^S is not necessarily perpendicular to the single-surface normal n_Γ, it is obvious that

$$[\![p^{\mathrm{FR}}]\!] = 0. \tag{88}$$

In the next step, (85) has to be applied to the fluid component φ^{FM}. Thus,

$$\left[\!\left[\left(\varepsilon^{\mathrm{FM}}+\tfrac{1}{2}\mathbf{v}_{\mathrm{FM}}\cdot\mathbf{v}_{\mathrm{FM}}\right)\rho^{\mathrm{FM}}\mathbf{w}_{\mathrm{FM}\Gamma}-\mathbf{T}^{\mathrm{FM}}\mathbf{v}_{\mathrm{FM}}+\mathbf{q}^{\mathrm{FM}}\right]\!\right]\cdot\mathbf{n}_{\Gamma}=0. \tag{89}$$

Following the same procedure as to obtain (71) from (70) with the gaseous phase of φ^{FM} only in B^+ and the liquid phase of φ^{FM} only in B^-, one obtains

$$\begin{aligned}&\left(\varepsilon^{\mathrm{G}}+\tfrac{1}{2}\mathbf{v}_{\mathrm{G}}\cdot\mathbf{v}_{\mathrm{G}}\right)\rho^{\mathrm{G}}\mathbf{w}_{\mathrm{G}\Gamma}\cdot\mathbf{n}_{\Gamma}-\left(\mathbf{T}^{\mathrm{G}}\mathbf{v}_{\mathrm{G}}-\mathbf{q}^{\mathrm{G}}\right)\cdot\mathbf{n}_{\Gamma}\\&\quad-\left(\varepsilon^{\mathrm{L}}+\tfrac{1}{2}\mathbf{v}_{\mathrm{L}}\cdot\mathbf{v}_{\mathrm{L}}\right)\rho^{\mathrm{L}}\mathbf{w}_{\mathrm{L}\Gamma}\cdot\mathbf{n}_{\Gamma}+\left(\mathbf{T}^{\mathrm{L}}\mathbf{v}_{\mathrm{L}}-\mathbf{q}^{\mathrm{L}}\right)\cdot\mathbf{n}_{\Gamma}=0.\end{aligned} \tag{90}$$

Applying $(73)_{1,2}$ to (90) yields

$$\begin{aligned}&\left(\varepsilon^{\mathrm{G}}+\tfrac{1}{2}\mathbf{v}_{\mathrm{G}}\cdot\mathbf{v}_{\mathrm{G}}\right)\hat{\varrho}_{\Gamma}^{\mathrm{G}}-\left(\mathbf{T}^{\mathrm{G}}\mathbf{v}_{\mathrm{G}}-\mathbf{q}^{\mathrm{G}}\right)\cdot\mathbf{n}_{\Gamma}\\&\quad+\left(\varepsilon^{\mathrm{L}}+\tfrac{1}{2}\mathbf{v}_{\mathrm{L}}\cdot\mathbf{v}_{\mathrm{L}}\right)\hat{\varrho}_{\Gamma}^{\mathrm{L}}+\left(\mathbf{T}^{\mathrm{L}}\mathbf{v}_{\mathrm{L}}-\mathbf{q}^{\mathrm{L}}\right)\cdot\mathbf{n}_{\Gamma}=0.\end{aligned} \tag{91}$$

Finally, this equation can be solved with the aid of $(73)_3$ yielding $\hat{\varrho}_{\Gamma}^{\mathrm{L}}=-\hat{\varrho}_{\Gamma}^{\mathrm{G}}$, such that

$$\hat{\varrho}_{\Gamma}^{\mathrm{L}}=\frac{\left(n^{\mathrm{G}}p^{\mathrm{GR}}\mathbf{v}_{\mathrm{G}}-n^{\mathrm{L}}p^{\mathrm{LR}}\mathbf{v}_{\mathrm{L}}+\mathbf{q}^{\mathrm{G}}-\mathbf{q}^{\mathrm{L}}\right)\cdot\mathbf{n}_{\Gamma}}{\varepsilon^{\mathrm{G}}-\varepsilon^{\mathrm{L}}+\tfrac{1}{2}\mathbf{v}_{\mathrm{G}}\cdot\mathbf{v}_{\mathrm{G}}-\tfrac{1}{2}\mathbf{v}_{\mathrm{L}}\cdot\mathbf{v}_{\mathrm{L}}}, \tag{92}$$

where $(26)_{2,3}$ has been used to substitute the partial stresses TGTG and TLTL.

In (91), the difference $\varepsilon^G-\varepsilon^L$ in internal energies can be substituted by the *Gibbs* energy (enthalpy) difference $\zeta^G-\zeta^L$, if one applies the *Legendre* transformation

$$\zeta^{\beta}=\varepsilon^{\beta}+\frac{p^{\beta\mathrm{R}}}{\rho^{\beta\mathrm{R}}}\quad\rightarrow\quad\varepsilon^{\beta}=\zeta^{\beta}-\frac{p^{\beta\mathrm{R}}}{\rho^{\beta\mathrm{R}}}. \tag{93}$$

Furthermore, since the effective pore pressure p^{FR} has been found to be jump-free, cf. (88), one can conclude to

$$[\![p^{\mathrm{FR}}]\!]=p^{\mathrm{FR}+}-p^{\mathrm{FR}-}=0\ \text{ with }\ \left\{\begin{array}{l}p^{\mathrm{FR}+}=s^{\mathrm{G}}p^{\mathrm{GR}}\\p^{\mathrm{FR}-}=s^{\mathrm{L}}p^{\mathrm{LR}}\end{array}\right.. \tag{94}$$

Following this, the partial pore pressures s^Gp^{GR} and s^Lp^{LR} and, as a result, the partial pressures n^Gp^{GR} and n^Lp^{LR} are equivalent, such that

$$n^{\mathrm{G}}p^{\mathrm{GR}}\mathbf{v}_{\mathrm{G}}-n^{\mathrm{L}}p^{\mathrm{LR}}\mathbf{v}_{\mathrm{L}}=n^{\mathrm{L}}p^{\mathrm{LR}}(\mathbf{v}_{\mathrm{G}}-\mathbf{v}_{\mathrm{L}})=n^{\mathrm{L}}p^{\mathrm{LR}}(\mathbf{w}_{\mathrm{G}}-\mathbf{w}_{\mathrm{L}}). \tag{95}$$

Inserting (93) and (95) in (91) finally yields

$$\hat{\varrho}_{\Gamma}^{\mathrm{L}}=\frac{\left[n^{\mathrm{L}}p^{\mathrm{LR}}(\mathbf{w}_{\mathrm{G}}-\mathbf{w}_{\mathrm{L}})+\mathbf{q}^{\mathrm{G}}-\mathbf{q}^{\mathrm{L}}\right]\cdot\mathbf{n}_{\Gamma}}{\Delta\zeta_{\mathrm{vap}}-\frac{p^{\mathrm{GR}}}{\rho^{\mathrm{GR}}}+\frac{p^{\mathrm{LR}}}{\rho^{\mathrm{LR}}}+\tfrac{1}{2}\mathbf{v}_{\mathrm{G}}\cdot\mathbf{v}_{\mathrm{G}}-\tfrac{1}{2}\mathbf{v}_{\mathrm{L}}\cdot\mathbf{v}_{\mathrm{L}}}, \tag{96}$$

where $\Delta\zeta_{vap}:=\zeta^G-\zeta^L$ is the latent heat or the enthalpy of evaporation.

Equation (92) is often simplified with the argument that differences in mass-specific pressures $\frac{p^{LR}}{\rho^{LR}}-\frac{p^{GR}}{\rho^{GR}}$ and in mass-specific kinetic energies $\frac{1}{2}\mathbf{v}_G\cdot\mathbf{v}_G-\frac{1}{2}\mathbf{v}_L\cdot\mathbf{v}_L$ are small in comparison with the latent heat $\Delta\zeta_{vap}$, cf. Morland and Gray (1995). Following this argumentation leads to

$$\hat{\varrho}_{\Gamma}^{\mathrm{L}}=\frac{\left[n^{\mathrm{L}}p^{\mathrm{LR}}(\mathbf{w}_{\mathrm{G}}-\mathbf{w}_{\mathrm{L}})+\mathbf{q}^{\mathrm{G}}-\mathbf{q}^{\mathrm{L}}\right]\cdot\mathbf{n}_{\Gamma}}{\Delta\zeta_{\mathrm{vap}}}. \tag{97}$$

Furthermore, in case that phase transitions are mainly induced by heat, the pressure-dependent term in the nominator of (97) is negligible compared to the heat conduction and (97) reduces to

$$\hat{\varrho}^{L}_{\Gamma} = \frac{(\mathbf{q}^{G} - \mathbf{q}^{L}) \cdot \mathbf{n}_{\Gamma}}{\Delta\zeta_{\text{vap}}}. \qquad (98)$$

This equation, however, is well known from classical thermodynamics, cf. e.g. Silhavy (1997).

Finally, the interfacial normal n_Γ, cf. Figure 2, has to be found. For this purpose, use is made of the fact that the interface is oriented perpendicular to the gradient grad$\rho^{\beta R}$ of the fluid densities. Thus, similar to the level-set method, grad$\rho^{\beta R}$ is taken and normalized to provide a simple way for the determination of n_Γ.

With the above equations, the set of constitutive relations is completed and can be applied together with the governing equations for the computation of initial-boundary-value problems of phase transitions in deformable porous media with heat transfer.

COMPUTATIONAL ISSUES

Proceeding from a non-isothermal, single-temperature triphasic formulation of partially saturated soil filled with two fluid phases of a single substance and undergoing phase-change processes, any computational procedure is based on a basic set of six primary variables given by the solid displacement u_S, the seepage velocities w^G and w^L, the effective pore-fluid pressures p^{GR} and p^{LR}, and the temperature θ. Under quasi-static conditions, one obtains a coupling between the seepage velocities and the effective liquid and gas pressures resulting from the individual fluid momentum balances and the constitutive setting yielding *Darcy*-like relations, cf. (62). Following this reduces the set of primary variables from six to four: the solid displacement u_S, the effective pressures p^{GR} and p^{LR}, and the temperature θ. The corresponding set of governing equations is then given by the vector-valued overall momentum balance corresponding to u_S, the scalar-valued gas and liquid mass balance equations corresponding to p^{GR} and p^{LR}, and the scalar-valued overall energy balance corresponding to θ:

overall momentum balance:

$$\mathbf{0} = -\,\text{grad}\, p^{FR} + \text{div}\, \mathbf{T}^{S}_{E} + \rho\, \mathbf{g} - \hat{\rho}^{G}(\mathbf{w}_{G} - \mathbf{w}_{L}),$$

gas and liquid mass balances:

$$(\rho^{\beta})'_{S} + \rho^{\beta} \text{div}\, \mathbf{v}_{S} + \text{div}\,(\rho^{\beta}\mathbf{w}_{\beta}) = \hat{\rho}^{\beta}, \quad \beta = \{G, L\},$$

overall energy balance:

$$\sum_{\alpha} \rho^{\alpha} (\varepsilon^{\alpha})'_{\alpha} = (\mathbf{T}^{S}_{E} - p^{FR}\,\mathbf{I}) \cdot \mathbf{L}_{S} - n^{L} p^{LR}\, \text{div}\, \mathbf{w}_{L} - n^{G} p^{GR}\, \text{div}\, \mathbf{w}_{G}$$
$$- \,\text{div}\, \mathbf{q} - \hat{\mathbf{p}}^{L} \cdot \mathbf{w}_{L} - \hat{\mathbf{p}}^{G} \cdot \mathbf{w}_{G} - \hat{\rho}^{G}\left[\varepsilon^{G} - \varepsilon^{L} + \tfrac{1}{2}(\mathbf{w}_{G} \cdot \mathbf{w}_{G} - \mathbf{w}_{L} \cdot \mathbf{w}_{L})\right]. \qquad (99)$$

The overall momentum balance is obtained by summing up the momentum balances $(25)_3$ for φ^S, φ^G and φ^L, where the partial stresses T^α and the solid extra stresses $\mathbf{T}^{S}_{E} \approx \boldsymbol{\sigma}^{S}_{E}$ are given by $(26)_{2,3}$ and (45)1,ρ1,ρ is given by (5), and $\sum_\alpha \hat{\mathbf{p}}^{\alpha}$ has been computed according to $(23)_1$ and $(24)_2$. The gas and liquid mass balances have been taken from $(25)_2$, where they have been derived such that the material time derivatives of the gas and liquid phases have been expressed by the solid time derivative and additional terms according to the modified *Euler*ian setting of the fluid components. Under the assumption of negligible external heat supplies (r^α=0), the overall energy balance has been taken from (69), where the same stresses have been included as in the momentum balance. The total heat influx has been defined by $\mathbf{q} = \sum_\alpha \mathbf{q}^{\alpha}$ with qα after (67) and (68), and $\hat{\mathbf{p}}^{S}$ and $\hat{\rho}^{\beta}$ have been substituted with respect to $(23)_1$ and $(24)_{1,2}$. Furthermore, note in passing that the kinetic part of the phase-change energy, $\hat{\rho}^{G}\,[\,\frac{1}{2}(\mathbf{w}_{G} \cdot \mathbf{w}_{G} - \mathbf{w}_{L} \cdot \mathbf{w}_{L})\,]$, can be neglected under creeping-flow conditions. Thus, this term is dropped in the weak form of the governing equations.

Given (99), one obtains the weak form of the governing equations by multiplication of (99)1−3with the test functions $\delta\mathbf{u}_S$, δp^{GR}, δp^{LR} and $\delta\theta$ and integration over the volume BB using the *Gauss*ian integral theorem. Thus,

- overall momentum balance:

$$\mathcal{G}_{\mathbf{u}_S} = \int_{\mathcal{B}} (\boldsymbol{\sigma}^{S}_{E} - p^{FR}\,\mathbf{I}) \cdot \delta\boldsymbol{\varepsilon}_{S}\, dv - \int_{\mathcal{B}} \rho\, \mathbf{g} \cdot \delta\mathbf{u}_{S}\, dv + \int_{\mathcal{B}} \hat{\rho}^{L}\,(\mathbf{w}_{L} - \mathbf{w}_{G}) \cdot \delta\mathbf{u}_{S}\, dv$$
$$- \int_{\mathcal{S}} (\boldsymbol{\sigma}^{S}_{E} - p^{FR}\,\mathbf{I})\, \mathbf{n} \cdot \delta\mathbf{u}_{S}\, da = 0, \qquad (100)$$

- gas and liquid mass balances:

$$\mathcal{G}_{p^{\mathrm{GR}}} = \int_{\mathcal{B}} \left[(\rho^{\mathrm{G}})'_{\mathrm{S}} + \rho^{\mathrm{G}} \operatorname{div} (\mathbf{u}_{\mathrm{S}})'_{\mathrm{S}} \right] \delta p^{\mathrm{GR}} \, \mathrm{d}v - \int_{\mathcal{B}} \rho^{\mathrm{G}} \, \mathbf{w}_{\mathrm{G}} \cdot \operatorname{grad} \delta p^{\mathrm{GR}} \, \mathrm{d}v$$
$$- \int_{\mathcal{B}} \hat{\rho}^{\mathrm{G}} \, \delta p^{\mathrm{GR}} \, \mathrm{d}v + \int_{\mathcal{S}} \rho^{\mathrm{G}} \, \mathbf{w}_{\mathrm{G}} \cdot \mathbf{n} \, \delta p^{\mathrm{GR}} \, \mathrm{d}a = 0,$$
$$\mathcal{G}_{p^{\mathrm{LR}}} = \int_{\mathcal{B}} \left[(\rho^{\mathrm{L}})'_{\mathrm{S}} + \rho^{\mathrm{L}} \operatorname{div} (\mathbf{u}_{\mathrm{S}})'_{\mathrm{S}} \right] \delta p^{\mathrm{LR}} \, \mathrm{d}v - \int_{\mathcal{B}} \rho^{\mathrm{L}} \, \mathbf{w}_{\mathrm{L}} \cdot \operatorname{grad} \delta p^{\mathrm{LR}} \, \mathrm{d}v$$
$$- \int_{\mathcal{B}} \hat{\rho}^{\mathrm{L}} \, \delta p^{\mathrm{LR}} \mathrm{d}v + \int_{\mathcal{S}} \rho^{\mathrm{L}} \, \mathbf{w}_{\mathrm{L}} \cdot \mathbf{n} \, \delta p^{\mathrm{LR}} \, \mathrm{d}a = 0, \qquad (101)$$

- overall energy balance:

$$\mathcal{G}_{\theta} = \int_{\mathcal{B}} \Big\{ \rho^{\mathrm{S}} \, (\varepsilon^{\mathrm{S}})'_{\mathrm{S}} + \rho^{\mathrm{L}} \, (\varepsilon^{\mathrm{L}})'_{\mathrm{L}} + \rho^{\mathrm{G}} \, (\varepsilon^{\mathrm{G}})'_{\mathrm{G}} - (\boldsymbol{\sigma}^{\mathrm{S}}_{\mathrm{E}} - p^{\mathrm{FR}} \, \mathbf{I}) \cdot (\boldsymbol{\varepsilon}_{\mathrm{S}})'_{\mathrm{S}}$$
$$+ \left[\, \hat{\mathbf{p}}^{\mathrm{L}}_{\mathrm{E}} - n^{\mathrm{L}} \operatorname{grad} p^{\mathrm{LR}} + p^{\mathrm{C}} \, (s^{\mathrm{G}} \operatorname{grad} n^{\mathrm{L}} - s^{\mathrm{L}} \operatorname{grad} n^{\mathrm{G}}) \right] \cdot \mathbf{w}_{\mathrm{L}}$$
$$+ \, (\hat{\mathbf{p}}^{\mathrm{G}}_{\mathrm{E}} - n^{\mathrm{G}} \operatorname{grad} p^{\mathrm{GR}}) \cdot \mathbf{w}_{\mathrm{G}} + \hat{\rho}^{\mathrm{G}} \, (\varepsilon^{\mathrm{G}} - \varepsilon^{\mathrm{L}}) \Big\} \, \delta\theta \, \mathrm{d}v$$
$$- \int_{\mathcal{B}} (\mathbf{q} + n^{\mathrm{L}} \, p^{\mathrm{LR}} \, \mathbf{w}_{\mathrm{L}} + n^{\mathrm{G}} \, p^{\mathrm{GR}} \, \mathbf{w}_{\mathrm{G}}) \cdot \operatorname{grad} \delta\theta \, \mathrm{d}v$$
$$+ \int_{\mathcal{S}} (\mathbf{q} + n^{\mathrm{L}} \, p^{\mathrm{LR}} \, \mathbf{w}_{\mathrm{L}} + n^{\mathrm{G}} \, p^{\mathrm{GR}} \, \mathbf{w}_{\mathrm{G}}) \cdot \mathbf{n} \, \delta\theta \, \mathrm{d}a = 0. \qquad (102)$$

In (100)–(102), the mass-transition term $\hat{\rho}^{\mathrm{G}}$ is given by (76), (77), and (96), and $\hat{\mathbf{p}}^{\mathrm{L}}_{\mathrm{E}}$ and $\hat{\mathbf{p}}^{\mathrm{G}}_{\mathrm{E}}$ are given by (61). To set an example on how the temperature change comes into play, consider the terms

$$\rho^{\mathrm{S}}_{0\mathrm{S}} (\varepsilon^{\mathrm{S}})'_{\mathrm{S}} - \boldsymbol{\sigma}^{\mathrm{S}}_{\mathrm{E}} \cdot (\boldsymbol{\varepsilon}_{\mathrm{S}})'_{\mathrm{S}} = \rho^{\mathrm{S}}_{0\mathrm{S}} \, c^{\mathrm{S}}_{\mathrm{V}} \, \theta'_{\mathrm{S}} - m^{\mathrm{S}} \Delta\theta \, (\boldsymbol{\varepsilon}_{\mathrm{S}})'_{\mathrm{S}} \cdot \mathbf{I}, \qquad (103)$$

cf. $(45)_1$ and (48), where, in the framework of small-strain approaches, $\rho^{\mathrm{S}} \approx \rho^{\mathrm{S}}_{0\mathrm{S}}$ has been used. Further terms yielding $\theta'_{\beta} = \theta'_{\mathrm{S}} + \operatorname{grad} \theta \cdot \mathbf{w}_{\beta}$ are included in $(\varepsilon^{\beta})'_{\beta}$.

Given the above theoretical framework, the finite-element method (FEM) can be applied for the description and the illustration of numerical examples.

Condensation of CO_2 in a Deformable Porous Rock

In order to present the potential of the derived model, a two-dimensional example of a condensation process of CO_2 in a porous rock is simulated. Choosing this example, we are aware that considering the pure substance is somehow academic, since in real processes, such as CO_2 sequestration, the dissolution of CO_2 in saline water or brine, respectively, comes into play. Thus, our example might be considered as a state, where the injected CO_2 has completely displaced the saline water. Furthermore, the choice of CO_2 as the fluid under consideration results from the good knowledge of its thermodynamical behavior and its low-temperature condensation point. Nevertheless, substituting CO_2 by water would also have been an option.

The initial state of the simulation domain of 10 m × 10 m is composed of a thermoelastic porous solid filled with gaseous CO_2, which is guaranteed by an initial pore pressure of p^{FR}=4.0pFR=4.0 MPa. This value is also applied as *Dirichlet* boundary condition at the upper boundary in order to simulate an open boundary with connection to a surrounding environment also containing gaseous CO_2. Note in passing that $s^L=0$ at the upper boundary yields $p^{FR}=p^{GR}$. The initial temperature in the whole system is set to 320 K, while the domain is horizontally confined at the left and right boundaries and vertically confined at the bottom, cf. Figure 4. Then, the blue-colored part of 2 m at the bottom is subjected to a temperature decrease from 320 K to 200 K over 500 s and held constant thereafter.

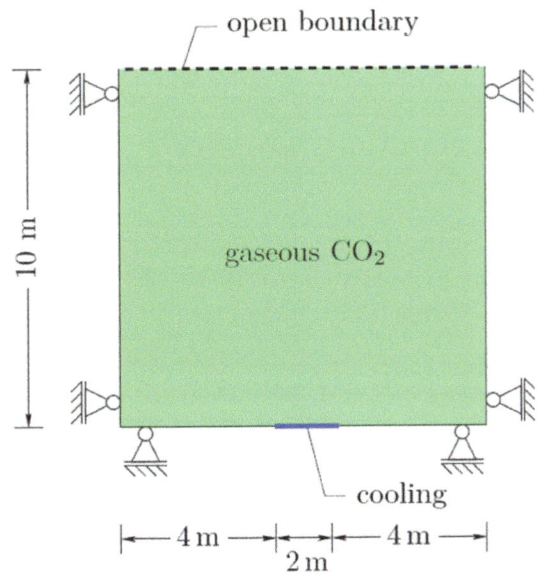

Figure 4: Simulation setup with constant pressure p^{FR}=4.0 MPa at the top boundary and cooling from 320 K to 200 K at the *blue* part of the bottom boundary.

For a realistic simulation, the solid parameters are resembling sandstone, which are included in Table 2. The thermodynamic parameters of CO_2 in Table 2 have been taken from Abbott and Ness (1989), and fluid and solid parameters from Graf (2008) and Rutqvist et al. (2010). The effective fluid shear viscosities are determined as a function of temperature and effective density given by Fenghour et al. (1998) via

$$\begin{aligned}\mu^{\beta R}(\theta,\, \rho^{\beta R}) &= \left[\mu_0^{\beta R}(\theta) + \Delta\mu^{\beta R}(\theta,\, \rho^{\beta R})\right] 10^{-6},\\ \mu_0^{\beta R}(\theta) &= \frac{1.00697\,\sqrt{\theta}}{e^{\sigma^*(\theta)}} \quad \text{with} \quad \sigma^*(\theta) = \sum_{i=1}^{m} a_i \left(\ln \frac{\theta}{251.196\,\text{K}}\right)^i,\\ \Delta\mu^{\beta R}(\theta,\, \rho^{\beta R}) &= \sum_{i=0}^{n} b_i(\theta)\,(\rho^{\beta R})^i \quad \text{with} \quad b_i(\theta) = \sum_{j=1}^{m} d_{ij} \left(\frac{\theta}{251.196\,\text{K}}\right)^{1-j},\end{aligned} \tag{104}$$

where the divergence of the viscosity around the critical point has been omitted, since we do not particularly describe the critical region here. However, for the formulation with respect to the critical region, the reader is referred to Vesovic et al. (1990).

In contrast to the original publication, (104)11 has been multiplied by 10^{-6}. This is due to the fact that *Fenghour et al.* have based their shear viscosity formulation on μPa s instead of Pa s, what is required in this publication. The coefficients a_i and d_{ij} included in $(104)_{2,3}$ can be taken from Table 3, where all not listed d_{ij} are vanishing. Finally, note that m=4 and n=8.

Table 2: Parameters used for the 2D simulation

Initial solid volume fraction	$n_{0S}^S = 0.9$
Effective densities	$\rho_{0S}^{SR} = 2650\ \text{kg/m}^3$
Intrinsic permeability, permeability parameter	$K_{0S}^S = 1.3 \times 10^{-10}\text{m}^2,\ \pi = 1$
Lamé parameters	$\mu^S = 2.5 \times 10^9$ Pa, $\lambda^S = 1.67 \times 10^9$ Pa
Thermal expansion coefficient	$\alpha^S = 1.2 \times 10^{-5}\,1/\text{K}$
Medial grain diameter	$d_{50} = 6 \times 10^{-5}\text{m}$
Initial temperature	$\theta_0 = 320$ K
Brooks and *Corey* parameters	$p^D = 2000$ Pa, $l = 1.3$
Residual saturations	$s_{\text{res}}^L = 0.01,\ s_{\text{res}}^G = 0.01$
Thermodynamic parameters for CO_2	$\bar{R}^{CO_2} = 188.91$ mJ/K, $\theta_{\text{crit}}^{CO_2} = 304.21$ K
	$p_{\text{crit}}^{CO_2R} = 7.38 \times 10^6$ Pa
Antoine parameters for CO_2	$A = 7.8101,\ B = 987.44,\ C = 290.9$
Specific heat capacities	$c_V^S = 700$ J/(kg K), $c_V^{LR} = 933.6$ J/(kg K)
	$c_V^{GR} = 790.65$ J/(kg K)
Thermal conductivities	$H^S = 2000$ W/(m K), $H^{CO_2R} = 0.26$ W/(m K)

Table 3: Parameters for the calculation of the shear viscosities. *Left* Coefficients aiai for the formulation of the zero-density viscosity. *Right*

Coefficients dijdij for the formulation of the excess viscosity. Both for CO_2, cf. Fenghour et al. (1998)

i	a_i
0	0.235156
1	−0.491566
2	5.211155×10^{-2}
3	5.347906×10^{-2}
4	-1.537105×10^{-2}
ij	d_{ij}
11	0.4071119×10^{-2}
21	0.7198037×10^{-4}
64	$0.2411697 \times 10^{-16}$
81	$0.2971072 \times 10^{-22}$
82	$-0.1627888 \times 10^{-22}$

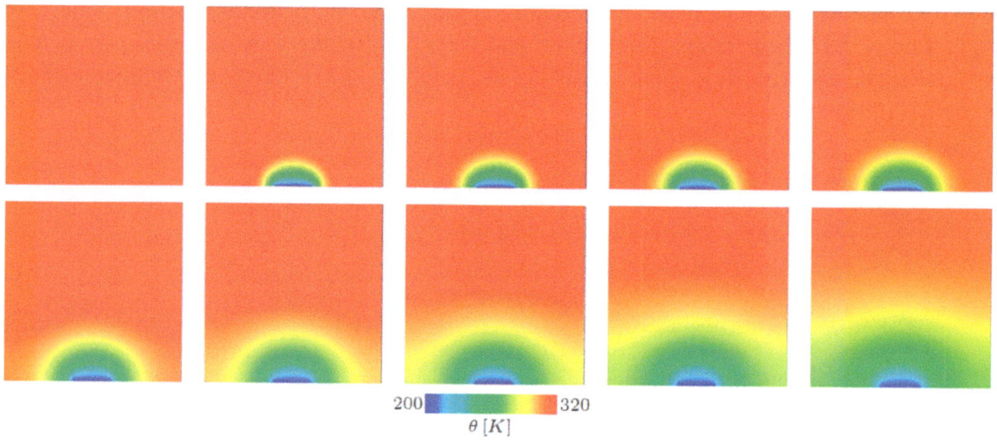

Figure 5: Temperature θ.

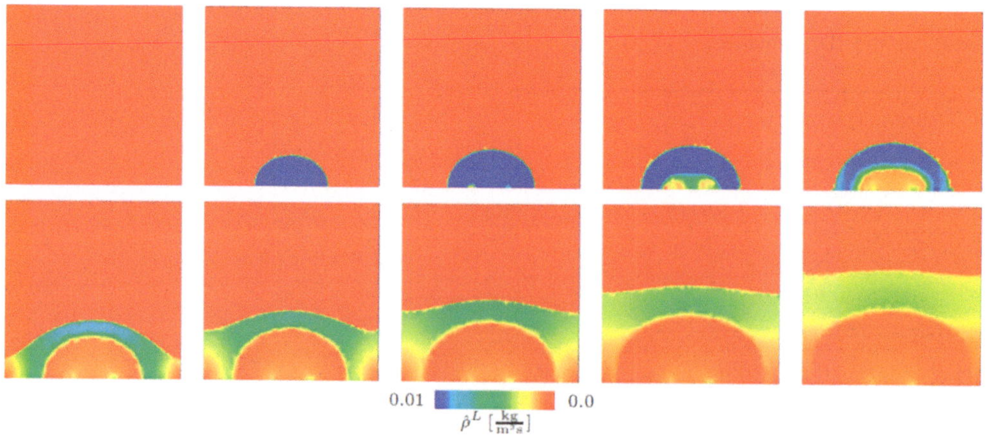

Figure 6: Liquid mass production $\hat{\rho}^L$.

After discretization, the strongly coupled system of partial differential equations given by (100)–(102) is solved monolithically with an unconditionally stable implicit time-integration scheme by use of the finite-element solver PANDAS.

The results of the simulation are depicted in Figs. 5, 6, 7, 8, 9, 10, and 11 and visualize the condensation of gaseous CO_2 due to cooling. For each parameter, ten snapshots are presented, taken during the simulation at times at 0 h, 6.6 h, 12.2 h, 17.5 h, 22.5 h, 35.0 h, 47.0 h, 59.7 h, 72.2 h, and 82.8 h. The first set of pictures in Figure 5 depicts the change in temperature due to the applied cooling condition. Figure 6 shows the liquid mass production $\hat{\rho}^L$, indicating the mass fraction of gaseous CO_2 transferred to liquid CO_2. It can be clearly observed that the mass transfer only appears in the transition zone, where both phases coexist. Note that this zone is indicated by the intermediate gas saturations s^G given in Figure 7. Thus, after complete transition of gaseous CO_2 to liquid CO_2, the mass production vanishes. The saturation plots, Figs. 7 and 8, also contain the gaseous and liquid seepage velocity vectors, respectively. It can be seen that gaseous CO_2 is replenished from the open boundary and liquid CO_2 is fanning-out along the transition zone. Finally, Figs. 9 and 10 show the partial pore densities $\rho_F^\beta := s^\beta \rho^{\beta R}$ of the fluids depicting the transition from gaseous CO_2 with a density of about 110 kg/m³ to liquid CO_2 with a maximum density of 1200 kg/m³. Consequently, this increase in density causes a drop in pore pressure p^{FR} that again affects the field of solid displacement vectors, represented in Figure 11 by black arrows exhibiting a settlement zone around the cooling region.

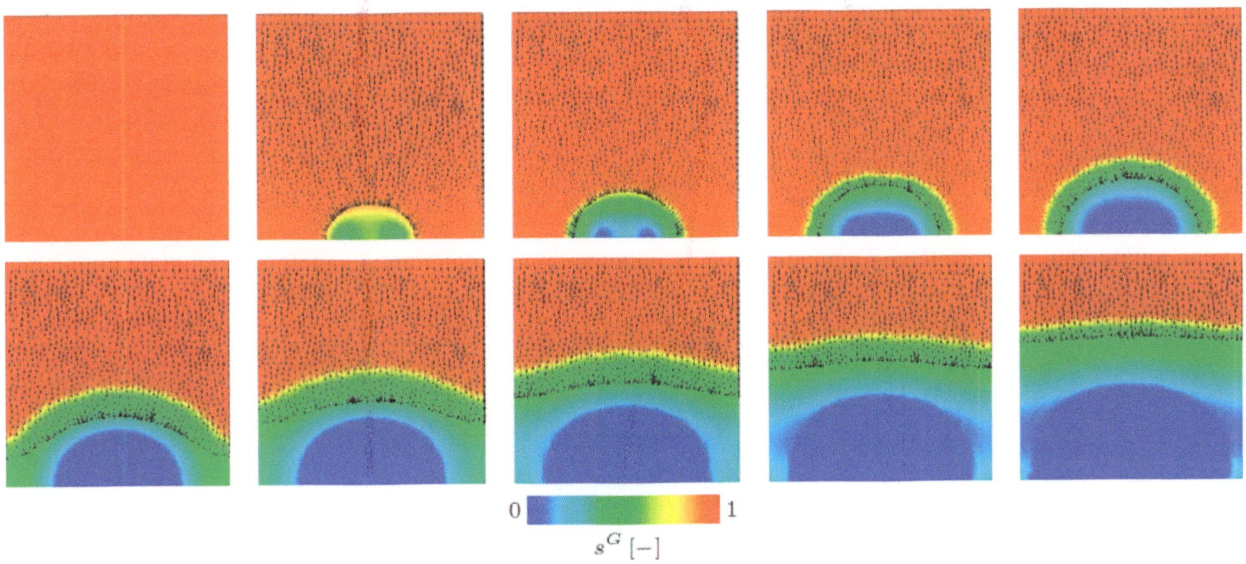

Figure 7: Gas saturation s^G and gaseous seepage-velocity vectors (*black arrows*).

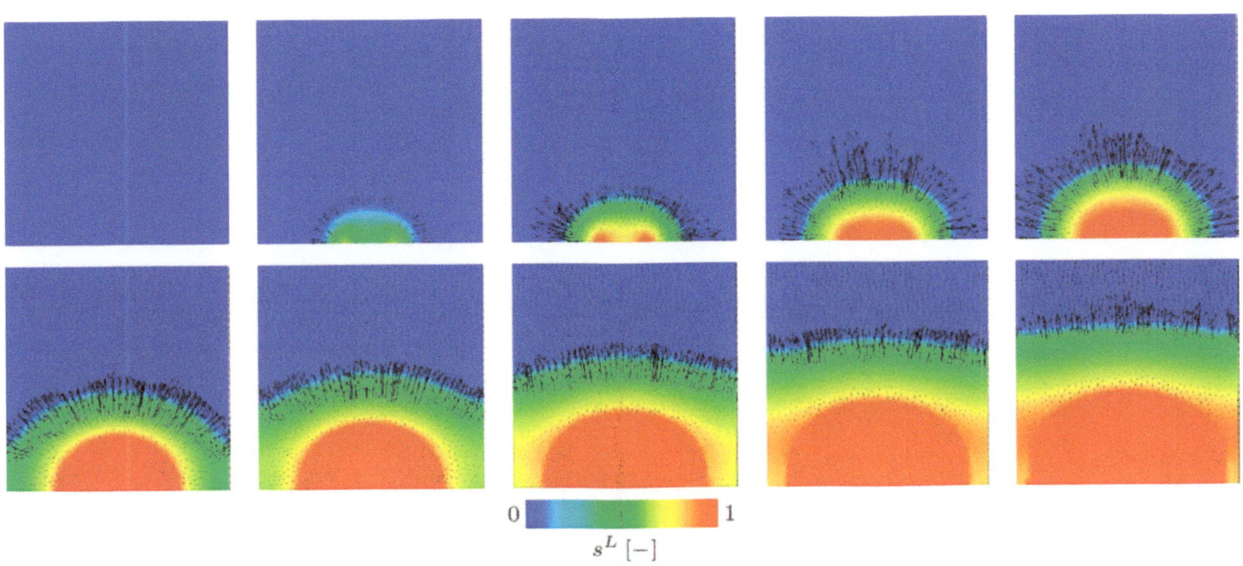

Figure 8: Liquid saturation s^L and liquid seepage-velocity vectors (*black arrows*).

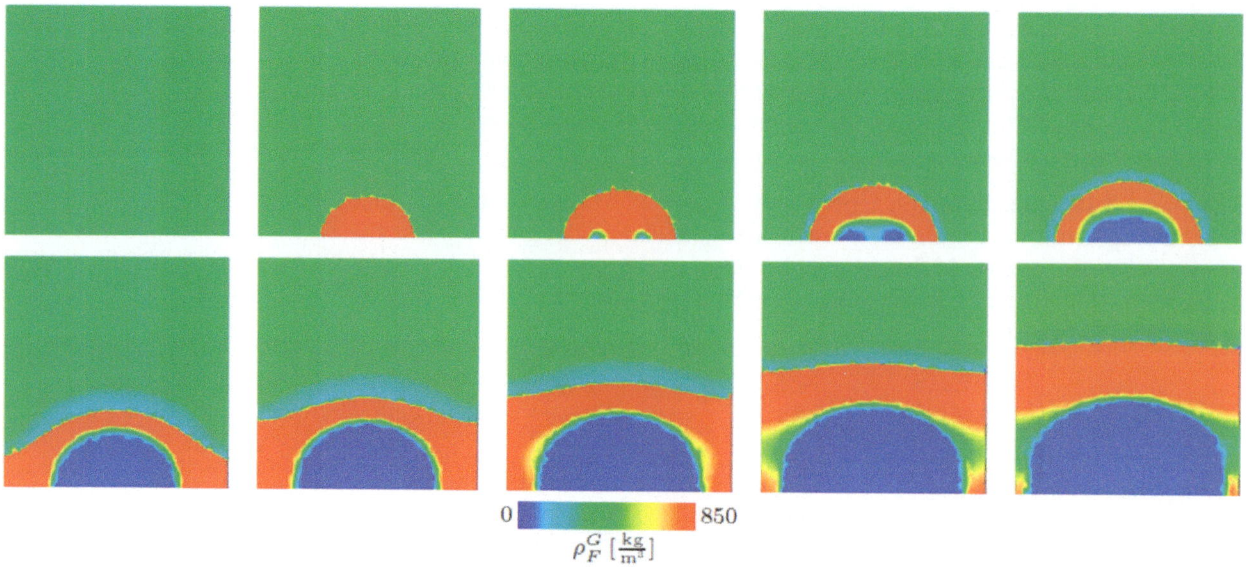

Figure 9: Partial pore density $\rho_F^{\mathrm{G}} = s^{\mathrm{G}} \rho^{\mathrm{GR}}$ of the gaseous phase.

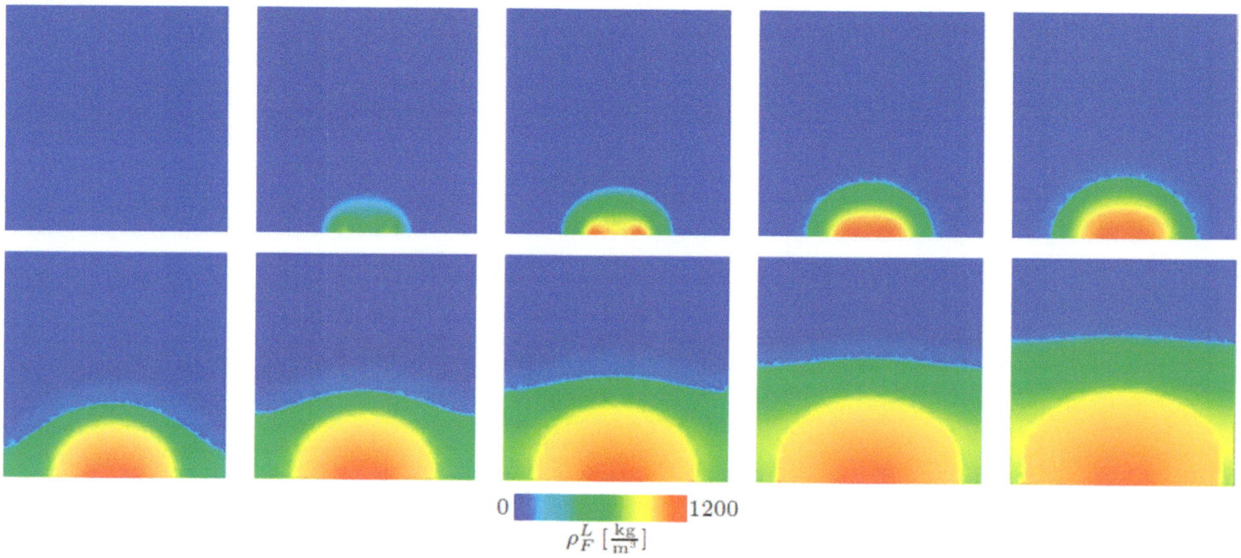

Figure 10: Partial pore density $\rho_F^{\mathrm{L}} = s^{\mathrm{L}} \rho^{\mathrm{LR}}$ of the liquid phase.

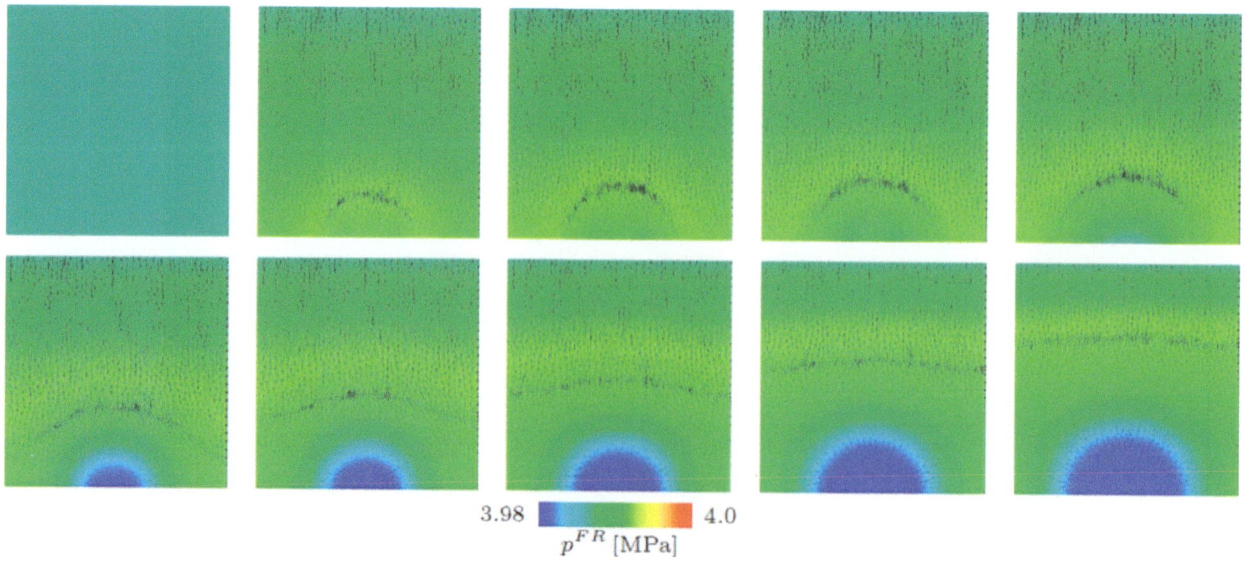

Figure 11: Pore pressure p^{FR} together with the solid displacement vectors (*black arrows*).

CONCLUSION

In this article, the phase-transition process between liquid and gaseous phases of the same fluid substance has been simulated by the mass transfer over a phase-change interface. Apart from the direct momentum productions $\hat{\mathbf{p}}^{\beta}$, the mass transfer couples the mass balance relations of the two fluid phases and influences the momentum and energy balances. To derive a thermodynamically consistent constitutive relation for the mass transfer, an immaterial, singular surface has been introduced representing the interface, across which jump conditions for the fluid constituents could be identified. Since the phase change did not affect the solid material, it has been assumed that the solid is continuous over the interface. The evaluation of the jump conditions of the balance equations led to a relation between the induced mechanical and non-mechanical power and the latent heat of vaporization, thus describing the mass transfer over the interface. After averaging the mass-transfer term over the REV by introducing a so-called interfacial area, the averaged mass-transfer $\hat{\rho}^{\mathrm{L}} = -\hat{\rho}^{\mathrm{G}}$ has been included in the global weak balance relations and, consequently, has been used for the simulation of gas–liquid phase transitions. Within a numerical example, the simulation of a condensation process of CO_2 in a thermoelastic porous rock was carried out showing reasonable results of the interaction between fluid thermodynamics, phase-change processes, fluid flow, and solid deformations with heat transfer.

The potential of the present model, which distinguishes from other more or less simple models, lies in the inclusion of solid deformations, compressible fluid phases based on the *van der Waals* equation, and an explicit constitutive relation of the interfacial mass production derived from the evaluation of balance relations at the interface, and not from additional assumptions.

NOMENCLATURE

The notation in this article follows the conventions commonly used in modern tensor calculus, compare, for example, Ehlers (2015), while the symbols used in the porous-media context follow the established nomenclature by de Boer (2000) and Ehlers (2002, 2009, 1989).

CONVENTIONS

Kernel conventions	
$(\cdot)$	Place holder for arbitrary quantities
$s, t, \ldots$ or $\sigma, \tau, \ldots$	Scalars (tensors of order 0)
$\mathbf{s}, \mathbf{t}, \ldots$ or $\boldsymbol{\sigma}, \boldsymbol{\tau}, \ldots$	Vectors (tensors of order 1)
$\mathbf{S}, \mathbf{T}, \ldots$ or $\boldsymbol{\Psi}, \boldsymbol{\Phi}, \ldots$	Tensors (tensors of order 2)
$\overset{n}{\mathbf{S}}, \overset{n}{\mathbf{T}} \ldots$	Higher-order tensors (tensors of order n)
Index and suffix conventions	
i, j, m, n	Indices as super- or subscripts range from 1 to N, where $N = 3$ indicates quantities of the usual three-dimensional (3D) space of our physical experience
$(\cdot)_\alpha$	Subscripts indicate kinematical quantities of a constituent within porous-media or mixture theories
$(\cdot)^\alpha$	Superscripts indicate the belonging of non-kinematical quantities to a constituent within mixture theories
$(\cdot)'_\alpha$	Material time derivative following the motion of a constituent α with the solid and fluid constituents $\alpha = \{\mathrm{S}, \mathrm{L}, \mathrm{G}\}$
$(\cdot)_{0\alpha}$	Initial value of a non-kinematical quantity with respect to the referential configuration of a constituent
$(\cdot)^{\mathrm{FM}}$, $(\cdot)_{\mathrm{FM}}$	Subscripts and superscripts "FM" indicate quantities of the fluid matter under consideration
$(\cdot)_{\mathrm{m}}$, $(\cdot)_\theta$	Subscripts "m" and "θ" indicate purely mechanical and purely thermal parts associated with thermoelastic solid kinematics
$(\cdot)_{\mathrm{crit}}$	Subscript "crit" indicates values at the critical point
$\mathrm{d}(\cdot)$	Differential operator
$\partial(\cdot)$	Partial derivative operator
$\delta(\cdot)$	Test functions of the respective degrees of freedom
$[\![(\cdot)]\!]$	Jump-related value on the discontinuity surface Γ
$(\cdot)^+, (\cdot)^-$	Quantity belonging to the pore gas ($\mathcal{B}^+ = \mathcal{B}^{\mathrm{G}}$) or pore liquid ($\mathcal{B}^- = \mathcal{B}^{\mathrm{L}}$)

Symbols

Symbol	Unit	Description
α		Constituent identifier in super- and subscript, i.e. $\alpha = \{S, L, G\}$
α^S	[1/K]	Coefficient of thermal expansion of φ^S
β		Fluid constituent identifier (here: $\beta = \{L, G\}$)
Γ		Interface between the fluid phases
ε^α	[J/kg]	Mass-specific internal energy of φ^α
$\hat{\varepsilon}^\alpha$	[J/m^3 s]	Volume-specific direct energy production of φ^α
ζ^α	[J/K m^3 s]	Mass-specific *Gibbs* energy (enthalpy) of φ^α
$\Delta\zeta_{vap}$	[J/K m^3 s]	Mass-specific latent heat or enthalpy of evaporation
η^α	[J/K kg]	Mass-specific entropy of φ^α
θ, θ^α	[K]	Absolute temperature of φ and φ^α
$\theta_0, \Delta\theta$	[K]	Initial temperature and temperature variation
κ	[–]	Exponent governing the deformation dependency of K^S
λ	[–]	Pore-size distribution index for *Brooks & Corey* law
λ^S	[N/m^2]	1st *Lamé* constant of φ^S
$\mu^{\beta R}$	[Pa s]	Effective dynamic fluid viscosity of φ^β
$\mu_0^{\beta R}, \Delta\mu^{\beta R}$	[µPa s]	Effective dynamic fluid viscosity of φ^β at zero density and excess part
μ^S	[N/m^2]	2nd *Lamé* constant of φ^S
π	[–]	Circle constant
ρ	[kg/m^3]	Density of the overall aggregate φ
$\rho^\alpha, \rho^{\alpha R}, \rho_F^\beta$	[kg/m^3]	Partial and effective (realistic) density of φ^α and partial pore density of φ^β
$\hat{\rho}^\alpha$	[kg/m^3 s]	Volume-specific mass production of φ^α
$\hat{\varrho}_\Gamma^\beta$	[kg/m^2 s]	Area-specific interfacial mass transfer of φ^β
σ^α		Scalar-valued supply terms of mechanical quantities
σ^*	[–]	Auxiliary term in the derivation of the shear viscosity
φ, φ^α		Overall aggregate and constituent α
ψ, ψ^α	[J/kg]	Mass-specific *Helmholtz* free energy of φ^α
Ψ^α	[·/m^3]	Volume-specific densities of scalar mechanical quantities
$\hat{\Psi}^\alpha$	[·/m^3]	Volume-specific productions of scalar mechanical quantities
$\boldsymbol{\sigma}^\alpha$		Vector-valued supply terms of mechanical quantities
$\boldsymbol{\phi}^\alpha$		General vector-valued mechanical quantities
$\boldsymbol{\Psi}^\alpha$	[·/m^3]	Volume-specific densities of vectorial mechanical quantities
$\hat{\boldsymbol{\Psi}}^\alpha$	[·/m^3]	Volume-specific productions of vectorial mechanical quantities
$\boldsymbol{\chi}_\alpha, \boldsymbol{\chi}_\alpha^{-1}$		Motion and inverse motion function of constituent φ^α
$\boldsymbol{\varepsilon}_S$	[–]	Linearised *Green–Lagrangean* solid strain tensor
$\boldsymbol{\Phi}, \boldsymbol{\Phi}^\alpha$		General tensor-valued mechanical quantities
$\boldsymbol{\sigma}_E^S$	[N/m^2]	Linearised 2nd *Piola–Kirchhoff* extra stress tensor of φ^S
a	[m^5/kg s^2]	Cohesion pressure, constant of the *van der Waals* equation
a_Γ	[1/m]	Volume-averaged interfacial area
b	[m^3/kg]	Co-volume, constant of the *van der Waals* equation
A, B, C	[–]	Empirical parameters of the *Antoine* equation
A^{GL}, A_Γ	[m^2]	Gas–liquid contact area in the volume-equivalent sphere, where $A^{GL} = A_\Gamma$
A^{SG}, A^{SL}	[m^2]	Solid–gas and solid–liquid contact areas of the volume-equivalent sphere
A_{REV}	[m^2]	Area of the REV
a_i, b_i, d_{ij}	[–]	Coefficients for the calculation of the shear viscosity
$\mathcal{B}, \mathcal{B}^\alpha$		Body of the overall aggregate and partial body of constituent φ^α
$\mathcal{G}$		Weak formulation of a governing equation related to a primary variable
$c_V^S, c_V^{\beta R}$	[J/kg K]	Solid and effective fluid-specific heat capacities at constant volume
d_{50}	[m]	Medial grain diameter of granular soil
$\mathrm{d}a_\Gamma, \mathrm{d}a_{\Gamma\,REV}$	[m^2]	Actual area element of the interface Γ and specific in the REV
$\mathrm{d}m^\alpha$	[kg]	Local mass element of φ^α
$\mathrm{d}v^\alpha$	[m^3]	Local volume element of φ^α

ACKNOWLEDGMENTS

The authors would like to thank the German Research Foundation (DFG) for financial support of this work within the International Graduate School for Non-linearities and Upscaling in Porous Media (NUPUS) at the University of Stuttgart.

REFERENCES

1. Abbott, M.M., van Ness, H.C.: Thermodynamics with Chemical Applications. McGraw-Hill, London (1989)
2. Alts, T., Hutter, K.: Continuum description of the dynamics and thermodynamics of phase boundaries between ice and water. Part I: surface balance laws and their interpretation in terms of three-dimensional balance laws averaged over the phase change boundary layer. J. Non-Equilib. Thermodyn. **13**, 221–257 (1988)
3. Antoine, C.: Tensions des vapeurs; nouvelle relation entre les tensions et les températures. Comptes Rendus des Séances de l'Académie des Sciences **107**, 681–684 (1888)
4. Bedeaux, D., Kjelstrup, S.: Transfer coefficients for evaporation. Phys. A **270**, 413–426 (1999)
5. Biot, M.A.: General theory of three dimensional consolidation. J. Appl. Phys. **12**, 155–164 (1941)
6. Bowen, R.M.: Theory of mixtures. In: Eringen, A.C. (ed.) Continuum Physics, III edn, pp. 1–127. Academic Press, New York (1976)
7. Bowen, R.M.: Incompressible porous media models by use of the theory of mixtures. Int. J. Eng. Sci. **18**, 1129–1148 (1980)
8. Bowen, R.M.: Compressible porous media models by use of the theory of mixtures. Int. J. Eng. Sci. **20**, 697–735 (1982)
9. Brooks, R.H., Corey, A.T.: Hydraulic Properties of Porous Media, Hydrology Papers, vol. 3. Colorado State University, Fort Collins (1964)
10. Chammari, A., Naon, B., Cherblanc, F., Bénet, J.C.: Water transport in soil with phase change. Mech. Model. Comput. Issues Civ. Eng. **23**, 135–142 (2005)
11. de Boer, R.: Thermodynamics of phase transitions in porous media. Appl. Mech. Rev. **48**, 613–622 (1995)
12. de Boer, R.: Theory of Porous Media. Springer, Berlin (2000)
13. de Boer, R., Bluhm, J.: Phase transitions in gas- and liquid-saturated porous solids. Transp. Porous Media **34**, 249–267 (1999)
14. de Boer, R., Kowalski, S.J.: Thermodynamics of fluid-saturated porous media with a phase change. Acta Mech. **109**, 167–189 (1995)
15. de Boer, R., Ehlers, W.: Theorie der Mehrkomponentenkontinua mit Anwendungen auf bodenmechanische Probleme. Forschungsberichte aus dem Fachbereich Bauwesen **40**, Universität-GH-Essen (1986)
16. Ehlers, W.: Poröse Medien – ein kontinuumsmechanisches Modell auf der Basis der Mischungstheorie. Habilitation thesis, Report No. II-22 of the Institute of Applied Mechanics (CE), University of Stuttgart 2012. Reproduction of Forschungsberichte aus dem Fachbereich Bauwesen der Universität-GH-Essen **47**, Essen (1989)
17. Ehlers, W.: Vector and Tensor Calculus. An Introduction. Lecture Notes, Institute of Applied Mechanics, University of Stuttgart 2015. http://www.mechbau.uni-stuttgart.de/ls2/Downloads/vektortensorskript_eng_ws1314.pdf
18. Ehlers, W.: Toward finite theories of liquid-saturated elasto-plastic porous media. Int. J. Plast.**7**, 433–475 (1991)
19. Ehlers, W.: Grundlegende Konzepte in der Theorie Poröser Medien. Tech. Mech. **16**, 63–76 (1996)
20. Ehlers, W.: Foundations of multiphasic and porous materials. In: Ehlers, W., Bluhm, J. (eds.) Porous Media: Theory, Experiments and Numerical Applications, pp. 3–86. Springer, Berlin (2002)
21. Ehlers, W., Graf, T., Ammann, M.: Deformation and localization analysis in partially saturated soil. Comput. Methods Appl. Mech. Eng. **193**, 2885–2910 (2004)
22. Ehlers, W.: Challenges of porous media models in geo- and biomechanical engineering including electro-chemically active polymers and gels. Int. J. Adv. Eng. Sci. Appl. Math. **1**, 1–24 (2009)
23. Ehlers, W., Graf, T.: Saturated elasto-plastic porous media under consideration of gaseous and liquid phase transitions. In: Schanz, T. (ed.) Theoretical and Numerical Unsaturated Soil Mechanics, Springer Proceedings in Physics 113, pp. 111–118. Springer, Berlin (2007)
24. Fenghour, A., Wakeham, W.A., Vesovic, V.: The viscosity of carbon dioxide. J. Phys. Chem. Ref. Data **27**, 31–44 (1998)
25. Ghadiani, S.R.: A Multiphasic Continuum Mechanical Model for Design Investigations of an Effusion-Cooled Rocket Thrust Chamber. Dissertation thesis, Report No. II-13 of the Institute of Applied Mechanics (CE), University of Stuttgart (2005)
26. Graf, T.: Multiphase Flow Processes in Deformable Porous Media under Consideration of Fluid Phase Transitions. Dissertation thesis, Report No. II-17 of the Institute of Applied Mechanics (CE), University of Stuttgart (2008)
27. Gray, W.G.: General conservation equations for multi-phase systems: 4. Constitutive theory including phase change. Adv. Water Resour. **6**, 130–140 (1983)
28. Hassanizadeh, S.M., Gray, W.G.: Mechanics and thermodynamics of multiphase flow in porous media including interphase boundaries. Adv. Water Resour. **13**, 169–186 (1990)
29. Huang, H., Lin, P., Zhou, W.: Moisture transport and diffusive instability during bread baking. SIAM J. Appl. Math. **68**, 222–238 (2006)
30. Jamet, D.: Diffuse interface models in fluid mechanics. GdR CNRS documentation, cf.http://pmc.polytechnique.fr/mp/GDR/docu/Jamet.pdf (2014)
31. Joekar-Niasar, V., Hassanizadeh, S.M., Leijnse, A.: Insights into the relationships among capillary pressure, saturation, interfacial area and relative permeability using pore-network modelling. Transp. Porous Media **74**, 201–219 (2008)
32. Juric, D., Tryggvason, G.: Computations of boiling flows. Int. J. Multiph. Flow **24**, 387–410 (1998)
33. Kowalski, S.J.: Toward a thermodynamics and mechanics of drying processes. Chem. Eng. Sci.**55**, 1289–1304 (2000)

34. Lozano, A.-L., Cherblanc, F., Cousin, B., Bénet, J.-C.: Experimental study and modelling of water phase change kinetics in soils. Eur. J. Soil Sci. **59**, 939–949 (2008)

35. Lozano, A.-L., Cherblanc, F., Bénet, J.-C.: Water evaporation versus condensation in a hygroscopic soil. Transp. Porous Media **80**, 209–222 (2009)

36. Lu, S.C.H., Pister, K.S.: Decomposition of deformation and representation of the free energy function for isotropic thermoelastic solids. Int. J. Solids Struct. **11**, 927–934 (1975)

37. Lykov, A.V.: On systems of differential equations for heat and mass transfer in capillary porous bodies. J. Eng. Phys. **26**, 11–17 (1974)

38. Markert, B.: A constitutive approach to 3-d nonlinear fluid flow through finite deformable porous continua. Transp. Porous Media **70**, 427–450 (2007)

39. Morland, L.W., Gray, J.M.N.T.: Phase change interactions and singular fronts. Contin. Mech. Thermodyn. **7**, 387–414 (1995)

40. Morland, L.W., Sellers, S.: Multiphase mixtures and singular surfaces. Int. J. Non-Linear Mech.**36**, 131–146 (2001)

41. Niessner, J., Hassanizadeh, S.M.: A model for two-phase flow in porous media including fluid–fluid interfacial area. Water Resour. Res. **44**, 1–10 (2008)

42. Niessner, J., Hassanizadeh, S.M.: Modeling kinetic interphase mass transfer for two-phase flow in porous media including fluid–fluid interfacial area. Transp. Porous Media **80**, 329–344 (2009a)

43. Niessner, J., Hassanizadeh, S.M.: Non-equilibrium interphase heat and mass transfer during two-phase flow in porous media—theoretical considerations and modeling. Adv. Water Resour.**32**, 1756–1766 (2009b)

44. Ruiz, T., Bénet, J.-C.: Phase change in a heterogeneous medium: comparison between the vaporisation of water and heptane in an unsaturated soil at two temperatures. Transp. Porous Media **44**, 337–353 (2001)

45. Rutqvist, J., Vasco, D.W., Myer, L.: Coupled reservoir-geomechanical analysis of CO 2 CO 2 injection and ground deformations at In Salah, Algeria. Int. J. Greenhouse Gas Control **4**, 225–230 (2010)

46. Sahimi, M.: Flow and Transport in Porous Media and Fractured Rock: From Classical Methods to Modern Approaches, Second, Revised and Enlarged edn. Wiley-VCH, Weinheim (2011)

47. Schrefler, B.A., Zhan, X.: A fully coupled model for water flow and air flow in deformable porous media. Water Resour. Res. **29**, 155–167 (1993)

48. Schrefler, B.A., Scotta, R.: A fully coupled model for two-pase flow in deformable porous media. Comput. Methods Appl. Mech. Eng. **190**, 3223–3246 (2001)

49. Silhavy, M.: The Mechanics and Thermodynamics of Continuous Media. Springer, Berlin (1997)

50. Tanguy, S., Ménard, T., Berlemont, A.: A level set method for vaporizing two-phase flow. J. Comput. Phys. **221**, 837–853 (2007)

51. Truesdell, C.: Rational Thermodynamics, 2nd edn. Springer, New York (1984)

52. van der Waals, J.D.: Over de Continuiteit van den Gas-en Vloeistoftoestand. Academisch proefschrift, Sijthoff, Leiden (1873)

53. van Genuchten, M.T.: A closed-form equation for predicting the hydraulic conductivity of unsaturated soils. Soil Sci. Soc. Am. J. **44**, 892–898 (1980)

54. Vesovic, V., Wakeham, W.A., Olchowy, G.A., Sengers, J.V., Watson, J.T.R., Millat, J.: The transport properties of carbon dioxide. J. Phys. Chem. Ref. Data **19**, 763–808 (1990)

55. Wang, Y., Oberlack, M.: A thermodynamic model of multiphase flows with moving interfaces and contact line. Contin. Mech. Thermodyn. **23**, 409–433 (2011)

56. Whitaker, S.: Simultaneous heat, mass, and momentum transfer in porous media: a theory of drying. Adv. Heat Transf. **13**, 119–203 (1977)

57. Wieners, C., Ammann, M., Graf, T., Ehlers, W.: Parallel Krylov methods and the application to 3-d simulations of a triphasic porous media model in soil mechanics. Comput. Mech. **36**, 409–420 (2005)

CHAPTER 13

THERMODYNAMICS OF POLARIZED RELATIVISTIC MATTER

Pavel Kovtun

Keywords: Holography and quark-gluon plasmas, Thermal Field Theory

INTRODUCTION

We would like to understand collective macroscopic behavior of matter subject to external fields. In the high-temperature limit this is often captured by classical hydrodynamics. The ingredients that go into writing down the hydrodynamic equations are: the identification of relevant variables (conserved densities, order parameters), the derivative expansion (small gradients near equilibrium), and symmetry constraints. The hydrodynamic equations are modified when the system is subject to external electric and magnetic fields. The latter will induce polarization (electric, magnetic, or both) in a fluid, and as a result the transport properties of the fluid will change. Our focus here will be on isotropic relativistic matter because a) electromagnetic fields are intrinsically relativistic, b) relativistic fluids have more symmetry than non-relativistic fluids, and c) relativistic fluids have been a subject of much recent attention in the literature due to their appearance in heavy-ion physics [1, 2], in gravitational physics, through the holographic duality [3], and even in condensed matter physics [4, 5]. The systematic description (including the derivative expansion) of polarized relativistic fluids is largely missing, and the present paper is a step in filling that gap.

In order to understand the hydrodynamics of matter subject to external fields, one needs to understand its thermodynamics first. In what follows we will describe the procedure for obtaining the energy-momentum tensor and the current density for stationary equilibrium polarized matter subject to external gravitational and electromagnetic fields. We will find simple expressions for "bound" currents, including equilibrium surface currents and surface momenta.

Let us start with the standard description of equilibrium thermodynamics without external fields. In the grand canonical ensemble at temperature $T_0 = 1/\beta_0$ and chemical potential μ_0, extensivity in the large-volume limit dictates that the logarithm of the grandcanonical partition function $Z[T_0, \mu_0]$ is proportional to the d-dimensional spatial volume,

$$\ln Z = \beta_0 \int d^d x\; P(T_0, \mu_0)\,,$$

where the pressure $P(T_0, \mu_0)$ is constant in equilibrium [6]. The partition function Z may be computed from a Euclidean path integral with a Euclidean (imaginary) time compactified with period β_0, see e.g. [7]. In the path integral action, the fundamental fields of the microscopic theory can then be coupled to time-independent external sources: the (Euclidean) metric $g^E_{\mu\nu}$ and the (Euclidean) gauge field A^E_μ. The gauge field couples to the conserved current, whose time component is the charge density corresponding to the chemical potential. See ref. [8] for a convenient parametrization of the Euclidean sources g^E and A^E. The Euclidean path integral gives rise to the partition function $Z = Z[T_0, \mu_0, g^E, A^E]$, where $T_0 = 1/\beta_0$ is the coordinate periodicity of the Euclidean time. We assume that the coupling to time-independent external sources leaves the system in equilibrium, so that no entropy is produced. The temperature and the chemical potential will be altered by the external sources and are not uniform any more. For example, the equilibrium temperature becomes $T(\mathbf{x}) = T_0/\sqrt{g^E_{00}(\mathbf{x})}$ [6]. Similarly, the chemical potential will be shifted by the time component of the external gauge field. We can write $W = -i\ln Z$ as

Citation: Pavel Kovtun, "Thermodynamics of polarized relativistic matter," doi:10.1007/JHEP07(2016)028.

$$W[T_0,\mu_0,g^E,A^E] = -i\beta_0 \int d^dx\, \sqrt{g^E}\, \mathcal{F}(T_0,\mu_0,g^E,A^E)\,, \tag{1.1}$$

where $\sqrt{g^E}$ is the square root of the determinant of $g^E_{\mu\nu}$, and F is the negative of the grand canonical free energy density. In flat space and without external gauge fields, F reduces to the pressure P, and in general F is a complicated function of the spatially varying external sources. In a slight abuse of terminology, we will refer to F as the free energy density, and to W as the free energy. Varying W with respect to a time-independent source gives rise to a zero-frequency insertion in the Euclidean path integral of the operator coupled to the source. The relevant operators are the energy-momentum tensor (coupled to the metric), and the conserved current (coupled to the gauge field). Thus W is the generating functional for zero-frequency correlation functions of the energy-momentum tensor and the current in equilibrium.

The Euclidean external sources $g^E_{\mu\nu}$ and A^E_μ may be "un-Wick-rotated" to Minkowski time to obtain the physical real-time external sources $g_{\mu\nu}$ and A_μ, for example $g^E_{00} = -g_{00}$, $g^E_{0k} = -ig_{0k}$, $A^E_0 = -iA_0$ etc. In what follows we will omit the dependence on T_0 and μ_0, and will denote the Euclidean generating functional with arguments continued to physical time as W[g, A], so that

$$W[g,A] = \int d^{d+1}x\, \sqrt{-g}\, \mathcal{F}(T_0,\mu_0,g,A)\,, \tag{1.2}$$

where $\int d^{d+1}x$ stands for $-i\beta_0 \int d^dx$. We may as well view R d d+1x as containing an integral over the physical time, as the argument of the integral does not depend on time anyway.

For a relativistic microscopic theory without gauge and gravitational anomalies, the generating functional is both gauge- and diffeomorphism-invariant. Let us further assume that all long-range interactions are screened due to a non-zero temperature T, so that the spatial correlations are local on scales longer than the screening length. The effective description of static correlations on such long scales will then be given by W[g, A], where the density F is a local function of the external sources. For external fields that vary slowly in space, the above locality implies that F may be written as a derivative expansion in the gradients of the external fields. See ref. [8] for a study of the local generating functional in the Euclidean form, and ref. [9] for the Minkowski form. Here we will use the Minkowski form, in which the underlying gauge and diffeomorphism invariance is manifested in a more straightforward way.

In order to implement the derivative expansion in practice, one needs to postulate the derivative scaling of the external sources $g_{\mu\nu}$ and A_μ. Physically, this amounts to deciding whether the external sources are taken as "strong" or "weak" on the scale of the spatial inhomogeneity in equilibrium. In refs. [8, 9], the external sources were taken as "weak" in the sense that both $g_{\mu\nu}$ and A_μ were assumed to be O(1) in the derivative expansion, so that both electric and magnetic fields appear at order $O(\partial)$ in the expansion. This choice of scaling makes the description of equilibrium polarization rather awkward: for example, in 3+1 dimensions, the thermodynamic response to constant homogeneous magnetic field B appears at the same order as the response to two derivatives of temperature, $B^2 \sim (\partial T)^2$.

In order to describe polarized matter in constant (or slowly varying in space) electric and magnetic fields, a different derivative counting scheme is more natural, one in which constant homogeneous electric and magnetic fields are taken to be O(1) in the derivative expansion, rather than order $O(\partial)$. This will be our goal here: to implement the derivative expansion of the free energy in the regime when the external gravitational field is still "weak" so that $g_{\mu\nu}$ is O(1), while the external electromagnetic field is "strong" so that the field strength $F_{\mu\nu} = \partial_\mu A_\nu - \partial_\nu A_\mu$ is O(1). Following the general approach of ref. [9], we will obtain simple expressions for the energy-momentum tensor and the conserved current in relativistic polarized matter subject to external fields.

FREE ENERGY

Thermodynamic Parameters

Let us first outline the starting point. The free energy W[g, A] is a gauge- and diffeomorphism-invariant functional of A_μ, $g_{\mu\nu}$, and their derivatives. Being in equilibrium means that that there is a Killing vector V, such that the Lie derivative with respect to V vanishes on all observables, $\pounds_V(\ldots)=0$. In suitable coordinates, $V^\mu = (1, 0)$. In the grand canonical ensemble, the equilibrium state is parametrized by the temperature, velocity, and the chemical potential. Their relation to the external sources is

$$T = \frac{1}{\beta_0\sqrt{-V^2}}\,, \qquad u^\mu = \frac{V^\mu}{\sqrt{-V^2}}\,, \qquad \mu = \frac{V^\mu A_\mu + \Lambda_V}{\sqrt{-V^2}}\,, \tag{2.1}$$

where β_0 is a constant setting the normalization of temperature, and Λ_V is a gauge parameter which ensures that μ is gauge-invariant. The constant μ0 is absorbed into Λ_V. Without external gauge fields, relations (2.1) are the covariant versions of the statement that $T\sqrt{-g_{00}}$ and $\mu\sqrt{-g_{00}}$ are constant in equilibrium [6]. The vector u^μ is the normalized ($u^2 = -1$) velocity of matter, and the coordinates in which $V^\mu = (1, 0)$ correspond to the matter "at rest". Both T and μ are gauge invariant and transform as scalars under diffeomorphisms. For a discussion of gauge and diffeomorphism covariance of the equilibrium parameters see section 5 of ref. [10]. For a system occupying a spacetime region M with a boundary ∂M, we assume that the generating functional can be separated into bulk and boundary contributions, and we take

$$W[g,A] = \int_{\mathcal{M}} d^{d+1}x\, \sqrt{-g}\, \mathcal{F} + \int_{\partial\mathcal{M}} d^dx\, \sqrt{\gamma}\, L\,. \tag{2.2}$$

Here the first term describes the bulk contribution, and the second term the boundary contribution. To leading order in the derivative expansion, F is the pressure, and L is the surface tension. In the bulk term, g is the determinant of $g_{\mu\nu}$, and F is a function of T, u^μ, μ, as well as of the sources A_μ, $g_{\mu\nu}$, and their derivatives. For the boundary with coordinates y a whose shape is specified by $x^\mu(y^a)$, the tangent vectors are $e^\mu_a = \partial x^\mu/\partial y^a$, and the projector onto the boundary is $P^{\mu\nu} = g^{\mu\nu} - n^\mu n^\nu$.

The induced metric on the boundary is $\gamma_{ab} = e^\mu_a e^\nu_b g_{\mu\nu}$. In the boundary term, γ is the determinant of the induced metric, while L in addition may depend on n^μ, the spacelike unit normal vector to the boundary.

Response to External Sources

The energy-momentum tensor and the current are defined as

$$\begin{aligned}\delta W[g,A] &= \frac{1}{2}\int_{\mathcal{M}} d^{d+1}x\,\sqrt{-g}\,T^{\mu\nu}\,\delta g_{\mu\nu} + \int_{\mathcal{M}} d^{d+1}x\,\sqrt{-g}\,J^\mu\,\delta A_\mu \\ &+\frac{1}{2}\int_{\partial\mathcal{M}} d^dx\,\sqrt{\gamma}\,T^{\mu\nu}_{\rm s}\,\delta g_{\mu\nu} + \int_{\partial\mathcal{M}} d^dx\,\sqrt{\gamma}\,J^\mu_{\rm s}\,\delta A_\mu \\ &+\int_{\partial\mathcal{M}} d^dx\,\sqrt{\gamma}\,K^{\mu\nu}_{\rm s}\,\pounds_n\delta g_{\mu\nu} + \int_{\partial\mathcal{M}} d^dx\,\sqrt{\gamma}\,K^\mu_{\rm s}\,\pounds_n\delta A_\mu + \dots\,,\end{aligned} \tag{2.3}$$

where $\pounds_n$ is the Lie derivative along the normal, the dots denote boundary terms with higher normal derivatives of the sources. The variations are performed at fixed V^μ and Λ_V. Here $T^{\mu\nu}$, J^μ are the bulk energy-momentum tensor and the current, and $T^{\mu\nu}_s$, J^μ_s are the boundary energy-momentum tensor and the current. The tensors $K^{\mu\nu}_s$ and K^μ_s describe interactions between bulk and boundary degrees of freedom. The gauge- and diffeomorphism-invariance of W[g, A] implies conservation laws

$$\nabla_\mu J^\mu = 0\,, \tag{2.4a}$$

$$\nabla_\mu T^{\mu\nu} = F^{\nu\lambda}J_\lambda\,. \tag{2.4b}$$

The derivative expansion for the free energy density F in the generating functional (2.2) gives rise to the derivative expansion for the equilibrium $T^{\mu\nu}$ and J^μ, as described in [9]. The boundary energy-momentum tensor and the current in (2.3) may be decomposed into the contributions tangential to and normal to the boundary,

$$\begin{aligned}T^{\mu\nu}_{\rm s}\,\delta g_{\mu\nu} &= T^{ab}_{\rm s}\,\delta\gamma_{ab} + \Pi^a_{\rm s}\,\delta g_a + \Pi_{\rm s}\,\delta g_{\mathbf{n}}\,,\\ J^\mu_{\rm s}\,\delta A_\mu &= J^a_{\rm s}\delta A_a + J_{\rm s}\,\delta A_{\mathbf{n}}\,.\end{aligned}$$

Here δg_a is the pullback of $P^\lambda_\mu\,\delta g_{\lambda\nu}n^\nu$ to the boundary, $\delta g_{\mathbf{n}} = n^\mu n^\nu\delta g_{\mu\nu}$, δA_a is the pullback of δA_μ to the boundary, and $\delta A_{\mathbf{n}} = n^\mu\delta A_\mu$. Similarly, one can vary the generating functional with respect to the field strength $F_{\mu\nu}$,

$$\begin{aligned}\delta_F W[g,A] &= \frac{1}{2}\int_{\mathcal{M}} d^{d+1}x\,\sqrt{-g}\,M^{\mu\nu}\,\delta F_{\mu\nu} + \frac{1}{2}\int_{\partial\mathcal{M}} d^dx\,\sqrt{\gamma}\,M^{ab}_{\rm s}\,\delta F_{ab}\\ &+\int_{\partial\mathcal{M}} d^dx\,\sqrt{\gamma}\,M^a_{\rm s}\,\delta F_a + \int_{\partial\mathcal{M}} d^dx\,\sqrt{\gamma}\,L^{\mu\nu}_{\rm s}\,\pounds_n\delta F_{\mu\nu} + \dots\end{aligned} \tag{2.5}$$

where again δF_{ab} is the pullback of $\delta F_{\mu\nu}$ to the boundary, δF_a is the pullback of $P^\lambda_\mu\,\delta F_{\lambda\nu}n^\nu$ to the boundary, and the dots denote boundary terms with higher normal derivatives of $\delta F_{\mu\nu}$. This defines the bulk polarization tensor $M_{\mu\nu}$, and the boundary polarization tensor M^{ab}_s . 1 The surface terms $J_{\rm s}, \Pi^a_{\rm s}, \Pi_{\rm s}, M^a_{\rm s}$ depend on how the equilibrium is set up, and what the boundary conditions on ∂M are, as determined by the nature of the phase separation at ∂M.

In all the above variations, we assume that the region M occupied by matter is unchanged. One could also consider the response of the generating functional to changing the shape of ∂M, however this will not be needed for our purposes. See ref. [11] for a recent discussion of surface terms in the Euclidean generating functional.

The polarization tensor contains both electric and magnetic components. We define the electric field as $E_\mu \equiv F_{\mu\nu}u^\nu$, the magnetic field as $B \equiv -\frac{1}{2}\epsilon^{\mu\alpha\beta}u_\mu F_{\alpha\beta}$ for d = 2, and $B^\mu \equiv \frac{1}{2}\epsilon^{\mu\nu\alpha\beta}u_\nu F_{\alpha\beta}$ for d = 3. In 1+1 dimensions, we define the "magnetic field" as $B \equiv \frac{1}{2}\epsilon^{\mu\nu}F_{\mu\nu}$, so that $F_{\mu\nu} = -B\epsilon_{\mu\nu}$. The Levi-Civita tensor is $\epsilon^{\mu\nu\alpha\beta} = \varepsilon^{\mu\nu\alpha\beta}/\sqrt{-g}$, with $\varepsilon^{0123} = 1$, and similarly in other dimensions. Both E_μ and B_μ are spacelike and orthogonal to u^μ. We have the following decomposition of the field strength:

$$\begin{aligned}d=1: &\quad F_{\mu\nu} = u_\mu E_\nu - u_\nu E_\mu\,,\\ d=2: &\quad F_{\mu\nu} = u_\mu E_\nu - u_\nu E_\mu - \epsilon_{\mu\nu\rho}u^\rho B\,,\\ d=3: &\quad F_{\mu\nu} = u_\mu E_\nu - u_\nu E_\mu - \epsilon_{\mu\nu\rho\sigma}u^\rho B^\sigma\,.\end{aligned}$$

The electric polarization vector p^{α} and the magnetization vector m^{α} (for d=3) are defined by rewriting the integrand in (2.5) as $\frac{1}{2}M^{\mu\nu}\delta F_{\mu\nu} = p^{\alpha}\delta E_{\alpha} + m^{\alpha}\delta B_{\alpha}$. For d=2, the variation is $\frac{1}{2}M^{\mu\nu}\delta F_{\mu\nu} = p^{\alpha}\delta E_{\alpha} + m\delta B$, which defines the magnetization m. The decomposition of the polarization tensor into the electric and magnetic parts is then

$$d=1: \qquad M_{\mu\nu} = m\epsilon_{\mu\nu}\,, \tag{2.6a}$$

$$d=2: \qquad M_{\mu\nu} = p_{\mu}u_{\nu} - p_{\nu}u_{\mu} - \epsilon_{\mu\nu\rho}u^{\rho}m\,, \tag{2.6b}$$

$$d=3: \qquad M_{\mu\nu} = p_{\mu}u_{\nu} - p_{\nu}u_{\mu} - \epsilon_{\mu\nu\rho\sigma}u^{\rho}m^{\sigma}\,, \tag{2.6c}$$

where $p^{\mu} \equiv u_{\lambda}M^{\lambda\mu}$, $m^{\mu} \equiv \frac{1}{2}\epsilon^{\mu\nu\alpha\beta}u_{\nu}M_{\alpha\beta}$ (for d = 3), and $m \equiv -\frac{1}{2}\epsilon^{\mu\alpha\beta}u_{\mu}M_{\alpha\beta}$ (for d = 2). Both p_{α} and m_{α} are transverse to u^{α}.

Equilibrium Relations

The equilibrium definitions (2.1) together with $£_V(\ldots) = 0$ give

$$\partial_{\lambda}T = -Ta_{\lambda}\,, \qquad \partial_{\lambda}\mu = E_{\lambda} - \mu a_{\lambda}\,, \tag{2.7}$$

where $a^{\mu} \equiv u^{\lambda}\nabla_{\lambda}u^{\mu}$ is the acceleration vector, $u_{\mu}a^{\mu} = 0$. These relations imply that $T\,\partial_{\lambda}(\mu/T) - E_{\lambda}$ vanishes in equilibrium. In other words, a system subject to an external electric field will develop a gradient of μ/T in order to compensate the applied field and ensure that the equilibrium is maintained.

This has implication for derivative counting. For "weak" electric fields $E_{\lambda} \sim O(\partial)$, the gradients of T and μ are $O(\partial)$ as well. For "strong" electric fields $E_{\lambda} \sim O(1)$, there will be an O(1) gradient of μ/T. How exactly this gradient is achieved depends on the nature of the microscopic degrees of freedom. Given that the chemical potential determines the number of charge carriers, we take "strong" electric fields to mean that both E and ∂_{μ} are O(1), while ∂T is still $O(\partial)$, so that $\frac{\partial\mu}{\mu} \gg \frac{\partial T}{T}$. In the generating functional, the derivatives of the chemical potential may then be traded for the electric field.

Similarly, the derivative of the velocity can be decomposed in equilibrium as

$$d=1: \qquad \nabla_{\mu}u_{\nu} = -u_{\mu}a_{\nu}\,, \tag{2.8a}$$

$$d=2: \qquad \nabla_{\mu}u_{\nu} = -u_{\mu}a_{\nu} - \frac{1}{2}\epsilon_{\mu\nu\alpha}\,u^{\alpha}\Omega\,, \tag{2.8b}$$

$$d=3: \qquad \nabla_{\mu}u_{\nu} = -u_{\mu}a_{\nu} - \frac{1}{2}\epsilon_{\mu\nu\alpha\beta}\,u^{\alpha}\Omega^{\beta}\,, \tag{2.8c}$$

The vorticity is $\Omega \equiv -\epsilon^{\mu\nu\lambda}u_{\mu}\nabla_{\nu}u_{\lambda}$ for d = 2, and $\Omega^{\mu} \equiv \epsilon^{\mu\nu\alpha\beta}u_{\nu}\nabla_{\alpha}u_{\beta}$ for d = 3. This velocity decomposition implies that both the expansion $\nabla_{\mu}u^{\mu}$ and the shear tensor $\sigma^{\mu\nu} \equiv (\Delta^{\mu\alpha}\Delta^{\nu\beta} + \Delta^{\nu\alpha}\Delta^{\mu\beta} - \frac{2}{d}\Delta^{\mu\nu}\Delta^{\alpha\beta})\nabla_{\alpha}u_{\beta}$ (where $\Delta^{\mu\nu} = g^{\mu\nu} + u^{\mu}u^{\nu}$ is the transverse projector) vanish in equilibrium. This is as it should be: out of equilibrium, the expansion would contribute to dissipation through bulk viscosity, and the shear tensor would contribute to dissipation through shear viscosity.

Combined with the electromagnetic "Bianchi identity" $\epsilon^{\mu\nu\alpha\beta}\nabla_{\nu}F_{\alpha\beta} = 0$ in 3+1 dimensions, the velocity decomposition (2.8) implies

$$\nabla{\cdot}B = B{\cdot}a - E{\cdot}\Omega\,,$$

$$u_{\mu}\epsilon^{\mu\nu\rho\sigma}\nabla_{\rho}E_{\sigma} = u_{\mu}\epsilon^{\mu\nu\rho\sigma}E_{\rho}a_{\sigma}\,.$$

These are the covariant versions of the familiar flat-space equilibrium relations $\nabla{\cdot}B = 0$ and $\nabla\times E = 0$. More generally, for the electric field in equilibrium we have

$$d=2: \qquad \epsilon^{\mu\alpha\beta}\nabla_{\alpha}E_{\beta} = \epsilon^{\mu\alpha\beta}E_{\alpha}a_{\beta}\,, \tag{2.9a}$$

$$d=3: \qquad \epsilon^{\mu\nu\rho\sigma}\nabla_{\rho}E_{\sigma} = \epsilon^{\mu\nu\rho\sigma}E_{\rho}a_{\sigma}\,, \tag{2.9b}$$

as a consequence of $£_V E_{\alpha} = 0$ and $E_{\alpha}u^{\alpha} = 0$.

Polarization Ambiguities

The electromagnetic Bianchi identity also implies that there is an ambiguity in the definition of the polarization tensor: in 3+1 dimensions, one can always add to the generating functional an identically vanishing term $W_{\varnothing} = \frac{1}{2}\int\sqrt{-g}\,C_{\mu}\,\epsilon^{\mu\nu\alpha\beta}\nabla_{\nu}F_{\alpha\beta}$, where C_{μ} can be a function of the field strength and its derivatives. Such a term shifts the polarization tensor by

$$M^{\alpha\beta} \to M^{\alpha\beta} + \epsilon^{\alpha\beta\mu\nu}\nabla_{\mu}C_{\nu}\,. \tag{2.10}$$

The polarization vectors correspondingly shift as

$$p^{\mu} \to p^{\mu} - \epsilon^{\mu\nu\rho\sigma}u_{\nu}\nabla_{\rho}C_{\sigma}\,, \tag{2.11a}$$

$$m_{\mu} \to m_{\mu} - \nabla_{\mu}(C{\cdot}u) - (C{\cdot}u)a_{\mu}\,. \tag{2.11b}$$

In 2+1 dimensions, we can add an identically vanishing term $W_{\varnothing} = \frac{1}{2}\int\sqrt{-g}\,C\epsilon^{\mu\alpha\beta}\nabla_{\mu}F_{\alpha\beta}$, where again C can be a function of the field strength and its derivatives. The polarization tensor then shifts by

$$M^{\alpha\beta} \to M^{\alpha\beta} - \epsilon^{\alpha\beta\mu}\,\nabla_{\mu}C\,. \tag{2.12}$$

The electric polarization vector correspondingly changes as

$$p^\mu \to p^\mu + \epsilon^{\mu\nu\rho} u_\nu \nabla_\rho C\,, \tag{2.13}$$

while the magnetic polarization m remains unchanged.

The variational derivatives of $W_\emptyset$ with respect to both $g_{\mu\nu}$ and A_μ vanish. As a result, the energy-momentum tensor and the current (both bulk and boundary) are not affected by such unphysical shifts.

Bound Charges and Bound Currents

For matter whose degrees of freedom carry gauge charges, it is conventional to separate the charge into the "free charge" and "bound charge" components. In the grand canonical ensemble, the chemical potential μ describes the coupling of the system to a reservoir of "free charges". Demanding local charge neutrality for free charges in the bulk would amount to demanding $\partial F/\partial\mu = 0$. Doing so would eliminate the contribution of free charges to polarization. One may refer to μ-dependent contributions as coming from "free charges", and μ-independent contributions as coming from "bound charges", though such a separation is somewhat artificial. We will not impose $\partial F/\partial\mu = 0$, and will keep the contribution to polarization from both free charges and bound charges.

The current J^μ admits a simple expression in terms of the polarization tensor to any order in the derivative expansion. Indeed, the free energy density F can be written as

$$\mathcal{F} = S_{(0)} + S_{(1)}^{\lambda\mu\nu} \nabla_\lambda F_{\mu\nu} + S_{(2)}^{\lambda\sigma\mu\nu} \nabla_\lambda \nabla_\sigma F_{\mu\nu} + \dots\,,$$

where the coefficients $S_{(n)}$ do not contain derivatives of the electromagnetic field strength. The derivative of the chemical potential can be traded for the electric field according to (2.7), hence we can take $S_{(n)} = S_{(n)}(T, \mu, F_{\alpha\beta}, \dots)$ where dots denote the arguments which do not depend on the gauge field. The polarization tensor can be easily found in terms of $S_{(n)}$ through integration by parts. It is then clear that the current extracted from the generating functional according to (2.3) is

$$J^\alpha = \rho u^\alpha - \nabla_\lambda M^{\lambda\alpha}\,, \tag{2.14}$$

to any order in the derivative expansion, where $\rho \equiv \partial F/\partial\mu$. The first term in the right hand side is the standard equilibrium current in the absence of polarization: to leading order in the derivative expansion the free energy density F is just the pressure P, and $\rho = \partial P/\partial\mu$ is the density of "free charges". The second term in the right-hand side is a total derivative of an anti-symmetric tensor. It therefore does not contribute to the conservation equation $\nabla_\mu J^\mu = 0$, and can be interpreted in terms of "bound" charges and "bound" currents. It is clear from the expression (2.14) that the unphysical polarization shifts (2.10) and (2.12) do not affect the current.

The current can be decomposed with respect to the velocity u^μ as

$$J^\mu = \mathcal{N} u^\mu + \mathcal{J}^\mu\,, \tag{2.5}$$

where $\mathcal{N} \equiv -u_\mu J^\mu$ is the charge density, and the spatial current $\mathcal{J}_\mu \equiv \Delta_{\mu\lambda} J^\lambda$ is transverse to u^μ. For the polarization tensor of the form (2.6), the definitions (2.1) together with $\pounds_V(\dots) = 0$ lead to the following equilibrium expressions for the charge density:

$$d = 2: \quad \mathcal{N} = \rho - \nabla_\mu p^\mu + p^\mu a_\mu - m\Omega\,, \tag{2.16a}$$

$$d = 3: \quad \mathcal{N} = \rho - \nabla_\mu p^\mu + p^\mu a_\mu - m_\mu \Omega^\mu\,. \tag{2.16b}$$

Consider the charge density in d = 3 spatial dimensions. The second term in the right-hand side is the familiar electrostatic bound charge density, which in flat space reduces to $-\nabla\cdot p$. The third term is the bound charge density induced by gravity: in the static Newtonian gravitational field it becomes $p\cdot\nabla\phi$, where ϕ is the gravitational potential. The last term is the bound charge density induced in magnetized matter which is rotating. For a system undergoing rotation with small (meaning $|\omega| R \ll 1$, where R is the size of the system) angular velocity ω, the last term in the right-hand side becomes $-2m\cdot\omega$.

Similarly, the definitions (2.1) together with $\pounds_V = 0$ lead to the following equilibrium expressions for the spatial current:

$$d = 2: \quad \mathcal{J}^\mu = \epsilon^{\mu\nu\rho} u_\nu \nabla_\rho m + \epsilon^{\mu\nu\rho} u_\nu a_\rho m\,, \tag{2.17a}$$

$$d = 3: \quad \mathcal{J}^\mu = \epsilon^{\mu\nu\rho\sigma} u_\nu \nabla_\rho m_\sigma + \epsilon^{\mu\nu\rho\sigma} u_\nu a_\rho m_\sigma\,. \tag{2.17b}$$

Consider the current density in d = 3 spatial dimensions. The first term in the right-hand side is the familiar bound current, which in flat space reduces to $\nabla\times m$. The second term is the bound current induced by the gravitational field: in the static Newtonian gravitational field it reduces to $(\nabla\phi)\times m$, where ϕ is the gravitational potential.

We emphasize that the above expressions for bound charges and bound currents are simply a consequence of thermal equilibrium. Equations (2.16), (2.17) do not assume any particular microscopic model of matter, and moreover they hold to any order in the derivative expansion.

Derivative Expansion

We close this section with a comment on the derivative expansion of the free energy. As a schematic example, consider the functional W[a, g] which depends on two sources a(x) and g(x) which both vary slowly in space. Assuming locality, the derivative expansion is

$$\begin{aligned} W[a,g] = &\int P_0(a,g) + \int P_1(a,g)\, a'(x) + \int P_2(a,g)\, g'(x) \\ &+ \int P_3(a,g)\, a''(x) + \int P_4(a,g)\, g''(x) + \int P_5(a,g)\, a'(x) g'(x) \\ &+ \int P_6(a,g)\, a'(x)^2 + \int P_7(a,g)\, g'(x)^2 + O(\partial^3)\,, \end{aligned}$$

and the boundary terms are implied.

Suppose now that a changes much faster than g, such that $|a'(x)/a(x)| \gg |g'(x)/g(x)|$. Naively, one may think that the terms containing the derivatives of a(x) are more important than those with derivatives of g(x), and there is a separate derivative counting associated with a(x) and g(x). This is not in general so: for example, integrating the P_2 term by parts gives rise to $\int(\partial P_2/\partial a)\, g(x) a'(x)$, which may be of the same order as the P_1 term. It is possible to count the derivatives of g differently from the derivatives of a if the "cross" terms ($\partial P_2/\partial a$) are in some sense small. For example, we could introduce two counting parameters ε and $\gamma \ll \varepsilon$ and count the derivatives as $a' \sim \varepsilon$, $g' \sim \gamma$, while $\partial P_2/\partial a$, $\partial P_4/\partial a$, $\partial P_5/\partial a$, $\partial P_7/\partial a$ are of order γ/ε.

Physically, a will be the external gauge potential $A\mu$, and g the external metric $g\mu\nu$. By "strong" electromagnetic fields we will mean the fields such that this derivative counting is valid, i.e. electromagnetism is more important than gravity. For such "strong" fields, by the leading order in the derivative expansion we will mean: i) setting γ to zero, ii) isolating terms polynomial in $a'(x)$, and iii) summing those terms into a single function $P(a, g, a')$.

LEADING ORDER IN THE DERIVATIVE EXPANSION

Weak Electromagnetic Fields

Let us start with "weak" electromagnetic fields. For the sources with $A\mu \sim O(1)$, $g_{\mu\nu} \sim O(1)$, there are only two gauge and diffeomorphism invariants at leading order in the derivative expansion in the bulk: T and μ. On the boundary, there is an extra invariant $u_n \equiv u^\mu n_\mu$. The static generating functional to leading order in the derivative expansion is then

$$W[g, A] = \int_{\mathcal{M}} d^{d+1}x\, \sqrt{-g}\, P(T, \mu) + \int_{\partial\mathcal{M}} d^d x\, \sqrt{\gamma}\, L(T, \mu, u_n)\,.$$

The definitions (2.3) give

$$T^{\mu\nu} = P g^{\mu\nu} + (Ts + \mu\rho) u^\mu u^\nu\,,$$
$$J^\mu = \rho u^\mu\,,$$

where $s = \partial P/\partial T$ is the bulk entropy density, $\rho = \partial P/\partial \mu$ is the bulk charge density. These are the standard expressions for the energy-momentum tensor and the current in a relativistic perfect fluid. The boundary energy-momentum tensor and the current are

$$T_s^{ab} = L\gamma^{ab} + (Ts_s + \mu\rho_s)\, u^a u^b\,,$$
$$J_s^a = \rho_s u^a\,,$$

and the other boundary terms are $\Pi_s^a = (Ts_s + \mu\rho_s)\, u_n u^a$, $\Pi_s = \frac{1}{2}(Ts_s + \mu\rho_s)\, u_n^2 + \varepsilon_s u_n$, and $J_s = \rho_s u_n$, where we have defined $s_s \equiv \partial L/\partial T$, $\rho_s \equiv \partial L/\partial\mu$, $\varepsilon_s \equiv \partial L/\partial u_n$. Again, these describe a perfect fluid on the boundary with pressure L. At leading order in the derivative expansion, both K_s^μ and $K_s^{\mu\nu}$ vanish.

Strong Electromagnetic Fields

Let us now consider "strong" electromagnetic fields, such that $F_{\mu\nu} \sim O(1)$ and $g_{\mu\nu} \sim O(1)$ in the derivative expansion. To leading order, the static generating functional is

$$W[g, A] = \int_{\mathcal{M}} d^{d+1}x\, \sqrt{-g}\, P(T, \mu, F_{\alpha\beta}) + \int_{\partial\mathcal{M}} d^d x\, \sqrt{\gamma}\, L(T, \mu, F_{\alpha\beta}, n_\alpha)\,. \tag{3.1}$$

The dependence on $F_{\alpha\beta}$ includes the dependence on electric and magnetic fields, and for the boundary part also on their normal components.

The bulk current is given by (2.14), with the polarization tensor $M^{\mu\nu} = 2\partial P/\partial F_{\mu\nu}$. In what follows we will express $M^{\mu\nu}$ in terms of electric and magnetic susceptibilities. In order to find the energy-momentum tensor, we need to be more specific about the dependence of P and L on the metric. Gauge and diffeomorphism invariance requires that $P = P(s^{(0)})$ is a function of scalars $s^{(0)}$, which are made out of the electromagnetic field strength (we will use the term "scalar" for both scalars and pseudo-scalars). The superscript signifies that we are working to leading order in the derivative expansion. The number of scalars $s^{(0)}$ depends on the dimension. To leading order in the derivative expansion, we choose to work with the following independent scalars:

$$d = 1:\ s^{(0)} = \{T,\ \mu,\ B\}\,, \tag{3.2a}$$

$$d = 2:\ s^{(0)} = \{T,\ \mu,\ B,\ E^2\}\,, \tag{3.2b}$$

$$d = 3:\ s^{(0)} = \{T,\ \mu,\ B^2,\ E{\cdot}B,\ E^2\}\,. \tag{3.2c}$$

Let us express the bulk energy-momentum tensor using the decomposition with respect to the velocity u^μ, as is often done in relativistic fluid dynamics,

$$T^{\mu\nu} = \mathcal{E} u^\mu u^\nu + \mathcal{P}\Delta^{\mu\nu} + \mathcal{Q}^\mu u^\nu + \mathcal{Q}^\nu u^\mu + \mathcal{T}^{\mu\nu}\,. \tag{3.3}$$

Here $\mathcal{E} \equiv u_\mu u_\nu T^{\mu\nu}$ is the energy density, $\mathcal{P} \equiv \frac{1}{d}\Delta_{\mu\nu}T^{\mu\nu}$ is the pressure, the momentum density $\mathcal{Q}_\mu \equiv -\Delta_{\mu\alpha}u_\beta T^{\alpha\beta}$ is transverse to u μ, and $\mathcal{T}^{\mu\nu} \equiv \frac{1}{2}(\Delta_{\mu\alpha}\Delta_{\nu\beta} + \Delta_{\nu\alpha}\Delta_{\mu\beta} - \frac{2}{d}\Delta_{\mu\nu}\Delta_{\alpha\beta})T^{\alpha\beta}$ is transverse to u μ, symmetric, and traceless. Given P as a function of the above scalars, the energy-momentum tensor can be read off from the definition (2.3).

For d = 1, we have P = P(T, μ, B). This gives the following energy-momentum tensor:

$$\mathcal{E} = -P + Ts + \mu\rho + mB\,, \tag{3.4a}$$

$$\mathcal{P} = P - mB\,, \tag{3.4b}$$

$$\mathcal{Q}^\mu = 0\,, \tag{3.4c}$$

$$\mathcal{T}^{\mu\nu} = 0\,. \tag{3.4d}$$

Here the "magnetization" density $m \equiv \partial P/\partial B$ determines the polarization tensor as

$$M^{\mu\nu} = m\epsilon^{\mu\nu} \,. \tag{3.5}$$

For d = 2, P = P(T, μ, B, E^2). This gives the following energy-momentum tensor:

$$\mathcal{E} = -P + Ts + \mu\rho + \chi_{\rm E} E^2 \,, \tag{3.6a}$$

$$\mathcal{P} = P - mB - \frac{1}{2}\chi_{\rm E} E^2 \,, \tag{3.6b}$$

$$\mathcal{Q}^\mu = m\,\epsilon^{\mu\rho\sigma} E_\rho\, u_\sigma \,, \tag{3.6c}$$

$$\mathcal{T}^{\mu\nu} = -\chi_{\rm E}\left(E^\mu E^\nu - \frac{1}{2}\Delta^{\mu\nu}E^2\right) . \tag{3.6d}$$

Here again s = ∂P/∂T is the entropy density, ρ = ∂P/∂μ is the charge density, m = ∂P/∂B is the magnetization density, and $\chi_E \equiv 2\partial P/\partial E^2$ is the electric susceptibility. They determine the polarization tensor as

$$M^{\mu\nu} = -m\, u_\lambda \epsilon^{\lambda\mu\nu} + \chi_{\rm E}\left(E^\mu u^\nu - E^\nu u^\mu\right) . \tag{3.7}$$

The dependence P = P(T, μ, B, E^2) implies that the electric polarization vector is $p^\mu = \chi_E E^\mu$, and the polarization tensor (3.7) coincides with the general expression (2.6), as it should.

For d = 3, P = P(T, μ, B^2 , E·B, E^2). This gives the following energy-momentum tensor:

$$\mathcal{E} = -P + Ts + \mu\rho + \chi_{\rm EE} E^2 + \chi_{\rm EB}\, E\cdot B \,, \tag{3.8a}$$

$$\mathcal{P} = P - \frac{2}{3}\chi_{\rm BB} B^2 - \frac{1}{3}\chi_{\rm EE} E^2 - \chi_{\rm EB}\, E\cdot B \,, \tag{3.8b}$$

$$\mathcal{Q}^\mu = -\chi_{\rm BB} S^\mu \,, \tag{3.8c}$$

$$\mathcal{T}^{\mu\nu} = -\chi_{\rm EE}\left(E^\mu E^\nu - \frac{1}{3}\Delta^{\mu\nu}E^2\right) + \chi_{\rm BB}\left(B^\mu B^\nu - \frac{1}{3}\Delta^{\mu\nu}B^2\right) . \tag{3.8d}$$

Here again s = ∂P/∂T, ρ = ∂P/∂μ are the entropy and charge densities, $S^\mu = \epsilon^{\mu\rho\sigma\lambda}u_\rho E_\sigma B_\lambda$ is the Poynting vector, χEE ≡ $2\partial P/\partial E^2$ is the electric susceptibility, χEB ≡ ∂P/∂(E·B) is the electro-magnetic susceptibility, χBB ≡ $2\partial P/\partial B^2$ is the magnetic susceptibility. They determine the polarization tensor as

$$M^{\mu\nu} = \chi_{\rm EE}(E^\mu u^\nu - E^\nu u^\mu) + \chi_{\rm BB}\epsilon^{\mu\nu\alpha\beta}B_\alpha u_\beta - \chi_{\rm EB} G^{\mu\nu} \,, \tag{3.9}$$

where $G^{\mu\nu} = \frac{1}{2}\epsilon^{\mu\nu\alpha\beta}F_{\alpha\beta}$ is the dual field strength. The dependence P = P(T, μ, B^2 , E·B, E^2), implies that the polarization vectors are

$$p^\mu = \chi_{\rm EE}E^\mu + \chi_{\rm EB}B^\mu \,,$$
$$m^\mu = \chi_{\rm BE}E^\mu + \chi_{\rm BB}B^\mu \,.$$

The magneto-electric susceptibility χ_{BE} is equal to the electro-magnetic susceptibility χ_{EB}, and the polarization tensor (3.9) coincides with the general expression (2.6), as it should.

So far we have presented T μν in terms of the decomposition (3.3) with respect to the velocity u $^\mu$, whose coefficients E, P, Q^μ , and T $^{\mu\nu}$ are expressed in terms of the electric and magnetic fields, and the susceptibilities. The same energy-momentum tensors (3.4), (3.6), (3.8) can be equivalently expressed in terms of the polarization tensor $M^{\mu\nu} = 2\partial P/\partial F_{\mu\nu}$ as

$$T^{\mu\nu} = Pg^{\mu\nu} + (Ts + \mu\rho)u^\mu u^\nu + T^{\mu\nu}_{\rm EM} \,, \tag{3.10a}$$

Where

$$T^{\mu\nu}_{\rm EM} = M^{\mu\alpha}g_{\alpha\beta}F^{\beta\nu} + u^\mu u^\alpha\left(M_{\alpha\beta}F^{\beta\nu} - F_{\alpha\beta}M^{\beta\nu}\right) \tag{3.10b}$$

is the "electromagnetic correction" to the perfect fluid form. Note that P, s, and ρ in (3.10a) are functions of the electric and magnetic fields. The tensor (3.10b) is symmetric; if we set the external electric field to zero (in two or three spatial dimensions), then $T^{\mu\nu}_{\rm EM}$ reduces to its first term, and is still symmetric. Note that $\nabla_\mu T^{\mu\nu}_{\rm EM}$ does not equal $F^\nu{}_\lambda(-\nabla_\nu M^{\nu\lambda})$.

The above expression for $T^{\mu\nu}_{\rm EM}$ was first derived by W. Israel [14], for a free gas of polarized relativistic particles. We emphasize that one does not need to assume any particular microscopic model of matter in order to arrive at the above energy-momentum tensor: expression (3.10) is a direct consequence of gauge and diffeomorphism invariance of the theory, to leading order in the derivative expansion.

We now turn to the boundary energy-momentum tensor and the current which follow from the generating functional (3.1). The boundary current may be expressed in terms of the boundary polarization tensor $m^{\mu\nu} \equiv 2\partial L/\partial F_{\mu\nu}$ (keeping T, μ, and n μ fixed). Upon integrating by parts on the boundary, the definition (2.3) gives the following boundary currents:

$$J^a_{\rm s} = M^a + (\rho_{\rm s}u^a - \nabla_b m^{ba}) \,, \tag{3.11}$$

Here M^a is the boundary current arising from integrating the variation of P(T, μ, $F_{\alpha\beta}$) by parts, $e^\mu_a M^a = n_\lambda M^{\lambda\mu}$. As one can see from the polarization tensor (2.6), in flat space in 3+1 dimensions the boundary current $n_\lambda M^{\lambda\mu}$ reduces to a vector whose time component (surface charge density) is p·n, while the spatial part (surface bound current) is m×n. These are the familiar expressions from electro- and magneto-statics. The other term in the boundary current, $(\rho_{\rm s}u^a - \nabla_b m^{ba})$, arises due to the presence of charged degrees of freedom on the boundary described by L, and mimics the bulk current (2.14), with ρs ≡ ∂L/∂μ. The other boundary currents, $J_s = \rho_s u_n$, and $K^\mu_{\rm s} = n_\lambda m^{\lambda\mu}$, emerge from L as well.

At leading order in the derivative expansion, the only contribution to the boundary energy-momentum tensor arises from the surface tension term in (3.1). In 2+1 dimensions, $L = L(T, \mu, u_n, B, E^2, E_n)$, where $u_n \equiv u^\mu n_\mu,\ E_n \equiv E_\mu n^\mu$. At leading order K μν s vanishes, and the definition (2.3) gives

$$T_s^{ab} = (Ts_s + \mu\rho_s - m_s B + \alpha_{s,E}E^2 + \chi_{n,E}E_n)\, u^a u^b + (L - m_s B)\gamma^{ab} + m_s(u^a S^b + u^b S^a) - \alpha_{s,E}E^a E^b .$$

Here $S_a = e_a^\alpha P_{\alpha\mu}\epsilon^{\mu\rho\sigma}E_\rho u_\sigma$, and the coefficients are $s_s \equiv \partial L/\partial T$, $\rho_s \equiv \partial L/\partial \mu$, $m_s \equiv \partial L/\partial B$, $\varepsilon_s \equiv \partial L/\partial u_n$, $\alpha_{s,E} \equiv 2\partial L/\partial E^2$, $\chi_n,E \equiv \partial L/\partial E_n$. The boundary energy-momentum tensor in 3+1 dimensions looks similar, and we won't write it down explicitly.

NEXT ORDER IN THE DERIVATIVE EXPANSION

We now proceed to the next (first) order in the derivative expansion, taking into account O(∂) terms in the generating functional (2.2). We will take the surface tension L to be constant for simplicity, and will focus on the bulk contributions to thermodynamics. The free energy density at first order in the derivative expansion is

$$\mathcal{F} = P(s^{(0)}) + \sum_n M_n(s^{(0)})\, s_n^{(1)} , \tag{4.1}$$

where P is the leading-order pressure. For weak electromagnetic fields, the leading order scalars are $s^{(0)} = \{T, \mu\}$, while for strong electromagnetic fields $s^{(0)}$ are given by eq. (3.2). The functions $M_n(s^{(0)})$ parametrize the thermodynamic response at first order, and are determined by the microscopic theory. The gauge- and diffeomorphism-invariant scalars s (1) n depend on T, u^μ, μ, and the sources A_μ, $g_{\mu\nu}$. The number of such first-order scalars depends on the dimension, and on whether the external electromagnetic fields are weak or strong. We will enumerate the scalars s (1) n in what follows.

The bulk current is still given by the general expression (2.14). At first order, the free energy density (4.1) may be equivalently rewritten as

$$\mathcal{F} = P(s^{(0)}) + P'(s^{(0)}, s^{(1)}) + \frac{1}{2}S^{\lambda\mu\nu}(s^{(0)})\nabla_\lambda F_{\mu\nu} , \tag{4.2}$$

where P' is O(∂), but contains no derivatives of $F_{\alpha\beta}$, while the last term parametrizes the static response to inhomogeneous electromagnetic fields. The polarization tensor is then

$$M^{\alpha\beta} = 2\frac{\partial P}{\partial F_{\alpha\beta}} + 2\frac{\partial P'}{\partial F_{\alpha\beta}} + \frac{\partial S^{\lambda\mu\nu}}{\partial F_{\alpha\beta}}\nabla_\lambda F_{\mu\nu} - \nabla_\lambda S^{\lambda\alpha\beta} . \tag{4.3}$$

The equilibrium relation $\partial_\lambda \mu = E\lambda + O(\partial)$ now implies that for strong electric fields the leading-order polarization tensor may receive contributions from subleading terms in the generating functional

$$M^{\alpha\beta} = 2\frac{\partial P}{\partial F_{\alpha\beta}} - E_\lambda \frac{\partial S^{\lambda\alpha\beta}}{\partial \mu} + O(\partial) .$$

The second term in the right-hand side describes a contribution of free charges to polarization. We will assume for simplicity that the effects of free charges are less important than those of bound charges, in the sense that $\partial M_n/\partial\mu \sim O(\partial)$. (Alternatively, the effects of the free charges may be lumped into the leading-order free energy, but in this case isolating their contribution becomes less straightforward.) The boundary currents are

$$J_s^a = M^a - \nabla_b S^{ba} , \tag{4.4a}$$

$$K_s^\mu = n_\lambda n_\alpha S^{\lambda\alpha\mu} , \tag{4.4b}$$

where $M^a e_a^\mu = n_\lambda M^{\lambda\mu}$ as before, and $e_a^\mu e_b^\nu S^{ab} = n_\lambda S^{\lambda\alpha\beta}P_\alpha^\mu P_\beta^\nu$. Equation (4.4a) shows that for strong electromagnetic fields beyond the leading order in the derivative expansion, the surface current is not determined solely by the bulk bound current any more, even in the absence of charged degrees of freedom on the boundary. Similarly, the energy-momentum tensor will differ from the simple form (3.10) beyond leading order.

Weak Electromagnetic Fields

For weak electromagnetic fields with $A_\mu \sim O(1)$, $g_{\mu\nu} \sim O(1)$, we have the following nonvanishing scalars at order O(∂) in the derivative expansion:

$$d = 1: \quad s^{(1)} = \{B,\ \Omega\} , \tag{4.5a}$$

$$d = 2: \quad s^{(1)} = \{B,\ \Omega\} , \tag{4.5b}$$

$$d = 3: \quad s^{(1)} = \{\} . \tag{4.5c}$$

In d = 1, we define the vorticity as $\Omega \equiv \epsilon^{\mu\nu}\nabla_\mu u_\nu = \epsilon^{\mu\nu}a_\mu u_\nu$. In d = 3, both the magnetic field and the vorticity are vectors, and there are no scalars at order O(∂).

Focussing on d = 2, the equilibrium generating functional is given by eq. (2.2), with the free energy density

$$\mathcal{F}(T, \mu, B, \Omega) = P + M_B B + M_\Omega \Omega$$

where P, MB, M_Ω are functions of T and μ. Note that both B and Ω are pseudoscalars, hence we are describing thermodynamics of a microscopic system which intrinsically violates parity. The bulk current is given by (2.14), with the polarization tensor $M^a e_a^\mu = -M_B\,\epsilon^{\mu\alpha\beta}u_\alpha n_\beta$. The boundary current is $J_s^a = M^a$, where $M^{\alpha\beta} = -M_B\, u_\mu \epsilon^{\mu\alpha\beta}$.

The bulk energy-momentum tensor can be expressed as a general decomposition (3.3), whose coefficients are

$$\mathcal{P} = P ,$$

$$\mathcal{E} = -\mathcal{F} + T\frac{\partial \mathcal{F}}{\partial T} + \mu\frac{\partial \mathcal{F}}{\partial \mu} - M_\Omega \Omega ,$$

$$\mathcal{Q}^\mu = g_1\,\epsilon^{\mu\alpha\beta}u_\alpha a_\beta + g_2\epsilon^{\mu\alpha\beta}u_\alpha E_\beta ,$$

$$\mathcal{T}^{\mu\nu} = 0 ,$$

where

$g_1 \equiv (2M_\Omega - T\frac{\partial M_\Omega}{\partial T} - \mu\frac{\partial M_\Omega}{\partial \mu})$, $g_2 \equiv (-M_B + \frac{\partial M_\Omega}{\partial \mu})$, see ref. [9, 15]. The boundary energy-momentum tensor can be expressed in terms of the $\ell^\mu \equiv M_\Omega\, \epsilon^{\mu\alpha\beta} n_\alpha u_\beta$ which is tangent to the boundary,

$$T_{\rm s}^{ab} = L\gamma^{ab} + \ell^a u^b + \ell^b u^a\,, \tag{4.6}$$

where again $\ell^a e_a^\mu = M_\Omega\, \epsilon^{\mu\alpha\beta} n_\alpha u_\beta$, and we have assumed that the surface tension L is constant. The other boundary momentum currents are $\Pi_{\rm s}^a = u_n \ell^a$. The vector ` a (the energy-momentum analogue of the boundary magnetization current) is the density of momentum flowing along the boundary in equilibrium, as is generically expected to happen in a parity-violating system. To sum up, the boundary current is determined by the magnetization MB, while the boundary momentum is determined by M_Ω.

Strong Electromagnetic Fields: 1+1 Dimensions

Now let us turn to strong electromagnetic fields, with $F_{\mu\nu} \sim O(1)$, $g_{\mu\nu} \sim O(1)$. In 1+1 dimensions, there are only two independent scalars in equilibrium at $O(\partial)$ in the derivative expansion, which may be taken to be

$$s_n^{(1)} = \{\epsilon^{\mu\nu}\nabla_\mu u_\nu,\ \epsilon^{\mu\nu}u_\nu\partial_\mu B\}\,.$$

The equilibrium generating functional is given by eq. (2.2), with the free energy density

$$\mathcal{F} = P + M_1\epsilon^{\mu\nu}\nabla_\mu u_\nu + M_2\,\epsilon^{\mu\nu}u_\nu\partial_\mu B\,,$$

where P, M_1, M_2 are functions of T, μ, and B. The polarization tensor is given by (2.6), and the magnetization m is a combination of the thermodynamic derivatives of M_1, M_2, and the scalars $s_n^{(1)}$. The boundary currents are given by eq. (4.4), where $S^{\lambda\mu\nu} = M_2\epsilon^{\mu\nu}\epsilon^{\lambda\sigma}u_\sigma$. The energy-momentum tensor may be derived in a straightforward way from the definition (2.3).

Table 1. Independent $O(\partial)$ invariants in 2+1 dimensions. The first row in the table is the number of the invariant, and the second row says what the invariant is. The rows labeled C, P, T indicate the eigenvalue of the invariant under charge conjugation, parity, and time reversal, respectively. Parity in 2+1 dimensions is defined as a reflection of one of the spatial coordinates. The row labeled W shows the weight w of the invariant under a local rescaling of the metric; the invariants which do not transform homogeneously are marked as "n/a".

n	1	2	3	4	5	6	7	8
$s_n^{(1)}$	$E^\mu\partial_\mu(\frac{B}{T^2})$	$E^\mu\partial_\mu(\frac{E^2}{T^4})$	$v^\mu\partial_\mu(\frac{B}{T^2})$	$v^\mu\partial_\mu(\frac{E^2}{T^4})$	$\nabla_\mu E^\mu$	$E^\mu\partial_\mu T$	$v^\mu\partial_\mu T$	$\epsilon^{\lambda\mu\nu}u_\lambda\partial_\nu u_\mu$
C	+	−	+	−	−	−	−	+
P	−	+	+	−	+	+	−	−
T	−	+	−	+	+	+	+	−
W	3	3	3	3	3	n/a	n/a	1

Strong Electromagnetic Fields: 2+1 Dimensions

In two spatial dimensions, there is a large number of $O(\partial)$ scalars. However, equilibrium relations such as (2.7), (2.8) reduce the number of independent non-zero invariants to just eight. One choice of the independent invariants is listed in table 1, where $v^\mu \equiv \epsilon^{\mu\alpha\beta}E_\alpha u_\beta$. The table indicates how the invariants transform under charge conjugation, parity, and time reversal.

The table also indicates the weight of the invariants under a Weyl rescaling of the metric, $g_{\mu\nu} \to \tilde{g}_{\mu\nu} = e^{-2\varphi}g_{\mu\nu}$, where ϕ satisfies $V^\mu\partial_\mu\varphi = 0$. A quantity Φ transforms homogeneously with weight w under the Weyl rescaling if $\Phi \to \tilde{\Phi} = e^{w\varphi}\Phi$. For a review of Weyl rescaling in relativistic hydrodynamics, see ref. [3]. Temperature T, chemical potential μ, velocity u^μ, and the electric field E^μ all have w = 1. The factors of T^2 and T^4 in the first four invariants in table 1 are inserted in order to ensure that the invariant has a well-defined weight. For the scalars which transform homogeneously, their weight w coincides with their mass dimension. The invariants $s_6^{(1)}$ and $s_7^{(1)}$ do not transform homogeneously and can not appear in a conformally invariant generating functional.

The first five invariants are in general already non-zero in flat space. For the static Newtonian gravitational field with potential ϕ, we have $s_6^{(1)} \sim E_i\partial_i\varphi$, $s_7^{(1)} \sim \epsilon^{ij}E_i\partial_j\varphi$. The last invariant is the vorticity, $s_8^{(1)} = \Omega$, which is non-zero if the system is rotating. The equilibrium generating functional is then given by eq. (2.2), with the free energy density

$$\mathcal{F} = P(T,\mu,B,E^2) + \sum_{n=1}^{8} M_n(T,\mu,B,E^2)\, s_n^{(1)}\,.$$

There are eight scalar functions M_n, in addition to pressure, which specify the thermodynamic response at first order. For a system whose microscopic dynamics is PT-invariant, the coefficients M_3, M_4, and M_7 must vanish, in order for the generating functional to be PT-invariant (none of the leading-order invariants are PT-odd). For a system whose microscopic dynamics is conformally invariant, the generating functional must be conformally invariant as well, hence the coefficients M_6 and M_7 must vanish.

While eight might seem like a large number, if one were to naively write down the constitutive relations directly for $T^{\mu\nu}$ and $M^{\mu\nu}$ in terms of all available $O(\partial)$ scalars, vectors, and tensors, doing so would involve introducing many more than eight unknown O(1) scalar functions, even in equilibrium. The generating functional, on the other hand, allows one to obtain the simplest expressions for the equilibrium quantities without overcounting the parameters.

At leading order in the derivative expansion, the electric polarization vector p^λ was simply proportional to the external electric field. At first order, electric polarization can also be induced by the gradients of T, B, and E^2. One finds

$$p^\lambda = \chi_E E^\lambda + \gamma_1 \nabla^\lambda T + \gamma_2 \nabla^\lambda B + \gamma_3 \nabla^\lambda E^2 + \epsilon^{\lambda\rho\sigma} u_\rho \left(\gamma_4\, \partial_\sigma T + \gamma_5\, \partial_\sigma B + \gamma_6\, \partial_\sigma E^2\right) \tag{4.7}$$

The susceptibility here is a function of the parameters M_n of the generating functional,

$$\chi_E = 2\frac{\partial P}{\partial E^2} + 2\sum_{n=1}^{8} \frac{\partial M_n}{\partial E^2} s_n^{(1)} - \frac{\partial M_5}{\partial \mu} - \frac{2}{T^4}\nabla_\lambda \left(M_2 E^\lambda + M_4 \epsilon^{\lambda\alpha\beta} E_\alpha u_\beta\right),$$

and the other coefficients are as follows:

$$\gamma_1 = M_6 - \frac{\partial M_5}{\partial T} - \frac{\mu}{T}\frac{\partial M_5}{\partial \mu} - \frac{2M_1 B}{T^3} - \frac{4M_2 E^2}{T^5},$$

$$\gamma_2 = \frac{M_1}{T^2} - \frac{\partial M_5}{\partial B}, \qquad \gamma_3 = \frac{M_2}{T^4} - \frac{\partial M_5}{\partial E^2},$$

$$\gamma_4 = M_7 - \frac{2M_3 B}{T^3} - \frac{4M_4 E^2}{T^5}, \qquad \gamma_5 = \frac{M_3}{T^2}, \qquad \gamma_6 = \frac{M_4}{T^4}.$$

The magnetization is

$$m = \frac{\partial P}{\partial B} + \sum_{n=1}^{8} \frac{\partial M_n}{\partial B} s_n^{(1)} - \frac{1}{T^2}\nabla_\lambda \left(M_1 E^\lambda + M_3 \epsilon^{\lambda\alpha\beta} E_\alpha u_\beta\right).$$

The above m and p μ give the polarization tensor according to eq. (2.6b), and thus determine the $O(\partial^2)$ contributions to equilibrium bound charges and bound currents, following (2.14). Finally, we note that the polarization ambiguities of section 2.4 allow one to simplify the polarization vector p^λ: adding to the free energy the WØ term with $\partial C/\partial B = -\gamma_5$, $\partial C/\partial E^2 = -\gamma_6$ eliminates the $\gamma_4, \gamma_5, \gamma_6$ contributions in (4.7), and adds the term $\partial C/\partial\mu\, \epsilon^{\lambda\alpha\beta} u_\alpha E_\beta$.

The energy-momentum tensor can be read off from the definition (2.3), however the general expressions are rather cumbersome, involving thermodynamic derivatives of all eight M_n's. It is easy to derive the energy-momentum tensor when the external electric field vanishes (in a certain set of coordinates), in which case $T^{\mu\nu}$ is only determined by P and M_8 (the electric polarization vector is still determined by M_1, M_3, M_5, M_6, and M_7). In the hydrodynamic-like decomposition (3.3), the components of the energy-momentum

$$\mathcal{E} = -\mathcal{F} + T\frac{\partial \mathcal{F}}{\partial T} + \mu\frac{\partial \mathcal{F}}{\partial \mu} - M_\Omega \Omega, \tag{4.8a}$$

$$\mathcal{P} = \mathcal{F} - B\frac{\partial \mathcal{F}}{\partial B} - M_\Omega \Omega, \tag{4.8b}$$

$$\mathcal{Q}^\mu = g_1\, \epsilon^{\mu\rho\sigma} u_\rho a_\sigma + \frac{\partial M_\Omega}{\partial B} \epsilon^{\mu\rho\sigma} u_\rho \partial_\sigma B, \tag{4.8c}$$

$$\mathcal{T}^{\mu\nu} = 0, \tag{4.8d}$$

where we have defined $M_\Omega \equiv M_8$ and $g_1 \equiv (2M_\Omega - T\frac{\partial M_\Omega}{\partial T} - \mu\frac{\partial M_\Omega}{\partial \mu})$, to mimic the notation in section 4.1, and F = P(T, μ, B) + M_Ω(T, μ, B)Ω after we have set the electric field to zero. Even in flat space and without external electric fields, there is an equilibrium energy flux, caused by the inhomogeneous magnetic field. The magnetization m, which determines the spatial bound current according to eq. (2.17), simplifies to m = ∂P/∂B + Ω ∂M_Ω/∂B. There is a surface momentum ` a flowing along the boundary in equilibrium, completely analogous to the expression in section 4.1,

$$T_s^{ab} = L\gamma^{ab} + \ell^a u^b + \ell^b u^a, \tag{4.9}$$

where $\ell^a e^\mu_a = M_\Omega(T,\mu,B)\, \epsilon^{\mu\alpha\beta} n_\alpha u_\beta$ may now depend on the external magnetic field. The other boundary momentum currents are $\Pi^a_s = u_n \ell^a$ and Πs = 0, as before. The energy-momentum tensors (4.8), (4.9) will receive extra contributions proportional to the external electric field when the latter is non-zero.

Strong Electromagnetic Fields: 3+1 Dimensions

In 3+1 dimensions, there is again a large number of O(∂) scalars, but many are not independent due to equilibrium constraints such as (2.7), (2.8). I counted twenty-one independent non-zero invariants. One choice is listed in table 2, where $S^\mu = \epsilon^{\mu\rho\sigma\lambda} u_\rho E_\sigma B_\lambda$ is the Poynting vector, $a_\mu = -\partial_\mu T/T$ is the acceleration, and $\Omega^\mu = \epsilon^{\mu\nu\alpha\beta} u_\nu \nabla_\alpha u_\beta$ is the vorticity. The notation in the table is the same as in the 2+1 dimensional case. The linear combinations in $s_8^{(1)}$, $s_9^{(1)}$, and $s_{10}^{(1)}$ are taken so that the invariant has a well-defined weight under Weyl rescaling.

The first fifteen invariants are in general non-zero already in flat space. The equilibrium generating functional is given by eq. (2.2), with the free energy density

$$\mathcal{F} = P(T,\mu,E^2,B^2,E{\cdot}B) + \sum_{n=1}^{21} M_n(T,\mu,E^2,B^2,E{\cdot}B)\, s_n^{(1)}\,. \tag{4.10}$$

There are twenty-one scalar functions M_n, in addition to pressure, which specify the thermodynamic response at first order. For a system whose microscopic dynamics is PT-invariant, the coefficients $M_9, \ldots, M_{15}$, M_{18}, and M_{21} must vanish, in order for the generating functional to be PT-invariant (none of the leading-order invariants are PT-odd). For a system whose microscopic dynamics is conformally invariant, the generating functional must be conformally invariant as well, hence the coefficients M_{16}, M_{17}, and M_{18} must vanish.

Table 2. Independent O(∂) invariants in 3+1 dimensions

n	1	2	3	4	5	6	7
$s_n^{(1)}$	$E^\mu\partial_\mu(\frac{B^2}{T^4})$	$E^\mu\partial_\mu(\frac{E^2}{T^4})$	$E^\mu\partial_\mu(\frac{E\cdot B}{T^4})$	$B^\mu\partial_\mu(\frac{B^2}{T^4})$	$B^\mu\partial_\mu(\frac{E^2}{T^4})$	$B^\mu\partial_\mu(\frac{E\cdot B}{T^4})$	$\nabla_\mu(TE^\mu)$
C	−	−	−	−	−	−	−
P	+	+	−	−	−	+	+
T	+	+	−	−	−	+	+
W	3	3	3	3	3	3	4

n	8	9	10
$s_n^{(1)}$	$B^\mu B^\nu\nabla_\mu E_\nu - B^2 E{\cdot}a + 3(E{\cdot}B)B{\cdot}a$	$S^\mu B^\nu\nabla_\mu E_\nu + 2(E{\cdot}B)S{\cdot}a$	$E^\mu S^\nu\nabla_\mu B_\nu + (E{\cdot}B)S{\cdot}a$
C	−	+	+
P	+	−	−
T	+	+	+
W	7	9	9

n	11	12	13	14	15
$s_n^{(1)}$	$S^\mu\partial_\mu(\frac{B^2}{T^4})$	$S^\mu\partial_\mu(\frac{E^2}{T^4})$	$S^\mu\partial_\mu(\frac{E\cdot B}{T^4})$	$\nabla_\mu(\frac{S^\mu}{T})$	$\epsilon^{\mu\nu\rho\sigma}u_\mu B_\nu\nabla_\rho B_\sigma$
C	+	+	+	+	+
P	+	+	−	+	−
T	−	−	+	−	+
W	5	5	5	4	5

n	16	17	18	19	20	21
$s_n^{(1)}$	$E{\cdot}a$	$B{\cdot}a$	$S{\cdot}a$	$E{\cdot}\Omega$	$B{\cdot}\Omega$	$S{\cdot}\Omega$
C	−	−	+	−	−	+
P	+	−	+	−	+	−
T	+	−	−	−	+	+
W	n/a	n/a	n/a	3	3	5

Analogously to what happens in 2+1 dimensions, polarization may be induced by the gradients of the applied fields. The electric polarization vector which follows from the free energy is

$$p^\lambda = \chi_{EE}E^\lambda + \chi_{EB}B^\lambda + \chi_{E\Omega}\Omega^\lambda + \chi_{ES}S^\lambda$$
$$+ \gamma_1\nabla^\lambda T + \gamma_2\nabla^\lambda B^2 + \gamma_3\nabla^\lambda E^2 + \gamma_4\nabla^\lambda(E\cdot B)$$
$$+ \gamma_5 X^{\lambda\alpha}\partial_\alpha T + \gamma_6 X^{\lambda\alpha}\partial_\alpha B^2 + \gamma_7 X^{\lambda\alpha}\partial_\alpha E^2 + \gamma_8 X^{\lambda\alpha}\partial_\alpha(E\cdot B) + \gamma_9 X^{\lambda\alpha}\Omega_\alpha$$
$$+ \gamma_{10}\Delta^\lambda_\rho S^\alpha\nabla_\alpha B^\rho + \gamma_{11}\Delta^\lambda_\rho S^\alpha\nabla^\rho B_\alpha + \gamma_{12}\Delta^\lambda_\rho B^\alpha\nabla_\alpha B^\rho$$
$$+ \gamma_{13}X^{\lambda\rho}B^\sigma\nabla_\rho E_\sigma + \gamma_{14}X^{\lambda\rho}E^\sigma\nabla_\sigma B_\rho\,, \tag{4.11}$$

where $X^{\alpha\beta} \equiv \epsilon^{\alpha\beta\rho\sigma}u_\rho B_\sigma$. The susceptibility coefficients χ_{EE} etc and γ_k are determined by thermodynamic derivatives of the coefficients M_n, and can be easily read off from the free energy density (4.10). However, as explained in section 2.4, polarization vectors only make sense up to certain redefinitions. For example, by choosing the arbitrary vector C_{μ} in (2.11) appropriately, one can eliminate $\chi_{E\Omega}$, and trade χ_{ES}, γ_5, γ_6, γ_7, γ_8 in favor of a single contribution proportional to $\epsilon^{\lambda\nu\rho\sigma}u_\nu\nabla_\rho B_\sigma$. The coefficients χ_{EE} and χ_{EB} suffer from similar ambiguities.

The magnetic polarization vector which follows from the free energy is

$$m^\lambda = \chi_{BB}B^\lambda + \chi_{BE}E^\lambda + \chi_{B\Omega}\Omega^\lambda + \chi_{BS}S^\lambda$$
$$+ \delta_1\nabla^\lambda T + \delta_2\nabla^\lambda B^2 + \delta_3\nabla^\lambda E^2 + \delta_4\nabla^\lambda(E\cdot B)$$
$$+ \delta_5 Y^{\lambda\alpha}\partial_\alpha T + \delta_6 Y^{\lambda\alpha}\partial_\alpha B^2 + \delta_7 Y^{\lambda\alpha}\partial_\alpha E^2 + \delta_8 Y^{\lambda\alpha}\partial_\alpha(E\cdot B) + \delta_9 Y^{\lambda\alpha}\Omega_\alpha$$
$$+ \delta_{10}\Delta^\lambda_\rho B^\mu\nabla^\rho E_\mu + \delta_{11}\Delta^\lambda_\rho B^\mu\nabla_\mu E^\rho + \delta_{12}Y^{\lambda\rho}B^\sigma\nabla_\rho E_\sigma + \delta_{13}Y^{\lambda\rho}E^\sigma\nabla_\sigma B_\rho$$
$$+ \delta_{14}\Delta^\lambda_\rho S^\mu\nabla_\mu E^\rho + \delta_{15}\Delta^\lambda_\rho E^\mu\nabla_\mu S^\rho + \delta_{16}\epsilon^{\lambda\mu\rho\sigma}u_\mu\nabla_\rho B_\sigma$$
$$+ \delta_{17}X^{\lambda\alpha}\partial_\alpha T + \delta_{18}X^{\lambda\alpha}\partial_\alpha B^2 + \delta_{19}X^{\lambda\alpha}\partial_\alpha E^2 + \delta_{20}X^{\lambda\alpha}\partial_\alpha(E\cdot B)\,, \tag{4.12}$$

where $Y^{\alpha\beta} \equiv \epsilon^{\alpha\beta\rho\sigma}u_\rho E_\sigma$. The susceptibility coefficients χ_{BB} etc and δ_k can be easily read off from the free energy density (4.10). The ambiguities (2.11) also affect the magnetic polarization: adding to the free energy the WØ term with $C^\mu = Cu^\mu$ shifts $\chi_{BE} \to \chi_{BE} + \partial C/\partial\mu$ (in addition to shifting δ_1, δ_2, δ_3, δ_4).

While the polarization vectors are ambiguous, the energy-momentum tensor and the current are not. As an example, consider the M_{21} term in the free energy. It gives rise to polarization vectors $p^\mu = M_{21}\epsilon^{\mu\nu\rho\sigma}u_\nu B_\rho\Omega_\sigma$ and $m^\mu = -M_{21}\epsilon^{\mu\nu\rho\sigma}u_\nu E_\rho\Omega_\sigma$ which do not suffer from polarization ambiguities. Such contributions to p μ and mμ only come from M21, and therefore the magneto-vortical response of the surface charge density is correlated with the electro-vortical response of the surface current. The corresponding boundary current is

$$J^\mu_s = \left(M_{21}n_\alpha\epsilon^{\alpha\beta\rho\sigma}u_\beta B_\rho\Omega_\sigma\right)u^\mu + M_{21}\left(\Omega_n E^\mu - E_n\Omega^\mu\right)\,,$$

where $\Omega n \equiv \Omega\cdot n$, $E_n \equiv E\cdot n$, and we have assumed $u_n = 0$.

As another example, let us set the electric field to zero (in a certain set of coordinates), while keeping the magnetic field non-zero. The equilibrium bulk energy-momentum tensor is then determined by only four functions M_4, M_{15}, M_{17}, and M_{20}, in addition to the leading-order pressure $P = P(T, \mu, B^2)$. The correction to the leading-order energymomentum tensor (3.8) is straightforward to derive, and we will not write it down explicitly. Both E, P, and T μν will receive derivative corrections, proportional to $\partial_u T$, $\nabla_u B_v$, and Ω^μ. In addition, the magneto-vortical term M_{20} will give rise to equilibrium energy currents Qμ proportional to $\epsilon^{\mu\nu\rho\sigma}u_\nu B_\rho a_\sigma$, and $\epsilon^{\mu\nu\rho\sigma}u_\nu B_\rho\partial_\sigma B^2$. There is also a nonvanishing boundary energy-momentum tensor $T_s^{\alpha\beta}$, defined by (2.3). In the decomposition $T_s^{\alpha\beta} = \mathcal{E}_s u^\alpha u^\beta + \mathcal{P}_s\Delta^{\alpha\beta} + (\mathcal{Q}_s^\alpha u^\beta + \mathcal{Q}_s^\beta u^\alpha) + \mathcal{T}_s^{\alpha\beta}$ as in (3.3) we have

$$\mathcal{E}_s = -\,B\cdot n\left(M_{17} + 4M_4B^2/T^4\right)\,, \tag{4.13a}$$

$$\mathcal{P}_s = -\frac{4}{3}B\cdot n\,M_4B^2/T^4\,, \tag{4.13b}$$

$$\mathcal{Q}_s^\alpha = M_{20}\,n_\rho B_\mu u_\nu\,\epsilon^{\rho\mu\nu\alpha}\,, \tag{4.13c}$$

$$\mathcal{T}_{\text{s}}^{\alpha\beta} = 2B{\cdot}n\, M_4/T^4 \left(B^\alpha B^\beta - \frac{1}{3}\Delta^{\alpha\beta}B^2\right)$$
$$+ M_{15} n_\rho u_\mu B_\nu \left(\epsilon^{\rho\mu\nu\alpha}B^\beta + \epsilon^{\rho\mu\nu\beta}B^\alpha\right) .$$
(4.13d)

Here n^μ is the unit normal vector the boundary as before, and we have omitted the surface tension L. One can see that in addition to the standard surface tension, even a uniform magnetic field generates energy density, pressure, energy current, and spatial stress on the boundary.

SUMMARY

Let us summarize. We have presented the equilibrium free energy of isotropic relativistic matter, in the regime when external electromagnetic fields are more important than external gravitational fields. From a technical point of view, this amounts to generalizing the analysis of ref. [9] by i) performing a partial summation of electromagnetic contributions, and ii) by taking into account surface terms in the generating functional. From a physical point of view, this amounts to describing the effects of polarization. The equilibrium electric current can be expressed in terms of the polarization tensor $M^{\mu\nu}$ to all orders, $J^\alpha = \rho u^\alpha - \nabla_\lambda M^{\lambda\alpha}$. The charge density and the spatial current are given by eqs. (2.16) and (2.17). In 3+1 dimensions in flat space3 they reduce to

$$n = \rho - \boldsymbol{\nabla}{\cdot}\mathbf{p} - \mathbf{p}{\cdot}\boldsymbol{\nabla}T/T - 2\mathbf{m}{\cdot}\boldsymbol{\omega}\,, \quad (5.1)$$

$$\mathbf{j} = \boldsymbol{\nabla}{\times}\mathbf{m} + \mathbf{m}{\times}\boldsymbol{\nabla}T/T\,, \quad (5.2)$$

where p is the electric polarization vector, m is the magnetic polarization vector, and ω is the angular velocity. These expressions generalize the familiar $n = \rho - \nabla{\cdot}p$ and $j = \nabla{\times}m$ in electro- and magneto-statics. At leading order in the derivative expansion, the surface current is $J^\mu_{\text{s}} = n_\lambda M^{\lambda\mu}$, which says that the surface charge density is p·n, and the surface spatial current is m×n.

The notion of polarization is ambiguous when the external fields vary in space. This is because polarization is defined as a response to electric and magnetic fields, which are not fundamental quantities, but are rather derived from the vector potential A_μ. In particular, the electro-vortical susceptibility $\chi_{E\Omega}$ is unphysical, as well as the magneto-electric susceptibility χ_{BE} in the presence of free charges. Nevertheless, most O(∂) contributions to polarization are not affected by this ambiguity and may be derived from the equilibrium free energy, as described in section 4. For example, in a parity-violating system, there is a contribution to the electric polarization vector p ∝ B × ω, and the contribution to the magnetic polarization vector m ∝ E × ω which do not suffer from this ambiguity. The corresponding surface charge density σ_s = c n·(B × ω) and the surface current js = c (E(ω·n) − ω(E·n)) are determined by the same coefficient c = 2M21.

When the external fields are non-uniform, the boundary charge and spatial current are no longer determined by polarization. This is not surprising: while the polarization vectors are ambiguous, the charge and the current are not. At first order in the derivative expansion the boundary current is

$$J^\mu_{\text{s}} = n_\lambda M^{\lambda\mu} - \nabla_\lambda\left(n_\alpha P^\lambda_\rho P^\mu_\sigma S^{\alpha\rho\sigma}\right) , \quad (5.3)$$

where $P^\lambda_\rho = \delta^\lambda_\rho - n^\lambda n_\rho$, and $S^{\alpha\rho\sigma}$ is defined by (4.2). For the generating functional (4.10) in 3+1 dimensions, there are 15 contributions to $S^{\alpha\rho\sigma}$. As an example, consider the effect of the M_2 term for non-rotating matter at constant temperature in flat space. For the boundary with vanishing extrinsic curvature ($\nabla_\mu n_\nu = 0$), the surface charge density is

$$\sigma_{\text{s}} = \mathbf{p}{\cdot}\mathbf{n} - \frac{2}{T^4}\partial_a(M_2 E^a E_n)\,,$$

where E_n is the normal component of the electric field. This describes the response of the boundary charge density to the changes of the external electric field along the boundary.

The equilibrium energy-momentum tensor to leading order takes a simple form (3.10) which we repeat here:

$$T^{\mu\nu} = P g^{\mu\nu} + (Ts + \mu\rho)u^\mu u^\nu + T^{\mu\nu}_{\text{EM}}\,, \quad (5.4a)$$

and

$$T^{\mu\nu}_{\text{EM}} = M^{\mu\alpha}g_{\alpha\beta}F^{\beta\nu} + u^\mu u^\alpha\left(M_{\alpha\beta}F^{\beta\nu} - F_{\alpha\beta}M^{\beta\nu}\right) . \quad (5.4b)$$

This expression is model-independent, and is a leading-order consequence of gauge invariance, diffeomorphism invariance, and locality (on scales longer than the screening length). Beyond the leading order in derivatives, the form of the equilibrium $T^{\mu\nu}$ is more complicated.

Equilibrium μ-independent contributions to the current J^α are usually referred to as "bound charges" and "bound currents". There exist analogous contributions to the equilibrium $T^{\alpha\beta}$, which one may similarly christen "bound energy", "bound pressure", "bound momentum", and "bound stress". Just like bound charges and bound currents, these live both in the bulk and on the surface. For matter subject to external magnetic field (and no electric field), there will be bulk energy currents Q ∝ B×∇T, Q ∝ B×∇B². The boundary energy current $Q_s = \chi_{B\Omega}$ B×n is determined by the same susceptibility $\chi_{B\Omega}$ which fixes the response of magnetization to rotation, m = χ_{BB}B + 2$\chi_{B\Omega}$ω + O(∂T, ∂B).

Finally, our discussion so far was restricted to the state of global equilibrium, i.e. to thermodynamics. It is straightforward to extend it to hydrodynamics of polarized relativistic matter, if one assumes that the external electromagnetic and gravitational fields are not dynamical. In order to do so, one promotes u μ and T to dynamical variables, and postulates the hydrodynamic equations in the form $\nabla_\mu T^{\mu\nu} = F^{\nu\lambda}J_\lambda,\ \nabla_\mu J^\mu = 0$, with the leading order energy-momentum tensor given by eq. (5.4). Beyond the leading order, the energymomentum tensor becomes much more involved as discussed in section 4, plus the extra transport coefficients such as viscosity make their way into the hydrodynamic equations.

If the electromagnetic fields are dynamical, the conservation equations for $T^{\mu\nu}$ and J^μ need to be supplemented by the evolution equations for the electromagnetic fields. These are usually taken to be Maxwell's equations, $\nabla_\nu F^{\mu\nu} = J^\mu$. Substituting the equilibrium current (2.14) gives

$$\nabla_\nu \left(F^{\mu\nu} - M^{\mu\nu}\right) = \rho u^\mu \,, \tag{5.5}$$

which is the standard covariant form of Maxwell's equations in matter, see e.g. [16]. In the right-hand side of (5.5), ρ is the density of free charges, while the effects of polarization are in the left-hand side. In the framework of (5.5), the derivative expansion in hydrodynamics can be implemented through the derivative expansion for $M^{\mu\nu}$, however eq. (5.5) itself will receive corrections, e.g. due to the electrical conductivity. We plan to return to the study of hydrodynamics of polarized relativistic matter in the future.

ACKNOWLEDGMENTS

I would like to thank Kristan Jensen and Adam Ritz for helpful conversations. This work was supported in part by NSERC of Canada.

REFERENCES

1. S. Jeon and U. Heinz, *Introduction to hydrodynamics*, *Int. J. Mod. Phys.* E 24 (2015) 1530010.
2. T. Schäfer and D. Teaney, *Nearly perfect fluidity: from cold atomic gases to hot quark gluon plasmas*, *Rept. Prog. Phys.* 72 (2009) 126001.
3. M. Rangamani, *Gravity and hydrodynamics: lectures on the fluid-gravity correspondence*, *Class. Quant. Grav.* 26 (2009) 224003
4. M. Müller, J. Schmalian and L. Fritz, *Graphene: a nearly perfect fluid*, *Phys. Rev. Lett.* 103 (2009) 025301.
5. Lucas, R.A. Davison and S. Sachdev, *Hydrodynamic theory of thermoelectric transport and negative magnetoresistance in Weyl semimetals*, arXiv:1604.08598 .
6. L.D. Landau and E.M. Lifshitz, *Statistical physics, Part I*, Pergamon Press, Oxford U.K. (1980).
7. L.S. Brown, *Quantum field theory*, Cambridge University Press, Cambridge U.K. (1992).
8. N. Banerjee et al., *Constraints on fluid dynamics from equilibrium partition functions*,*JHEP* 09 (2012) 046.
9. K. Jensen, M. Kaminski, P. Kovtun, R. Meyer, A. Ritz and A. Yarom, *Towards hydrodynamics without an entropy current*, *Phys. Rev. Lett.* 109 (2012) 101601.
10. K. Jensen, R. Loganayagam and A. Yarom, *Anomaly inflow and thermal equilibrium*, *JHEP* 05 (2014) 134.
11. J. Armas, J. Bhattacharya and N. Kundu, *Surface transport in plasma-balls*, *JHEP* 06(2016) 015 .
12. X.G. Wen, *Theory of the edge states in fractional quantum Hall effects*, *Int. J. Mod. Phys.*B 06 (1992) 1711.
13. S. Bhattacharyya, S. Jain, S. Minwalla and T. Sharma, *Constraints on superfluid hydrodynamics from equilibrium partition functions*, *JHEP* 01 (2013) 040.
14. W. Israel, *The dynamics of polarization*, *Gen. Rel. Grav.* 9 (1978) 451 .
15. K. Jensen, M. Kaminski, P. Kovtun, R. Meyer, A. Ritz and A. Yarom, *Parity-violating hydrodynamics in 2 + 1 dimensions*, *JHEP* 05 (2012) 102 .
16. S.R. de Groot and L.G. Suttorp, *Foundations of electrodynamics*, Horth-Holland, Netherlands (1972).

CHAPTER 14

THERMODYNAMICS AND FLUCTUATIONS FAR FROM EQUILIBRIUM

John Ross and Alejandro Fernández Villaverde

Keywords: Nonequilibrium thermodynamics, irreversible thermodynamics, stochastic processes, diffusion in chemical reaction kinetics, chemical reactions

INTRODUCTION

In the last twenty years we have developed a mesoscopic formulation of thermodynamics and fluctuations both far from, and near to, equilibrium [1]. We briefly review some of this work here. The word mesoscopic denotes an approach based on the master equation for probability distributions [2]. This is not as detailed as a statistical mechanical analysis based on averages over mechanical motions, but is related to probability distributions and macroscopic chemical rate, and transport, coefficients. We treat primarily chemical reaction systems, but also transport processes and electrochemical kinetic systems.

We establish thermodynamic state functions, dependent on the irreversible processes, that provide necessary and sufficient conditions for stability and instability of systems and provide the fluctuations in the system; these state functions can be measured and have simple relations to the concept of work. For systems with multiple stable stationary states the thermodynamic functions provide measures of relative stability and equistability. A number of experiments are cited and discussed which substantiate the theory. The review consists of chosen selections, shortened and edited, from the book *Thermodynamics and Fluctuations far from Equilibrium* by John Ross [1].

ONE-VARIABLE SYSTEMS

We begin this section with the presentation of the linear one-variable case, for which a thermodynamic state function is obtained in 2.1, and proceed to the more general nonlinear case in 2.2. These first two subsections deal with establishing reference states. We then discuss dissipation in section 2.3; connect the thermodynamic theory with stochastic theory in 2.4, and finally comment on the relative stability of multiple stationary stable states in 2.5.

Linear One-variable Systems

Consider the reaction sequence:

$$A \Leftrightarrow X \Leftrightarrow B \qquad (1)$$

with rate coefficients k_1 and k_2 for the forward and reverse reaction of the first reaction ($A \Leftrightarrow X$) and k_3 and k_4 the corresponding rates for the second reaction. In this sequence A is the reactant, X the intermediate, and B the product. Let the reactions occur in the schematic apparatus of Figure 1 at constant temperature.

Let us assume ideal gases and mass action kinetics; in that case the deterministic rate equation is:

$$\frac{dp_X}{dt} = \left(k_1 p_A + k_4 p_B\right) - \left(k_2 + k_3\right) p_X \qquad (2)$$

Hence at a stationary state the pressure of X is $p_X^{ss} = \frac{k_1 p_A + k_4 p_B}{k_2 + k_3}$. We denote the first term on the RHS (right-hand side) of (2) by t_X^+ and the second term by t_X^- [3]. At the stationary state, t_X^+ is constant, and the pressure p_X is given by:

$$\frac{p_X^s}{p_X} = \frac{t_X^+}{t_X^-} = \frac{t_X^{+s}}{t_X^-} \qquad (3)$$

Citation: John Ross and Alejandro Fernández Villaverde, "Thermodynamics and Fluctuations Far From Equilibrium," Entropy 2010, 12(10), 2199-2243; doi:10.3390/e12102199.

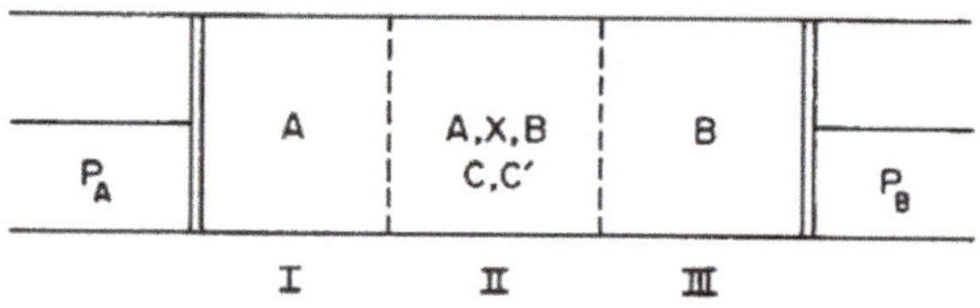

Figure 1. Two piston model. The reaction compartment (II) is separated from a reservoir of species A (I) by a membrane permeable only to A and from a reservoir of species B (III) by a membrane permeable only to B. The pressures of A and B are held fixed by constant external forces on the pistons. Catalysts C and C' are required for the reactions to occur at appreciable rates and are contained only in region II.

We use the hypothesis of *local equilibrium*: it is assumed that at each time there exists a temperature, a pressure, and a chemical potential for each chemical species. These quantities change on shorter time scales than the changes in pressure or concentration of chemical species due to chemical reaction. Thus the *chemical potential* is $\mu_X = \mu_X^0 + RT \ln p_X$, where μ_X^0 is the chemical potential in the standard state. Hence we have:

$$\mu_X - \mu_X^s = -RT \ln \frac{t_X^+}{t_X^-} \tag{4}$$

We define a thermodynamic state function ϕ [3]:

$$\phi(p_X) = V_{II} \int (\mu_X - \mu_X^s) dp_X \tag{5}$$

where V_{II} is a volume shown in Figure 1. This function has many important properties. At the stationary state of this system, ϕ is zero. If we start at the stationary state and increase p_X then $dp_X \geq 0$ and the integrand is larger than zero; hence ϕ is positive. Similarly, if we start at the stationary state and decrease p_X, then dp_X and the integrand are both negative and ϕ is positive. Hence ϕ is an extremum at the stable stationary state, a minimum.

Nonlinear One-Variable Systems

Let us consider now a nonlinear model of a stoichiometric equation, also occurring in the apparatus of Figure 1, as follows:

$$A + (r-1)X \underset{k_2}{\overset{k_1}{\rightleftharpoons}} rX,$$
$$(s-1)X + B \underset{k_3}{\overset{k_4}{\rightleftharpoons}} sX. \tag{6}$$

Since this isothermal system has chambers I and III at constant pressure and chamber II at constant volume, the proper thermodynamic function for the entire system is a linear sum of Gibbs free energies for I and III and the Helmholtz free energy for II. If in (6) we set r = 3 and s = 1 then we have the Schlögl model [4]:

$$A + 2X \Leftrightarrow 3X \tag{7}$$

$$X \Leftrightarrow B \tag{8}$$

with the rate coefficients k_1 and k_2 for the forward and reverse reaction in (7), and k_3 and k_4 in (8). For this system there exists the possibility of multiple stationary states for given constraints of the pressures p_A and p_B. The kinetic equation for p_X is:

$$\frac{dp_X}{dt} = k_1 p_A p_X^2 + k_4 p_B - \left(k_2 p_X^3 + k_3 p_X\right) \tag{9}$$

which is cubic in p_X and hence may have three stationary states (RHS of (9) equals zero), as shown in Figure 2.

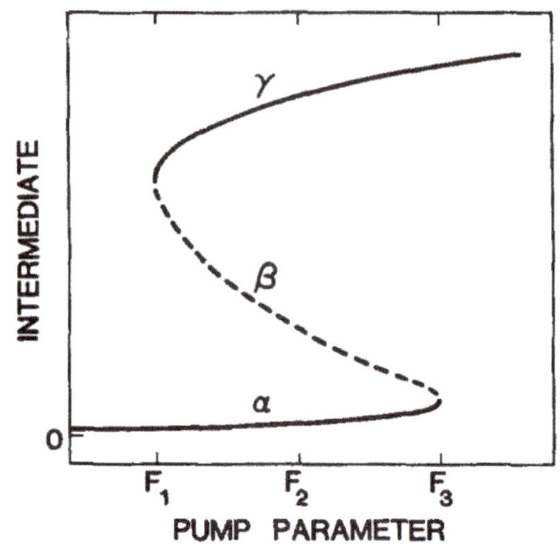

Figure 2. Stationary states of the Schlögl model with fixed reactants and products pressures. Plot of the pressure of the intermediate p_X *vs.* the pump parameter (p_A/p_B). The branches of stable stationary states are labeled α and γ and the branch of unstable stationary states is labeled β. The marginal stability points are at F_1 and F_3 and the system has two stable stationary states between these limits. The equistability point of the two stable stationary states is at F_2.

The deterministic kinetic equation for the Schlögl model is:

$$\frac{dp_X}{dt} = k_1 p_A p_X^2 + k_4 p_B - \left(k_2 p_X^3 + k_3 p_X\right) \tag{10}$$

The first two positive terms on the RHS of (10) are again given the symbol t_X^+ and the two negative terms the symbol t_X^-; their ratio is:

$$\frac{t_X^+}{t_X^-} = \frac{k_1 p_A p_X^2 + k_4 p_B}{k_2 p_X^3 + k_3 p_X} \tag{11}$$

which we use to define the quantity p_X^*, the pressure in a reference state for which the following holds:

$$\frac{t_X^+}{t_X^-} = \frac{p_X^*}{p_X} \tag{12}$$

Thus p_X^* is:

$$p_X^* = \frac{k_1 p_A p_X^2 + k_4 p_B}{k_2 p_X^2 + k_3} \tag{13}$$

It is useful to compare the linear model with the Schlögl model in the following way: assign to each model the same values of p_A, p_B, T, V_I, V_{II}, V_{III}, and equilibrium constants for the A⇔X and B⇔X reactions. Then the two model systems are *instantaneously thermodynamically equivalent*. If furthermore t_X^+ and t_X^- have the same values in the two systems at each point in time, the two systems are *instantaneously kinetically indistinguishable*. If we define p_X^* as the pressure of X in the instantaneously indistinguishable linear system at stationary state, we may write from (4) and (5):

$$\mu_X - \mu_X^* = RT\ln\frac{p_X}{p_X^*} = -RT\ln\frac{t_X^+}{t_X^-} \quad (14)$$

This equation relates the chemical driving force towards a stable stationary state (LHS) to the ratio of sums of fluxes of X (RHS). We write the analog of (5) for our chosen thermodynamic function:

$$\phi^*(p_X) = \int(\mu_X - \mu_X^*)dp_X \quad (15)$$

where the integrand is a species specific activity which plays a crucial role. The function in (5) is an *excess work*: the work of moving the system from a stable stationary state to an arbitrary value p_X compared to the work of moving the system from the stationary state of the instantaneous indistinguishable linear system to p_X. If A and B in (6) are chosen such that the ratio of their pressures equals the equilibrium constant, then $\phi^* = \Delta G$ and p_X^* =ps.

With the species specific activity and the thermodynamic state function ϕ^* we can state the following conditions:

$$\frac{d\phi^*}{dp_X} = 0 \quad (16)$$

at each stationary state,

$$\frac{d^2\phi^*}{dp_X^2} \geq 0 \quad (17)$$

at each stable stationary state, with the equality sign holding at marginal stability,

$$\frac{d^2\phi^*}{dp_X^2} \leq 0 \quad (18)$$

at each unstable stationary state, with the equality sign holding at marginal stability.

Equations (16,17) are necessary and sufficient conditions for the existence and stability of nonequilibrium stationary states.

Dissipation

For a spontaneously occurring chemical reaction at constant pressure, p, and temperature, T, the Gibbs free energy change gives the maximum work, other than $p{\cdot}V$ work, that can be obtained from the reaction. For systems at constant V,T it is the Helmholtz free energy change that yields that measure. If no work is done by the reaction then the respective free energy changes are dissipated, lost. For reactions of ideal gases run in the apparatus in Figure 1, we can define a *hybrid free energy, M*:

$$M = \mu_A(n_A^I + n_A^{II}) + \mu_B(n_B^{II} + n_B^{III}) + \mu_X n_X^{II} - RT(n_A^{II} + n_B^{II} + n_X^{II}) \quad (19)$$

The time rate of change of M is:

$$dM/dt = \mu_A dn_A^I/dt + \mu_B dn_B^{III}/dt + \mu_X dn_X^{II}/dt \quad (20)$$

if there is no depletion of the reservoirs I and III. According to conservation of mass we have:

$$0 = \mu_X^* dn_A^I/dt + \mu_X^* dn_B^{III}/dt + \mu_X^* dn_X^{II}/dt \quad (21)$$

and therefore we may write:

$$dM/dt = (\mu_A - \mu_X^*)dn_A^I/dt + (\mu_B - \mu_X^*)dn_B^{III}/dt + (\mu_X - \mu_X^*)dn_X^{II}/dt \quad (22)$$

Hence we write for the dissipation D:

$$D = -dM/dt = -dM_{res}/dt - dM_X/dt \quad (23)$$

where the first term on the RHS is the dissipation due to the conversion of A to X at the pressure p_X^* and at the rate $-\frac{dn_A^I}{dt}$ and the conversion of X to B at the same pressure of X and the rate $\frac{dn_B^{III}}{dt}$. The second term on the RHS of (23) is a species-specific dissipation, D_X:

$$-dM_X/dt = -(\mu_X - \mu_X^*)dn_X^{II}/dt = RT(t_X^+ - t_X^-)\ln(t_X^+/t_X^-) \equiv D_X \quad (24)$$

From this last equation it is clear that DX=−dMX/dt≥0 for all p_X, regardless of the reaction mechanism; the equality holds only at the stationary state.

As we shall discuss later, the total dissipation D is not an extremum at stationary states in general, but there may be exceptions. D_X is such an extremum and the integral:

$$\phi^* = \int(\mu_X - \mu_X^*)dn_X \quad (25)$$

is a Lyapunov function in the domain of attraction of each stable stationary state.

The dissipation in a reaction can range from zero, for a reversible reaction, to its maximum of ΔG when no work is done in the surroundings. Hence the dissipation can be taken to be a measure of the efficiency of a reaction in regard to doing work. There is more on this subject in Section 7.

Connection of the Thermodynamic Theory with Stochastic Theory

The deterministic theory of chemical kinetics is formulated in terms of pressures, for gases, or concentrations of species for gases and solutions. These quantities are macroscopic variables and fluctuations of theses variables are neglected in this approach. In stochastic theory, however, one assumes that fluctuations do occur, say in the number of particles of a given species *X*, that there is a probability distribution *P*(*X*,*t*) for that number of particles at a given time, and that changes in this distribution occur due to chemical reactions. The transition probabilities of such changes are assumed to be given by macroscopic kinetics. We shall show that the non-equilibrium thermodynamic functions ϕ (for linear systems), ϕ^* (for nonlinear systems), the excess work, determines the stationary, timeindependent, probability distribution, which leads to a physical interpretation of the connection of the thermodynamic and stochastic theory. At equilibrium the probability distribution of fluctuations is determined by the Gibbs free energy change at constant *T*,*p*, which is the work other than $p \cdot V$ work. We restrict the analysis at first to reaction mechanisms for which the number of molecules of species *X* changes by ±1 in each elementary step. We take the probability distribution to obey the master equation. For the cubic Schlögl model ((6) with r = 3, s = 1) the master equation is [2,3]:

$$\partial P(X,t)/\partial = t^+(X-1)P(X-1,t)+t^-(X+1)P(X+1,t)-[t^+(X)+t^-(X)]P(X,t) \quad (26)$$

The first two terms on the RHS yield an increase in *X*, the last two terms a decrease in *X*. The fluxes in this equation are:

$$t^+(X)=c_1AX(X-1)/2!+c_4B$$
$$t^-(X)=c_2X(X-1)(X-2)/3!+c_3X \quad (27)$$

with the parameters c_i related to the rate coefficients k_i by $k_i = V^{m_i-1}(c_i/n_i!)$ for $1 \le i \le 4$, where m_i is the molecularity of the ith step and n_i the molecularity in *X*.

From the master equation we can derive the result that the average concentration, the average number of X in a volume V, obeys the deterministic rate equation in the limit of large numbers of molecules.

The time-independent solution of the master equation is:

$$P_S(X)=P_S(0)\prod_{i=1}^{X}\frac{t^+(i-1)}{t^-(i)} \quad (28)$$

which by retention of only the leading term in the Euler-MacLaurin summation formula reduces to

$$P_S(X)=N\exp\left[-\frac{1}{RT}\int^{x}(\mu_X-\mu_X^*)dX\right] \quad (29)$$

and *N* is a normalization constant. The connection between the thermodynamic and stochastic theory is established with the use of (14) to give:

$$P_S(X)=N\exp\left[-\frac{1}{RT}\int^{x}(\mu_X-\mu_X^*)dX\right] \quad (30)$$

The Lyapunov function ϕ^* (15) is both the thermodynamic driving force towards a stable stationary state and determines the stationary probability distribution of the master equation. The stationary distributions (29,30) are non-equilibrium analogs of the Einstein relations at equilibrium, which give fluctuations around equilibrium. There is another interesting connection [3]. We define $P(X_1,t_1;X_0,t_0)$ to be the probability density of observing X_1 molecules in V at time t_1 given that there are X_0 molecules at t_0. This function is the solution of the master equation (26) for the initial condition $P(X,t=t_0)=\delta(X-X_0)$. The probability density can be factored into two terms [3]:

$$P(X_1,t_1;X_0,t_0)=F_1(X_0\to X_1)F_2(X_1,X_0,t_1-t_0) \quad (31)$$

in which the first term on the RHS is independent of the path from X_0 to X_1 and independent of the time interval (t_1-t_0). To the same approximation with which we obtained (29) we can reduce the first term to:

$$F_1(X_0\to X_1)=\exp\left[\frac{1}{2}\int_{X_0}^{X_1}(\ln t^+/t^-)dX\right] \quad (32)$$

and find it to be of the same form as the probability distribution (29). It contains the irreversible part of the probability density (31).

Relative Stability of Multiple Stationary Stable States

For systems with multiple stable stationary states there arises the issue of relative stability of such states. We treat systems with a single intermediate and stoichiometric changes in X are limited to ±1.

In regions of multistability the stationary probability distribution is bimodal and is shown in Figure 3 for the cubic Schlögl model. Stable stationary states are located at maxima, labeled 1 and 3, and unstable stationary states at minima, labeled 2.

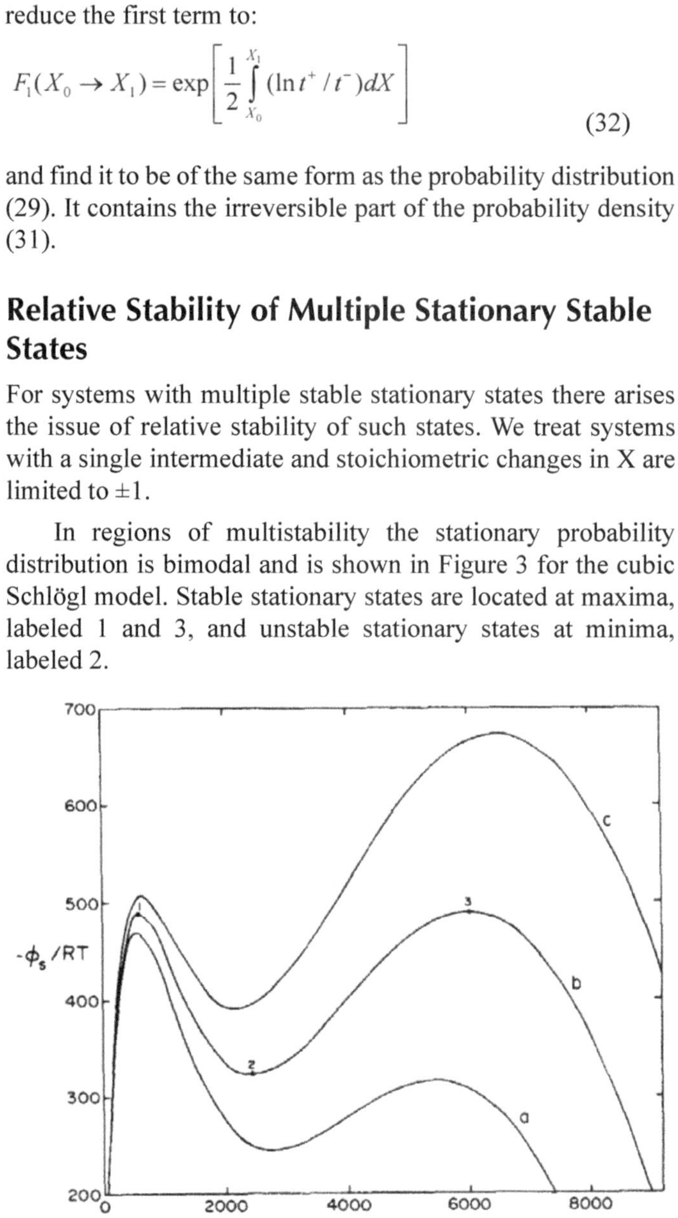

Figure 3. Plot of the integral in (30), marked ϕs *vs.* *X* for the Schlögl model with parameters: $c_1 = 3.10^{-10}$ s^{-1}; $c_2 = 1.10^{-7}$ s^{-1}; $c^3 = 0.33$ s^{-1}; $c_4 = 1.5.10^{-4}$ s^{-1}; and A = B. For curve (a) B = $9.8.10^6$; for curve (b) B = $1.01.10^6$; curve (c) B = $1.04.10^6$. Curve (b) lies close to the equistability of the stable stationary states 1 and 3; 2 marks the unstable stationary state.

Consider now the ratio of the probability density (31), for a given transition from X_1 to X_2 to that of the reverse transition:

$$\frac{P(X_2,t_2;X_1,t_1)}{P(X_1,t_2;X_2,t_1)} = \exp\left[\int_1^2 \ln\frac{t_X^+}{t_X^-}dX\right] \tag{33}$$

Equistability of two stable stationary states, labeled now 1 and 3 to correspond to Figure 3, is defined by:

$$\frac{P(X_3,t;X_1,0)}{P(X_1,t;X_3,0)} = 1 \tag{34}$$

which with the use of (24) we may also express as:

$$\int_2^3 D_X\,dt = \int_2^1 D_X\,dt \tag{35}$$

The integral of the species-specific dissipation from the unstable stationary state 2 to the stable stationary state 3 equals, at equistability, to the integral of the dissipation from the unstable stationary state 2 to the stable stationary state 1. Whereas the integral of the total dissipation for the limits in (35) goes to infinity, that of the species specific dissipation is finite. We can restate (35) in terms of the excess work:

$$\int_2^3 \left(\mu_X - \mu_X^*\right)dn_X^{II} = \int_2^1 \left(\mu_X - \mu_X^*\right)dn_X^{II} \tag{36}$$

These results constitute a fluctuation-dissipation relation, rediscovered in a statistical mechanical approach by Evans and coworkers (see [5] and references therein). At equistability the integral of the excess work from 2 to 3 equals the integral of the excess work from 2 to 1. Equations (34–36) provide necessary and sufficient conditions of equistability of stable stationary states.

The master equation has been investigated for a sequence of unimolecular (non-autocatalytic) reactions based on moment generating functions [6]; these yield Poissonian stationary distribution for single intermediate systems in terms of the number of particles X of species X, with X^{ss} that number in the stationary state, $P_S(X) = \left[(X^{SS})^X / X!\right]\exp(-X^{SS})$. Our results are consistent with (36), as can be seen from the use of (15) and (30), a change of variables to particle numbers X, and the use of Stirling's approximation:

$$P_S(X) = \wp\, exp\left[-(k_B T)^{(-1)}\int^{nx}(\mu_X - \mu_X^{SS})dn'_X\right] = \wp\, exp(X lnX^{SS} - X lnX + X) = \wp\left[\frac{(X^{SS})^X}{X!}\right] \tag{37}$$

Here $\wp$ is the normalization constant $\exp(-X^{ss})$. The formulation given in this section has the advantage of the physical interpretation in terms of species-specific thermodynamic driving forces and in terms of Lyapunov functions; further our formulation is generalizable to autocatalytic systems and many variable systems.

THERMODYNAMIC STATE FUNCTION FOR MULTIVARIABLE SYSTEMS

In the previous section we have obtained thermodynamic state functions for single variable systems, both linear (5) and nonlinear (15). In this section we obtain a thermodynamic state function for *multivariable* systems. To this end we need to consider fluctuations [7]. We start with the master equation [2]:

$$\frac{\partial}{\partial t}P_X(\mathbf{X},t) = \sum_{\mathbf{r}}\left[W(\mathbf{X}-\mathbf{r},\mathbf{r})P_X(\mathbf{X}-\mathbf{r},t) - W(\mathbf{X},\mathbf{r})P_X(\mathbf{X},t)\right] \tag{38}$$

in which P_X is the probability distribution of finding $\boldsymbol{X}$ particles (molecules) in a given volume, and $W(\boldsymbol{X},\boldsymbol{r})$ is the transition probability due to reaction from $\boldsymbol{X}$ to $\boldsymbol{X+r}$ particles. Now we do a Taylor expansion of the term W(X−r,r)P(X−r,t)around $\boldsymbol{X}$:

$$W(\mathbf{X}-\mathbf{r},\mathbf{r})P_X(\mathbf{X}-\mathbf{r},t) = W(\mathbf{X},\mathbf{r})P_X(\mathbf{X},t) + \sum_{m=1}^{\infty}\frac{(-1)^m}{m!}(\mathbf{r}\cdot\nabla_X)^m\left[W(\mathbf{X},\mathbf{r})P_X(\mathbf{X},t)\right] \tag{39}$$

and introduce the concentration vector x=X/V; then we have the reduced relations:

$$\nabla_X = \frac{1}{V}\nabla_x,\ P_X(\mathbf{X},t)\Delta\mathbf{X} = P_x(\mathbf{x},t)d\mathbf{x},\ \Delta\mathbf{X} = 1,\ w(\mathbf{x},\mathbf{r};V) = \frac{1}{V}W(\mathbf{x}V,\mathbf{r}) \tag{40}$$

where V is the volume of the system. We substitute these relations into (39) and obtain:

$$W(\mathbf{X}-\mathbf{r},\mathbf{r})P_X(\mathbf{X}-\mathbf{r},t) = d\mathbf{x}W(\mathbf{x}V,\mathbf{r})P_x(\mathbf{x},t) + d\mathbf{x}\sum_{m=1}^{\infty}\frac{(-1)^m}{m!}\left(\frac{1}{V}\mathbf{r}\cdot\nabla_x\right)^m\left[W(\mathbf{x}V,\mathbf{r})P_x(\mathbf{x},t)\right] \tag{41}$$

Next we introduce the momentum operator, p^=−1V∇X, with which we can write:

$$1+\sum_{m=1}^{\infty}\frac{(-1)^m}{m!}\left(\frac{1}{V}\mathbf{r}\cdot\nabla_X\right)^m \ldots = 1+\sum_{m=1}^{\infty}\frac{1}{m!}(\mathbf{r}\cdot\hat{\mathbf{p}})^m \ldots = exp(\mathbf{r}\cdot\hat{\mathbf{p}}) \tag{42}$$

The master equation becomes:

$$\frac{1}{V}\frac{\partial}{\partial t}P_x(\mathbf{x},t) = \sum_{\mathbf{r}} w(\mathbf{x},\mathbf{r};V)\left[\exp(\mathbf{r}\cdot\hat{\mathbf{p}})-1\right]P_x(\mathbf{x},t) = \hat{H}_+(\mathbf{x},\hat{\mathbf{p}})P_x(\mathbf{x},t) \tag{43}$$

where we have defined the Hamiltonian operator [8,9]:

$$\hat{H}_+(\mathbf{x},\hat{\mathbf{p}})\ldots = \sum_{\mathbf{r}} w(\mathbf{x},\mathbf{r};V)\left[\exp(\mathbf{r}\cdot\hat{\mathbf{p}})-1\right] \tag{44}$$

Thus we have formally, and exactly, converted the master equation to a Schrödinger equation. This has the substantial advantage that we can apply well-known approximations in quantum mechanics to obtain solutions to the master equation. In particular we refer to the W.K.B. approximation valid for semi-classical cases, those for which Planck's constant formally approaches zero. The equivalent limit for (43) is that of large volumes (large numbers of particles). Hence we seek a stationary solution of (43), that is the time derivative of $P_X(\boldsymbol{X},t)$ is set to zero, of the form:

$$P_S^{(n)}(\mathbf{X}) = C^{(n)}\exp\left[-VS_n(\mathbf{X})\right] \tag{45}$$

where S_n will be shown to be the classical action of a fluctuational trajectory accessible from the nth stable stationary state. We substitute (45) into the stationary part of (43) and obtain:

$$H\left[\mathbf{x}, \frac{\partial S_n(\mathbf{x})}{\partial \mathbf{x}}\right] = 0 \tag{46}$$

the equation satisfied by $S_n(\mathbf{X})$ with the Hamiltonian function (not operator):

$$H(\mathbf{x},\mathbf{p}) = \sum_r w(\mathbf{x},\mathbf{r})[\exp(\mathbf{r}\cdot\mathbf{p}) - 1] \tag{47}$$

and the boundary condition:

$$S_n(\mathbf{x}_n^S) = 0. \tag{48}$$

These equations show that it is the classical action S_n that satisfies the Hamiltonian-Jacoby equation (46) with coordinate **x**, momentum $p=\partial S_n(x)/\partial x$, and Hamiltonian equal to zero (stationary condition). The Hamiltonian equations of motion for the system are:

$$\frac{d\mathbf{x}}{dt} = \sum_r \mathbf{r}\omega(\mathbf{x},\mathbf{r})\exp(\mathbf{r}\cdot\mathbf{p}) \tag{49}$$

and:

$$\frac{d\mathbf{p}}{dt} = -\sum_r [\exp(\mathbf{r}\cdot\mathbf{p}) - 1]\frac{\partial\omega(\mathbf{x},\mathbf{r})}{\partial\mathbf{x}} \tag{50}$$

From these relations we determine the *action*:

$$S_n(\mathbf{x}) = \int_{-\infty}^{0} \mathrm{d}t\, \mathbf{p}\cdot \mathrm{d}\mathbf{x}/\mathrm{d}t \tag{51}$$

for the fluctuational trajectory starting at the nth stable stationary state $\mathbf{x}_n^s$ with $\mathbf{p} = 0$ at $t = -\infty$ and ending at **x** at $t = 0$.

Linear Multi-variable Systems

Let us consider the following linear reaction system [7]:

$$A \underset{k_2}{\overset{k_1}{\rightleftarrows}} X \underset{k_4}{\overset{k_3}{\rightleftarrows}} Y \underset{k_6}{\overset{k_5}{\rightleftarrows}} B \tag{52}$$

run in an apparatus as in Figure 1, where the pressures of A and B are held constant. The action is:

$$S_n(\mathbf{x}) = \int_S^{X,Y} [\ln(X/X_S)dx + \ln(Y/Y_S)dy] \tag{53}$$

where X_s, Y_s are stable stationary state concentrations. We see that the integrand is an exact differential and hence the action is independent of the path of integration in concentration space. The action is a state function [8,9].

The physical interpretation of the action comes from consideration of the free energy M for the three compartments in Figure 1 (19):

$$M = G^I + A^{II} + G^{III} \tag{54}$$

For differential changes in A,X,Y,B the differential change in M is:

$$dM = \mu_A dn_A + \mu_X dn_X + \mu_Y dn_Y + \mu_B dn_B \tag{55}$$

The differential excess free energy change $d\phi$ is the difference between dM, the system with arbitrary concentrations of X and Y, and dM_s, the system in the stationary state. Hence we have:

$$d\phi = (\mu_X - \mu_X^S)dn_X + (\mu_Y - \mu_Y^S)dn_Y \tag{56}$$

When we compare Equation (56) with (53) we see that:

$$d\phi/kT = V\, dSq \tag{57}$$

This important physical result was first given in [7]: the mathematical concept of the action can be identified with the thermodynamic excess work.

On a fluctuational trajectory the differential excess free energy is positive and zero at a stable stationary state. We show this by considering the differential action:

$$\left.\frac{dS}{dt}\right|_{fl} = \mathbf{p}\cdot d\mathbf{x}/dt\big|_{fl} - H(\mathbf{x},\mathbf{p}) = \sum_r \omega(\mathbf{x},\mathbf{r})[(\mathbf{r}\cdot\mathbf{p})\exp(\mathbf{r}\cdot\mathbf{p}) - \exp(\mathbf{r}\cdot\mathbf{p}) + 1] \tag{58}$$

The transition probabilities $\omega(\mathbf{x},\mathbf{r})$ are all positive and the square bracket is larger than zero except for $\mathbf{p} = 0$, that is at the stable stationary state. Therefore we have:

$$\left.\frac{dS}{dt}\right|_{fl} \geq 0 \tag{59}$$

and hence the excess differential free energy $d\phi$ is positive in general and zero at stationary states.

Suppose we prepare this system at a given (x,y) and let it proceed along the deterministic trajectory back to the stationary state. Along this path $d\phi$ is negative which follows from the deterministic variation of the action in time:

$$\left.\frac{dS}{dt}\right|_{\det} = \nabla S\left.\frac{d\mathbf{x}}{dt}\right|_{\det} = \mathbf{p}\left.\frac{d\mathbf{x}}{dt}\right|_{\det} = \mathbf{p}\left.\frac{d\mathbf{x}}{dt}\right|_{\det} - H(\mathbf{x},\mathbf{p}) = \sum_r \omega(\mathbf{x},\mathbf{r})[(\mathbf{r}\cdot\mathbf{p}) - \exp(\mathbf{r}\cdot\mathbf{p}) + 1] \tag{60}$$

which holds since the Hamiltonian is zero. For all real values of $\mathbf{r}\cdot\mathbf{p}$ the square bracket in (60) is negative unless $\mathbf{p} = 0$. And therefore:

$$\left.\frac{dS}{dt}\right|_{\det} \leq 0 \tag{61}$$

with the equality holding only at a stationary state. Hence an excess work is required to move a system from a stable stationary state, and excess work can be done by a system relaxing towards a stable stationary state. The action and the excess work are both Lyapunov functions; they serve for nonequilibrium systems the role the Gibbs free energy serves for systems going to equilibrium .

We note here that (59) and (61) hold for nonlinear multi-variable systems as well; no assumption of a linear reaction mechanism was made in their derivation.

Nonlinear Multi-variable Systems

Let us consider now a non-linear multi-variable system such as the following example:

$$A+(m-1)X \underset{k_2}{\overset{k_1}{\rightleftharpoons}} mX,$$
$$rX+(s-1)Y \underset{k_4}{\overset{k_3}{\rightleftharpoons}} (r-1)X+sY,$$
$$nY \underset{k_6}{\overset{k_5}{\rightleftharpoons}} (n-1)Y+B. \qquad (62)$$

The stationary distribution is given by (45) and (51), with **p** and **dx**/dt obtained from solutions of Hamilton's equations. We now choose our reference by using the equations:

$$p_x = \ln(X/X^0)$$
$$p_y = \ln(Y/Y^0) \qquad (63)$$

Equation (62) yields unique values of (X^0, Y^0) in the absence of certain crossings of fluctuational trajectories in the (X, Y) space, called 'caustics', see [10]. There may be more than one fluctuational trajectory which starts at **p** = 0 at a stable stationary state and passes through a given (X, Y). These trajectories will have different values of **p** and the one with the lowest value of **p** will determine the action in the thermodynamic limit, the contributions from other trajectories vanishing in that limit.

Hence we find for the action the expressions:

$$S_n(\mathbf{x}) = \int_{-\infty}^{0} dt\left[\ln(X/X^0)\frac{dx}{dt}\bigg|_{fl} + \ln(Y/Y^0)\frac{dy}{dt}\bigg|_{fl}\right] = \int_S^{X,Y}\left[\ln(X/X^0)dx + \ln(Y/Y^0)dy\right]$$
$$= 1/RT\int_S^{X,Y}\left[(\mu_X-\mu_X^0)dx + (\mu_Y-\mu_Y^0)dy\right] = (1/VkT)\int_S^{X,Y} d\phi^0 \qquad (64)$$

where the reference state used here (X^0, Y^0) replaces the starred reference state of section 2, see Equation (13). The important point is that the action and the excess work in (64) are state functions for single and multi-variable systems. Both X^0 and $Y^0$are functions of X and Y in general, but the integrand in (64) is an exact differential, because **p** is the gradient of the action, (51). For the starred reference state the excess work is a state function only for single variable systems.

The fluctuational trajectory away from a stationary state to a given point in concentration space (X, Y) in general differs from the deterministic path from that point back to the stationary state for systems without detailed balance. We show this in some calculations for the Selkov model [11]; in (62) we take m = n = r = 1, s = 3; other parameters are given in [7], p. 4555.Figure 4 gives some results of these calculations.

For one-variable systems the fluctuational trajectory away from the stationary state is the same as the deterministic trajectory back to the stationary state. Therefore for such systems ϕ^* equals ϕ^0 (64). In summary, we define the state functionϕ^0 with the use of (64):

$$\phi^0 = \int_S^{X,Y} d\phi^0 \qquad (65)$$

and list the following results (compare with the results listed in section 2 for single variable systems). ϕ^0 is a thermodynamic state function. It is a potential for the stationary probability distribution of the master equation, and is a Lyapunov function in the domain of each stable stationary state. See also [10,11,12,13,14,15,16,17]. It is directly related to the excess work necessary to remove a system from a stable stationary state, and the work obtainable from a system in its return to such a state. It is an extremum at stationary states; a minimum (zero) at stable stationary states, a maximum at unstable stationary states (59). For a fluctuational trajectory ϕ^0 increases away from the stable stationary state (59); for a deterministic trajectory towards a stable stationary state it decreases, (60). The first derivative of ϕ^0 is larger than zero at each stable stationary state, smaller than zero at each unstable stationary state. The function ϕ0 provides necessary and sufficient criteria for the existence and stability of stationary states. ϕ^0 serves to determine relative stability of multi-variable homogeneous systems in exactly the same way as shown in (33) for single variable systems. Comparison with experiments on relative stability requires consideration of space-dependent (inhomogeneous) systems and that subject is discussed in the next section.

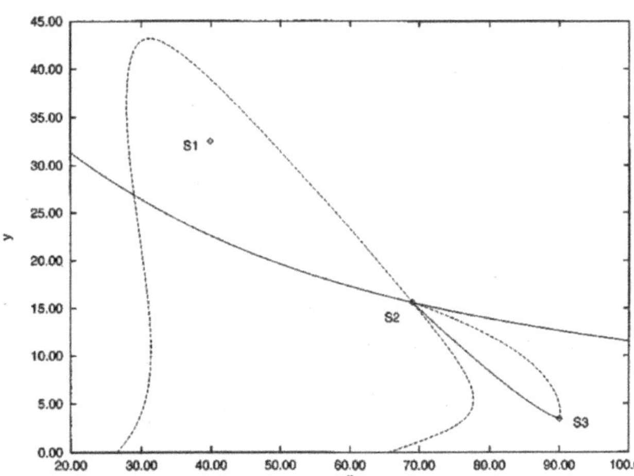

Figure 4. From [7]. S1 and S3 are stable stationary states (stable foci); S2 denotes an unstable stationary state. The solid line from S2 to S3 indicates the deterministic trajectory. The other solid line through S2 is the deterministic separatrix, that is the line that separates deterministic trajectories, on one side going towards S2 and on the other side going towards S3. The dotted lines are fluctuational trajectories: one from S3 to S2 and the others proceeding from S2 in two different directions. The fluctuational trajectory need not differ so much from the reverse of the deterministic trajectory.

The specification of the reference state X^0, Y^0 requires solution of the master equation for a particular reaction mechanism. This in general demands numerical solutions, which can be lengthy. However, if we do not need to obtain a thermodynamic function, this is not necessary and the deterministic approach of section 2 may be used instead.

The state function ϕ^0 can be determined from macroscopic electrochemical measurements, as well as other measurements, see section 6.3.

THERMODYNAMIC AND STOCHASTIC THEORY OF REACTION-DIFFUSION SYSTEMS

In the previous sections we have studied

only *homogeneous* reaction systems. In the more general, inhomogeneous case, concentrations are not only function of time but also of space. Hence there may be concentration gradients in space and diffusion will occur. In this section we formulate a thermodynamic and stochastic theory for such systems [18]. First we analyze one-variable systems and then multi-variable systems with multiple stable stationary states; then we apply the theory to study relative stability of these multiple stationary states.

Systems with One Intermediate

We begin with a one-variable system such as the Schlögl model, which is described by (7,8). If the system is homogeneous its deterministic kinetics is:

$$\frac{dX}{dt} = t^{+}(X) - t^{-}(X) \tag{66}$$

where t^{+}, t^{-} are kinetic fluxes. If the system is inhomogeneous, the concentrations are functions of time and spatial coordinates. Let us consider a one-dimensional system with spatial coordinate z and discretize the space into many boxes labeled with $\ldots i-1$, i, $i+1$, as in Figure 5. The change in the number of particles X_i is due to reaction and diffusion into and out of box i and can be written:

$$\frac{dX}{dt} = t^{+}(X) - t^{-}(X) + t^{+}_{Di} - t^{-}_{Di} \tag{67}$$

where the fluxes of diffusion can be expressed as a function of a constant diffusion coefficient, d:

$$t^{+}_{Di} = d(X_{i-1} + X_{i+1})$$
$$t^{-}_{Di} = 2dX_i \tag{68}$$

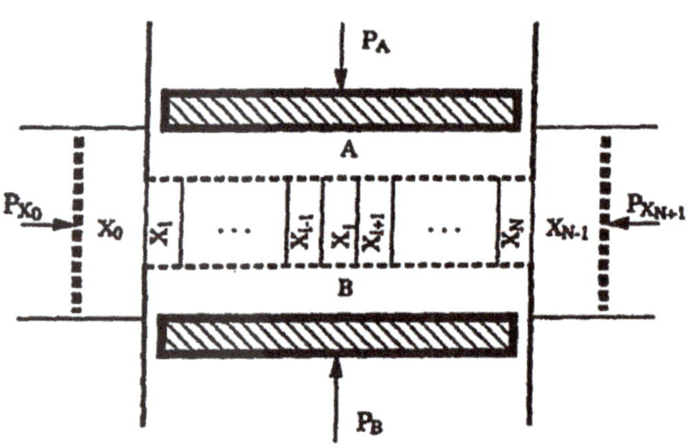

Figure 5. Schematic apparatus for reaction-diffusion system in one spatial dimension. The boxes 1 to N are separated from a constant-pressure reservoir of *A* by a membrane permeable only to *A*, and similarly for the reservoir of *B* [18].

Let us write the diffusivity of the system as $D = l^2 d$, where l is the length of a box. Then in the continuous limit we have:

$$\frac{dx(z)}{dt} = t^{+}[x(z)] - t^{-}[x(z)] + D\frac{\partial^2 x(z)}{\partial z^2} \tag{69}$$

Since changes in *A* and *B* take place at constant temperature and pressure, and in X at constant temperature and volume, the differential free energy for a box i, dM_i, is:

$$dM_i = \mu(A_i)dA_i + \mu(X_i)dX_i + \mu(B_i)dB_i \tag{70}$$

A linear system thermodynamically and kinetically equivalent to (67) is:

$$\frac{dX_i^L}{dt} = a_i - b_i X_i^L + t_{Di}^{+L} - t_{Di}^{-L} \tag{71}$$

where the superscript L stands for linear. For the instantaneous equivalencies we require $a_i = t^{+}(X_i)$ and $b_i = \frac{t^{-}(X_i)}{X_i}$. The stationary solution is:

$$X_i^{LS} = X_i^L \frac{a_i + d(X_{i-1}^{LS} + X_{i+1}^{LS})}{b_i X_i^L + t_{Di}^{-L}} \tag{72}$$

The excess work for a box i of the equivalent linear system is the difference between the free energy change dM at an arbitrary state X_i^L and that at the stationary state X_i^{LS}:

$$\tilde{d}\phi_i(X_i^L) = dM_i - dM_i^S = \{\mu(X_i^L) - \mu(X_i^{LS})\}dX_i^L = kT\ln\frac{X_i^L}{X_i^{LS}}dX_i^L \tag{73}$$

where the notation d˜ indicates inexact differential. The driving force towards the stationary state is

$$\{\mu(X_i^L) - \mu(X_i^{LS})\} \tag{74}$$

In a nonlinear system the driving force is the potential difference between state X_i and a reference state X_i^*, which is the stationary state of the equivalent linear system at the specified value of X_i. Thus the reference state is:

$$X_i^* = X_i^{LS} = X_i \frac{t^{+}(X_i) + d(X_{i-1}^* + X_{i+1}^*)}{t^{-}(X_i) + t_{Di}^{-}} \tag{75}$$

The excess work for a box i of the nonlinear system is:

$$\tilde{d}\phi_i(X_i) = \{\mu(X_i) - \mu(X_i^*)\}dX_i = kT\ln\frac{X_i}{X_i^*}dX_i \tag{76}$$

and the total excess work of all the boxes is:

$$\phi[\{X_i\}] = kT\sum_i\int^{x_i}\ln\frac{X_i'}{X_i'^*}dX_i' = kT\sum_i\int^{x_i}\ln\frac{t^{-}(X_i') + t_{Di}^{-}}{t^{+}(X_i') + t_{Di}^{+*}}dX_i' \tag{77}$$

Since ϕ is not a state function, a path of integration must be specified in order to evaluate it. We can easily check two limiting cases: the homogeneous limit (no diffusion) and the limit of no reaction (diffusion only). In the homogeneous limit there is only one box for which $t_D^{+} = t_D^{-} = 0$, so:

$$\phi(X) = kT\int^{x}\ln\frac{X'}{X'^*}dX' = kT\int^{x}\ln\frac{t^{-}(X')}{t^{+}(X')}dX' \tag{78}$$

which agrees with (14,15). In the limit of no reaction we have

$t^+(X_i) = t^-(X_i) = 0$, and hence

$$\phi[\{X_i\}] = kT\sum_i \int^{X_i} \ln\frac{X_i'}{X_i'^*}dX_i' = kT\sum_i \int^{X_i} \ln\frac{t_{Di}^-}{t_{Di}^{+*}}dX_i' \tag{79}$$

If the spatial variable z is continuous and S is the area perpendicular to the z axis, (77) becomes:

$$\phi[x(z)] = S\int dz\left[\int^{x(z)} dx'\{\mu(x') - \mu(x'^*)\}\right] \tag{80}$$

Since the number of X molecules in each box, X_i, is an independent variable, a reaction-diffusion system is isomorphic to a multivariable homogeneous reaction system, which can be linearized around a stable stationary state. At that state we have $\left.\frac{d\phi}{dX_i}\right|_{det}^{S} = 0$, where 'det' means that the deterministic path is the path of integration. The time derivative of ϕ satisfies $\frac{d\phi}{dt} \le 0$. ϕ is a minimum at stable stationary states and a Lyapunov function in their vicinity.

Systems with Two Intermediates

Now we extend this approach to systems with two intermediates and multiple stationary states. Let us consider first a *linear* system such as the one in (52). By carrying out the same calculations as in the previous subsection, we obtain the following expression for the total excess work:

$$\phi[\{X_i,Y_i\}] = \sum_i\left(\int^{X_i} dX_i'\{\mu(X_i') - \mu(X_i'^S)\} + \int^{Y_i} dY_i'\{\mu(Y_i') - \mu(Y_i'^S)\}\right) \tag{81}$$

which is a state function for linear systems, since the integral is path-independent. It can be seen that the derivative of ϕ with respect to time is negative semidefinite, so the system tends towards the minimum of ϕ, that is towards the stable stationary state. Hence, ϕ is a Lyapunov function. Further, ϕ satisfies the stationary solution of the master equation in the thermodynamic limit. These properties assure that ϕ provides necessary and sufficient conditions for the existence and stability of stationary states.

Let us consider now *nonlinear* systems such as the reaction mechanism of (62). For each set of (X_i, Y_i) we can uniquely map the non-linear system to a thermodynamically and kinetically equivalent linear system. In this case the total excess work is:

$$\phi = \sum_i\int^{X_i,Y_i} \tilde{d}\phi_i = \sum_i\int^{X_i,Y_i} \{\mu(X_i) - \mu(X_i^*)\}dX_i + \{\mu(Y_i) - \mu(Y_i^*)\}dY_i = kT\sum_i\int^{X_i,Y_i} \ln\frac{X_i}{X_i^*}dX_i + \ln\frac{Y_i}{Y_i^*}dY_i \tag{82}$$

which depends on the path of integration. ϕ is zero at the stationary state and so are its first derivatives. Since the nonlinear system is indistinguishable from the instantaneously equivalent linear system, their time derivatives are equal. It can be seen that $\left.\frac{d\phi}{dt}\right|_{nonlinear} \le 0$, with the equality holding only at the stationary state (which is a minimum). Hence ϕ is a Lyapunov function and it provides necessary and sufficient conditions for the existence and stability of stationary states.

Relative Stability of Stable Stationary States

Let us consider a reaction-diffusion system with two phases numbered 1 and 3, each at a different stable stationary state. If both phases are placed in contact under the same external constraints, reaction and diffusion will occur and a reaction diffusion front may be formed in an interphase region. This front will travel into the less stable state, and the more stable state will displace the less stable one.

For systems at equilibrium the Gibbs free energy serves as a criterion of equistability; for systems far from equilibrium it is the excess work which serves that purpose. If the excess work required to form the reaction diffusion front from phase 1 is less than that necessary to form the front from phase 3, then phase 1 is more stable than phase 3. If both excess works are equal then we may expect equistability.

The excess work can be calculated as follows. We divide the interphase region into N boxes and integrate numerically the $2N$ ordinary differential equations:

$$\frac{dX_i}{dt} = t_x^+(X_i,Y_i) - t_x^-(X_i,Y_i) + d_x(X_{i+1} + X_{i-1} - 2X_i)$$

$$\frac{dY_i}{dt} = t_y^+(X_i,Y_i) - t_y^-(X_i,Y_i) + d_y(Y_{i+1} + Y_{i-1} - 2Y_i), \tag{83}$$

From (82), the excess work as a function of the reference states (X_i^*, Y_i^*) is:

$$\Delta\phi(1\to 3) = kT\sum_i\int_{phase\,1}^{phase\,2} \ln\frac{X_i}{X_i^*}dX_i + \ln\frac{Y_i}{Y_i^*}dY_i \tag{84}$$

which may be written as:

$$\Delta\phi(1\to 3) = kT\sum_i\int_{phase\,1}^{St.Fr.}\left\{\ln\frac{X_i}{X_i^*}dX_i + \ln\frac{Y_i}{Y_i^*}dY_i\right\} - kT\sum_i\int_{phase\,3}^{St.Fr.}\left\{\ln\frac{X_i}{X_i^*}dX_i + \ln\frac{Y_i}{Y_i^*}dY_i\right\}$$
$$= \Delta\phi(1\to St.Fr.) - \Delta\phi(3\to St.Fr.) \tag{85}$$

where "St.Fr." stands for stationary front. The direction of propagation of the interface is predicted by the theory as a function of the sign of Δϕ. If we have Δϕ (1→St.Fr.)>Δϕ(3→St.Fr.), then the more stable phase is 3 and the interface region moves in the direction which annihilates phase 1. If in the previous expression we have < instead of >, the case is the opposite. If both sides of the expression are equal (=), we have equistability and the interface does not move.

This can be illustrated with an example taken from [18]: the Selkov model, which is obtained from (62) with m = n = r = 1, s = 3. Figure 6 shows an example of the solutions of the reaction diffusion equations for a certain selection of parameter values.

For a different selection of values a stationary front can be obtained and the interface propagates with zero velocity. This is shown in Figure 7, which plots the curve of zero velocity as a function of two parameters: k_6, which is the rate coefficient corresponding to the transformation from B to Y, and the ratio of diffusivities of X and Y, $\delta = D_y/D_x$. Finally, we compare the predictions of the theory with the numerical results in Figure 8. The theoretical results approach the numerical ones as the length of the interface region is increased.

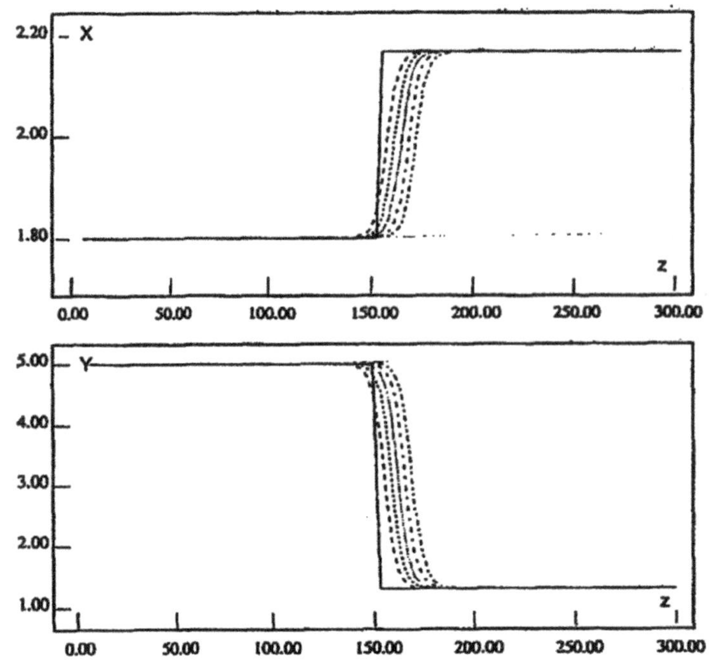

Figure 6. Plots of concentration profiles of *X* and *Y vs.* distance *z* during the front propagation in the Selkov model. The solid line is the initial concentration profile; the dotted lines are the concentration profiles with uniform time spacing.

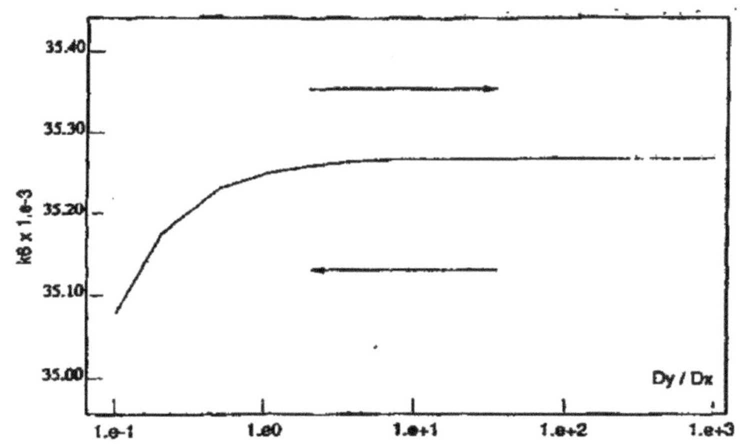

Figure 7. Selkov model: the solid line is a plot of zero velocity of the interface between phase 1 and 3. Above the solid line the interface moves to the right, below it to the left.

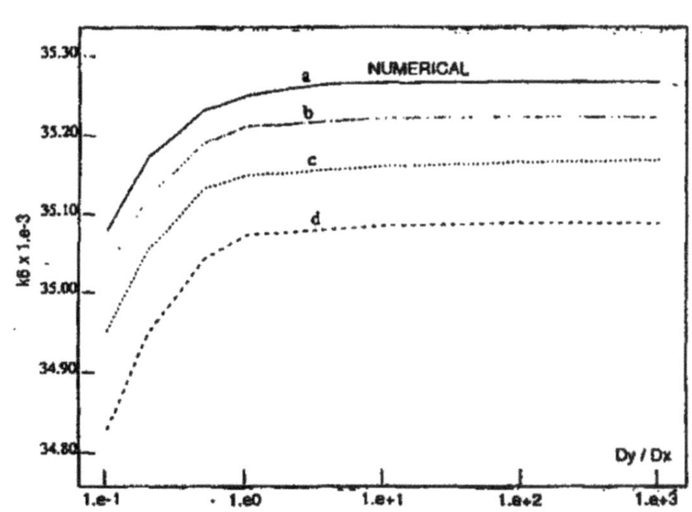

Figure 8. Comparison of predictions of equistability from the thermodynamic theory (b, c, d) with the numerical solution (a) repeated from Figure 7. The theoretical solutions correspond to different lengths of the interface region L: (b) 6L, (c) 2L, (d) L. The theoretical curves approach the numerical calculations as the length of the interface region, and the number of boxes, are increased.

Stability and Relative Stability of Reaction-diffusion Systems Related to Fluctuations

In Section 4 we have discussed the thermodynamics of reaction-diffusion systems by means of deterministic kinetic equations. This can also be carried out on the basis of consideration of fluctuations, as was done in section 3 for the homogeneous systems previously discussed in section 2. In this subsection we follow again this approach. In parallel withsection 3, we connect the stationary solution of the master equation and the thermodynamic excess work, a state function, ϕ^0. This presentation is based on the results in [17,20]. For a system with two intermediates (x,y) we can write:

$$ds = \frac{1}{kTV} d\phi^0 = \left(\mu_x - \mu_{x^0}\right) dn_{x,fl} + \left(\mu_y - \mu_{y^0}\right) dn_{y,fl} \tag{86}$$

where (x^0,y^0) refer to a reference state given by px=ln(x/x0), py=ln(y/y0), which hold for an equivalent linear system. The displacements on the RHS of (86) are along the most probable fluctuational path. The momentum p is the gradient of the action, and therefore ds and $d\phi^0$ are exact differentials. Let us divide the interphase region into N boxes; then the state function of the total excess work is the sum of that in each box:

$$\phi^0(X,Y) = \int_s^{(X,Y)} \sum_{i=1}^{N} kTVd\phi_i^0 = kTV \sum_{i=1}^{N} \left[\int_s^{(X,Y)} dX_i \ln(X_i / X_i^0) + dY_i \ln(Y_i / Y_i^0) \right] \tag{87}$$

The reference state is determined as for the homogeneous case, but now in the full 2N dimensional case. Figure 9 shows the developed interface region, the dotted line. At equistability this line does not move, and translation of the dotted line does not change ϕ^0. If the stationary state 1 (SS1) is slightly more stable than the stationary state 3 (SS3) then the deterministic motion of the front is a translation to the right. Since ϕ^0 is a Lyapunov function $\Delta\phi^0$ for this process must be negative. Similarly for the opposite case, 3 slightly more stable than 1, $\Delta\phi^0$ is also negative. Hence at equistability the limiting value of $\Delta\phi^0$ for a translation along the position z must be zero.

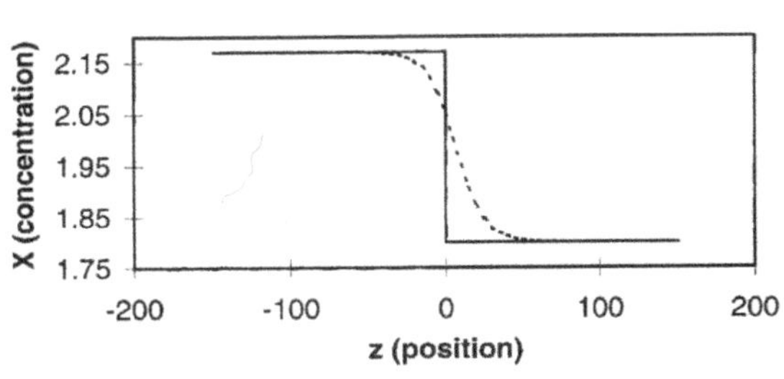

Figure 9. Concentration (X) *versus* position (z) for the Selkov model. The initial concentration profile is shown by the solid line; the space with negative z is filled initially with stationary state 1, the space with positive z is filled initially with stationary state 3. The dotted line denotes the interphase region.

The change in the excess work for establishing the stable front (SF) from SS1 equals at equistability the change in excess work of establishing the stable front from SS3:

$$\Delta\phi^0(SS1 \to SF) = \Delta\phi^0(SS3 \to SF) \tag{88}$$

Thus the stationary probability distribution of the master equation in the eikonal approximation [20] is a Lyapunov function which gives necessary and sufficient conditions of the existence and stability of non-equilibrium stationary states and provides a measure of relative stability on the basis of inhomogeneous fluctuations (88).

An Experiment on Relative Stability of Multiple Stationary States

In order to put to test the theory presented in this section we consider a multi-variable system, the bromate oxidation of ferroin, which exhibits bistability. We assume the NFT reaction mechanism considered in [22] and discuss the experimental results reported in [23].

In the experimental setup, two solution mixtures in different stable stationary states are pumped through a laminar flow reactor. They are brought in contact with each other on one edge of each solution; there is hardly any mixing due to the flow. When the flows are stopped diffusion occurs. The goal of the experiment is to measure the front propagation of one stable state to the other. The experimental results shown on Figure 10 yield an average value of $(6.1 \pm 0.6) \times 10^{-3}$ s^{-1} for the flow rate coefficient at zero velocity of propagation. This experimental result was compared with a calculation based on a further simplification of the NFT mechanism to a two-variable system [24]. Numerical solution of the corresponding deterministic reaction-diffusion equations yields the value of 12.2×10^{-3} s^{-1}, while the thermodynamic theory of section 4.3predicts a value of 12.45×10^{-3} s^{-1}. In view of the limited precision of the experiments and the use of a very simple model of the reaction mechanism, the agreement of the experiment with the theory and the calculations is satisfactory.

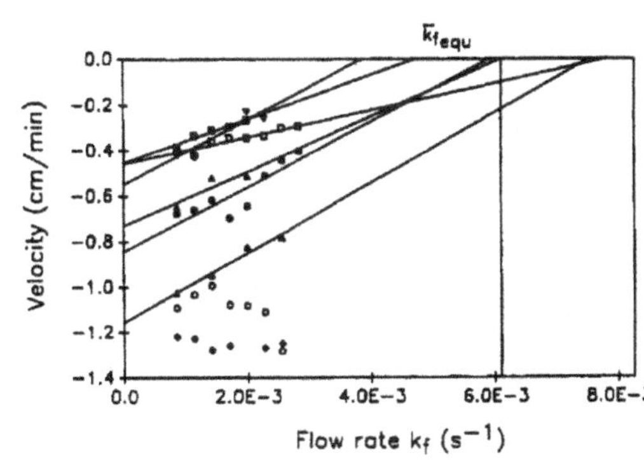

Figure 10. Plots of the measured dependency of the velocity of front propagation of one stationary state into the other on the flow rate, k_f. The eight different symbols correspond to eight experiments. Two of the plots are shifted from their original positions for the purpose of better display: circles by −0.6 cm/min in V; diamonds by −0.7 cm/min in V. Lines are fitted to each set of points for purpose of extrapolation to zero velocity of front propagation. The precision of the points at the largest velocities is insufficient to permit extrapolation.

THERMODYNAMIC AND STOCHASTIC THEORY OF TRANSPORT PROCESSES

We present a brief description of the thermodynamic and stochastic theory of simple transport processes, linear and nonlinear: diffusion, thermal conduction, and viscous flow. We select thermal conduction as a specific example, and indicate some advanced work on hydrodynamic equations and some interesting experiments.

Linear Thermal Conduction

We consider a simple schematic of the apparatus depicted in Figure 11. It consists of two heat reservoirs, each of infinite heat capacity, one at temperature T_1 and the other at T_3, with $T_1 > T_3$. Volume 2 is between the two thermal reservoirs and is of small width so that its temperature T is uniform within it. The flow of heat occurs with conservation of energy and no work done.

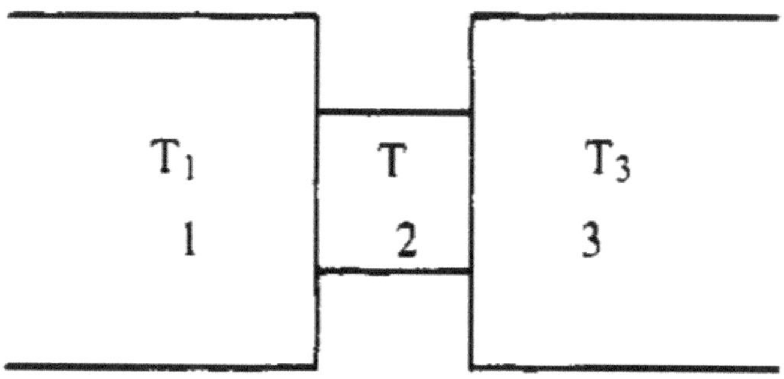

Figure 11. Schematic apparatus for study of thermal conduction.

We write the 'mixed' thermodynamic function M as a function of the entropies S:

$$dM = dS_1 + dS + dS_3 \tag{89}$$

or equivalently, $dM = \frac{dQ_1}{T_1} + \frac{dQ}{T} + \frac{dQ_3}{T_3}$. At the stationary state we have $dM_S = \frac{dQ_1}{T_1} + \frac{dQ}{T_S} + \frac{dQ_3}{T_3}$; hence the driving force towards the stationary state is:

$$T_S^{-1} d\phi = -(dM - dM_S) = \left(\frac{1}{T_S} - \frac{1}{T}\right) dQ \tag{90}$$

for the same changes dQ_1, dQ, dQ_3. The integral of $d\phi$ is:

$$\phi(T) = \phi(T) - \phi(T_S) = C_V \left(T - T_S - T_S \ln\left(\frac{T}{T_S}\right)\right). \tag{91}$$

This is an excess work as seen from the last two equations.

The macroscopic transport equation for this thermal conduction process is: $dT/dt = k_1(T_1 - T) + k_3(T_3 - T) = F(T)$

(92)

where k_1 and k_3 are proportional to thermal conduction coefficients for the interface of the system with the heat reservoirs 1 and 3, respectively. The temperature of the stationary state is $T_S = \frac{k_1 T_1 + k_3 T_3}{k_1 + k_3}$, and hence we may write $dT/dt = (k_1 + k_3)(T_S - T)$. If we divide by dt in (90) and use $\frac{dQ}{dt} = C_V dT$, we obtain for the time derivative of ϕ:

$$\dot{\phi} = C_V \left(1 - \frac{T_S}{T}\right) 2k(T_S - T) \le 0; \quad 2k = k_1 + k_3 \tag{93}$$

Since $\phi(T) \ge 0$ and lower-bounded it is a Lyapunov function. Its first derivative with respect to T is:

$$\frac{d\phi}{dT} = C_V \left(1 - \frac{T_S}{T}\right) \begin{cases} > 0 & T > T_S \\ = 0 & T = T_S \\ < 0 & T < T_S \end{cases} \tag{94}$$

and the second derivative with respect to T is:

$$\frac{d^2\phi}{dT^2} = \frac{C_V}{T^2} \ge 0 \tag{95}$$

so that ϕ is a minimum at the stationary state T_S.

The total dissipation is the sum of the dissipation of the reservoirs in the stationary state, and $\dot{\phi}$. Next we seek a stochastic equation for the distribution of fluctuations of the macroscopic temperature T for which the excess work (91) is its stationary distribution. This is a Fokker-Planck type equation:

$$\frac{\partial}{\partial t} P(T,t) = -\frac{\partial}{\partial t}(f(T)P(T,t)) + \frac{\partial^2}{\partial T^2}\{(f(T)P(T,t))\} \tag{96}$$

in which the probability diffusion coefficient $f(T)$ is cT with $c = k(k_1 T_1 + k_3 T_3)/C_V$, which is constant. The stationary solution of the stochastic equation is:

$$P_S(T) = P_0 \frac{T_S}{T} \exp\left(-\frac{1}{kT_S}\phi\right). \tag{97}$$

We rescale the temperature τ=CV/kTST and use Stirling's approximation for lnX!=XlnX−X and obtain for the stationary distribution:

$$P_S(\tau) = \frac{P_0'}{\tau} \exp\left(-\tau + \tau_S + \tau_S \ln\left(\frac{\tau}{\tau_S}\right)\right) \approx \frac{\tilde{P}_0}{\Gamma(\tau_S)} \tau^{\tau_S^{-1}} e^{-\tau} \tag{98}$$

where Γ is the gamma function. If we expand this solution around T_S we arrive at the expected quadratic form:

$$\phi(T) = C_V \left(T - T_S - T_S \ln\left(\frac{T}{T_S}\right)\right) \approx \frac{C_V}{2} \frac{(T - T_S)^2}{T_S}. \tag{99}$$

and the Gaussian stationary distribution:

$$P_S(T) = P_0 \exp\left(-\frac{\phi}{kT_S}\right) = P_0 \exp\left(-\frac{C_V}{2kT_S^2}(T - T_S)^2\right). \tag{100}$$

This is exactly the same form as the equilibrium probability distribution for the fluctuations in temperature 2 at the equilibrium temperature $T_S = T_{equ}$. For an ideal gas we have $E =$

$C_V T$ and the fluctuations are in the Gaussian form in energy:

$$\phi(E) = \frac{1}{2}\frac{(E - E_S)^2}{C_V T_S} \qquad (101)$$

The generalization to a system with a discreet distribution of temperature follows closely the development for diffusion.

In the next subsection we present an experimental test of the theory for thermal conduction.

An Experiment on Optical Bistability

Here we discuss measurements of the relative stability of two stable stationary states in a bistable system, that of an optically bistable ZnSe interference filter. A thin ZnSe rectangle is illuminated with an argon laser (514 nm) and bounded by nonilluminated regions at room temperature. This produces optical bistability in certain ranges of power of the irradiating light. Part of the incoming light is absorbed by the filter and turned into heat, resulting in an increase in temperature. Two possible temperature profiles [25] are shown on Figure 12, where temperature T is plotted against distance z.

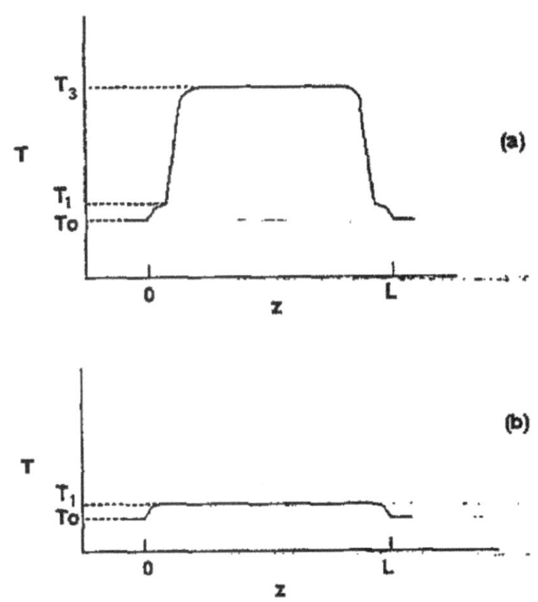

Figure 12. Two possible stable stationary states of an optically bistable interference filter. The region of irradiation is the length L; the ambient, lower, and upper temperatures are T_0, T_1, and T_3.

The upper and lower stable stationary states decay differently on stopping the irradiation. The upper state is annihilated by fronts moving inward from the boundaries; the lower state by the simultaneous decay of all regions at T_3 toward T_0. Calculations are shown on the left side of Figure 13 [26]; they are confirmed by the experimental measurements on the right side. The reaction-diffusion theory is amended for a problem in thermal conduction [25], yielding a calculation of the irradiation power at equistability can of 1.375 mW. This shows a very good agreement with the experimental measurement, which was 1.395 ± 0.015 mW.

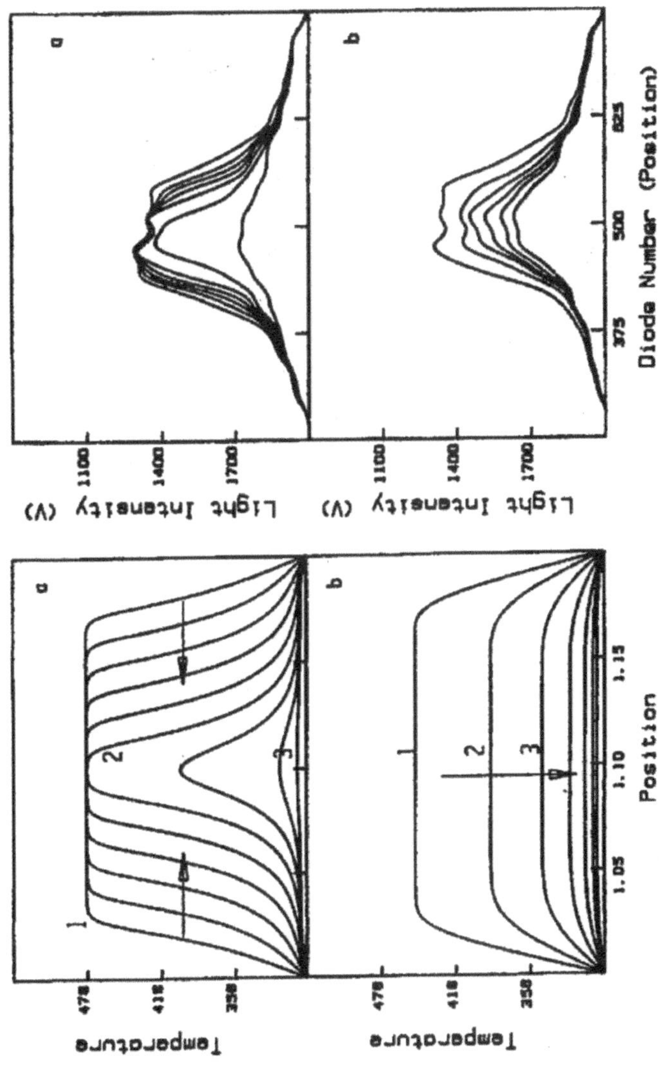

Figure 13. Calculated (left) and measured (right) plots of the decay of temperature profiles on stoppage of irradiation of the upper stationary state (**a**) and the lower stationary state (**b**).

Coupled Transport Processes: An Approach to Thermodynamics and Fluctuations in Hydrodynamics

Lorenz Equations and an Interesting Experiment

Coupled transport processes are described macroscopically by hydrodynamic equations, the NavierStokes equations [27]. These are difficult, highly nonlinear coupled partial differential equations; they are frequently approximated. One such approximation consists of the Lorenz equations [28,29], which are obtained from the Navier-Stokes equations by Fourier transform of the spatial variables in those equations, retention of first order Fourier modes, and restriction to small deviations from a bifurcation of an homogeneous motionless stationary state (a conductive state) to an inhomogeneous convective state in Rayleigh-Benard convection [30]. The Lorenz equations have been applied successfully in various fields ranging from meteorology to laser physics.

The aims of a theory of thermodynamics for hydrodynamics are the establishment of evolution criteria (Lyapunov functions) with physical significance, such as the *excess* work; the work and power available from a transient decay to a stationary state; macroscopic necessary and sufficient criteria of stability of stationary states; thermodynamic criteria for bifurcations from one type of stationary state to another type; thermodynamic criteria of relative stability, that is thermodynamic criteria of state selection; and a connection of the thermodynamic theory to fluctuations.

Attempts in these directions have a long history going back to Helmholtz, Korteweg, and Rayleigh, which we shall not review here; for a comprehensive account see Lamb's classical treatise [27]. The emphasis was on the total dissipation, which however does not provide thermodynamic evolution criteria far from equilibrium. Glansdorff and Prigogine [27,31] considered the second variation of the entropy as a criterion for evolution and stability. Their approach is limited to small deviations from a stationary state, provides only sufficient not necessary conditions, and has no connections with work or excess work, nor with fluctuations (master equation). Keizer formulated a stochastic approach for the relaxation to stationary states and fluctuations around a single stationary state [32]; he assumed Gaussian fluctuations, limited to small fluctuations related to linearized kinetics (for chemical kinetics). There are several approaches to the statistical mechanics of stationary states and fluctuating hydrodynamics (see [33] and references therein); some consist of the addition of Gaussian fluctuations to the linearized Navier-Stokes equations [34,35]. Here the thermodynamics may be sufficient for systems approaching equilibrium but not for stationary states far from equilibrium. In most of these studies no connections are made to work or power, nor to Lyapunov functions, nor to issues of relative stability when several states are available.

We developed our thermodynamic theory for the Lorenz equations, obtained with approximations from the Navier-Stokes equations (we present almost no mathematics here; that is given in detail in [36]). The Lorenz equations are:

$$\dot{X} = P(Y - X),$$
$$\dot{Y} = -XZ - Y + rX,$$
$$\dot{Z} = XY - bZ, \qquad (102)$$

where X represents the amplitude of the stream function of the macroscopic velocity of the fluid; Y the reduced temperature mode of the thermal conduction; and Z the reduced temperature mode related to the vertical flow in the liquid layer. P is the Prandtl number (the kinematic viscosity divided by the thermal conductivity). The parameters r and b are$r = Rq^2/(\pi^2+q^2)^3$, $b = 4\pi^2/(\pi^2+q^2)$, where R is the Rayleigh number, $R = \rho\alpha g h^3 \Delta T/\eta\kappa$, and the density, the thermal expansion coefficient, g the gravitational constant, h the height of the fluid layer, ΔT the temperature difference across the layer of liquid, is the viscosity, and the thermal conductivity. These are the variables and parameters of this system.

One solution of the Lorenz equations is $(X,Y,Z) = (0,0,0)$. When the control parameter r is less than unity, that is the Rayleigh number is less than its critical value R_c, then the zero solution is unique and stable, and it corresponds to the motionless conductive state of the fluid. At the bifurcation point $r = 1$ this solution becomes unstable, and a new solution becomes stable corresponding to convective modes. These solutions can be used to construct an excess work function, just as we did for single transport properties.

Now we return to a fascinating experiment in alignment with our theory. Zamora and Ray de Luna [37] carried out Rayleigh-Benard experiments in an apparatus that could be inverted 180° in the gravitational field. In their experimental arrangement the bottom and top sides of the fluid are connected to heat reservoirs at different temperatures. Suppose a stationary state is homogeneous and conductive, achieved by setting the low temperature reservoir below the one at high temperature in the gravitational field. Next the entire apparatus is turned in the gravitational field; then the conductive state becomes unstable and a stable convective state appears. Throughout the experiment the reservoir temperatures are held fixed and the average temperature of the fluid is nearly constant. They measured the heat fluxes into and out from the reservoirs during all transients, that is the relaxations from a conductive to a convective state, and the reverse. For example, say the fluid is in a convective state; invert the apparatus; the convective state is unstable and a transient change to a conductive state takes place. Measure the heat fluxes in and out of the reservoirs. The integration of the differences of the heat fluxes in and out during this transient process is the total energy change accompanying the destruction of the convective rolls. Similarly they measured the total energy change accompanying the formation of the convective rolls. The results show that the heat releases for both the destruction and formation of convective structures are *positive*, which means that the system *always* releases energy in the form of heat when it approaches a stable stationary state, either the convective state or the motionless conductive state. In an auxiliary experiment they found that the change in average temperature of the system is very small; the change in internal energy due to this small temperature is less than 10% of the heat release during the relaxation processes, so this can not be the reason for the experimental observations. Moreover, the change in internal energy cannot explain the observed heat release in both the destruction and the formation of a convective structure.

Our theory based on the concept of excess work accounts for these experiments, at least qualitatively. According to our theory, when the system approaches a stable stationary state, either convective or conductive, there is a decrease in ϕ, the excess work, and a positive excess work is released, which will be dissipated and released as heat. This is shown in a calculation from the theory in Figure 14, from [36]. Figure 14a is for the transition from a conductive to a convective state, and Figure 14b for the reverse transition. In each case the integral of $d\phi / dt$ over time is negative, that is excess work is dissipated and heat is generated. Further we note that the entropy production in the convective state is always larger than that in the conductive state, and no explanation of heat release in the transition from one state to the other can thereby be derived. The theoretical results, based on the Lorenz model, agree with experiments qualitatively in that the total excess work change is of the same order of magnitude as the heat release measured in the experiments, and this is a major confirmation of our theory.

Rayleigh Scattering in a Fluid in a Temperature Gradient

In a simple fluid, in an imposed temperature gradient, light is scattered due to fluctuations in temperature and due to fluctuations in the transverse hydrodynamic velocity. Excellent experiments have been made on the measurement of this light scattering. The problem has been studied theoretically as well, by means of fluctuating hydrodynamics [34,35], valid for small fluctuations around a conductive state with a constant temperature gradient, which can be close to or far from equilibrium. Theory and experiment are in very good agreement [38,39,40,41].

In [42] we developed the relation of our studies on the thermodynamic and stochastic theory of transport properties to the reported research on this topic. There we showed that the deterministic excess work, as formulated in Section 2 for reactions and in this section for transport processes, provides a thermodynamic interpretation of fluctuations around a stationary state, either close to or far from equilibrium, for the case of Raleigh scattering from fluctuations in a fluid with an imposed temperature gradient. The stationary probability distribution is determined by a quantity proportional to the excess deterministic work. From the probability distribution we obtain, in the Gaussian approximation for small fluctuations, the matrix of correlations derived from fluctuating hydrodynamics, Equation (51) in [42]. Thus in this limit of small fluctuations there is agreement between the theory of fluctuating hydrodynamics and our theory of the thermodynamics and fluctuations in transport properties.

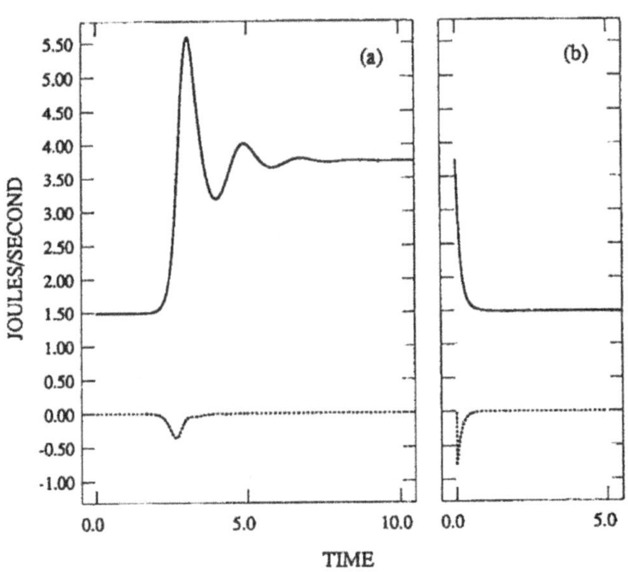

Figure 14. The product of the total rate of dissipation times temperature (solid line) in J/s and the time derivative of excess work (dashed line) *vs.* time in the following processes for the Lorenz model: (a) Gravity is initially set in the direction along which the temperature decreases, and the system is at a stable motionless conductive stationary state; at $t = 0$, invert the direction of gravity; the motionless conductive state becomes unstable and the system approaches the convective stationary state. (b) The reverse process. The temperature difference is $|\Delta T|$=4K in both cases.

It should be noted that the presentations in the prior sections have been limited to ideal systems, either gases or solutions. The thermodynamic and stochastic theory can be extended for non-ideal systems; the interested reader is referred to [43].

ELECTROCHEMICAL EXPERIMENTS IN SYSTEMS FAR FROM EQUILIBRIUM

In chemical systems at equilibrium, Gibbs free energy changes and other thermodynamic quantities can be determined from electrochemical experiments [44]. At equilibrium the Nernst equation relates the Gibbs free energy change of a chemical reaction, ΔG, to the voltage (the electrochemical potential E) generated by that reaction run in an electrochemical cell:

$$\Delta G = -Nfe \qquad (103)$$

where n is the number of equivalents of electrons transferred from one electrode to the other.

In the preceding sections we have presented a thermodynamic and stochastic theory of chemical reactions and transport processes in non-equilibrium stationary and transient states approaching nonequilibrium stationary states. We established a state function ϕ, which for systems approaching equilibrium reduces to ΔG. Since Gibbs free energy changes can be determined by macroscopic electrochemical measurements, we seek a parallel development for the determination of ϕ by macroscopic electrochemical and other measurements. We discuss two kinds of experiments in Section 6.1 and Section 6.2; then in Section 6.3 we turn to the development of the thermodynamic and stochastic theory to connect with these experiments.

Measurement of Electrochemical Potentials in Non-equilibrium Stationary States

When chemical species come into equilibrium with an electrode in an open circuit, the potential between the electrode and a reference electrode is related to the potential difference of the half reaction occurring at the electrode. If no other reactions are occurring then this potential is related to the Gibbs free energy difference of the half reaction at the electrode. If there are other reactions occurring then the species may be in non-equilibrium states, even though they are in equilibrium with the electrode, and the potential is that of a non-equilibrium stationary state. If local equilibrium holds then the potential is the Gibbs free energy difference; if it does not hold, in that there are degrees of freedom, such as the reactions, which are explicitly held away from equilibrium, then deviations from the Gibbs free energy difference may occur. We shall speak of Nernstian (103) and non-Nernstian contributions to the electrochemical potential. There is one prior measurement of the type to be described and that is by Keizer and Chang [45] following a suggestion by Keizer [46,47,48] that there should be a nonNernstian contribution to the electrochemical potential in nonlinear reaction systems approaching to, or in, stationary states far from equilibrium. They reported a very small non-Nernstian contribution in a Fe(II)/Fe(III) reaction system in a non-equilibrium stationary state.

We studied the autocatalytic minimal bromate reaction, which can be oscillatory, but was studied in a bistable regime. A mechanism for this reaction was proposed in [49]. The net reaction is the oxidation of Ce(III) to Ce(IV) by bromate. In the bistable regime there is a state, where essentially no reaction occurs, which coexists with a state in which a percentage of Ce(III) is oxidized to Ce(IV). In this system we measured [50] at the same time the optical density which gives concentrations of Ce(IV) by Beer's law, and hence also the concentration of Ce(III) by conservation, and the emf of a Pt electrode which at equilibrium follows the Nernst equation (103). The experiment consisted of the measurement of the emf of the Ce(III)/Ce(IV) half reaction at a redox (Pt-Ag/AgCl) electrode under equilibrium and stationary non-equilibrium conditions. From these measurements we determined that there exists a non-Nernstian contribution in a non-equilibrium stationary state as shown in Table 1. The concentration of $[Ce(III)]_{ss}$ in the stationary state is obtained from the absorption measurement. The local equilibrium emf, the third column in Table 1 is calculated from the ratio Ce(III)/Ce(IV) and the Nernst equation. The measured emf in the fourth column of the table is that measured by the Pt electrode. The difference is small, about 1% of the emf at the largest inflow concentration of $Ce(III)_0$ and decreases for smaller inflow concentrations.

Table 1. Results from the Minimal Bromate Experiment with Various Concentrations of Ce(III) in the combined feedstreams into the Reactor, $[Ce(III)]_0$. From [50]

$[Ce(III)]_0$ (M)	$[Ce(III)]_{ss}$	IE emf (mV)	M emf (mV)
1.500×10^{-3}	1.393×10^{-3}	1167.0	1180.2
1.397×10^{-3}	1.361×10^{-3}	1176.0	1183.0
1.297×10^{-3}	1.277×10^{-3}	1187.5	1189.7
8.333×10^{-4}	8.700×10^{-4}	1223.7	1223.8

Kinetic and Thermodynamic Information Derived from Electrochemical Measurements

We study now the electrochemical displacement of a nonlinear chemical system from nonequilibrium stationary states and from equilibrium. Then in Section 6.3 we shall relate such measurements to the thermodynamic and stochastic theory of potentials governing fluctuations in electrochemical systems in stationary states far from, near to, and at equilibrium.

We study again the minimal bromate reaction, where we measure the Ce(III)/Ce(IV) potential (for further details about the experimental setup see [51]). The contents of three reservoirs are pumped separately into a CSTR; the three reservoirs contain 0.00450 M Ce^{III}, 0.0100 M BrO_2^-, and 1.00 $\times 10^{-6}$ M Br^-, and each reservoir contains also 0.72 M H_2SO_4. To run the reaction at equilibrium the three solutions are mixed and allowed to react for a day prior to being pumped into the CSTR. First we measure the Ce(III)/Ce(IV) potential with zero imposed current from an external current source; then we impose various currents and displace the equilibrium mixture in the CSTR from equilibrium. A non-equilibrium stationary state is achieved by flowing the reacting solutions into the CSTR at given flow rates, that is given residence times in the reactor. For a residence time of 175 seconds, Figure 15 shows the measured voltages plotted against the imposed current, as well as the Ce(IV) concentration, and the product of the measured voltage minus the stationary state voltage multiplied by the imposed current. The left hand part of Figure 15 shows results for an equilibrium mixture, while the right hand part shows displacements from a non-equilibrium stationary state.

It is interesting to compare the equilibrium displacement plot with the plot of displacements from nonequilibrium stationary states. To achieve a given displacement, either in Ce(IV) concentration or in the potential from its stationary value at zero current, a larger imposed current is necessary in the nonequilibrium case than in the equilibrium case. We further note that the plot of $(V - V_{ss}) \times I$ in the equilibrium case is nearly symmetric, but that for the non-equilibrium cases it is not.

The plots of power input *vs.* imposed current can be obtained by a simple, nearly dimensional argument from our theory. For a one-variable linear system the excess work is, as seen in section 2, $\phi = \int (\mu_X - \mu_X^S) dX$; hence $d\phi = (\mu_X - \mu_X^S) dX$ and $\dot{\phi} = (\mu_X - \mu_X^S)\dot{X}$; so that with Nernst's equation we have:

$$\dot{\phi} = -(V - V_S) I \qquad (104)$$

The time derivative of the excess work, which is that part of the dissipation due to the variation in *X*, equals the power input necessary to maintain the system away from its stationary state. In the next subsection we outline the theory in more detail.

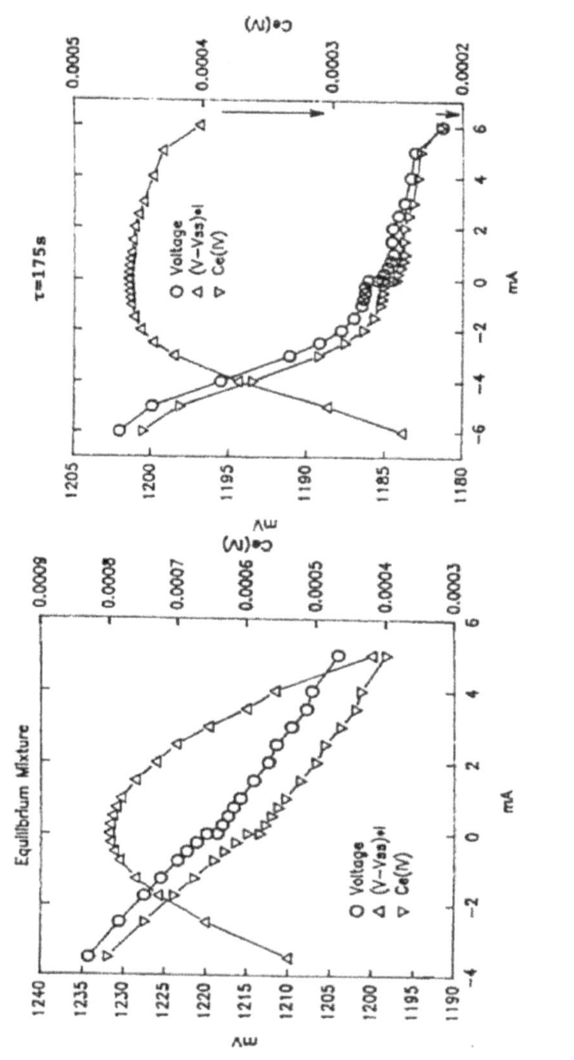

Figure 15. Plot of voltage V, the power input $(V - V_{ss}) \times I$, and the Ce(IV) concentration*versus* the imposed current I. V_{ss} is the measured voltage at a non-equilibrium stationary state at zero imposed current. The plot on the left corresponds to an equilibrium stationary state; the residence time in the CSTR is 200 seconds. The plot on the right hand corresponds to a non-equilibrium stationary state at zero imposed current and displacements from that state with imposed currents; the residence time is 175 s. The arrows indicate transitions to other stationary states. From [51].

Theory of Determination of Thermodynamic and Stochastic Potentials from Macroscopic Measurements

First we present an outline to an approach to the determination of thermodynamic and stochastic potentials, ϕ, for non-equilibrium systems from electrochemical measurements; and second, a parallel development for neutral (not ionic) systems in general. We conclude with some suggestions for testing the consistency of the master equation with measurements.

For the first purpose we choose a chemical reaction system with some ionic species, as for example the minimal bromate reaction, for which we presented some experiments in chapter X. We consider a simple representation of a reaction:

$$R_+ + Q_+ \underset{k_{+1}^-}{\overset{k_{+1}^+}{\rightleftarrows}} A_1 + B_2 \underset{k_0^-}{\overset{k_0^+}{\rightleftarrows}} B_1 + A_2 \underset{k_{-1}^-}{\overset{k_{+1}^+}{\rightleftarrows}} R_- + Q_- \qquad (105)$$

where the signs + or − indicate ionic species and the other species are neutral.

The reactions system may be in equilibrium or in a non-equilibrium stationary state. An ion selective electrode is inserted into the chemical system and connected to a reference electrode. The imposition of a current flow through the electrode connection drives the chemical system (CS), written above, away from its initial stationary state to a new stationary state of the combined chemical and electrochemical system (CCECS), analogous to driving the CS away from equilibrium in the same manner. A potential difference is generated by the imposed current, which consists of a Nernstian term dependent on concentrations only, and a non-Nernstian term dependent on the kinetics. We shall relate the potential difference to the stochastic potential; for this we need to know the ionic species present and their concentrations, but we do not need to know the reaction mechanism of the chemical system, nor rate coefficients.

The differential of the stochastic potential for the chemical system in one of its stationary states, $d\phi_c$, is (see Section 3):

$$d\phi_c = \sum_i (\mu_i - \mu_i^0) dn_i \qquad (106)$$

with the index i extending over A_1, B_1, taken to be neutral, and the species A_2, B_2 taken to be negatively charged. The differential $d\phi_c$ is exact and any path of integration suffices to obtain ϕ. The exponential of the integral in (106) is a formal representation of the eikonal approximation for the stationary solution of the master equation of the chemical system.

We choose the same variables for the chemical as for the electrochemical system. The reactions at the electrodes are sufficiently fast that the measured potential is the equilibrium potential; fluctuations in that potential and in the imposed current are neglected. This is analogous to neglecting fluctuations in concentrations of species in equilibrium with mass reservoirs. For systems for which equilibrium is the only stable attractor, the chemical potential of each chemical species, say that of A_2, is $\mu_{A_2} + E_{A_2}NF$, where E_{A2} is the potential for a given imposed current, N is the number of equivalents in the half-cell reaction for A_2, and F is the Faraday constant. We postulate that we may write $d\phi_E$ for the combined chemical and electrochemical system in a parallel way:

$$\begin{aligned} d\phi_E &= (\mu_{A_1} - \mu_{A_1}^0) dn_{A_1} + (\mu_{B_1} - \mu_{B_1}^0) dn_{B_1} + (\mu_{A_2} + E_{A_2}NF - \mu_{A_2}^0 - E_{A_2}^0 NF) dn_{A_2} \\ &\quad + (\mu_{B_2} + E_{B_2}NF - \mu_{B_2}^0 - E_{B_2}^0 NF) dn_{B_2} \\ &= d\phi_C + (E_{A_2} - E_{A_2}^0) NF dn_{A_2} + (E_{B_2} - E_{B_2}^0) NF dn_{B_2} \end{aligned} \qquad (107)$$

where the first line is for the neutral species and the next two line for the ionic species. The fourth line gives the relation of the stochastic potential of the combined systems to that of the chemical system.

This postulate is consistent to given approximations with an expansion of the master equation or an equivalent Hamilton-Jacobi equation. The study of stochastic equations of electrochemical systems is in its infancy; we know of no prior work on this subject. Further intensive studies will be necessary to fully substantiate the postulate of (107).

At a stationary state of the combined system the differential $d\phi_E = 0$ and therefore we have for $d\phi_c$ the result:

$$\begin{aligned} d\phi_C &= -NF\left[\left(E_{A_2} - E_{A_2(s)}\right) - \left(E_{A_2}^0 - E_{A_2(s)}\right)\right] dn_{A_2} \\ &\quad - NF\left[\left(E_{B_2} - E_{B_2(s)}\right) - \left(E_{B_2}^0 - E_{B_2(s)}\right)\right] dn_{B_2}, \end{aligned} \qquad (108)$$

where we added and subtracted the potentials $E_{A2(s)}$ of A and $E_{B2(s)}$ of B in the stationary state of the system of the chemical system. The first term in each square bracket depends on concentrations only and thus is the Nernstian contribution to the measured electrochemical potential. The second term in each square bracket depends on the kinetics of the chemical system and thus is the non-Nernstian contribution the electrochemical potential. Measurements of this potential, say with ion-specific electrodes, yield the slopes $\frac{\partial \phi_c}{\partial n_{A_2}}$ and $\frac{\partial \phi_c}{\partial n_{B_2}}$; thus with measurements of the macroscopic concentrations of A_1, A_2, B_1, and B_2 at a sufficient number of displacements from the stationary state of the chemical system we can determine the stochastic potential of that system from macroscopic measurements. To obtain these results no direct use of any master equation has been made and no model of the reaction mechanism was necessary.

Determination of the Stochastic Potential in Chemical Systems with Imposed Fluxes

Consider the chemical system in (105) with the species being either ions or neutrals; the system is in a reaction chamber in a non-equilibrium stationary state. We impose a flux of species A_1, $J = k'A_1'$, into the reaction chamber with Q_+ and Q_- held constant and thereby move the chemical system to a different non-equilibrium stationary state with different concentrations of the reacting species A_1, B_1, A_2, B_2. This procedure allows the sampling of different combinations of the reacting species by means of the imposition of different fluxes of these reactants. These combinations represent different nonstationary states in the absence of imposed fluxes, but with the imposed fluxes they are stationary states and hence measurements may be made without constraints of time. If we would attempt to measure concentrations in non-stationary states then the measurement technique would have to be fast compared to the time scale of change of the concentrations due to chemical reactions.

Now we impose a flux of a chemical species present in the system and inquire on the effect of that imposition on the stochastic potential of the system. For that we need to go from the deterministic kinetic equations to a stochastic equation, say the lowest order eikonal approximation to the chemical master equation. The response measurements to the imposed flux provide an indirect determination of the stochastic potential, one that depends on the use of the master equation, an assumed reaction mechanism, and assumed rate coefficients.

This procedure is easy for a one-variable system because we know the solution of the stationary master equation to this approximation. For example, for the one variable Schlögl model we have the elementary reaction steps:

$$A+2X \underset{k_2}{\overset{k_1}{\rightleftharpoons}} 3X, \quad X \underset{k_4}{\overset{k_3}{\rightleftharpoons}} B \tag{109}$$

with the concentrations of A and B held constant. The kinetic equation without the imposed flux is:

$$\frac{dX}{dt} = k_1AX^2 + k_4B - (k_2X^3 + k_3X) \equiv t^+ - t^- \tag{110}$$

Let the imposed flux be $J = k'X'$. The stationary solution of the lowest order eikonal approximation of the master equation for the system Equation (15) with the imposed flux is $P(X) \approx \exp(-\phi'/(kBT))$, where:

$$-\frac{1}{k_BTV}\frac{\partial \phi'}{\partial X} = \ln\left(\frac{t^+ + J}{t^-}\right) = \ln\left(\frac{t^+}{t^-}\right) + \ln\left(1+\frac{J}{t^-}\right). \tag{111}$$

In the absence of an imposed flux the solution reduces to $-\frac{1}{k_BTV}\frac{\partial \phi}{\partial X} = \ln\left(\frac{t^+}{t^-}\right)$. At a stationary state of the system with imposed flux we have $\frac{\partial \phi'}{\partial X} = 0$ and hence from (111) we obtain:

$$-\frac{1}{k_BTV}\frac{\partial \phi}{\partial X} = \ln\left(\frac{t^+}{t^-}\right) = -\ln\left(1+\frac{J}{t^-}\right). \tag{112}$$

Thus from measurements with imposed flux we obtain the derivative of the stochastic potential for the system without imposed flux, but we need kinetic information, the rate coefficients in t^-, as well. For multivariable systems this approach is more difficult; the determination of the stochastic potential requires sufficient measurements to determine rate coefficients and then the numerical solution of the stationary form of the master equation.

Suggestions for Experimental Tests of the Master Equation

A direct test of the master equation for systems in non-equilibrium stationary states comes from the measurements of concentration fluctuations; there have been a few of such measurements. Some other tests of the master equation are possible based on the earlier sections in this chapter, where we can compare measurements of the stochastic potential with numerical solutions of the master equation (which requires knowledge of rate coefficients and the reaction mechanism of the system).

There are other indirect methods. Consider a one-variable system (or an effectively one variable). Let the system have multiple stationary states and in Figure 16, taken from [52], we show a schematic diagram of the hysteresis loop in such systems.

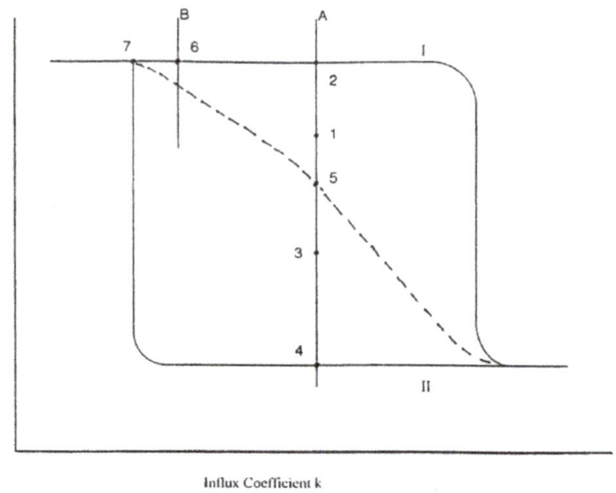

Figure 16. Typical hysteresis loop for a one-variable system with a cubic kinetic equation: plot of concentration c *vs.* influx coefficient. *Solid lines*, stable stationary states (nodes); *broken line*, unstable stationary state. For a discussion of lines A and B and numbers, see text.

Several experiments can be suggested to test aspects of the predictions of the master equation. To construct a diagram as in Figure 16 from the master equation we need to know or guess rate coefficients and the reaction mechanism of the system. For the experiments we need to measure the concentration c of a given species as the influx coefficient is varied. Thus we establish the solid lines by experiment. If we can form a combined chemical and electrochemical system, as discussed in subsection 6.3, then we can locate the combined system at point 1 on line A by imposing a given current flow. This point is a stable stationary state of the combined system. If the imposed current is stopped (the electrochemical system is disconnected) then the chemical system will return deterministically to the nearest stable stationary state of the chemical system, that is point 2 on line A and on the stable branch I. We can repeat this experiment by locating the combined chemical and electrochemical system, say on point 3 on line A; then on stopping the imposed current the chemical system will return to point 4 on line A and on the stable branch II. By means of such experiments we can locate the branch of unstable stationary states, the separatrix, the dotted line in Figure 16, and compare it with predictions of the master equation. The same approach works for the displacement of a system by imposition of an influx of a given species.

Equistability of a homogeneous stable stationary state on the upper branch of the hysteresis loop, labeled I in Figure 16, with a homogeneous stable stationary state on the lower branch, labeled II, occurs at one value of the influx coefficient $k$within the loop. Say that point occurs at the location of line A. The predictions of the stationary solution of the stochastic master equation are: (a) the minimum of the bimodal stationary probability distribution is located on the separatrix, and (b) at equistability the probability of fluctuations $P(c)$ obeys the condition:

$$\int_4^5 P(c)dc = \int_2^5 P(c)dc. \tag{113}$$

Approximately, at equistability the height of the probability peak at point 2 equals that at point 4. To either side of the value of k at equistability, the peak of the more stable stationary state is higher than the other peak.

A comparison of deterministic and stochastic calculations (not experiments) has been discussed in a different context, that of viewing the stochastic potential as an excess work [7,53]. A point at the end of a hysteresis loop, such as point 7 in Figure 16, is called a marginal stability point. Near such points, such as 6, critical slowing occurs. After a perturbation of the system in the stable stationary state at 6 the system returns to 6, but increasingly more slowly for values of the influx coefficient to the left of 6 but to the right of 7, and increasingly faster to the right of 6. This effect has been observed experimentally in several systems [54]. Critical slowing down manifests itself in the stationary solution of the master equation near marginal stability points. Hence a quantitative comparison of experiment and theoretical predictions leads to a test of the master equation and the assumed parameters.

DISSIPATION IN IRREVERSIBLE PROCESSES

The entropy production rate is a measure of the dissipation in an irreversible process. A popular 'principle' in the literature states that, if a stationary state is close enough to equilibrium, then the entropy production rate has an extremum at the steady state [55]. This is known as the principle of minimum entropy production. In this section we show that neither this principle nor the principle of maximum entropy production is valid. To this end we follow [56] and begin with the exact solution for thermal conduction.

Exact Solution for Thermal Conduction

We consider an example of thermal conduction, which we shall analyze exactly, without approximations. Consider a macroscopic, homogeneous system with cylindrical shape of length l and cross-sectional area A; and time-dependent temperature T. Heat is transported across the boundary of the cylinder at each end in contact with two thermal baths with temperature T_1 and T_2; we assume $T_1 > T_2$. We take the conduction of heat to be given by Newton's equations:

$$lA\rho c\frac{dT}{dt} = k\left(\frac{T_1 - T}{l}\right) + k\left(\frac{T_2 - T}{l}\right) \tag{114}$$

where k is the thermal conductivity, the density of the system taken to be constant, and c the mass specific heat capacity, also taken to be constant. The solution of this equation is:

$$lA\rho c\frac{dT}{dt} = k\left(\frac{T_1 - T}{l}\right) + k\left(\frac{T_2 - T}{l}\right) \tag{115}$$

where the relaxation rate is $2\varepsilon = 2k/(\rho c A l^2)$ and the stationary state temperature is $\lim_{t\to\infty} T = T_{st} = (T_1 + T_2)/2$. The rate of entropy production is the product of the heat flux times the conjugated force:

$$\sigma = Alk\frac{T_1 - T}{l}\frac{1}{l}\left(\frac{1}{T} - \frac{1}{T_1}\right) + Alk\frac{T_2 - T}{l}\frac{1}{l}\left(\frac{1}{T} - \frac{1}{T_2}\right) = \frac{kA}{TT_1T_2l}\left[T^2(T_1 + T_2) - 4T_1T_2T + T_1T_2(T_1 + T_2)\right] \geq 0 \tag{116}$$

and is always positive. Note that the flux is not proportional to the force. The derivative of the entropy production with respect to temperature is:

$$\frac{d}{dT}\sigma(T) = \frac{Ak(T^2 - T_1T_2)(T_1 + T_2)}{lT^2T_1T_2} \tag{117}$$

and at the stationary state we have:

$$\sigma(T_{st}) = Ak(T_1 - T_2)^2 / 2lT_1T_2$$
$$\frac{d}{dT}\sigma(T_{st}) = \frac{Ak(T_1 - T_2)^2}{lT_1T_2(T_1 + T_2)} \geq 0 \tag{118}$$

Since the derivative is always positive *the rate of entropy production is never an extremum at a stationary state, whether close to or far from equilibrium, except at equilibrium.* At equilibrium (T1=T2), the entropy production rate has an extremum which is a minimum.

Invalidity of the Principles of Minimum and Maximum Entropy Production

Glansdorff and Prigogine [30] stated the Minimum Entropy Production principle as follows: "... if the steady states occur*sufficiently close to equilibrium states* they may be characterized by an extremum principle according to which *the entropy production has its minimum value at the steady state compatible with the prescribed conditions* (constraints)." Since then it has been frequently presented as a fundamental law of nature with a deep meaning.

We have shown by a counterexample that the entropy production rate does not have an extremum for a steady state, not even in the vicinity of thermodynamic equilibrium. Two additional counterexamples are given in [56], for Fourier's law of heat conduction and for a chemical reaction network with mass-action kinetics. In fact, the entropy production rate has a minimum at the steady state only if the flux of the transport process is strictly proportional to the force. The common error of accepting the Minimum Entropy Production principle is caused by ignoring the noncommutativity of two different operations: (1) the truncation of a Taylor series for the entropy production rate in terms of a set of parameters which express the distance from equilibrium, and (2) the differentiation of entropy production rate with respect to temperature.

A related albeit less known theorem is the Maximum Entropy Production principle [57], which states that irreversible processes proceed in a direction which produces maximum entropy production. If there is a choice then the path with the highest entropy production has the fastest rate. For chemistry this principle is generally invalid; the rates of reactions are governed by the Gibbs free energy of activation, not by the rate of entropy production.

EFFICIENCY OF OSCILLATORY REACTIONS

State variables, such as concentrations of chemical species, are constant in a stationary state, whether at equilibrium or nonequilibrium; or they may be oscillatory in a non-equilibrium state, such as a limit cycle. In an oscillatory chemical reaction system the Gibbs free energy change of the reaction (ΔG) is oscillatory (usually not sinusoidally) and the rate of the overall reaction is oscillatory. The dissipation of the system is the product of ΔG and the rate of the reaction if no external work is done by the reaction.

There now appears a new quantity, the phase relation of the oscillations of ΔG to the oscillations of the rate. As in electric alternating current systems the phase between the current (rate) and the voltage (ΔG) determines the power output of the system. We illustrate this point with experiments on the oscillatory horse radish peroxidase (HRP) reaction:

$$NADH + H^+ + \frac{1}{2}O_2 \xrightarrow{HRP} NAD^+ + H_2O \tag{119}$$

In Figure 17 we plot ΔG, calculated from measurements of concentrations, and the rate *vs.* time for constant influx of oxygen, the flat part of the top line, and oscillatory influx of oxygen, the oscillatory part of the top line. The phase between the oscillation of ΔG and the rate is changed from the constant input of oxygen and the oscillatory input, as show in greater detail in Figure 18.

This change in phase changes the dissipation, or the power output, and we see that we can affect, control, these important quantities by controlling external inputs (influxes).

In [1] other examples are discussed: an oscillatory reaction with constant influx of reactants; the efficiency of pumped chemical reactions; and the efficiency of a proton pump.

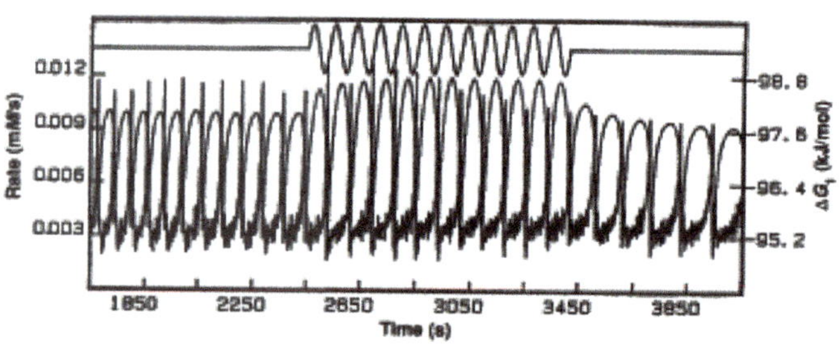

Figure 17. Plot of ΔG and the rate of the HRP reaction *vs.* time. See text for further explanation. From [58].

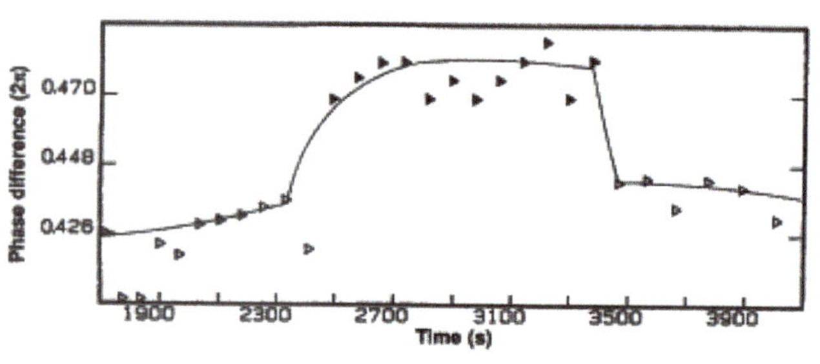

Figure 18. Plot of the phase difference between the Gibbs free energy change and the rate of the HRP reaction in units of 2π. From [58].

FLUCTUATION-DISSIPATION RELATIONS

We discuss briefly the relations between dissipative, deterministic kinetics and fluctuations for the purpose of an introduction to the interesting topic of fluctuation-dissipation relations. This subject has a long history, more than 100 years [59,60,61,62,63]; [59,60,61] are a classical review with many references to fundamental earlier work. A brief reminder of one of the early examples, that of Brownian motion, may be helpful.

The dynamics of a Brownian particle of mass m velocity v is described by a Langevin equation [44]:

$$m dv/dt + \gamma v = F(t), \tag{120}$$

where the first term is the force due to acceleration of the particle, the second term the frictional force on the particle, and $F(t)$ is a delta correlated Gaussian random force for which the first two moments are:

$$\langle F(t)\rangle = 0, \quad \langle F(t)F(t')\rangle = 2\lambda\delta(t-t'), \tag{121}$$

in which λ is a parameter that gives the strength of the fluctuations that affect the Brownian particle. In a derivation due to Langevin, given in [44], it is shown that the friction coefficient γ in (120), which characterizes the dissipative process, and λ, which characterizes the fluctuations, are related by:

$$\lambda\gamma = kT. \tag{122}$$

Equation (122) is a fluctuation-dissipation relation.

Now consider a one-variable open chemical system, either linear:

$$A \underset{k_2}{\overset{k_1}{\longleftrightarrow}} X \underset{k_4}{\overset{k_3}{\longleftrightarrow}} B \tag{123}$$

or non-linear:

$$A + 2X \underset{k_2}{\overset{k_1}{\longleftrightarrow}} 3X, \quad X \underset{k_4}{\overset{k_3}{\longleftrightarrow}} B. \tag{124}$$

We need two definitions, one for the net rate of the reaction:

$$t(x) = t^+(x) - t^-(x) \tag{125}$$

and the other for the total rates in the forward and reverse reaction:

$$D(x) = [t^+(x) + t^-(x)]/2. \tag{126}$$

With that we write a fluctuation–dissipation relation:

$$t(x) = 2D(x)\tanh\left[-\frac{1}{2Vkt}\cdot\left(\frac{\partial\phi}{\partial x}\right)\right] \tag{127}$$

$$t(x) = 2D(x)\tanh[-A(x)/2kT] \tag{128}$$

with the species specific affinity $A(x)$:

$$A(x) = \mu_X(x) - \mu_X(x_s), \tag{129}$$

for the linear case, and:

$$A(x) = \mu_X(x) - \mu_X(x^*) \tag{130}$$

for the nonlinear case. By substituting (125,126) into (127) and writing out the tanh term, $\tanh x = [\exp x - \exp(-x)]/$

[expx+exp(−x)], you obtain consistently the derivative of (5) for the linear reaction mechanism, and (15) for the non-linear reaction mechanism.

In (127) the macroscopic kinetics (dissipation) is connected to the thermodynamic potential which yields the fluctuations. What are the advantages of this formulation? The term *t(x)* is the net flux of the deterministic kinetics, (125), and the derivative of the state function ϕ is the species specific affinity, $(\mu_x - \mu^s_x)$ for the linear case or $(\mu_x - \mu^*_x)$ for the non-linear case, the driving force for the reaction towards a stationary state. Thus we have a flux–driving force relation. Secondly, the formulation is symmetric with respect to $t^+(x)$ and $t^-(x)$, which is not the case with other formulations. Third, the state function ϕ determines the probability distribution of fluctuations in *x* from its value at the stationary state, see (30). Further, as we shall show shortly, the term *D(x)* is a measure of the strength of the fluctuations of the total number of reaction events (total in the forward and reverse reaction). We emphasize that (127) is non-linear and holds for arbitrarily large fluctuations. Expressions for different regimes of affinity and dissipation can be easily obtained [64,65,66].

We return to the interpretation of the quantity *D(x)* in the fluctuation-dissipation relation, (127). Consider the concept of "reaction event" [64,65,66], that is a generation or consumption of *X* in reactions (123) or (124). Let the total number of reaction events be labeled *q(t)* which occur in the time interval (0,t). Further we take *q(t)* to be a non-decreasing *random* function of time even at a stationary state and at equilibrium. The variable *q* is a discreet time scale; the chemical reaction is a clock, the time being measured by the number of reaction events. Since the reaction events are independent the distribution of reaction events obeys Poisson statistics [7]. The first and second moments of the Poisson distribution are:

$$\langle q(x,t)\rangle = 2VD(x)t,$$
$$\langle \Delta q^2(x,t)\rangle = 2VD(x)t. \qquad (131)$$

The second of the equations (131) is similar to the Einstein equation for the mean square displacement of a Brownian particle in one dimension:

$$\langle \Delta X^2\rangle = 2Dt. \qquad (132)$$

Here *X* is the displacement, in one variable, and *D* is the diffusion coefficient of the Brownian particles. Hence we can identify *D(x)* in (131) as a probability diffusion coefficient in concentration space. The dispersion of *q*, the second equation in (131), is proportional to *D(x)* and hence *D(x)* is a measure of the strength of the fluctuations of *q* at a stationary state. For further analysis of this subject see [1].

ACKNOWLEDGEMENTS

A.F.V. acknowledges financial support from Spanish Ministry of Research and Innovation (MICINN) Grant MultiSysBio DPI2008-06880-C03-02. J.R. is grateful for partial support of this work by the National Science Foundation.

REFERENCES

1. Ross, J. *Thermodynamics and Fluctuations Far from Equilibrium*; Springer: Berlin, Germany, 2008.
2. Van KampenN.G.*Stochastic Processes in Physics and Chemistry*; North-Holland: Amsterdam, The Netherlands, 1981.
3. Ross, J.; Hunt, K.L.C.; Hunt, P.M. Thermodynamics far from equilibrium: Reactions with multiple stationary states. *J. Chem. Phys.* **1988**, *88*, 2719–2729.
4. Schlögl, F. On thermodynamics near a steady state. *Z. Phys.* **1971**, *248*, 446–458.
5. Evans, D.J.; Searles, D.J.; Williams, S.R. On the fluctuation theorem for the dissipation function and its connection with response theory. *J. Chem. Phys.* **2008**, *128*, 014504.
6. Glansdorff, P.; Nicolis, G.; Prigogine, I. The thermodynamic stability theory of non-equilibrium states. *Proc. Natl. Acad. Sci. U. S. A.* **1974**, *71*, 197–199.
7. Peng, B.; Hunt, K.L.C.; Hunt, P.M.; Suárez, A.; Ross, J. Thermodynamic and stochastic theory of nonequilibrium systems: Fluctuation probabilities and excess work. *J. Chem. Phys.* **1995**, *102*, 4548–4562.
8. Oppenheim, I.; Shuler, K.E.; Weiss, G.H. *Stochastic Processes in Chemical Physics: The Master Equation*; MIT: Cambridge, MA, USA, 1977.
9. Gardiner, C.W. *Handbook of Stochastic Methods*; Springer: New York, NY, USA, 1990.
10. Nicolis, G.; Babloyantz, A. Fluctuations in open systems. *J. Chem. Phys.* **1969**, *51*, 2632–2637.
11. Selkov, E.E. Self-oscillations in glycolysis. *J. Biochem.* **1968**, *4*, 79–86.
12. Graham, R.; Tél, T. Nonequilibrium potential for coexisting attractors. *Phys. Rev. A* **1986**, *33*, 1322–1337.
13. Maier, R.S.; Stein, D.L. Transition-rate theory for nongradient drift fields. *Phys. Rev. Lett.* **1992**, *69*, 3691–3695.
14. Crandall, M.G; Evans, L.C.; Lions, P.L. Some Properties of Viscosity Solutions of HamiltonJacobi Equations. *Trans. AMS* **1984**, *282*, 487–502.
15. Jauslin, H.R. Nondifferentiable potentials for nonequilibrium steady states. *Physica A* **1987**, *144*, 179–191.
16. Jauslin, H.R. Melnikov's criterion for nondifferentiable weak-noise potentials. *J. Stat. Phys.* **1986**, *42*, 573–585.
17. Freidlin, M.I.; Wentzell, A.D. *Random Perturbations of Mechanical Systems*; Springer: Berlin, Germany, 1984.
18. Chu,X.;Ross,J.;Hunt,P.M.;Hunt,K.L.C.Thermodynamic and stochastic theory of reactiondiffusion systems with multiple stationary status. *J. Chem. Phys.* **1993**, *99*, 3444–3454.
19. Hansen, N.F.; Ross, J. Lyapunov functions and relative stability in reaction-diffusion systems with multiple stationary states. *J. Phys. Chem.* **1996**, *100*, 8040–8043.
20. Hansen, N.F.; Ross, J. Relative stability of multiple stationary states related to fluctuations. *J. Phys. Chem.* **1998**, *102*, 7123–7126.

21. Glauber, R.J. *Lectures in Theoretical Physics*; Interscience Publishing: New York, NY, USA, 1959.

22. Noyes, R.M.; Field, R.J.; Thompson, R.C. Mechanism of reaction of bromine (V) with weak oneelectron reducing agents. *J. Am. Chem. Soc.* **1971**, *93*, 7315–7316.

23. Foerster, Y.-X.; Zhang, J.; Ross, J. Experiments on relative stability in the bistable multivariable bromate-ferroin reaction. *J. Phys. Chem.* **1993**, *97*, 4708–4713.

24. Bar-Eli, K; Geiseler, W. Mixing and relative stabilities of pumped stationary states. *J. Phys. Chem.* **1981**, *85*, 3461–3468.

25. Wolff, A.N.; Hjelmfelt, A.; Ross, J.; Hunt, P.M. Tests of thermodynamic theory of relative stability in one-variable systems. *J. Chem. Phys.* **1993**, *99*, 3455–3460.

26. Harding, R.H.; Ross, J. Experimental measurement of the relative stability of two stationary states in optically bistable ZnSe interference filters. *J. Chem. Phys.* **1990**, *92*, 1936–1946.

27. Lamb, H. *Hydrodynamics*, 6th ed.; Dover Publications: New York, NY, USA, 1945.

28. Saltzman, B. Finite amplitude free convection as an initial value problem-I. *J. Atmos. Sci.* **1962**, *19*, 329–341.

29. Lorenz, E.N. Deterministic nonperiodic flow. *J. Atmos. Sci.* **1963**, *20*, 130–141.

30. Getling, A.V. *Rayleigh-Bénard Convection. Structure and Dynamics*; World Scientific Pub. Co.: London, UK, 1997.

31. Glansdorff, P.; Prigogine, I. *Thermodynamic Theory of Structure, Stability, and Fluctuations*; Wiley: New York, NY, USA, 1971.

32. Keizer, J. *Statistical Thermodynamics of Nonequilibrium Processes*; Springer-Verlag: New York, NY, USA, 1987.

33. Attard, P. The second entropy: A general theory for non-equilibrium thermodynamics and statistical mechanics. *Annu. Rep. Prog. Chem. Sect. C Phys. Chem.* **2009**, *105*, 63–173.

34. Ronis, D.; Procaccia, I. Nonlinear resonant coupling between shear and heat fluctuations in fluids far from equilibrium. *Phys. Rev. A* **1982**, *26*, 1812–1815.

35. Fox, R.F. Gaussian stochastic processes in physics. *Phys. Rep.* **1978**, *48*, 179–283.

36. Velarde, M.G.; Chu, X.-L.; Ross, J. Toward a thermodynamic theory of hydrodynamics: The Lorenz equations. *Phys. Fluids.* **1994**, *6*, 550–563.

37. Zamora, M.; Rey de Luna, A. Energy of a system formed by a convective fluid and its container. *J. Fluid Mech.* **1986**,*167*, 427–437.

38. Law, B.M.; Sengers, J.V. Fluctuations in fluids out of thermal equilibrium. *J. Stat. Phys.* **1989**, *57*, 531–547.

39. Law, B.M.; Segrè, P.N.; Gammon, R.W.; Sengers, J.V. Light-scattering measurements of entropy and viscous fluctuations in a liquid far from thermal equilibrium. *Phys. Rev. A* **1990**, *41*, 816–824.

40. Segrè, P.N.; Gammon, R.W.; Sengers, J.V.; Law, B.M. Rayleigh scattering in a liquid far from thermal equilibrium.*Phys. Rev. A* **1992**, *45*, 714–724.

41. Li, W.B.; Segrè, P.N.; Gammon, R.W.; Sengers, J.V. Small-angle Rayleigh scattering from nonequilibrium fluctuations in liquids and liquid mixtures. *Physica A* **1994**, *204*, 399–435.

42. Suárez, A.; Ross, J. Thermodynamic and stochastic theory of coupled transport processes: Rayleigh scattering in a fluid in a temperature gradient. *J. Phys. Chem.* **1995**, *99*, 14854–14863.

43. Ross, J.; Chu, X.-L. Thermodynamic and stochastic theory for nonideal systems far from equilibrium. *J. Chem. Phys.***1993**, *98*, 9765–9770.

44. Berry, R.S.; Rice, S.A.; Ross, J. *Physical Chemistry*, 2nd ed.; Oxford University Press: New York, NY, USA, 2000.

45. Keizer, J.; Chang, O.-K. The nonequilibrium electromotive force. I. Measurements in a continuously stirred tank reactor. *J. Chem. Phys.* **1987**, *87*, 4064–4073.

46. Keizer, J. The nonequilibrium electromotive force. II. Theory for a continuously stirred tank reactor. *J. Chem. Phys.***1987**, *87*, 4074–4087.

47. Keizer, J. Thermodynamics at isothermal, isobaric steady states: vapor pressure, colligative properties, and the electromotive force. *J. Phys. Chem.* **1989**, *93*, 6939–6943.

48. Keizer, J. Heat, work, and the thermodynamic temperature at nonequilibrium steady states. *J. Chem. Phys.* **1985**, *82*, 2751–2771.

49. Field, R.J.; Burger, M. *Oscillations and Traveling Waves in Chemical Systems*; Wiley: New York, NY, USA, 1985.

50. Hjelmfelt, A.; Ross, J. Electrochemical experiments on thermodynamics at nonequilibrium steady states. *J. Phys. Chem.***1994**, *98*, 9900–9902.

51. Hjelmfelt, A.; Ross, J. Kinetic and thermodynamic information derived from electrochemical measurements on stationary states. *J. Phys. Chem. B* **1998**, *102*, 3441–3444.

52. Ross, J.; Hunt, K.L.C.; Vlad, M.O. Determination of thermodynamic and stochastic potentials in non-equilibrium system from macroscopic measurements. *J. Phys. Chem. A* **2002**, *106*, 10951–10960.

53. Ross, J.; Hunt, K.L.C.; Hunt, P. Thermodynamics and stochastic theory for non-equilibrium systems with multiple reactive intermediates: The concept and role of excess work. *J. Chem. Phys.* **2002**, *96*, 618–629.

54. Kramer, J.; Ross, J. Stabilization of unstable states, relaxation, and critical slowing down in a bistable system. *J. Chem. Phys.* **1985**, *83*, 6234–6241.

55. Prigogine, I. Moderation et transformations irreversibles des systemes ouverts. *Bull. Cl. Sci., Acad. R. Belg.* **1945**, *31*, 600–606.

56. Ross, J.; Vlad, M.O. Exact solutions for the entropy production rate of several irreversible processes. *J. Phys. Chem. A***2005**, *109*, 10607–10612.

57. Martyushev, L.M.; Seleznev, V.D. Maximum entropy production principle in physics, chemistry, and biology. *Phys. Rep.* **2006**, *426*, 1–45.

58. Lazar, J.G.; Ross, J. Changes in mean concentration, phase shifts, and dissipation in a forced oscillatory

reaction.*Science* **1990**, *247*, 189–192.

59. Lax, M. Fluctuations in a nonequilibrium stable state. *Rev. Mod. Phys.* **1960**, *32*, 25–64.
60. Lax, M. Classical noise III: Nonlinear markoff processes. *Rev. Mod. Phys.* **1966**, *38*, 359–379.
61. Lax, M. Classical noise IV: Langevin methods. *Rev. Mod. Phys.* **1966**, *38*, 541–566.
62. Reif, F. *Fundamentals of Statistical and Thermal Physics*; McGraw-Hill: New York, NY, USA, 1965; Chapters 1, 15.
63. Chandrasekhar, S. Stochastic Problems in Physics and Astronomy. *Rev. Mod. Phys.* **1943**, *15*, 1–89.
64. Vlad, M.O.; Ross, J. Fluctuation-dissipation relations for chemical systems far from equilibrium. *J. Chem. Phys.* **1994**,*100*, 7268–7278.
65. Vlad, M.O.; Ross, J. Random paths and fluctuation-dissipation dynamics for one-variable chemical systems far from equilibrium. *J. Chem. Phys.* **1994**, *100*, 7279–7294.
66. Vlad, M.O.; Ross, J. Thermodynamic approach to nonequilibrium chemical fluctuations. *J. Chem. Phys.* **1994**, *100*, 7295–7309.
67. Nicolis, G.; Prigogine, I. *Self-organization in Nonequilibrium Systems*; Wiley: New York, NY, USA, 1977.
68. Tolman, R.C. *The Principles of Statistical Mechanics*; Oxford University Press: London, UK, 1938.

SECTION III: APPLICATIONS

CHAPTER 15 STATISTICAL MECHANICS AND THERMODYNAMICS OF VIRAL EVOLUTION

Barbara A. Jones, Justin Lessler, Simone Bianco, James H. Kaufman

INTRODUCTION

Viruses are microscopic subcellular objects that infect cells of living organisms across all six kingdoms of life [1]. Because viruses require host cellular machinery to replicate [2], a common set of steps must occur for the reproduction of most viruses. First, the virus must enter the cell, which can occur through membrane fusion, endocytosis, or genetic injection [3]. During the replication process, tens to thousands of progeny are produced [2]. While the fidelity of the replication process varies between viruses, for most, particularly RNA viruses, the mutation rate is quite high [2]. Finally, progeny exit the cell (via budding, apoptosis, or exocytosis), in many cases killing the cell in the process [2]. The generally high levels of genetic variability created during replication lead to rapid "exploration" of genetic sequence space, allowing the virus to evade the host immune system, overcome environmental challenges such as antiviral drugs, and perhaps even adapt to new host species [4–6]. While even single cell organisms have an innate immune response, viral evolution becomes particularly important when viruses attempt to evade the adaptive immune system of humans and other vertebrates [2,7]. Successful viruses all must survive host defense mechanisms, compete to infect host cells, reproduce, and eventually pass to other hosts [2,8], though an immense variety of strategies are used to accomplish these tasks.

Recent technical advances in genome sequencing have revealed the enormous genetic diversity of RNA virus populations during infection [9], which is triggered by large population size and low replication fidelity. Information about mutation distributions during evolution has proven to be helpful in assessing the intricate mechanisms of viral reproduction [10,11]. Moreover, new insights into the tradeoff between mutational robustness, loosely defined as the invariance of the phenotype to genetic insults, and evolvability, the capability of the viral species to adapt to new environments, are emerging, supported by a wealth of new data [12,13].

However, it may not always be necessary, or even advisable, to capture the full intricacies of this system in useful models of viral evolution and dynamics. Highly simplified models may still reveal important principles about the behavior of viral populations. For example, Alonso and Fort measured thermodynamic observables to analyze a phase transition observed in a model of RNA virus error catastrophe [14–16]. In analogy to Bose condensation they derive an order parameter to characterize two phases separated by the error catastrophe phase transition. The error catastrophe literature demonstrates the importance of mutation rate and reveals a phase transition due to information loss at large rate [14–22]. Nowak and May consider the transition from sustained infection to elimination of infection as a function of basic reproductive ratios for various mutant strains [16]. In the current work we demonstrate a way to compute the steady state solutions for evolving viral quasispecies on a fitness landscape determined by two independent and competing energy terms, temperature and immunity.

Statistical mechanics allows physicists to describe the macroscopic characteristics of a multi-particle system based on microscopic properties [23–25]. Given a large collection of molecules or atomic particles, it is possible to use probability theory to define macroscopic properties in terms of thermodynamic quantities such as system heat, energy, and entropy [23–25]. These macroscopic properties are determined by an "ensemble" of all "microstates" of the collection, along with the probabilities associated with each microstate. If a simple model of viral replication, transmission and evolution

Citation: Jones BA, Lessler J, Bianco S, Kaufman JH (2015) Statistical Mechanics and Thermodynamics of Viral Evolution. PLoS ONE 10(9): e0137482. doi:10.1371/journal.pone.0137482.

can be developed that lends itself to such analysis, it may serve as a foundation on which to develop a powerful theory to describe the general behavior of viral systems using the same key concepts used in statistical mechanics.

METHODS

In this paper we present a model of viral replication and evolution within a single host and the analytic theory required to find steady-state solutions for this system. We then describe the steps used to solve the analytic equations and the methods used. Finally, we study the thermodynamics and statistical mechanics of the viral evolution model. We calculate thermodynamic quantities such as entropy, an order parameter, specific heat, energy, and properties of viral population dynamics such as host cell occupancy and viral load in the environment.

Viral Infection as Energy Barriers

Viruses replicate and transmit by a complex multi-step process. For an influenza virion to infect a cell and replicate, it must bind to receptors on the cellular membrane; induce endocytosis; release ribonucleoprotein (vRNP) complexes into the cytoplasm; vRNPs must be imported into the nucleus where replication can occur; and viral offspring must leave the cell through viral budding [26]. At each step there is some probability of failure, and the more fit the virus, the lower this probability [27,28]. We abstract this process as crossing two symmetric barriers, one for infecting the cell, and one for replication and exit.

The virus has some fitness for crossing these barriers (Fig 1), characterized by a probability of successes which depends on viral fitness and system "temperature", and is computed using an activated Arrhenius form [23]:

$$e_i = \exp(f_i / T) \tag{1}$$

where e_i is the probability of successfully crossing the barrier, f_i is viral fitness, and T is the system temperature. At this point one can view T as a parameter that governs how discriminating the barrier is between viruses with different numbers of "matches" to some target receptor. In classical chemistry the barrier height is a function of both reactants and products, while the temperature is a property of the reactants only (viruses in this case). When there is a distribution of energies for the reactants, k_BT is the average energy of the most probable distribution. In this model temperature changes the discrimination of the barriers to infection and reproduction. At low temperature viruses with a high match to the target are favored. At very high temperature, virtually all viruses of any match are able to infect and reproduce. As we will see, raising the temperature increases the fraction of viable virions within a quasispecies distribution. We will later demonstrate that temperature in this model is not only a tuning parameter, but also the thermodynamic temperature for the system, providing a distribution of energies for the viruses, which naturally form quasispecies distributions. We will also derive the effective Boltzmann constant relating temperature and the observed energy scales [23–26].

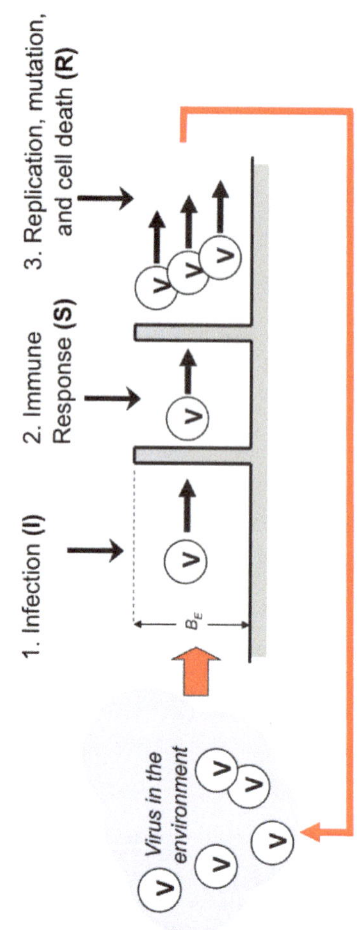

Figure 1. Model of an Idealized Virus Life Cycle. The barrier height (B_E) is equal to the number of mismatches of virus to a target. Viruses with different numbers of genetic matches will see barriers of different height. The probability of a virus passing is based on an activated Arrhenius model.

In viral entry, a protein creates a receptor on the viral envelope that binds to complementary receptors on a target cell membrane. We do not intend to model the complex interactions between these receptors. Instead, we abstract the fitness of a virus by how well the genetic letters (an abstraction of amino acids) in a virus's genome match an idealized target genetic sequence. The receptor in a real virus is a sub-region of the binding protein (providing many degrees of freedom). To capture these degrees of freedom we require that a sub-region of a virus "gene" matches a complementary genetic (receptor) target in the cell. Specifically, we define a target sequence of 50 letters in length associated with the host cell, and each virus is assigned a genome of 100 letters (i.e., 300 bases). As with real amino acids, letters are the phenotypic representation of a codon of three underlying bases (A,C,G, and U/T) in a redundant genetic code. We define m as the number of matches between host and target sequences at the alignment that minimizes the total number of mismatches (but still completely overlays the target). This model is not intended to capture the complex interactions between proteins based on their real shapes. Instead, viral fitness in this abstraction is completely characterized by the difference between the number of matches and the length of the genome (i.e., $f_i = f_m = -(50 - m)$). Once the number of matches to the target is determined for a particular virion, it is fixed for the life of that virion.

Then, given Eq (1), the probability of a successful barrier crossing is:

$$e_m = \exp(-(50 - m)/T) \quad \textbf{(2)}$$

On each replication there is some probability of mutation in a given base, allowing the distribution of viruses to change or evolve over time.

The Viral Life Cycle

In our simplified model of viral infection and replication the system of viruses passes through three stages in discrete generations (Fig 2). Free viruses first infect cells, passing into the post-infection stage, I. Some proportion of infected cells are then "killed" by the immune system, and instantly replaced by uninfected cells, and we enter the post-immunity stage, Ξ. Finally, viruses replicate and exit the cell, and we enter the post-reproduction/pre-infection stage, R. The system state in each stage can be described completely by two interacting sets of variables: the occupation of the host cells, and the distribution of "free" viruses in the environment. The self-consistent (steady state) solution for the virus life cycle is one in which each state remains unchanged after completing a full cycle.

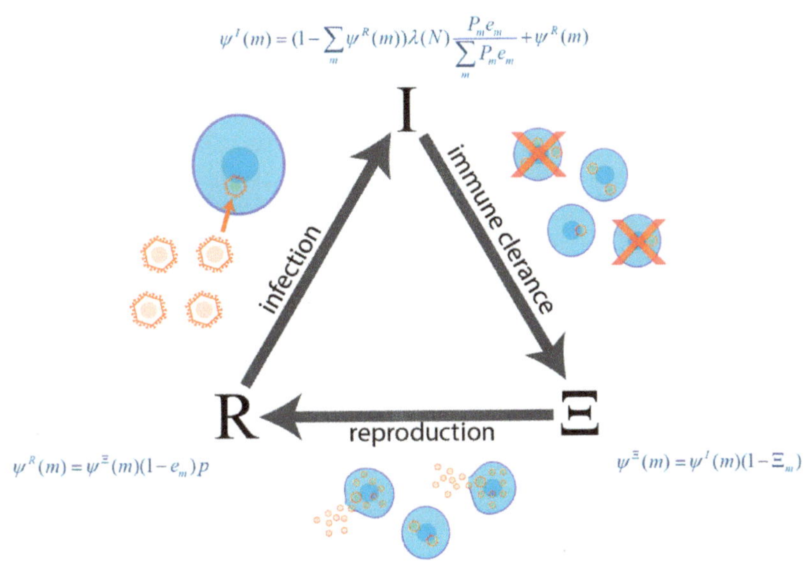

Figure 2. Virus Life Cycle. The changing states of all viruses must be computed self-consistently over the entire virus life cycle. The figure shows three important stages of the model virus life cycle: (I) Infection (entering the cell), (Ξ) Immune Clearance, and (R) Reproduction and exiting the cell. Also shown are the equations for cell occupancy at each stage.

We denote the probability that a cell in our model has a virus with given number of matches for each stage as $\Psi^I(m)$, $\Psi^\Xi(m)$, and $\Psi^R(m)$. Before reproduction, each cell is considered to contain at most a single virion. The number of free viruses is denoted by N and the proportion with m matches to the target is denoted as P_m. Both N and P_m are only defined in the post-reproduction stage, after the free virus population is replenished from those intra-cellular viruses that survive reproduction. The equations for cell occupancy at each stage are:

$$\psi^I(m) = (1 - \sum_m \psi^R(m))\lambda(N)\frac{P_m e_m}{\sum_m P_m e_m} + \psi^R(m)$$
$$\psi^\Xi(m) = \psi^I(m)(1 - \Xi_m)$$
$$\psi^R(m) = \psi^\Xi(m)(1 - e_m)p \quad \textbf{(3)}$$

where λ(N) is the overall infection rate, Ξ_m is the probability that the host immune response kills any virus in the cell with m matches, and p is the probability that a cell infected by a virus that does not reproduce survives until the next round of replication (cells that reproduce are considered to die, and all dying cells are considered to be instantly replaced with uninfected cells). Eq (3) represent the following: First, if the cell is empty, with probability $(1 - \sum_m \psi^R(m))$, the cell becomes infected. Next, the cell remains occupied if the virus survives the immune response with probability ($1-\Xi_m$). Finally, the cell only remains occupied if the virus does not reproduce with probability ($1-e_m$)p. The virus that remains, $\Psi^R(m)$, continues to occupy the cells when the next round of infection occurs.

In our model there are a finite number of identical target host cells available to infect at any one time, with the infection process proceeding as follows:

1. At any time at most one virus can infect each cell.
2. Each free virus successively attempts to infect the unoccupied cells with a success probability of each attempt of e_m.
3. Competition continues until either all cells are occupied or all free viruses have made an attempt.

With these criteria we can analytically derive the overall infection rate, λ(N), as a function of the number of target host cells and the number of free viruses in the environment N (see below).

The Immune Responses

Vertebrate hosts defend themselves from viral infections using both innate and adaptive immune responses [21]. In our model the innate immune response can be considered to be captured by the barrier that viruses must cross to infect and replicate in cells, while we explicitly model the adaptive immune response. In an adaptive immune response, the immune system develops an increasingly strong and specific response to infecting viruses by producing cells and antibodies which recognize and respond to specific viral epitopes (i.e., short sequences of amino acids that identify the virus) [21]. Here we assume that all parts of the virus are exposed to the immune system. Furthermore, we are interested in a steady state solution where the immune system has learned to recognize the target epitopes (not the entire viral genome). In steady state a virus genome matches some part of the target genome. In analogy to adaptive response to a specific set of epitopes, we use this matching sub-region to determine efficiency of immune response in steady state. In particular we represent the ability of the adaptive immune system to kill infected cells as a function of the match between a virion and the target as:

$$\Xi_m = A/(1+e^{-(m-v)/2}) \quad \textbf{(4)}$$

where $0 \leq A \leq 1$ is the maximum immune response, and v is the number of matching codons at which the virus achieves 50% efficiency when A = 1. In real viruses, the length of epitopes targeted by immune effector cells can vary widely, from as few as 3–4 of critical amino acids for B-cell conformational epitopes, to 8–11 amino acids for T cell epitopes (for example) [29]. Here we chose an intermediate value and v = 6 as a typical epitope length. The results and conclusions are not highly sensitive to this choice.

This abstraction is meant to model the steady state response of an adaptive antibody-mediated immunity. In this paper we explore the full range of A and m. Eq (4), together with Eq (2) and Eq (3), defines the two dimensional fitness landscape in this model.

Viral Reproduction and Mutation

Viral offspring differ from their parent through mutation of individual bases during the replication process. The resulting evolution is an important component of the survival strategy for many viruses, allowing them to evade the immune system and respond to changes in their environment (e.g., the introduction of chemotherapeutic agents). When a virus reproduces, the actual reproductive rate (number of offspring per parent) defines the "fecundity". In our model replication occurs with a fecundity, φ, and offspring have one, and only one, codon mutation per offspring. Mutation to the same amino acid is allowed.

As in the Moran model, single mutation can either reduce the maximal match length by one (Δm = −1), increase the maximal match length by one (Δm = +1) or leave the maximal match length unchanged (Δm = 0) [18,19]. Consider a virus with a maximum of m_0 codons matching the organism target genome. We find the sub-region(s) on the virus genome that contain the maximum number of matches, m_0, to the target by counting matches for each possible alignment. For a mutation to decrease m, two conditions must be true. First, the mutation must occur at a currently matched position in the matching region and must change the expressed amino acid (i.e., must not be a same-sense or silent mutation). Second, there must not be two or more alignments with the same maximal match. For a mutation to increase m, it must occur at a non-matching codon in a maximally matching alignment and result in a change to the amino acid at that position to one that matches the target.

The mutation operators take the distribution of viruses that survive the immune response process, and transform it into a distribution of free viruses that exist after reproduction and mutation (Fig 2). In order to calculate the probability distribution, P(m), for the viral state after mutation, we need to calculate the general "transition probability" matrix for changes in the number of matches as a function of m. Assume a virus with a given number of initial matches, m. To calculate the probability that a mutation on this virus will cause a +1, -1, or 0, change in the number of matches we first studied some limiting cases. The transition probability and its slope can be derived analytically at two limits: no (or few) matches and perfect or near perfect matches. Depending on the degeneracy of the target, the values for these limiting cases can vary considerably and have the most variability for viruses with low numbers of matches before mutation. In the case of a highly degenerate target, that is, one with a very limited number of distinct or unique codons, the probability of a single mutation keeping the same number of matches approaches zero in the low initial match limit. Conversely, the probability of a mutation increasing the number of matches goes rapidly (to one) in this limit.

In the opposite case of a target with no repeating codons, the results depend on whether the alphabet is much larger even than the size of the target, or whether the target uses almost all the possible codons. If there are a large number of distinct codons possible in the viruses, beyond the number already in the target, the probability of a mutation keeping the same number of matches approaches 1 in the low-match limit, while the probability of a mutation increasing the number of matches is low for all initial numbers of matches. If almost all the available codons appear in the target, and the viruses contain essentially the same selection of codons that the target does, then the probabilities for keeping the same number of matches and increasing the number of matches by 1 become nearly equal at 1/2 each in the low match limit.

The probability that a mutation decreases the number of matches starts at zero for zero initial matches, and typically rises to a value near 0.5 for a complete match with the target, independent of target degeneracy. The probability of a mutation causing no change in the perfectly matching virus is also near 0.5 regardless of target degeneracy.

These general cases are shown in Fig 3. It should be noted that these results hold for matches determined by complete overlay of the target binding sites by the virus. We do not consider alignments where the target extends past the end of the virus giving only partial overlay of the target binding sites. This alternative method would yield different limiting cases, as well as different expressions below for the mutation. For clarity, in our model there exist only v-t+1 allowed alignments, where v and t are the length of virus and target, respectively. (This is in contrast to the 'extended alignment' method, not implemented here, which has v+t-1 alignments.) The limiting cases shown in Fig 3 are derived for the specific choice of a virus genome segment exactly twice as long as the target segment (our model system).

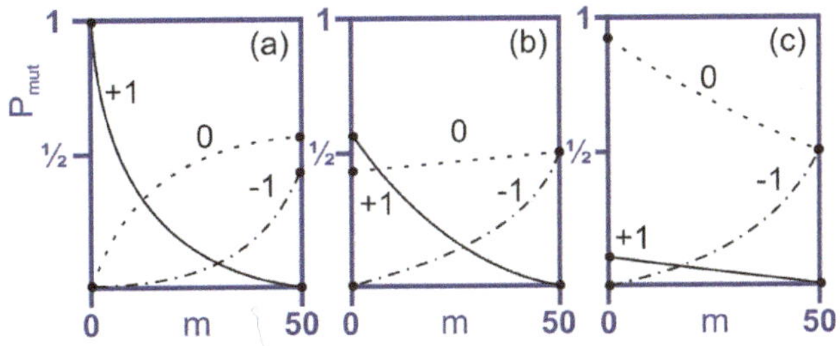

Figure 3. Three Limiting Cases of the Effect of Viral Mutation in a Single Codon. The transition probabilitiy (P_{mut}) as a function of the number of matches, m (pre-mutation). Curves labeled '+1' represent the probability of a mutation increasing the number of matches between the virus and target genomes, '-1' the probability of decreasing the number of matches, and '0' the probability of no change. Given alphabet length 'a,' the figure shows the limiting behavior for: (a) Small a, highly degenerate target; (b) a ≈ target+1, i.e., medium degeneracy; (c) Large a, i.e., low target degeneracy.

For specificity, the target genome segment was defined as:

THISISTHEENTRYTARGETFORREGULAR-CELLSWAYOFENTERINGIT

This example target uses 16 of the 26 possible codons. It has a maximum degeneracy of 7, and a length of 50. For a virus length of 100, and given this particular target genome segment, we were able to calculate analytically the behavior near the two end points and numerical continuation in between. The mutation probabilities are:

$$P_{mut}(\Delta m = -1) = \omega \frac{m}{100}\left(\frac{1}{1+e^{-(m-10)/2}}\right)$$

$$P_{mut}(\Delta m = +1) = \frac{\omega}{235.45}\left(e^{4.709(1-m/50)} - 1\right)$$

$$P_{mut}(\Delta m = 0) = 1 - P_{mut}(\Delta m = +1) - P_{mut}(\Delta m = -1) \quad (5)$$

where P_{mut} is the probability that a mutation of a virus with m matches will result in a change in m of $\Delta m = 0, \pm 1$. The variable ω, which we take to be 0.7867, represents the redundancy in the underlying genetic code due to the multiplicity of three-codon combinations that define an amino acid.

Note that the sum of the three terms is always one, corresponding to conservation of virus in the mutation process. We plot these mutation probabilities in Fig 4.

We note that our particular target is at the low degeneracy limit, and is part-way between the cases using all the available alphabet vs. those using only a small fraction of it. It thus can be viewed as fairly "challenging" to the virus. We find that changing the length of the virus genome, changing the length of the target, or changing the degeneracies in the codon "alphabet" (without going into trivial regimes or limiting cases of no degeneracy) only rescales temperature without changing the overall behavior or conclusions. In future work it is straightforward to explore other cases in detail.

Note also that many viruses not only mutate, but also recombine. Such viruses may infect individual cells multiple times. This model does not capture these processes. In addition, neither matches nor mismatches in this model should be interpreted as "replication error." In contrast to the quasispecies model(s) of viral evolution used to study viral error catastrophe [14–22], evolution in this model is a function of two independent and competing fitness pressures.

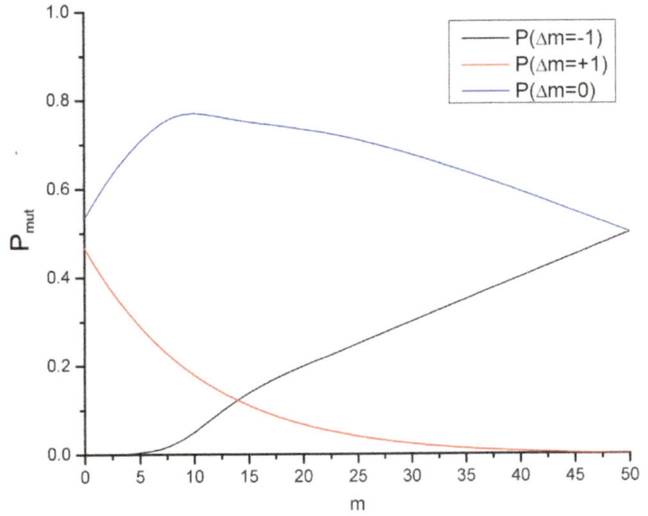

Figure 4. Viral Mutation Probabilities. For our model target, the transition probability (P_{mut}) as a function of the number of matches, m (pre-mutation). The figure shows the probability, P_{mut}, of a viral mutation causing an increase, decrease, or stasis in the number of matching codons between the virus and the actual target genome (Δm = +1,0,-1).

Self-Consistent Solutions

We solve self-consistently the probability functions of the virus in the host cells {Ψ^I, Ψ^Ξ, Ψ^R}, the virus distribution in the environment, and the total number of viruses. The 51x51 matrix of inter-related Ψ(m), Eq (3), can be diagonalized analytically to obtain the following:

$$\psi^I(m) = \frac{\lambda(N)[P_m e_m/E]}{[1+\sum_{m'} K_{m'}][1-(1-\Xi_m)(1-e_m)p]}$$

$$\psi^\Xi(m) = \frac{\lambda(N)[P_m e_m/E](1-\Xi_m)}{[1+\sum_{m'} K_{m'}][1-(1-\Xi_m)(1-e_m)p]} \quad (6)$$

$$\psi^R(m) = \frac{\lambda(N)[P_m e_m/E](1-\Xi_m)(1-e_m)p}{[1+\sum_{m'} K_{m'}][1-(1-\Xi_m)(1-e_m)p]}$$

With

$$K_m = \frac{\lambda(N)P_m e_m(1-\Xi_m)(1-e_m)p}{E[1-(1-\Xi_m)(1-e_m)p]}$$

where $E = \sum P_m e_m$ and all other terms are defined as in Eqs

(2–4). Eq (6) give the probability that a cell is occupied by virus with m matches at each stage of the life cycle (I, Ξ, and R). The stable solutions to Eq (6) hold for any given P_m and λ(N), which must also be solved self-consistently.

The limiting case of p = 0 has no virus remaining in the cells after each cycle so it decouples the solution of the occupation of the virus in the cells from the previous iteration ($\Psi^I(m)$ from $\Psi^R(m)$). If $\Psi^R(m)$ is zero in the equation for the infection rate, the solution of the virus in the cells becomes trivial. We have analyzed the model represented by the solutions in Eqs (6–10) for the full range of probability p, from p = 0 to 1. We find that the effect varying p, the probability that the virus remains in the cell if not cleared by the immune response, is only a slight rescaling of the temperature parameter, demonstrating universality in the solution described below. We also note that the addition of p breaks the symmetry between reproduction and infection but does not change the results. Since a value of p = 1 represents the most strongly coupled case of interaction between cells and virus, we present those results below.

Solution for the viral genetic states.

Eq (6) (solutions to Eq 3) provide the viral occupation (or load) in the cells as a function of the distribution of virus in the environment. We next solve for the steady state distribution of virus in the environment. Imposing self-consistency on the reproduction and mutation processes, we derive the following equation:

$$MD_mP_m = P_m\left[\sum_m D_mP_m\right] \tag{7}$$

Where $\boldsymbol{M}$ is a matrix of probabilities formed from Eq (5) and:

$$D_m = \frac{e_m^2(1-\Xi_m)}{[1-(1-\Xi_m)(1-e_m)p]} \tag{8}$$

Solving Eq (7) gives the steady-state viral probability distributions of the system. Eq (7) can be recognized as an eigenvalue equation where every valid eigenstate P_m of matrix MD_m must have eigenvalue $\sum_m D_mP_m$. It can be proven that any eigenvector solution of $MD_mP_m = \lambda_mP_m$ has an eigenvalue λ_m equal to $\sum_m D_mP_m$ as long as the eigenvectors P_m are normalizable as probability vectors (i.e., $\sum_m P_m = 1$) (note that D_m is expressed as a diagonal matrix).

Eq (7) defines a complex effective fitness landscape for the quasispecies population. We note that this fitness landscape is not purely multiplicative. Tripathi et al. (and references therein) modeled RNA virus evolution in a non-multiplicative fitness landscape where they allow for epistatic interactions [20]. However, the landscape considered here cannot simply be expressed as a sum of independent selection effects with cross terms.

Number of Viruses

For each solution P_m of Eq (7), the number, N, of viruses in the environment is equal to the total probability that a cell has a virus that successfully reproduces, times the number of target host cells, c, times the fecundity, φ, defined above:

$$N = c\varphi\sum_m e_m\psi^{\Xi}(m) \tag{9}$$

N can then be found as the solution of a pair of coupled transcendental equations, one for N as a function of the infection rate λ, and the other for λ as a function of N. Substituting Ψ^{Ξ} into Eq (9) from Eq (6), and using the definition of D_m from Eq (8), we obtain:

$$N = \lambda(N)\frac{c\varphi\sum_m D_mP_m}{E[1+\lambda(N)\sum_m K'_m]} \tag{10a}$$

With

$$K'_m = \frac{P_me_m(1-\Xi_m)(1-e_m)p}{E[1-(1-\Xi_m)(1-e_m)p]}$$

$$\lambda(N) = \sum_{n=1}^{c}\frac{n}{c}(1-E)^{(c-n)(N-n)}\prod_{i=0}^{n-1}\frac{(1-(1-E)^{N-i})(1-(1-E)^{c-i})}{(1-(1-E)^{i+1})}\,\Theta(N-n) \tag{10b}$$

With E, defined in the context of Eq (6), (1-E) is the probability of a cell not being infected in a single viral pass given the distribution of virus in the environment P_m. Further detail concerning the infection rate, λ(N), in Eq (10b). In the calculations in this paper, the number of target host cells, c, was taken to be 5 and the fecundity, φ, was 20, giving a maximum N of 100 viruses replicated from the cells into the environment.

These coupled nonlinear equations were solved using Newton-Raphson methods. It can be shown that the functional form above results in one and only one solution for N, for each choice of initial parameters and input viral distribution P_m.

Numerical Solutions

There are 51 roots of the eigenvalue Eq (7), corresponding to the vector size of P_m, (which is indexed by the different m, m = 0 to 50). More generally, given a genetic target of length G, there would exist G+1 roots. We employed a number of tests to determine which of the eigenstates are physical. Each eigenvector element represents a probability. The eigenvectors should not have imaginary elements, and after normalization they should not have negative elements. The Perron-Frobenius theorem states that a real square matrix with positive entries has a unique largest real eigenvalue and that eigenvector has strictly positive components. Our matrix meets the criteria required by Perron-Frobenius so the all-positive eigenvector corresponding to the largest eigenvalue provides the equilibrium solution. The number of viruses corresponding to an eigenstate (Eq 10) must also be greater than or equal to zero. With these conditions, only one nontrivial physical eigenstate was found for any set of initial conditions (temperature, immunity, etc.). The trivial zero state (no virus) is always a stable solution. We looked for dynamic solutions numerically and did not find any for the system defined in this paper. The dynamic solutions always converged to the analytically derived steady state result.

RESULTS AND DISCUSSION

Evolution of the Virus

The probability of a virus having a given number of matches, m, at a specific temperature and maximum immune response (Eq 4), A, is the normalized eigenstate, P_m (Fig 5). Each P_m can be thought of as the steady state quasispecies distribution, the peak of which represents the most "robust" virus type in the quasistates [14–16,20–22,30]. The width of each distribution reflects the accessible states and can be viewed as an indicator of evolvability or adaptive genetic diversity [4].

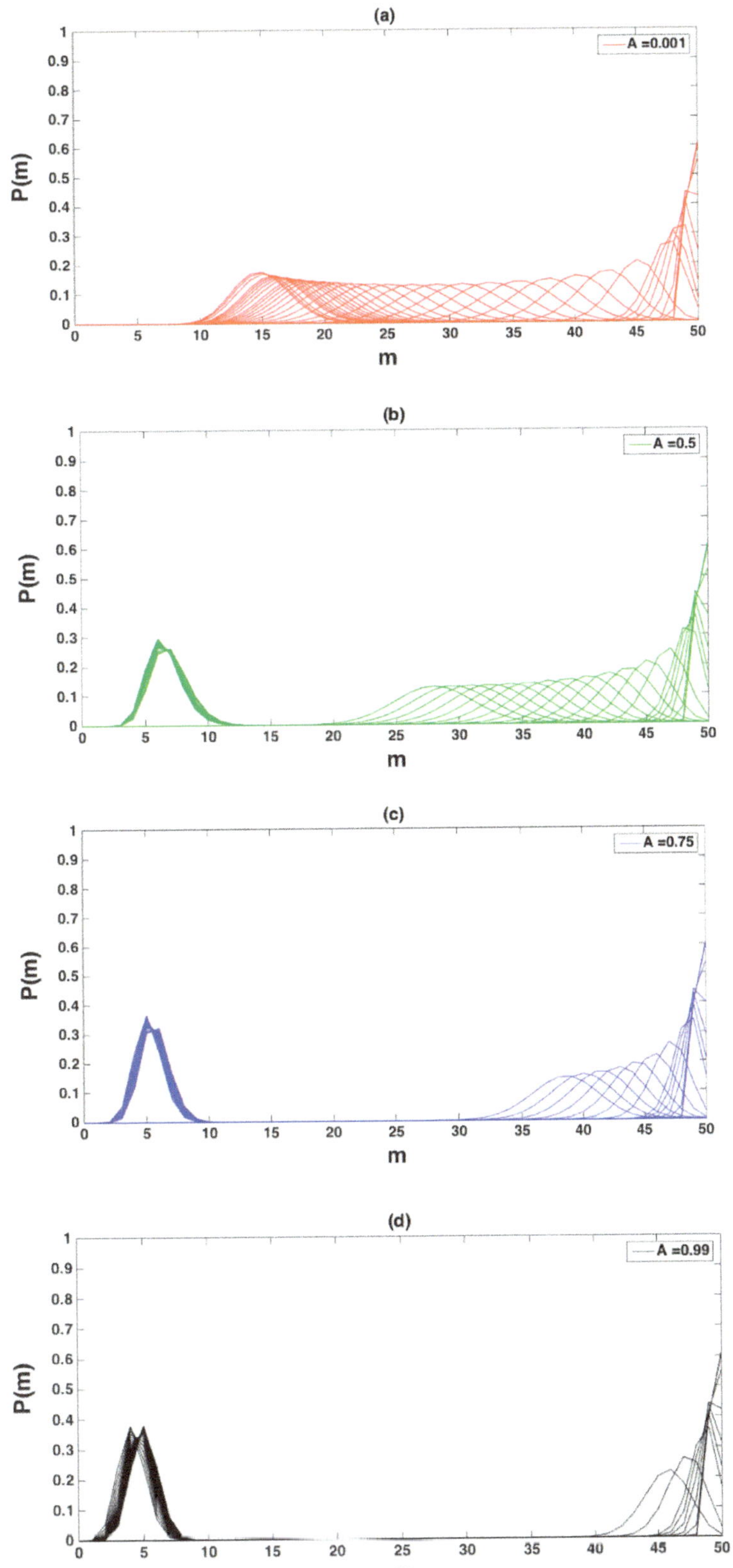

Figure 5. Eigenstates of the System. The figure shows the normalized eigenstates, P(m), vs. the number of matches, m, as a function of temperature and maximum immune response, A. Each distribution, P(m), represents the quasispecies distribution (i.e. probability of a given number of matches) at fixed A and temperature. The temperatures shown here are 0.01,0.03,0.05,0.1,0.3,0.5,1,2,3,4,5,10,…,100*,110,120,130,140,150,20

0,250,300 (*step by 5) for each immunity indicated. Low temperature is represented by the narrow distribution at high match (at far right).

For all temperatures and immunities studied (with the exception of T, A = 0, see below), only one stable (non-trivial, non-zero) eigenstate was found. At very low temperature, the virus must closely match the target genome (m ~ 50). As temperature is raised, the mean number of matches of the quasistate decreases, eventually excluding the perfect match as an important component of the solution (i.e., the state depins from m = 50). Two distinct behaviors are observed as a function of immunity. At low immune amplitude (Fig 5A), as T increases, the mean of the distributions moves smoothly from high match to low match (m~14.5). At higher immune amplitude (Fig 5B–5D), the quasistates distribution jumps from higher to lower m with increasing T. This is most pronounced in Fig 5D where all eigenstates are found only near higher or lower m regions.

At low temperatures the virus must be well adapted to the host as reflected in the high codon match. Conversely, at very high T, the barrier is less important (i.e., entry into the cell is thermally "activated") allowing the viruses to more easily avoid the immune system through greater genetic variation. We call distributions with mean near the perfect match "ordered states" of the virus, and distributions with low mean (m<10 in Fig 5D) "disordered states". This suggests that the mean of the eigenstate distributions may serve as a measure of an order parameter for the system, that is:

$$M_{env} = \frac{1}{50}\sum_{m} mP(m) \tag{11}$$

An equivalent order parameter for virus inside the cells is defined.

An order parameter, M, near 1.0 represents an ordered state and low M represents a disordered state. For low values of A, the order parameter decreases smoothly and continuously as temperature is raised (Fig 6). We will refer to this as the regime of normal replication. For high values of A, the order parameter jumps discontinuously from high to low as temperature is raised. This discontinuity suggests a first order phase transition in T at high immunity between the regime of normal replication and the disordered phase of the virus. The phase transition reported here reflects the competition between different natural "strategies" to resisting two different pressures. The first is immune response, and the second is the thermal barrier. Viruses responding to either of these pressures will have different characteristic energies. A first order phase transition occurs, by definition, when these energies cross. This phase transition is different from the phase transition that occurs in the Eigen and Schuster model, which reflects a loss in information associated with low fidelity of replication. It is also different from the very large literature on viral error catastrophe [14–22]. Error catastrophe is a phase transition that occurs in a dynamic model as mutation rate is increased past a critical rate. This is distinct from the phase transition observed above in the current steady state model. The phase transition observed here is a consequence of two competing energy terms. It is also distinct from the dynamic transition discussed by Nowak and May as a function of basic reproductive ratios for various mutant strains [16]. In our model even for increasing immunity in the disordered phase the viral population does not collapse.

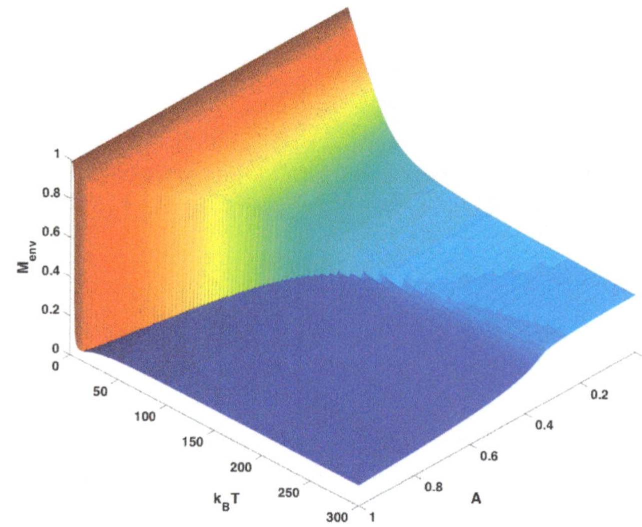

Figure 6. Order Parameter. The order parameter, M_{env}, as determined by sampling virus in the environment to measure the average fraction of matching codons, as a function of temperature and maximum immune response, A. The order parameter is defined in Eq (11). Sampling virus inside the cells yields almost exactly the same result

Fig 7 shows the occupancy of the cells after infection and immune response (before virus reproduction). One can view the occupancy as a measure of viral fitness. The occupancy fraction of the cells is between zero and 1. With zero immunity, A = 0, the cell occupancy is 1.0 for all T. As immunity is raised the occupancy decreases (approximately linearly in A) until reaching the phase boundary separating the regime of normal replication from the disordered phase. At high temperature and immune response the virus is in the disordered phase and cell occupancy plateaus at ~ 50%. These are viruses that never completely clear, but have low occupancy, low match, and evoke low immune response.

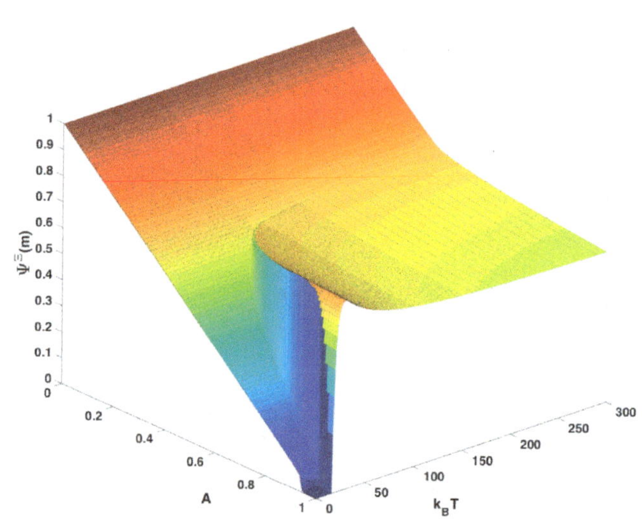

Figure 7. Occupancy of Cells. The occupancy of the cells, $\Psi^{\Xi}(m)$, derived from the steady-state solution (Eq 6), is shown as a function of temperature and maximum immune response, A. The same phase transition observed as in Fig 6 is evident here.

At low temperature the virus never enters the disordered phase and cell occupancy decreases linearly with increasing A, eventually falling to zero. The region of phase space with zero virus (viruses that clear in steady state) appears small but for an individual with full immunity, it actually extends to T =

16 degrees which is 32% of the maximum mismatch energy (50) scaled by the effective Boltzmann constant. Likewise the region of zero virus is bounded by maximum immune response A> = 0.94. In the regime of large A and low temperature, one would expect a dynamic process with basic reproductive ratio below unity. The discontinuity observed in Figs 6 and 7 is also evident in the order parameter measured for virus occupying the cells. These discontinuities suggest one or more phase transitions.

Thermodynamics and Statistical Mechanics

To understand the possible phase transitions (Figs 6 and 7) we now study the thermodynamics of the system. To do so we must first define a temperature. So far, we have used parameter Tas an "effective" temperature. At this point we go further and posit that T is in fact the natural temperature of the system. Systems at finite "natural" temperature do not stay in one equilibrium microstate. Rather, they sample all accessible states with a probability based on the Boltzmann distribution. We will estimate the effective Boltzmann constant and test the degree to which T acts as a real temperature below.

In order to determine the correct statistical thermodynamic ensemble of our system, we must identify the constant thermodynamic variables. In our model, the total number of cells, the size of the generic alphabet, and the length of the virus and the target genomes are all constant. To do thermodynamics, we need conservation of energy, and for this we need to define an energy. To be consistent with our definition of temperature, at T = 0 the system must enter a zero energy "ground state". In our model, due to the Arrhenius form with temperature in the denominator of the exponential, at T = 0 the probabilities of infection and reproduction become delta functions at the maximum number of matches. Only viruses with a perfect match will successfully reproduce. We thus assign energy E = 50 − m. With this definition, at T = 0, only the E = 0 state of the virus will be present. Any multiple of E would also serve. A Boltzmann constant must relate the energy and temperature scales. That is, with this definition of energy, our denominator in Eq (3) is actually $k_B T$.

In general, the expectation value of the energy of a viral state with N total viruses in the environment at a given temperature and immunity is:

$$E = N\sum_{m=0}^{50}(50 - m)P(m) \quad \textbf{(12)}$$

We calculate this and find the energy is zero all along the T = 0 axis for all immunities (as required), and increases monotonically with temperature.

So far we have discussed our model in terms of viruses in the cells and in the environment. It is clear that as temperature and immunity are changed, both the energy and the number of virions change. Energy and number are both conserved only if we imagine that our cells and environment are both in contact with a third reservoir or bath that includes all possible viruses in thermal and "chemical" equilibrium with the rest of the system. Chemical equilibrium in our model requires conservative flow of virus between the reservoir and the system. Classically, for particle number to have an associated chemical potential, chemical potential of the system must be conserved during the internal dynamics of the system, and only able to change when the system exchanges particles with an external reservoir. This is the classic definition of a macro-canonical or grand canonical ensemble (Fig 8). This ensemble is the natural statistical ensemble for modeling any system of viruses. It ensures conservation of both number and energy. Any viruses not in cells or the environment (e.g., those eliminated by immune response) are in the reservoir, and any new viruses entering the system (e.g., mutated offspring) are drawn from the reservoir.

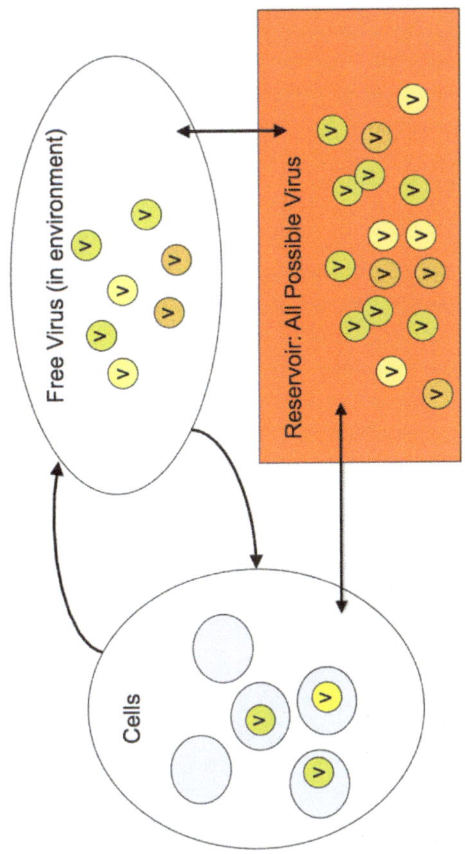

Figure 8. The Grand Canonical Ensemble for a System of Viruses. The three thermodynamic elements of the system are shown. The "Reservoir" of all possible virus is usually referred to as the "thermal bath".[23] In this case the bath of possible viral sequences is very large (effectively infinite). Free virus in the environment, at steady state, is populated from the reservoir with a distribution based on temperature and immunity. In the infection phase, virus that successfully infect cells are drawn from the environment. Virus that fails to infect are returned to the reservoir. Immunity may remove virus (from the cells back to the reservoir), and reproduction draws new offspring from the reservoir and repopulates the environment (emptying the cells). The double arrows indicate population from and return to the reservoir. The curved arrows show the virus life cycle.

Consider an initial (fully occupied) state of the reservoir with no viruses in the cells or the environment. In this state the bath includes enough copies of all possible virus sequences to populate any possible system state. That is for each of the 26^{100} possible viral sequences there must be a number of copies equal to fecundity times the number of cells. Reproduction is then a process of drawing new viruses from the theoretical reservoir constrained by the rules of mutation. Since the total number of viruses is fixed, the total energy is fixed, thus assuring conservation of energy.

Given this ensemble it is possible to use the methodologies of thermodynamics and statistical mechanics to calculate any thermodynamic quantity of interest. For example, given the expectation value of the energy, the specific heat, C(T), is defined by:

$$C(T) = \left(\frac{d\langle E\rangle}{dT}\right)_V \qquad \textbf{(13)}$$

where the derivative is taken at constant volume (here clearly maintained) and <E> is the average energy. Measurement of specific heat, or heat capacity, is another indicator of the type of phase transition. We observe a sharp maximum in specific heat (i.e., a latent heat) in the vicinity of the apparent first order phase transition seen in Fig 6. We do not observe a power law singularity in that part of phase space where the system smoothly transitions between states, suggesting there is no second order phase transition.

From the specific heat we also calculate entropy, a measure of the number of degrees of freedom of the system (Fig 9).

$$\Delta S = \int_a^b \frac{C(T)}{T} dT \qquad \textbf{(14)}$$

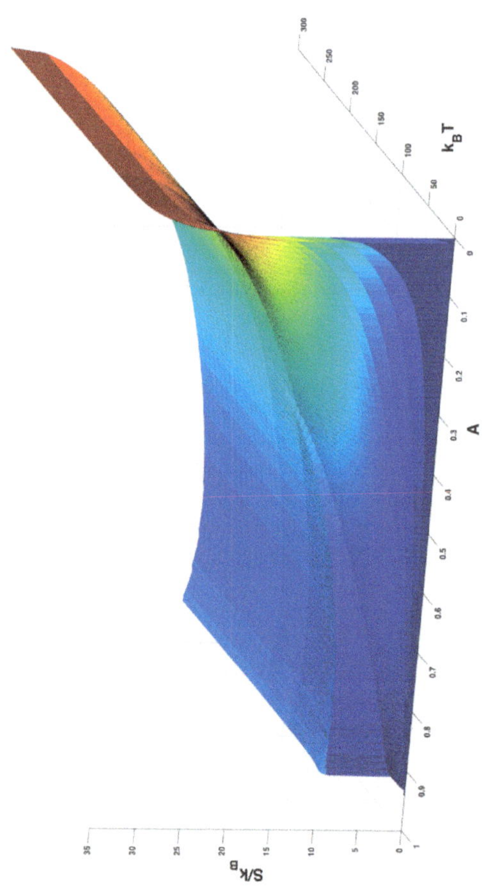

Figure 9. Entropy. The figure shows the entropy of virus, $S/k_B = \ln \Omega$, while in the cells as a function of temperature and maximum immune response, A, where k_B is the effective Boltzmann constant.

Formally, the entropy is defined as $S = k_B \ln\Omega$, where Ω is the effective number of degrees of freedom in the system, and k_B is the effective Boltzmann constant. Here Ω is a measure of the genetic variability of the viruses. One can see from Fig 9, that the number of degrees of freedom is very large (as large $e^{35} \sim 10^{15}$ even for our small genome). The number of degrees of freedom increases smoothly with decreasing peak immunity, A.

In addition to calculating the entropy, we also obtained the width, σ, of each quasistate as a function of temperature and maximum immune response, A. As discussed above this width is a measure of evolvability or adaptive diversity. The order parameter, which corresponds to the most abundant number of matches in the quasistates is a measure of robustness, m_{robust} or the number of amino acids that can change without changing the match number or phenotype. In Fig 10 we plot $m_{robust}/50$ as a function of evolvability for every temperature and immunity. Interestingly, we find that all of the data collapses onto a single universal curve. The evolvability is lowest for values of T and A that lead to quasispecies where m_{robust} closely matches the target as well as for quasispecies where the m_{robust} has almost no matches. The curve has a maximum for $m_{robust}/50 = 0.54$. The largest evolvability corresponds to quasistates near the phase transition where the curve breaks apart.

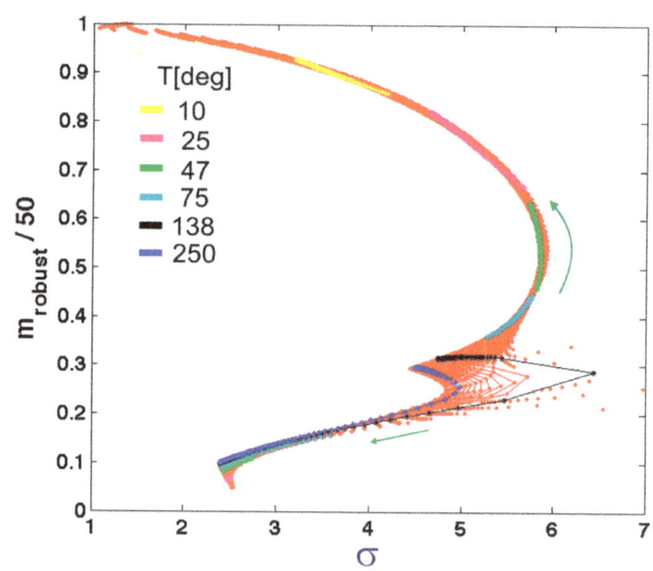

Figure 10. Robustness vs. Evolvability. The robustness ($m_{robust}/50$) as a function of "evolvability", σ, for each quasispecies, at all studied temperatures and immunities, A (red points). Here the robustness is defined by the order parameter, or average <m>, for each quasispecies distribution P(m). For a symmetric distribution this corresponds to the most probable m. The curve is nearly universal, breaking apart only near the phase transition. The colored segments and the green arrows indicate the trajectories as function of increasing immunity (for temperatures represented).

In Fig 10 the segment in yellow represents the trajectory along the universal curve where immunity, A, is varied at fixed T (T = 10 degrees). Counterintuitively, as immune pressure is increased, evolvability decreases and m_{robust} increases. This behavior provides an explanation for the phase transition. In this model viruses must survive two types of pressure. Low temperature selects for phenotypes best adapted to infect the host. Immune response puts pressure on phenotypes that most closely match the target. The direction in which quasistate distributions shift in response to these combined pressures depends on the relative steepness of each energy term as a function of T and A. In the example at T = 10 degrees, with increasing immunity the most robust virus shifts to even better match the target thus lowering temperature driven barriers to reproduction. While this leads to a slightly higher immune pressure, the immune response function has nearly plateaued for matches above 15 codons, so there is diminished benefit from lowering the match to avoid immune response. As immune pressure is increased still further, there comes a tipping point where increasing visibility to the immune response becomes too much for the virus, and there is a jump in population characteristics–the phase transition–favoring a much lower match to the target. In general, for all of the states inFig 10, all trajectories move away from the phase transition observed in Figs 5–8 (as indicated by the arrows in Fig 10). This behavior is further demonstrated by the shift in eigenstates as a function of immunity and temperature.

The relationship between viral robustness and evolvability explored in Fig 10 deserves to be understood more in depth. Viruses, especially RNA viruses, use a number of strategies to preserve genetic information during replication. Neutral (synonymous) mutations, large population sizes, co-infections and molecular chaperones are just few of these mechanisms. On the other hand, adaptation through new mutations to harsh environments is paramount for the evolution of the virus. The apparent antithetic nature of these two necessary mechanisms implies the need for a tradeoff between them. Thus, a relationship like the one obtained in Fig 10 may to be necessary for the evolution of the virus. The observation that evolvability is at a minimum when robustness is very low or very high and at a maximum for intermediate robustness was first reported by Draghi et al. in a dynamic genotype-phenotype network model [12]. Stern et al. further confirmed both theoretically and experimentally the relationship between evolvability and robustness and observed this proposed universal behavior in polio virus [13].

Thermodynamic Temperature

All of the analyses discussed above relate thermodynamic variables to strength of the immune system and an effective temperature. The question remains: how does our effective temperature relate to a real thermodynamic temperature? To determine this relationship, we calculate how the (genetic) states of the virus are distributed in energy.

$$\frac{1}{k_B T_{thermo}} \equiv \beta \equiv \frac{\partial \ln\Omega(E)}{\partial E} \qquad \textbf{(15)}$$

where Ω (E) is the number of accessible states at energy E. The accessible states represent the entire cohort of N viruses. In the previous sections we calculated the equilibrium viral state and its properties as a function of effective temperature (T) and immune strength (A). At a given effective T and A, each state has a well-defined number of viruses, N, and a probability distribution, P_m, representing the number of genetic matches (and mismatches) between the virus and target. While N and P_m are sufficient to calculate average properties (e.g., average energy), in order to calculate thermodynamic temperature one must enumerate the complete set of realizations of all systems with N viruses, and probability of match distributed as P_m. In order to calculate Ω (E), we need to do a careful counting of states as a function of energy.

We transform the probability distribution as a function of matches m, P_m, into a probability of finding a virus at an energy E, P(E), using the definition of Energy in Eq (12). Note that contributions to the probability of a virus at a given energy can be from several different quasistates. Details of how P(E) is calculated.

With the determination of P(E), we can define the accessible states in energy as:

$$\ln\Omega(E) = \ln P(E) + \frac{\sum_{j=1}^{n_E} P_j(E)\ln D_j(E) - P_j(E)\ln P_j(E)}{P(E)} \qquad \textbf{(16)}$$

With

$$\ln D_j(E) = \sum_{i=1}^{w_j} n_{ij}[\ln\Omega_o(m = m_i)] \qquad \textbf{(17)}$$

where Ω_o(m) is the number of distinguishable configurations of the codons for a virus with m matches. This very large number depends on the number of matches, length of the virus and target genome length, the size of the alphabet, and the number of codons used (and not used) in the target. In addition, to be accessible, the states must be connected by permissible mutations. In practice this limits Ω_o(m) from being the maximal value obtained by permutation alone. We have computed Ω_o(m) numerically and find ln Ω_o(m) ~ 47 for all m, given our definition of genomes, codon alphabet, and mutations.

In thermodynamics the formal relation between entropy and number of states is:

$$S = k_B \ln\Omega \qquad \textbf{(18)}$$

Note that in Eq (17) each contribution to lnΩ is of the form p lnp, which is the information theoretic entropy [31]. With these definitions we show below our calculated thermodynamic temperature as a function of the temperature parameter in our model, T_{model}. From Eq (15) the effective $k_B T_{thermo}$ is the inverse slope derived from a plot of lnΩ(E) vs. E.

For T_{model} less than the critical temperature (Figs 5 and 6), the system is in a regime of normal replication. In this phase, Fig 11 demonstrates that the thermodynamic temperature is defined, positive, and approximately linearly related to T_{model}. The constant of proportionality is the effective Boltzmann constant. Observe that the temperature scale set by the entropy and the number of states is tied to the genetic properties of the virus+host target pair. A different type of virus with a protein receptor of different length or different degeneracy, or a different target (host) receptor would change the entropy and the energy and therefore the corresponding temperature scale. Lowering temperature can cause some virions to fail to infect any cell (in a particular host) which might otherwise been able to infect a cell. This is analogous to changing immunity, which can cause some virions to die, which may not have otherwise died. Changing temperature changes a virion's fitness and shifts the steady state quasispecies distribution.

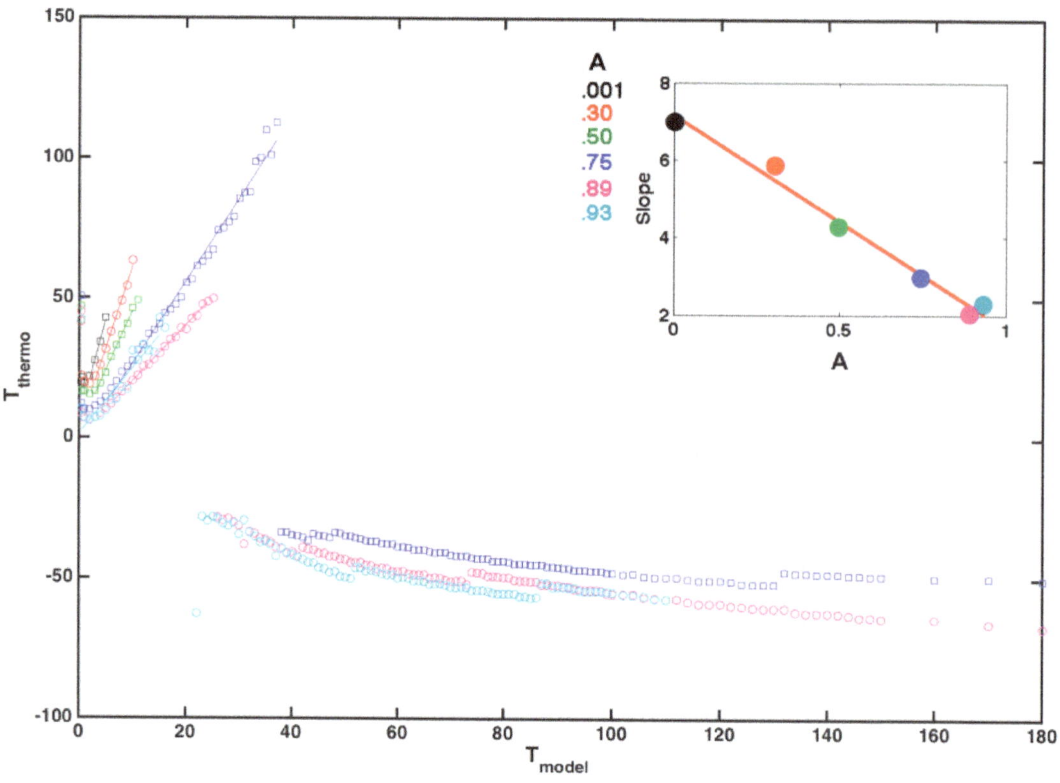

Figure 11. Testing for a Thermodynamic Temperature. The figure shows thermodynamic temperature T_{thermo} vs T_{model}. For T_{model} below the phase transition, the relationship is linear with slope k_B. From Eq (15) the effective $k_B T_{thermo}$ is the inverse slope derived from a plot of $\ln\Omega(E)$ vs. E. Above the phase transition a negative temperature is observed as expected. In the inset we plot the slope of T_{thermo} vs T_{model} for the values of maximum immune response, A, shown in the color legend, and observe that they all fall on a single line. This suggests that the effective k_B decreases linearly with increasing immune amplitude, A, and shows that immune strength rescales temperature.

Viruses compete to infect a finite number of cells. Fitness of a quasispecies distribution is affected by temperature, immunity, and also the number of possible states at each match. There are many more virus genetic configurations with 15 matches than there are with 50 matches. So entropy plays a critical role in defining the thermodynamic temperature (and the steady quasispecies distributions). As illustrated in Fig 5 (A) with near zero immune response (for example) changing temperature causes the quasispecies distribution to shift.

In the regime of normal replication we find that the effective Boltzmann constant also depends on the immune strength, A. In the inset to Fig 11,we plot the slope of the thermodynamic temperature, T_{thermo} vs T_{model} for the immunities shown in the color legend, and observe that they all fall on a single line. This suggests that the effective k_B decreases linearly with increasing maximum immune response, A, and shows that immune strength rescales temperature in the regime of normal replication. This rescaling is approximately linear with $k_B = -5.5A + 7.2$, where A is maximum immune response. We note with interest that for $A = 1$ we have $k_B \sim 2$ implying the thermodynamic and model temperature scales are not far apart.

At the critical temperature, T_c, there is a phase transition and the system switches from the regime of normal replication to a disordered phase for $T_{model} > T_c$. For reference, at high temperature and high immunity, the order parameter approaches zero and is nearly flat in the disordered phase. In fact, temperature is negative in the disordered phase. This negative temperature phase exists because at sufficiently high T_{model} there is less advantage to configurations with many matching codons, and at high immunity there is a survival penalty for eigenstates with high matches. Although it is possible to increase T_{model} to arbitrarily large values, the number of mismatches can never exceed the length of the target genome and the degeneracy of a state with maximal mismatch is constrained by the finite length codon alphabet. In classic textbook examples [23], $\Omega(E)$ is a rapidly increasing function of E. In this system, however, $\ln\Omega(E)$ has a maximum near the phase transition and then decreases. This occurs because as temperature increases past T_c there are actually fewer accessible states in the disordered phase. This gives rise to a negative slope of $\ln\Omega(E)$ vs E and, therefore, a negative temperature at high T_{model}. Physically, negative temperature occurs any time a finite system has both an upper and lower bound to the possible energies. This is precisely the case in any system with finite length genomes and a finite codon alphabet. Theoretically this should also be true of real biological viruses but it remains to be seen if any examples exist.

Negative temperate defines the highest energy state(s) of a system. The current biologically inspired model provides an easy to understand example of why a state with negative temperature is hotter than a state at positive temperature. Temperature is defined not only by a kinetic energy but also by the total entropy of the system. In an infinite system, entropy increases as temperature is raised. In this finite biological system, as energy is increased past the critical point, entropy actually decreases because the number of possible states or configurations with no matching codons is always less than the number of possible states at lower energy with (e.g.,) one matching codon. A fully disordered state cannot use any of the codons found in the target so it has lower entropy. In the limit of very high energy (and negative temperature) the disordered phase represents a state with a cohort of viruses, some with no codons that match the target genome. Due to mutation the cohort must contain some offspring in the environment with some matching codons.

CONCLUSIONS

In this paper we explored a simplified model of viruses and their life cycle. Within the model, the process of viral transmission is characterized by a series of energy barriers. A virus's ability to cross these barriers is defined by its genetic similarity to an idealized target sequence for the host. The genetic properties of the viruses evolve, through natural selection, to a steady-state distribution of genetic states best adapted to an environment at each fixed temperature and immune response. The immune response represents the host's ability to clear a virus based on both viral genetics and host immune memory. Viral evolution, in this case, is simply an operation on the genetic code of the multiple offspring of a parent virus.

The diversity of viral sequences in the extant population depends on the temperature of the system, T, and the strength of the immune response. At each temperature and immunity we find one stable quasi-state with a diverse distribution of viral sequences. The average of this distribution has a characteristic number of codons which match the target. This average match (M) defines an order parameter, and is found in our thermodynamic analysis to be related to the system energy. The width of the quasi-state distributions is a measure of the diversity (and evolvability) of the extant population. We find a nonlinear function relating evolvability to robustness that collapses all data at all temperatures and immune function to a single universal curve in agreement with previous theoretical and experimental literature [12,13].

We determined all equilibrium states of this model system, as well as the probability distribution describing the matches of those viruses as a function of temperature and immune response. The stable quasi-states and resulting virus phases as a function of immune response reflect the "strategies" a virus may take to efficiently infect a host cell while avoiding removal by the immune system. The order parameter based on the number of matches reveals two regimes. To understand these regimes we applied the machinery of thermodynamics and statistical mechanics. Enumerating the states of all possible viruses (those able to infect and reproduce in cells, off spring found in the environment, and a "reservoir" or "bath" of all remain states with their respective probabilities) we used the grand canonical ensemble to derive all of the thermodynamic variables for the system including thermodynamic temperature, immune suppression, entropy, specific heat, and total energy. The grand canonical ensemble is the natural statistical ensemble for modeling any system of viruses.

In response to temperature and immune pressure we observe a phase transition between a positive temperature regime of normal replication and a negative temperature "disordered" phase of the virus. In this model viruses must survive two types of pressure. Low temperature selects for phenotypes best adapted to infect the host. Conversely, immune pressure is strongest on phenotypes that most closely match the target. The direction in which quasistate distribution shifts in response to these combined pressures depends on the relative steepness of each energy term as a function of T and A. At some temperatures and immunities increasing immunity causes the virus quasistates to shift to even better match the target thus lowering temperature driven barriers to reproduction. As immune pressure is increased still further, there comes a tipping point where increasing visibility to the immune response becomes too much for the virus, and there is a jump in population characteristics–a phase transition–favoring a much lower match to the target.

The phase transition separates a regime of normal reproduction from a disordered regime with negative temperature. The negative temperature regime requires a scenario wherein a virus with few matching segments is still able to enter the cell. In real viruses there are many cases where we see large genetic diversity in individual genes, often those that are important to the immune response (e.g., surface proteins on hepatitis C virus and influenza). The action of these genes may be functionally more like a diffusion process, allowing greater diversity in the genes than would be expected in the regime of normal replication.

We have demonstrated that this simple model of viral replication has a real thermodynamic temperature linearly related to the effective model temperature where temperature is positive (thus defining the effective Boltzmann constant). Many important relevant modifications to the model and its parameters simply rescale the temperature. This suggests that if the model can be extended to capture the dynamics of true biological systems, complex aspects of such systems may similarly be understood using the formalisms of thermodynamics and statistical mechanics, thus greatly simplifying their analysis. Microbiological experiments systematically measuring the functional sensitivity of particular genes to changes in sequence may help to define the temperature scale of those genes and serve as an important step in adapting this model to real systems.

ACKNOWLEDGMENTS

We would like to thank Dr. Niina Haiminen, Dr. Charles Stevens, Professor Donald Burke, Dr. Kenneth Clarkson and Judith Douglas.

AUTHOR CONTRIBUTIONS

Conceived and designed the experiments: BAJ JHK. Performed the experiments: BAJ JHK. Analyzed the data: BAJ JL SB JHK. Contributed reagents/materials/analysis tools: BAJ JHK. Wrote the paper: BAJ JL SB JHK.

REFERENCES

1. Koonin EV, Senkevich TG, Dolja VV. The ancient virus world and evolution of cells. Biol Direct. 2006; 1: 1–27. doi: 10.1186/1745-6150-1-29. pmid:16542032
2. Knipe DM, Howley PM, Griffin DE. Fields virology. 2 vols. Philadelphia: Wolters Kluwer Health; 2007.
3. Marsh M, Helenius A. Virus entry: open sesame. Cell. 2006; 124: 729–740. pmid:16497584 doi: 10.1016/j.cell.2006.02.007

4. Van Nimwegen E, Crutchfield JP, Huynen M. Neutral evolution of mutational robustness. Proc Natl Acad Sci USA 1999; 96: 9716–9720. doi: 1073/pnas96.17.9716 pmid:10449760 doi: 10.1073/pnas.96.17.9716
5. Wilke CO, Wang JL, Ofria C, Lenski RE, Adami C. Evolution of digital organisms at high mutation rates leads to survival of the flattest. Nature. 2001; 412: 331–333. doi: 10.1038/35085569. pmid:11460163
6. Domingo E, Holland J. RNA virus mutations and fitness for survival. Ann Rev Microbiol. 1997; 51: 151–178. doi: 10.1146/annurev.micro.51.1.151
7. Janeway CA, Travers P, Walport M, Sclomchik MJ. Immunobiology. 6th edn. New York, NY: Garland Science; 2004.
8. Anderson RM, May RM. Infectious diseases of humans: dynamics and control. Oxford, UK: Oxford University Press; 1992.
9. Acevedo A, Brodsky L, and Andino R. Mutational and fitness landscapes of an RNA virus revealed through population sequencing. Nature 2014; 505.7485: 686–690. doi: 10.1038/nature12861. pmid:24284629
10. Schulte MB, Draghi JA, Plotkin JB, Andino R. Experimentally guided models reveal replication principles that shape the mutation distribution of RNA viruses. eLife. 2015; 4: e03753.09. doi: 10.7554/elife.03753
11. Schulte MB, Andino R. Single-cell analysis uncovers extensive biological noise in poliovirus replication. J Virol. 2014 Jun; 88(11): 6205–6212. doi: 10.1128/JVI.03539-13. Epub 2014 Mar 19. pmid:24648454
12. Draghi JA, Parsons TL, Wagner GP, Plotkin JB. Mutational robustness can facilitate adaptation. Nature. 2010; 463(7279): 353–355 doi: 10.1038/nature08694. pmid:20090752
13. Stern A, Bianco S, Te Yeh M, Wright C, Butcher K, Tang C, et al. Costs and benefits of mutational robustness in RNA viruses. Cell Rep. 2014 Aug 21;8(4):1026–1036. doi: 10.1016/j.celrep.2014.07.011. Epub 2014 Aug 7. pmid:25127138
14. Alonso J, Fort H. Error catastrophe for viruses infecting cells: analysis of the phase transition in terms of error classes. Philo Trans R Soc A. 2010; 368(1933): 5569–5582. doi: 10.1098/rsta.2010.0274
15. Hart GH, Ferguson AL. Error catastrophe and phase transition in the empirical fitness landscape of HIV. Phys Rev E. 2015 Mar; 91(3–1):032705. doi: 10.1103/PhysRevE.91.032705.
16. Nowak M, May RM. Virus dynamics: mathematical principles of immunology and virology. Oxford, UK: Oxford University Press; 2000.
17. Wright S. Evolution and the genetics of populations. Vol.1-4. Chicago, IL: University of Chicago Press; 1984.
18. Ramsey CL, De Jong KA, Grefenstettc JJ, Wu AS, Burke DS. Genome length as an evolutionary self-adaptation. Lecture Notes in Computer Science, 1498, 1998. p. 345–353. doi: 10.1007/bfb0056877
19. Hartl D, Clark A. Principles of population genetics. 4th ed. Stamford, CT: Sinauer Associates; 2007. p. 112. Tian JP. Evolution algebras and their applications. Lect Notes Math. 2008; 1921. Berlin: Springer-Verlag. doi: 10.1007/978-3-540-742845.
20. Tripathi K, Balagam R, Vishnoi NK, Dixit NM. Stochastic simulations suggest that HIV-1 survives close to its error threshold. PLoS Comp Biol. 2012; 8.9: e1002684. doi: 10.1371/journal.pcbi.1002684
21. Summers J, Litwin S. Examining the theory of error catastrophe. J Virol. 2006; 80 (1):20–26. pmid:16352527 doi: 10.1128/jvi.80.1.20-26.2006
22. Shekhar K, Ruberman CF, Ferguson AL, Barton JP, Kardar M, Chakradborty AK. Spin models inferred from patient-derived viral sequence data faithfully describe HIV fitness landscapes. Phys Rev E. 2013; 88(6): 062705. doi: 10.1103/physreve.88.062705
23. Reif F. Fundamentals of statistical and thermal physics. New York, NY: McGraw-Hill; 1965. p. 66.
24. Levine RD. Molecular reaction dynamics. Cambridge, UK: Cambridge University Press; 2005.
25. Landau LD, Lifshitz EM. Statistical physics. Oxford, UK: Pergamon Press; 1980. pp. 9ff.
26. Lakadamyali M, Rust MJ, Zhuang X. Endocytosis of influenza viruses. Microbes Infect. 2004; 6:929–936. doi: 10.1016/j.micinf.2004.05.002. pmid:15310470
27. Burch CL, Lin C. Evolution by small steps and rugged landscapes in the RNA virus □6. Genetics 1999; 151: 921–927. pmid:10049911
28. Novella IS, Duarte EA, Elena SF, Moya A, Domingo E, Holland JJ. Exponential increases of RNA virus fitness during large population transmissions. Proc Natl Acad Sci USA. 1995; 92: 5841–5844. pmid:7597039 doi: 10.1073/pnas.92.13.5841
29. Alberts B, Johnson A, Lewis J, Raff M, Roberts K, Walter P. Molecular biology of the cell. New York: Garland Science; 2002.
30. Eigen M, Schuster P. The hypercycle: a principle of natural self-organization. Berlin: Springer-Verlag; 1979.
31. Shannon CE, Weaver W. The mathematical theory of communication. Urbana and Chicago, IL: University of Illinois Press; 1949.

CHAPTER 16

ENERGETICS OF AN RF SQUID COUPLED TO TWO THERMAL RESERVOIRS

B. Gardas, J. Łuczka, A. Ptok, J. Dajka

INTRODUCTION

Current tendency toward miniaturization requires a better understanding of phenomena and processes in small space and time domain. In particular, one can notice an upsurge of interest in energy transport phenomena in micro- and nanostructures like molecular junctions, suspended nanotubes, quantum point contacts, nono-wires, phononic metamaterials, etc. [1,2]. Recent progress and rapid developments in technology will, sooner rather than later, allow to convert research findings into potential applications of functional thermal devices such as quantum heat engines [3], thermal rectifiers [4, 5], thermal transistors [6, 7], controllers of local temperature [8], heat—voltage converters, thermal circuits [9, 10], nano-refrigerators [11] and last but not least hybrid thermal circuits and microelectromechanical machines where energy transfer can be exploited to move microscopic devices.

In this paper we study energy transport in a temperature-biased Josephson junction ring, i.e. a superconducting loop interrupted by a Josephson junction (such a system is also called a rf SQUID). Josephson junctions form a reach family which exhibits interesting both classical and quantum phenomena. They are elements of many low-temperature devices and offer a number of beneficial applications. They are applied in quantum computing devices as single units of information (qubits). Circuits made of Josephson junctions have been turned into artificial atoms which can be manipulated and measured by methods developed in atomic physics. Of special interest are novel heat transport phenomena study of which still is at the initial stadium. A pioneering work in this subject was performed in Ref. [12] and in a series of papers by Guttmanet al. [13–15]. A more complete and recent literature survey is presented in Refs. [16, 17].

To some extend our studies have been inspired by coherent caloritronics and experiments with the dc SQUID [16]. Here, we consider a purely quantum model of the rf SQUID. However, this model is much more general: it is a quantum nonlinear oscillator widely studied in various contexts in nonlinear quantum optics, condensed matter physics, statistical physics and cold atoms physics. In a general context, we study energy transport passing through the nonlinear oscillator which is coupled to two thermal Markovian reservoirs of different temperatures. The applied theoretical approach provides information on the non-equilibrium states of the system, its dynamics, transient and stationary phenomena and influence of the initial state of the system. Therefore this method has the distinct advantage over other methods in use. We examine energy current as it evolves in time from the transient regime to the long-time stationary state. We find several different regimes in the steady-state as well as in the transient time-domain.

The sketch of the paper is as follows. First we briefly present main elements of quantum modelling for the rf SQUID. Next we formulate the corresponding model of Markovian dynamics for an open system weakly coupled to two thermostats of different temperature. Further we present general remarks on energy flow in the system and analyze in detail stationary energy flow as a function of control parameters. In particular, the dependence on the external magnetic flux exhibits periodic behavior and agrees with what was already known in literature and confirmed in experiments [16]. Remaining part of the work is devoted to transient effects: how energy flow depends on time. Finally we provide a summary and conclusions.

Citation: Gardas B, Łuczka J, Ptok A, Dajka J (2015) Energetics of an rf SQUID Coupled to Two Thermal Reservoirs. PLoS ONE 10(12): e0143912. doi:10.1371/journal.pone.0143912.

MATERIALS AND METHODS

Quantum Model of rf SQUID

Now, we present details of the model, see Fig 1. It is a superconducting loop of inductance Lcomprising a single Josephson tunnel junction with critical current I_c. The loop is threaded with a magnetic flux ϕ. The phase difference ψ of the Cooper pair wave function across the junction is related to the magnetic flux ϕ via the relation [18]

$$\psi = 2\pi(n - \phi/\phi_0), \tag{1}$$

where $2\pi n$ is the phase change per cycle around the loop and in the SI units $\phi_0 = h/2e$ is the superconducting flux quantum (the factor of two arises from electron pairing). When the external magnetic field is applied, the total flux consists of two parts, namely,

$$\phi = \phi_e + LI, \tag{2}$$

where ϕ_e is the external static flux and I is the total current flowing in the loop. We model the system in terms of the Hamiltonian [19]

$$\hat{H} = \frac{Q^2}{2C} + \frac{1}{2L}(\phi - \phi_e)^2 - E_J \cos\left(2\pi\frac{\phi}{\phi_0}\right). \tag{3}$$

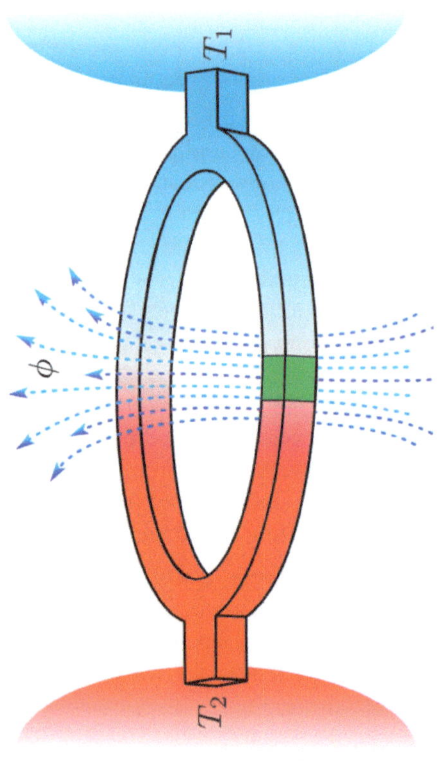

Figure 1. Schematic representation of an rf SQUID in contact with two thermal reservoirs of different temperature T_1 **and** T_2. The loop is threaded with a magnetic flux ϕ.

This Hamiltonian contains three energy scales entering charging energy E_C, inductive energy E_L and Josephson energy E_J, respectively. The junction capacitor charge Q is canonically conjugated to the flux ϕ and the commutation relation for the quantum operators Q and ϕ assumes the form $[\phi, Q] = i\hbar$. The Josephson energy E_J is related to the critical Josephson current I_J via the relation $E_J = \phi_0 I_J/2\pi$.

We transform Hamiltonian Eq (3) to the form

$$H = U^\dagger \hat{H} U = \frac{Q^2}{2C} + \frac{\phi^2}{2L} - E_J \cos\left(2\pi\frac{\phi + \phi_e}{\phi_0}\right), \tag{4}$$

where the unitary operator $U = \exp(-i\phi_e Q/\hbar)$. Such form of Hamiltonian allows to distill the role of the external flux ϕ_e piercing the SQUID.

In the limit $E_J \to 0$, Eq (4) reduces to the equation of motion for a harmonic oscillator of mass M = C and frequency $\omega = 1/\sqrt{LC}$. By analogy to the harmonic oscillator, we introduce a corresponding Hamiltonian in the form [21]

$$\begin{aligned} H &= \hbar\omega\left(a^\dagger a + \frac{1}{2}\right) \\ &- E_J \cos\left[\frac{2\pi}{\phi_0}\sqrt{\frac{\hbar}{2C\omega}}(a^\dagger + a) + 2\pi\frac{\phi_e}{\phi_0}\right], \end{aligned} \tag{5}$$

where the annihilation and creation bosonic operators are defined as [21]

$$a = \sqrt{\frac{C\omega}{2\hbar}}\left(\phi - \frac{i}{C\omega}Q\right), \quad a^\dagger = \sqrt{\frac{C\omega}{2\hbar}}\left(\phi + \frac{i}{C\omega}Q\right). \tag{6}$$

The dimensionless form of the Hamiltonian Eq (5) reads

$$\mathcal{H} = \frac{H}{\hbar\omega} = a^\dagger a + \frac{1}{2} - \mathcal{E}_J \cos\left[2\sqrt{\mathcal{E}_C}(a^\dagger + a) + 2\pi\Phi_e\right], \tag{7}$$

where the dimensionless parameters

$$\mathcal{E}_J = \frac{E_J}{\hbar\omega}, \quad \mathcal{E}_C = \frac{e^2/2C}{\hbar\omega}, \quad \Phi_e = \frac{\phi_e}{\phi_0} \tag{8}$$

have obvious meanings: ε_J is the dimensionless Josephson energy, ε_C is the dimensionless charging energy and Φ_e is the external flux in units of the magnetic flux quantum ϕ_0.

SQUID in Contact with Two Reservoirs

We now assume that the system under consideration is an open system, i.e. one side of the Josephson junction is kept at temperature T_1 and the other—at T_2, see Fig 1. These temperatures are controlled by two independent thermal reservoirs 1 and 2 of temperature $T_1$and T_2, respectively. If temperatures are different, there is heat flow between two reservoirs mediated by the SQUID. The influence of the SQUID on the process of heat flow is an interesting and difficult problem. In order to describe it, one should model two reservoirs in the framework of microscopic quantum many body systems a la Calderia-Legget Hamiltonian [20] and solve a Liouville-von Neumann equation for the total closed system: the SQUID + two reservoirs. However, it is impossible to solve it and therefore some approximations have to be applied. In the paper, we do not analyze the heat flow between two reservoirs but only energy change of the SQUID mediated by two reservoirs. We consider the system weakly interacting with reservoirs 1 and 2. In the sequel, we adopt the theory of quantum open systems in the weak-coupling regime [26]. This theory is valid under the condition that the relaxation time of heat baths are much shorter than the characteristic time of the system which can be identified with the inverse of the frequency ω. In such a case the dynamics is Markovian and governed by a completely positive linear map acting on the reduced density operator ρ(t) of the system. We follow the idea of Alicki [27] who studied a quantum open system coupled to several reservoirs as a model of a heat engine. In his approach, the reduced dynamics of the system alone is described by a master equation [27]

$$\frac{d\rho(t)}{dt} = -i[\mathcal{H}, \rho(t)] + D_1[\rho(t)] + D_2[\rho(t)]. \tag{9}$$

The first term $-i[\mathcal{H}, \rho(t)]$ describes the unitary evolution of the density matrix ρ(t) governed by the dimensionless Hamiltonian H defined in Eq (7) and the dimensionless time t = ωτ, where τ is the dimensional (original) time. The dissipative generators D_i (i = 1, 2) describe the influence of thermal reservoirs. In the microscopic formulation, the form of dissipators D_i should be derived from the original physical picture by exploiting e.g. the Davies theory [28]. To apply this theory, the spectral decomposition of the Hamiltonian Eq (7) is needed. However, its explicit form is not known. Therefore we make the next approximation and choose the dissipators in the form well-known in quantum optics [26], namely,

$$\begin{aligned} D_i(\rho) &= \gamma_i(N_i+1)(2a\rho a^\dagger - a^\dagger a\rho - \rho a^\dagger a) \\ &+ \gamma_i N_i(2a^\dagger \rho a - a a^\dagger \rho - \rho a a^\dagger), \quad i = 1, 2, \end{aligned} \tag{10}$$

where N_i is related to temperature T_i and the frequency ω_B of the i-th monochromatic reservoir via the expression

$$N_i = \left[\exp\frac{\hbar\omega_B}{k_B T_i} - 1\right]^{-1}. \tag{11}$$

The parameter γ_i is a coupling constant of the system to its i-th reservoir.

We are aware that a single mode monochromatic reservoirs are to crude for analysis of the heat flow between two reservoirs. However, for the energy change of the system alone it can be admissible to give insight into oscillation and relaxation phenomena. Moreover, a similar approach has been used in Ref. [21] to demonstrate the manner in which decoherence affects the quantum states of the loop. Lately [22], the same dissipators have been applied with success for analysis of the quantum dynamics of two electromagnetic oscillators coupled in series to a voltage-biased Josephson junction. The similar scheme has been proposed for a quantum thermal machine made by atoms interacting with a single nonequilibrium electromagnetic field [23]. The next system is a realistic device to reversibly extract work in a battery of finite energy: a hybrid optomechanical system [24]. Such devices consist of an optically active two-level quantum system interacting strongly with a nano-mechanical oscillator that provides and stores mechanical work, playing the role of the battery. Again, the authors of Ref. [24] exploite the same form of dissipators. Finally, we would like to mention the paper [25] in which time-resolved statistics of nonclassical light in Josephson photonics has been studied in the framework of the Lindblad master equation. The above examples show that the dissipators Eq (10) are universal and in many cases can correctly describe the influence of environment.

Energy Flow

The change of averaged energy $E(t) = \mathrm{Tr}[\mathcal{H}\rho(t)]$ of the SQUID is determined by the masterEq (9) and reads [26]

$$\frac{dE(t)}{dt} = J(t) = J_1(t) + J_2(t). \tag{12}$$

The energy current $J_1(t)$ is related to the change of energy of the SQUID caused by the coupling to the reservoir 1 and is given by the relation

$$J_1(t) = \mathrm{Tr}\{\mathcal{H}D_1[\rho(t)]\}. \tag{13}$$

In turn, the energy current $J_2(t)$ is related to the change of energy due to the coupling of the SQUID to the reservoir 2 and has the form

$$J_2(t) = \mathrm{Tr}\{\mathcal{H}D_2[\rho(t)]\}. \tag{14}$$

The dimensionless Hamiltonian Eq (7) is a periodic function of the magnetic flux Φ_e with period L = 1, i.e. $\mathcal{H}(\Phi_e) = \mathcal{H}(\Phi_e + 1)$. In consequence, averaged values of observables are also periodic. In particular, the energy currents are periodic with respect to Φ_e.

In the thermalization process, no work is performed and heat flows between reservoirs 1 and 2through the SQUID. In the stationary state, when t→∞, the statistical operator ρ(t)→ρ. In consequence dE(t)/dt = 0 and from Eq (12) it follows that the total energy current J(t)→J = J_1 +J_2 = 0. Hereafter, we will consider the case $T_2 > T_1$. Therefore the stationary energy current $J_2 > 0$ (heat is flowing from the reservoir 2 to the SQUID) and $J_1 < 0$ (heat is flowing from the SQUID to the reservoir 1).

We have solved the master Eq (9) numerically by truncation the infinite—dimensional Hilbert space at $\dim_{max}$ = 100. Next, we have solved a remaining system of differential equations using a variable—order and variable—step Adams method [29]. We have tested this procedure and ascertained that for considered parameters it allows to obtain converged and stable results. Parts of our so obtained results are presented next.

RESULTS AND DISCUSSION

Stationary Regime

Let us study generic properties of the long-time stationary energy change of the SQUID as a function of selected parameters of the model. In Figs 2 and 3 one can monitor the energy current approach to the stationary state. Because the total stationary energy current J = J_1 + J_2= 0, we analyze only the current $J_2 > 0$ (under the assumption $T_2 > T_1$). It is a periodic function of the magnetic flux, $J_2(\Phi_e) = J_2(\Phi_e + 1)$. There are regimes where Φ_e is a relevant control parameter. In Fig 2 we display the dependence of the stationary current J_2 on Φ_e. Let us notice that the energy current can be reduced by an external magnetic field by more that 50%. This opens a door for applications as a energy controller in small devices: by continuous changing of the external magnetic field, one can heat up or cool down the system. But this is not always the case because there are regimes where J_2 weakly depends on Φ_e and then Φ_e is an irrelevant control parameter, i.e. the energy current depends so weakly on Φ_e that it would be difficult to detect this behavior.

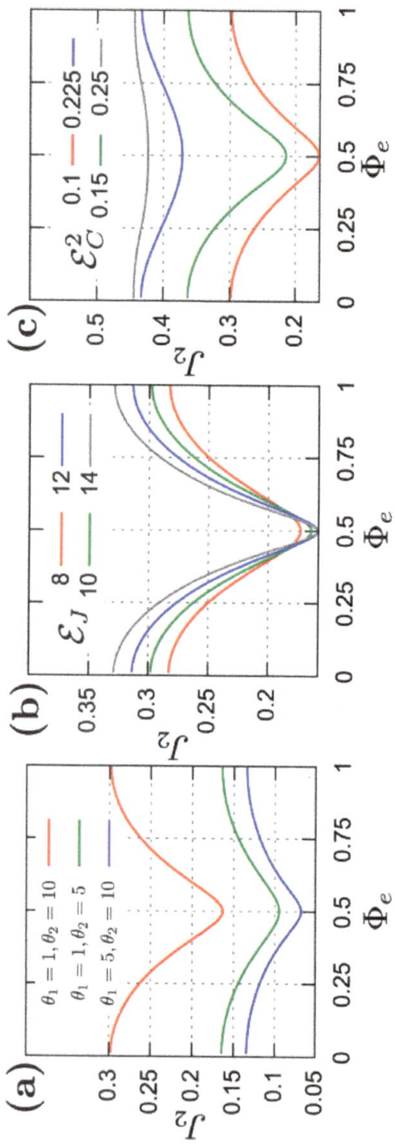

Figure 2. The stationary energy current J_2 **as a function of the external magnetic flux Φ_e**. Panel (a) for various temperatures of reservoirs: $(\{\theta_1, \theta_2\} = \{1, 10\}, \{1, 5\}, \{5, 10\}$. Panel (b) for $\mathcal{E}_C^2 = 0.1$ and various Josephson energy: $\varepsilon_J = 8, 10, 12, 14$. Panel (c) for $\varepsilon_J = 14$ and various charging energy $\mathcal{E}_C^2 = 0.1, 0.15, 0.225, 0.25$.

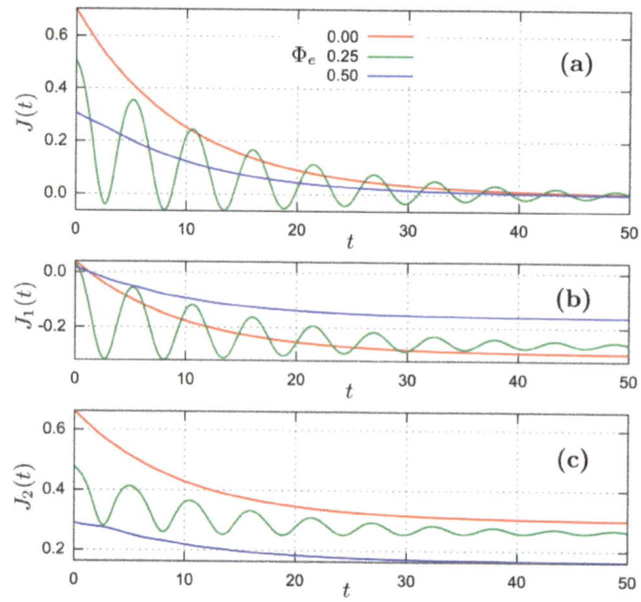

Figure 3. Short-time regime of (a) the total energy current J(t), (b) the energy current $J_1(t)$ induced only by the reservoir of temperature T_1 **and (c) the energy current** $J_2(t)$ induced by the reservoir of temperature T_2 **for selected values of the rescaled external magnetic flux Φ_e**. The reservoir rescaled temperatures are $\theta_1 = k_B T_1/\hbar\omega_B = 1$, $\theta_2 = k_B T_2/\hbar\omega_B = 10$. Remaining parameters are fixed as: $\varepsilon_J = 10$, $\mathcal{E}_C^2 = 0.1$ and $\gamma_1 = \gamma_2 = 0.05$. The dimensionless time t is rescaled by the oscillator frequency ω.

In Fig 2 we also show how the stationary energy current depends on the scaled Josephson energy ε_J and the scaled charging energy ε_C. The increase of ε_J leads to a greater amplitude of oscillations of J_2. It seems to be related to the form of the Hamiltonian Eq (7) in which ε_J plays a role of amplitude in the cosine part. The greater amplitude of oscillations in J_2 allows for a greater reduction of energy current by Φ_e. In turn, the change of the charging energy ε_C is not so evident. From panel (c) of Fig 2 one can notice that the variation of ε_C can increase the amplitude of the J_2-oscillations in a narrow interval of ε_C. Outside this interval, monotonic changes of ε_C result in flattening of modulation of J_2 and the interference effects are masked and cannot be detectable.

Time-Dependent Regime

In this section we analyze an approach to the stationary regime. To this aim, we have solved numerically the master Eq (9) with an initial condition $\rho(0)$. As an initial state of the system Eq (7), we consider the vacuum state $\rho(0) = |0\rangle\langle 0|$. Therefore both the total current J(t) and partial currents $J_1(t)$ and $J_2(t)$ are positive for very short time $t > 0$. It means that initially energy is pumped from both reservoirs to the SQUID. Next, one can observe several distinct evolution of the energy currents: (A) a monotonically decreasing function of time (as in the case $\Phi_e = 0$ inFig 3) or non-monotonic approach to the stationary flow. In turn, the non-monotonic time evolution can exhibit: (B) damped oscillations (the case $\Phi_e = 0.25$ in Fig 3) or: (C) pulse-type evolution (Fig 4). In some regimes, the external magnetic flux Φ_e has no qualitative but only small quantitative influence on the amplitude of energy currents (as in Fig 3). Most interesting is the regime (B). By varying the magnetic field one can cause drastic changes in energy flow, as it is visualized in Fig 3. For three values of the external magnetic flux Φ_e = 0, 1/4 and 1/2, the transition form monotonic to oscillatory and next again to monotonic flow is detected. If next one increases Φ_e from the value 1/2 through 3/4 to 1, the reversed behavior is observed. Because$T_2 > T_1$ the energy flows from the reservoir 2 through the SQUID to the reservoir 1. In a generic case, the averaged energy E(t) of the SQUID should grow in time to a constant stationary value E. In the case presented in Fig 3, the total current J(t) takes both positive and negative values. It means that there are time intervals in which the energy E(t) increases and there are time intervals in which the energy E(t) decreases. The phenomenon is quasi-periodic in time. It is an evident signature of the quantum interference effects in energy transport through the SQUID.

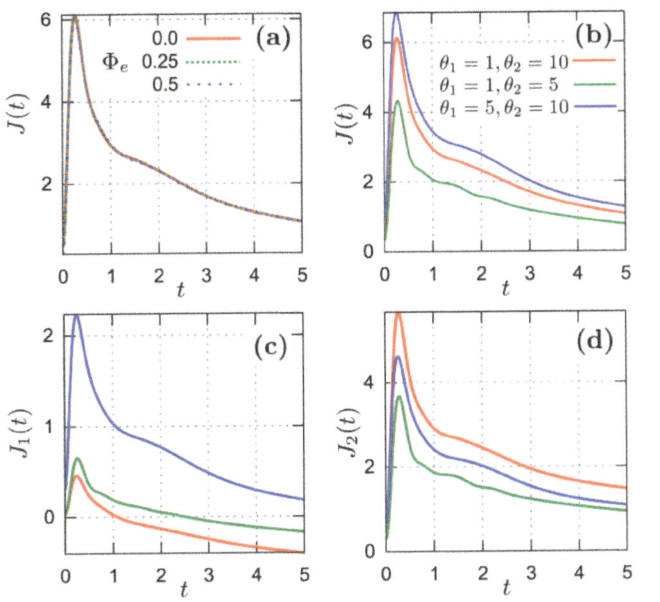

Figure 4. Pulse regime. (a) The total energy current J(t) = $J_1(t) + J_2(t)$ as a function of time for three values of the external magnetic flux Φ_e = 0, 0.25, 0.5. The reservoir rescaled temperatures are θ_1= 1, θ_2 = 10. (b) J(t) is shown for three sets of reservoirs temperatures $\{\theta_1, \theta_2\}$ = {1, 10}, {1, 5}, {5, 10}. (c) $J_1(t)$ and (d) $J_2(t)$ for the same set of temperatures as in panel (b). In panels (b)-(d) the flux Φ_e = 0. Remaining parameters are fixed as: ε_J = 4, ε_C^2 = 1.5 and$\gamma_1 = \gamma_2$ = 0.05. The dimensionless time t is rescaled by the oscillator frequency ω.

In Fig 4 we illustrate the pulse-type regime (C): at short time the energy current grows rapidly and after reaching its maximal value at some characteristic time it starts to decrease until a steady state is reached. In this case we find that the energy flow is robust against variation of the magnetic flux Φ_e. The magnitude of the pulse depends rather on both temperatures T_1 andT_2 and not on the temperature difference $T_1 - T_2$. Let us notice that in all three figures for J(t),$J_1(t)$ and $J_2(t)$, the sequence of colors is different. E.g. the maximal magnitude of the pulse for J(t) and $J_1(t)$ is noticed for the set $\{T_1 = 5, T_2 = 10\}$; for the current $J_2(t)$—for the set $\{T_1 = 1, T_2 = 10\}$. In turn, the minimal magnitude of the pulse for J(t) and $J_2(t)$ is observed for the set $\{T_1 = 1, T_2 = 5\}$ while for $J_1(t)$ it is for the set $\{T_1 = 1, T_2 = 10\}$.

Ending this section, we have to discuss the question of applicability of the Markovian master equation to time-dependent regimes. Starting from a giving initial state, for very short time its evolution is non-Markovian. It can be approximated by a Markovian process for time longer than the relaxation time of the system. We extrapolate results for short time but we are not able to prove that it is compatible with the description considerd in the paper. Only comparison with results obtained from the exact reduced dynamics could answer this question. However, the exact reduced dynamics is not known and the short-time regimes should be treated with limited trust.

SUMMARY

We studied the energy flow in a microscopic, fully quantum-mechanical dissipative model of an rf SQUID coupled to two heat baths. The results have implications beyond this system because the Hamiltonian Eq (3) describes several other quantum systems. In the stationary regime, the energy flow can be periodically modulated by Φ_e exemplifying the rf SQUID as a quantum energy interferometer. We have found three different time-dependent regimes of the energy current: monotonic decay, damped oscillations and pulse-type behavior. The first two regimes are sensitive to change of Φ_e. This feature allows to manipulate and control the energy flow by the external magnetic field. The total current can take both positive and negative values and in consequence there are time intervals in which the energy of the SQUID increases and there are time intervals in which the energy decreases. The phenomenon is quasi-periodic in time. It is an evident signature of the quantum interference effects in energy transport through the SQUID. The pulse regime is robust against variation of Φ_e. The magnitude and to some extent also the position of the pulse can be controlled by parameters of the model. Recent progress in novel experimental techniques makes the verification of our findings quite realistic. Nowadays, the system can be realized [16] and researchers can observe heat processes over time scales ranging between a few picoseconds and a few nanoseconds [30].

ACKNOWLEDGMENTS

This work was supported by the NCN (Poland) under grants UMO-2011/01/N/ST3/02473 (B. G) and Mobility Plus 1060/MOB/2013/0 (B. G), N202 052940 (J. Ł.) and DEC-2013/09/B/ST3/01659 (J. D.)

AUTHOR CONTRIBUTIONS

Conceived and designed the experiments: BG JL AP JD. Analyzed the data: BG JL AP JD. Contributed reagents/materials/analysis tools: BG JL AP JD. Wrote the paper: BG JL AP JD.

REFERENCES

1. Dubi Y, Di Ventra M (2011) Colloquium: Heat flow and thermoelectricity in atomic and molecular junctions. Rev Mod Phys 83: 131–155. doi: 10.1103/RevModPhys.83.131.

2. Li N, Ren J, Wang L, Zhang G, Hänggi P, et al. (2012) Colloquium: Phononics: Manipulating heat flow with electronic analogs and beyond. Rev Mod Phys 84: 1045–1066. doi: 10.1103/RevModPhys.84.1045.

3. Park JJ, Kim KH, Sagawa T, Kim SW (2013) Heat engine driven by purely quantum information. Phys Rev Lett 111: 230402. doi: 10.1103/PhysRevLett.111.230402. pmid:24476235

4. Chang CW, Okawa D, Majumdar A, Zettl A (2006) Solid-state thermal rectifier. Science 314: 1121–1124. doi: 10.1126/science.1132898. pmid:17110571

5. Li B, Wang L, Casati G (2004) Thermal diode: Rectification of heat flux. Phys Rev Lett 93: 184301. doi: 10.1103/PhysRevLett.93.184301. pmid:15525165

6. Saira OP, Meschke M, Giazotto F, Savin AM, Möttönen M, et al. (2007) Heat transistor: Demonstration of gate-controlled electronic refrigeration. Phys Rev Lett 99: 027203. doi: 10.1103/PhysRevLett.99.027203. pmid:17678252

7. Ben-Abdallah P, Biehs SA (2014) Near-field thermal transistor. Phys Rev Lett 112: 044301. doi: 10.1103/PhysRevLett.112.044301. pmid:24580455

8. Lee KH, McRae TG, Harris GI, Knittel J, Bowen WP (2010) Cooling and control of a cavity optoelectromechanical system. Phys Rev Lett 104: 123604. doi: 10.1103/PhysRevLett.104.123604. pmid:20366533

9. Wang L, Li B (2007) Thermal logic gates: Computation with phonons. Phys Rev Lett 99: 177208. doi: 10.1103/PhysRevLett.99.177208. pmid:17995368

10. Wang L, Li B (2008) Thermal memory: A storage of phononic information. Phys Rev Lett 101: 267203. doi: 10.1103/PhysRevLett.101.267203. pmid:19437667

11. Kafanov S, Kemppinen A, Pashkin YA, Meschke M, Tsai JS, et al. (2009) Single-electronic radio-frequency refrigerator. Phys Rev Lett 103: 120801. doi: 10.1103/PhysRevLett.103.120801. pmid:19792419

12. Maki K, Griffin A (1965) Entropy transport between two superconductors by electron tunneling. Phys Rev Lett 15: 921–923. doi: 10.1103/PhysRevLett.15.921.

13. Guttman GD, Nathanson B, Ben-Jacob E, Bergman DJ (1997) Phase-dependent thermal transport in josephson junctions. Phys Rev B 55: 3849–3855. doi: 10.1103/PhysRevB.55.3849.

14. Guttman GD, Nathanson B, Ben-Jacob E, Bergman DJ (1997) Thermoelectric and thermophase effects in josephson junctions. Phys Rev B 55: 12691–12700. doi: 10.1103/PhysRevB.55.12691.

15. Guttman GD, Ben-Jacob E, Bergman DJ (1998) Interference effect heat conductance in a josephson junction and its detection in an rf squid. Phys Rev B 57: 2717–2719. doi: 10.1103/PhysRevB.57.2717.

16. Giazotto F, Martínez-Pérez MJ (2012) The josephson heat interferometer. Nature 492: 401. doi: 10.1038/nature11702. pmid:23257882

17. Golubev D, Faivre T, Pekola JP (2013) Heat transport through a josephson junction. Phys Rev B 87: 094522. doi: 10.1103/PhysRevB.87.094522.

18. Barone A, Paterno G (1982) Physics and applications of the Josephson effect. Wiley, New York.

19. Makhlin Y, Schön G, Shnirman A (2001) Quantum-state engineering with josephson-junction devices. Rev Mod Phys 73: 357–400. doi: 10.1103/RevModPhys.73.357.

20. Caldeira A, Leggett A (1983) Path integral approach to quantum brownian motion. Physica A 121: 587–616. doi: 10.1016/0378-4371(83)90013-4.

21. Everitt MJ, Clark TD, Stiffell PB, Vourdas A, Ralph JF, et al. (2004) Superconducting analogs of quantum optical phenomena: Macroscopic quantum superpositions and squeezing in a superconducting quantum-interference device ring. Phys Rev A 69: 043804. doi: 10.1103/PhysRevA.69.043804.

22. Armour AD, Kubala B, Ankerhold J (2015) Josephson photonics with a two-mode superconducting circuit. Phys. Rev. B 91: 184508

23. Leggio B, Bellomo B, Antezza M (2015) Quantum thermal machines with single nonequilibrium environments. Phys. Rev. A 91: 012117 doi: 10.1103/PhysRevA.91.012117.

24. Elouard C, Richard M, and Aufféves A (2015) Reversible work extraction in a hybrid opto-mechanical system. New. J. Phys. 17: 055018 doi: 10.1088/1367-2630/17/5/055018.

25. Dambach S, Kubala B, Gramich V, Ankerhold J (2015) Time-resolved statistics of nonclassical light in Josephson photonics. Phys. Rev. B 92: 054508 doi: 10.1103/PhysRevB.92.054508.

26. Breuer HP, Petruccione F (2002) The Theory of Open Quantum Systems. Oxford University Press, Oxford.

27. Alicki R (1979) The quantum open system as a model of the heat engine. J Phys A: Math Gen 12: L103. doi: 10.1088/0305-4470/12/5/007.

28. Davies E (1974) Markovian master equations. Comm Math Phys 39: 91–110. doi: 10.1007/BF01608389.

29. Butcher JC (2003) Numerical Methods for Ordinary Differential Equations. John Wiley.

30. Volz S (2007) Topics in Applied Physics 107. Springer, Berlin.

CHAPTER 17

PROCESS EFFICIENCY IN POLYMER EXTRUSION: CORRELATION BETWEEN THE ENERGY DEMAND AND MELT THERMAL STABILITY

Chamil Abeykoon, Adrian L. Kelly, Javier Vera-Sorroche, Elaine C. Brown, Phil D. Coates, Jing Deng, Kang Li, Eileen Harkin-Jones, Mark Price

Keywords: Polymer extrusion; Process monitoring; Melt viscosity; Energy demand; Thermal stability; Energy efficiency

INTRODUCTION

Polymeric materials are widely used all over the world primarily due to their superior properties such as high strength to weight ratio; high temperature/chemical/corrosive resistance; non-conductivity; high clarity; re-processability; low cost and so forth. More importantly, polymeric materials are showing a great potential of saving energy consumption in aerospace, automotive, marine and transport sector. Moreover, they are quite easy to form into complex shapes compared to other conventional materials. Also, the energy requirement for processing of polymers is considerably lower than other conventional materials such as steel and glass. Perhaps, these may be among the major reasons for growing popularity of polymeric materials in diverse industrial sectors. Although the processing of polymers demands a less amount of energy compared to other materials [1], many polymer processes operate at poor energy efficiency. Usually, the specific energy consumption (SEC) in polymer extrusion reduces as processing speed increases [2] and [3]. However, the thermal fluctuations of the melt flow are increased as the process speed is increased [4], [5], [6], [7], [8] and [9]. Therefore, the majority of polymer processes are run at conservative rates to avoid thermal fluctuations at higher processing speeds. Currently, the polymer sector is under pressure to cut down excessive use of energy due to the gradual increase of energy prices in the world over the last few decades [10].

Usually, polymer extrusion is an unpredictable process and hence it is highly prone to fluctuations in nature. Moreover, the process parameters are complexly coupled each to other and hence difficult to set-up and control [7]. Therefore, the typical relationship between process thermal stability and energy efficiency may differ depending on the processing conditions; material and machine being used while the quality of the process monitoring and control also may have considerable effects. Among the polymer processing extruders, single screw continuous extruders are the most commonly used in industry [11] and that is mainly due to their low purchase and maintenance costs, simplicity of operation and the ability to generate the required pressure [12]. Therefore, this study is focused on single screw extrusion processes. More details on the process mechanisms, operational requirements, and the available process analytical techniques (PAT) of polymer extrusion can be found in the literature [13], [14], [15] and [16].

Efficiency of an Extruder

An illustration of an extruder based on its energy conservation can be illustrated as shown in Figure 1.

Citation: Chamil Abeykoon, Adrian L. Kelly, Javier Vera-Sorroche, Elaine C. Brown, Phil D. Coates, Jing Deng, Kang Li, Eileen Harkin Jones, Mark Price, Process efficiency in polymer extrusion: Correlation between the energy demand and melt thermal stability, Applied Energy, Volume 135, 15 December 2014, Pages 560-571, ISSN 0306-2619, http://dx.doi.org/10.1016/j.apenergy.2014.08.086.

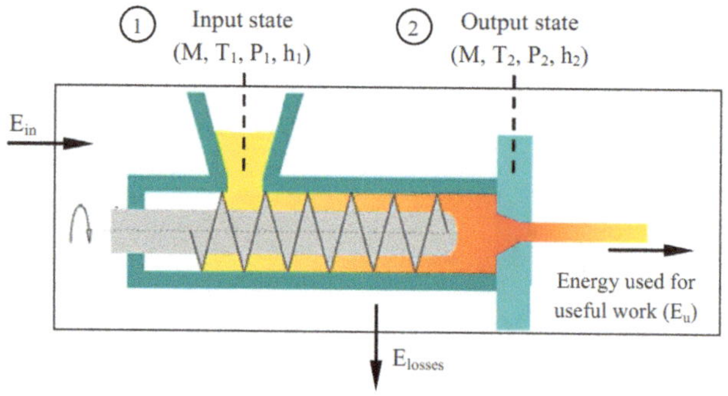

Figure 1. An illustration of an extruder based on its energy.

Then the energy used for useful work (E_u) from an extruder (i.e. the energy used for material melting and forming through the die) can be given as:

$$E_u = E_{in} - E_{losses} \tag{1}$$

where E*in* is the total energy supplied to the extruder and E*losses* is the total amount of energy wasted without involving in any useful work. Therefore, the extruder energy efficiency (η*extruder*) is given by:

$$\eta_{extruder} = \frac{E_{in} - E_{losses}}{E_{in}} \times 100\% \tag{2}$$

Here, the energy inputs (E*in*) are from the electrical energy supplied to the devices such as drive motor, motor cooling fan, barrel/die haters, barrel cooling fans, instruments in the control panel, and water pump. The energy losses which may occur in these devices and other mechanical/functional systems such as transmission, forced/natural cooling come under E *losses*. Of the energy consuming devices, drive motor and barrel/die heaters are likely to consume more than 90% of the total energy supply while these are also responsible for the highest energy losses. In extrusion there is a little potential of useful recovery of rejected energy as these are largely released to air or water. Drury [17] argues that over 40% from the energy input to the small scale extruder is wasted through drive/transmission losses, radiation, convection, conduction losses, etc.

The thermodynamic efficiency of an extruder can be determined by comparing the actual energy consumed by the extruder to the theoretical energy required to transform the polymer from initial (input) stage to the desired/output stage. Usually, the thermodynamic efficiency is calculated under the following assumptions:

- The extruder is under steady state operation.
- The process operates in uniform temperature, pressure and mass flow rate.
- The feed material entering to the extruder is assumed to be under uniform temperature and pressure.

The theoretical energy required for melting and forming (E_T) of material in polymer extrusion [18] is:

For semi-crystalline (SC) materials:

$$E_{T,SC} = \dot{M}\int_{T_1}^{T_2} C_p \times dT + \dot{M} \times H_f + \dot{M}\int_{P_1}^{P_2} \left(\frac{\partial h}{\partial P}\right)_T \times dP \tag{3}$$

For amorphous (AM) materials:

$$E_{T,AM} = \dot{M}\int_{T_1}^{T_2} C_p \times dT + \dot{M}\int_{P_1}^{P_2} \left(\frac{\partial h}{\partial P}\right)_T \times dP \tag{4}$$

where M is the mass flow rate, T_1 and T_2 are the feedstock temperatures at the input and output states respectively, P_1 and P_2 are the pressures at the input and output states respectively, C_p is the specific heat capacity of the material, H_f is the enthalpy of heat of fusion of the materials (this is zero for amorphous material), and ρ is the material density. The first and second terms of the right side of Eq. (3) are related to the energy required for melt preparation. The third term is the energy required for forming the material through the die and this is usually considered to be less than 5% of the E_T.

According to the first law of thermodynamics, the minimum energy required to go from state 1 to state 2 (see Figure 1) can be calculated from the enthalpy changes (Δh) occur in melting (assumes that these occur at constant pressure) and forming (assumes that this occur at constant temperature) operations:

$$\Delta h = (h_2 - h_1) = \Delta h_1 + \Delta h_2 \tag{5}$$

$$\Delta h_1 = \int_{T_1}^{T_2} C_p \times dT + H_f = \bar{C}_p \times (T_2 - T_1) + H_f \tag{6}$$

$$\Delta h_2 = \int_{P_1}^{P_2} \left(\frac{\partial h}{\partial P}\right)_T \times dP = \frac{(P_2 - P_1)}{\bar{\rho}} \tag{7}$$

where h_1 and h_2 are the specific enthalpy values of the materials at the input and output states respectively, Δh_1 and Δh_2 are the enthalpy changes occur at melting and forming operations respectively, $\bar{C}_p$ is the average specific heat capacity of the polymer and $\bar{\rho}$ is the average density of the material. Then, the thermal efficiency ($\eta_{extruder,thermo}$) of an extruder is given by:

$$\eta_{extruder,thermo} = \frac{\dot{M} \times \Delta h}{E_{in}} \times 100\% \tag{8}$$

The above details describe the energy efficiency of an extruder. However, the process should be efficient in both energy and thermal terms to ensure high quality products at a low production cost. A good thermal efficiency means that the process melt output should be uniform in temperature (i.e. the temperature uniformity of the melt over the time and across the melt flow cross-section) throughout the process operation. Moreover, the level of the melt temperature should be at the desired level and the material should have achieved the complete melting when it reaches the die. Therefore, it is better to define the overall efficiency of an extruder as a combination of both energy and thermal efficiencies.

Previous Studies on Thermal Stability and Energy Consumption in Polymer Extrusion

Since the 1950s, numerous previous studies can be found in the literature on melting/thermal issues in polymer extrusion. However, it seems that not much attention has been paid to studies of energy related issues until the 1990s. Traditionally, polymeric materials based manufactures primarily concerned about the thermal quality of the melt which is the key to from high quality products but they did not pay much attention to process energy efficiency. As world energy prices have been gradually increasing over the last few decades, manufacturing companies start to search for methods to cut down their energy bills by making their production lines energy efficient. However, it is challenging to achieve both the required melt quality and the energy efficiency at the same time in industrial applications.

Process cooling (e.g. screw cooling, barrel cooling) is a common operation in extrusion to maintain the thermal stability of the process. Also, some amount of the process heat is naturally released to the surroundings. Therefore, a considerable portion of the supplied energy to the process is taken away by the cooling water and air [17]. In fact, it is clear that a considerable amount of the supplied energy has to be sacrificed to maintain the process thermal stability. Obviously, the process energy and thermal efficiencies are likely to be inter-related. However, there is no point of making the process more energy efficient if the melt output is not of the required thermal quality. Therefore, thermal stability and energy efficiency should be achieved consecutively. Here, the level of melt temperature fluctuations can be taken as a measure of the process thermal stability. Rasid and Wood[19] observed the die melt temperature profile and effects of barrel set temperature on extruder power consumption. They found that the metering zone set temperature has the greatest influence on the level of melt temperature while the solids conveying zone set temperature has the greatest influence on the level of extruder power consumption. Moreover, they attempted to explore the effects of melt pressure on melt temperature profile however, no significant effect was observed. No attempt was made to investigate correlation/s between extruder power and melt temperature or melting stability.

Previous work reported by Sorroche et al. [20] measured extruder energy consumption and melt temperature dynamics in parallel. They found that the extruder specific energy consumption reduced as screw speed increased despite changes to process settings and screw geometry. Melt temperature measured across an extruder output melt flow showed that the magnitude of temperature fluctuations increased with the screw speed. Such an increasing trend of thermal fluctuations may arise due to the fact that the residence time of the material reduces with the increasing speed and hence there will be less time for melting and mixing of the material.

Previous work by the author [4], [5], [6] and [7] found that the temperature across the melt flow significantly varied over the processing conditions. The magnitude of the thermal fluctuations were found to increase with screw speed. Barrel set temperature also showed some influences on melt thermal variations but to a lesser extent. Moreover, process thermal fluctuations differed depending on the screw geometry. In general, the thermal fluctuations were found to be slow in nature and these were below 0.5 Hz over the different processing situations tested (i.e. from 10 to 90 rpm over 3 different barrel set temperature conditions).

Other work by the authors [4] and [21] reported an attempt to predict the process thermal stability inferentially. The correlations between the screw load torque, melt pressure and melt temperature fluctuations were examined by analyzing experimentally measured signals. However, no strong correlations between these signals could be observed. It was found that as the screw load torque signal is dominated by the solids conveying torque, it was not sensitive enough to identify unstable melting issues. Pressure fluctuations had slight correlations with melt temperature fluctuations particularly at low screw speeds. However, none of these signals showed sufficiently good performance for them to be used as a tool to monitor the process thermal stability inferentially.

According to the authors' knowledge, no previous work has attempted to correlate process thermal stability and energy consumption to explore possible ways of optimizing both of these parameters in parallel. Increasing both energy and thermal efficiencies may be achieved by improving the machine design, particularly by modifying screws to have better conveying, melting and mixing of the material. Also, direct drive extruders, improved heater/barrel design for reduced heat losses to the surroundings, and advanced vector control alternating current (AC) drives are some of the current topics in the extrusion field in terms of achieving an improved overall process efficiency. Ideally, for a given machine and material, thermal and energy efficiencies should be achieved through proper selection of process settings. Prior knowledge on the relationship/s between thermal and energy related parameters would therefore be invaluable.

In this work, both the extruder energy consumption and melt temperature (i.e. at the number of different radial positions across the melt flow) were observed at the same time over a number of different processing conditions. Then, these signals were used to identify any possible correlation/s between these parameters. The major aim was to identify possible ways to achieve optimal energy and thermal efficiencies at high throughput rates. A single screw extruder was used in the experiments as it is the most commonly used type in industrial polymer processing applications. Three different screw geometries and set temperature conditions were used for collecting the experimental data with single polymer material.

EXPERIMENTAL

All measurements were carried out on a highly instrumented 63.5 mm diameter (D) single screw extruder (Davis Standard BC-60) at the IRC laboratories of the University of Bradford. A gradual compression screw with 3:1 compression ratio; a tapered rapid compression screw with 3:1 compression ratio; and a barrier flighted screw with a spiral Maddock mixer and a 2.5:1 compression ratio were used to process the materials. Geometrical specifications of these screws are shown in Figure 2.

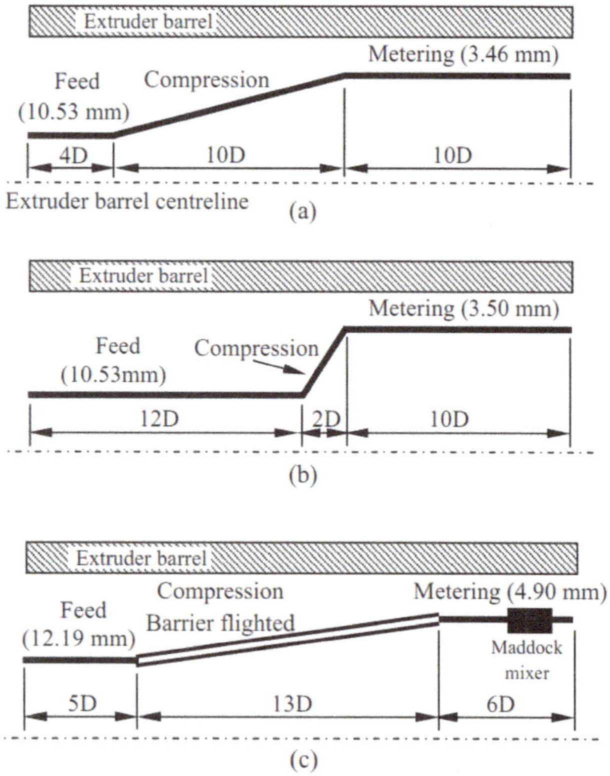

Figure 2. Details of the screws used in experiments.

From here onwards, these three screws are refereed as the GC screw, RC screw and BF screw, respectively. The extruder was fitted with a 38 mm diameter adapter by using a clamp ring prior to a short 6 mm diameter capillary die as shown in Figure 3.

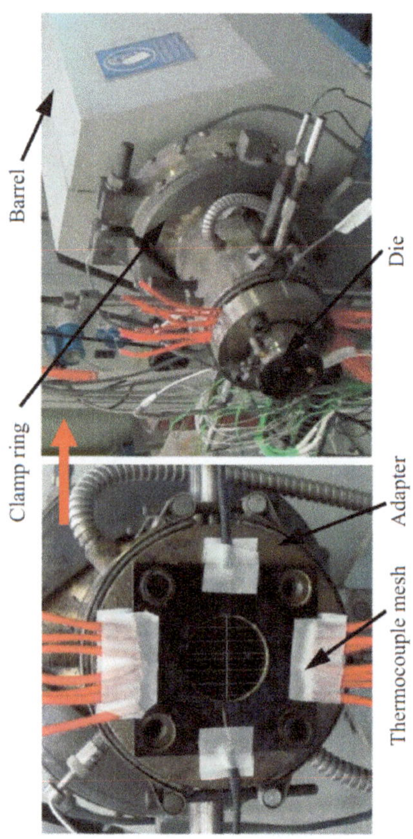

Figure 3. Arrangement and dimensions of the apparatus.

The extruder barrel has four separate temperature zones (each with a heater of 4 kW) and another three separate temperature zones at the clamp ring (with a heater of 0.9 kW), adapter (with a heater of 1.4 kW) and die (with a heater of 0.2 kW). All of these temperature zones are equipped with Davis Standard Dual-Therm controllers and the set temperature can be controlled individually. The extruder drive is a horizontal type separately excited direct current (SEDC) motor which has ratings: 460 Vdc, 50.0 hp (30.5 kW), at speed 1600 rpm. The motor and screw are connected through a fixed gearbox with a ratio of 13.6:1, and according to the manufacturers' information the gearbox efficiency is relatively constant at all speeds (~96%). The motor speed is controlled by a speed controller (MENTOR II) based on speed feedback obtained through a direct current (d.c.) tachometer generator. Moreover, cooling of the feed throat of the extruder is achieved via a water supply with a constant flow rate and also forced air cooling blowers have mounted along the extruder barrel.

Melt pressure at the adapter was recorded by a Dynisco TPT463E pressure transducer. The total extruder power and motor power were measured using a Hioki 3-phase power meter and an Acuvim IIE 3-phase power meter, respectively. Melt temperatures of the different radial locations of the melt flow at the end of the adapter (denoted as die melt temperatures throughout this paper) were measured using a thermocouple mesh placed in-between the adapter and the die as shown in Figure 4. As it was previously confirmed by Kelly et al. [22] and [23], the die melt temperature measurements are symmetrical across the thermocouple mesh centerline when averaged over a significantly long period of time. Therefore, seven thermocouple junctions (distance from the melt flow centerline to each radial position (rp): Trp-0: 0 mm, Trp-2.5: 2.5 mm, Trp-5: 5 mm, Trp-8: 8 mm, Trp-11: 11 mm, Trp-14: 14 mm, Trp-16.5: 16.5 mm) were placed asymmetrically across the melt flow along the diameter of the mesh as shown in Figure 4, and this asymmetric placement of wires gave the opportunity to increase the number of effective temperature measurements across the melt flow. Moreover, the temperature of the melt close to the die wall was measured from a flush mounted insulated wall thermocouple and this measurement was considered as the melt temperature of the 19 mm die radial position (Trp-19). Additionally, an infrared (IR) temperature sensor (Dynisco MTX 922-6/24) was used to make bulk melt temperature measurements (T-IR) of melt near the screw tip.

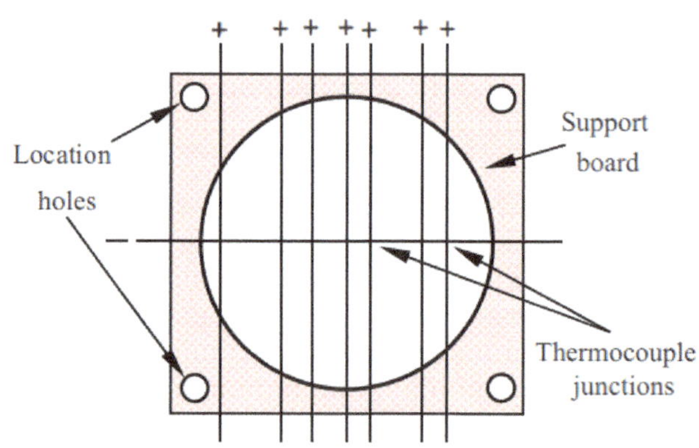

Figure 4. The thermocouple mesh arrangement.

A data acquisition programme developed in LabVIEW was used to communicate between the experimental instruments and a PC. All signals (screw speed, melt temperature, melt pressure, power) were acquired at 10 Hz using a 16-bit DAQ card, National Instruments (NI) PCI-6035E, through a NI TC-2095 thermocouple connector box and a NI low-noise SCXI-1000 connector box.

Materials and Experimental Conditions

Experimental trials were carried out on a virgin Polystyrene (PS), Styrolution PS 124N (density: 1.040 g/cm^3 and MVR: 12 cm^3/10 min). The volume melt-flow rate (MVR) and density values are presented according to the ISO 1133 (@ 200 °C, 5 kg) and ISO 1183 standards, respectively. The extruder temperature settings were fixed under three different set conditions denoted as A (low temperature), B (medium temperature) and C (high temperature) as described in Table 1.

Table 1. Extruder barrel temperature settings

Temperature settings	**Set temperatures (°C)**						
	Barrel zones				Clamp ring	Adapter	Die
	1	2	3	4			
A	130	155	165	180	180	180	180
B	140	170	185	200	200	200	200
C	150	185	200	220	220	220	220

Nine different experimental trials were carried out with the three screw geometries and three set temperature conditions, and the data was collected at 0 rpm for a small time period. Then, the screw speed was adjusted from 10 rpm to 90 rpm in steps of 20 rpm. All the data was recorded continuously whilst the extruder was allowed to stabilise at each screw speed.

RESULTS AND DISCUSSION

Rheological Analysis of the Materiel (Polystyrene)

A twin bore capillary rheometer (Rosand RH10, Malvern Instruments, UK) was used to measure the rheology of Polystyrene at extrusion process rates and temperatures. The experiments were carried out over a shear rate range of 10–1000 s^{-1} using capillary dies of 2.0 mm diameter with an entrance angle of 180°; a long die of length 32.0 mm and an orifice die with land length close to 0. Shear viscosity was calculated from measured wall shear stress at a range of shear strain rates. The Bagley ends correction [24] was applied to the results to correct for pressure loss due to convergence into the die entrance. Also the Rabinowitsch correction [25] was used for representing the pseudoplastic nature of the polymer melt. Tests were performed at three set temperatures (i.e. equivalent to the extruder set temperatures shown in Table 1):180 °C, 200 °C and 220 °C. Measured shear viscosity at each of the three set temperatures is shown in Figure 5. The measured rheology followed temperature dependent shear thinning behaviour typical of thermoplastic polymers. The Carreau model [26] and [27] (which is given by Eq. (9)) was fitted to the measured rheological data at each of the set temperatures, as shown in Figure 5. Figure legends are in the format of EXP/PRE-set temperature condition and the terms EXP and PRE are used to denote experimental and model predicted conditions, respectively.

$$\mu(\gamma, T) = \frac{\mu_0 \times f(T)}{[1 + (\lambda \times \gamma \times f(T))^a]^{\frac{n-1}{a}}} \quad (9)$$

where μ is the viscosity, μ_0 is the viscosity at zero shear rate, γ is the shear rate, λ is the relaxation time, *a* is the shift factor, f(T) is the temperature function and *n* is the power law index. This model describes the shear rate dependent behaviour and is able to fit both the linear power law section of the data and predict the transformation to a Newtonian plateau at lower strain rates. Temperature dependence of the measured viscosity was modelled using the Williams–Landel–Ferry (WLF) equation [28] and [29]which is an empirical equation associated with time-temperature superposition and this is given in Eq. (10).

$$f(T) = \frac{C_1 \times (T_r - T_s)}{(C_2 + T_r - T_s)} - \frac{C_1 \times (T - T_s)}{(C_2 + T - T_s)} \quad (10)$$

where T_r is a reference temperature chosen to construct the compliance master curve and C_1, C_2, T_s are empirical constants adjusted to fit the values of the superposition parameter.

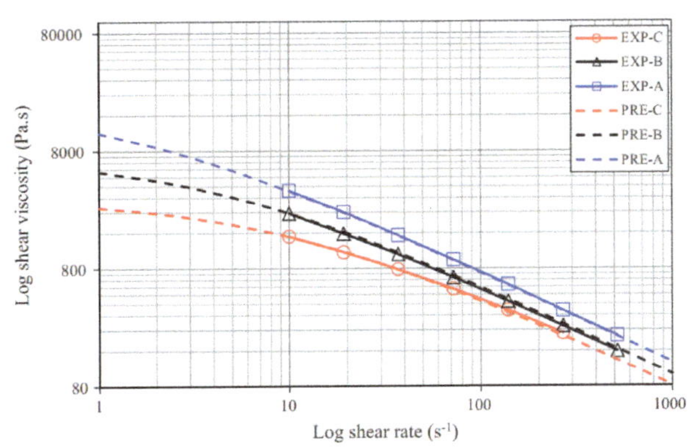

Figure 5. Measured and predicted shear viscosity of polystyrene at set temperature conditions A, B, and C.

The purpose of modelling the measured rheological data was to enable calculation of viscosity in the channel of the extruder screw metering section at each of the experimental conditions (screw rotation speed and melt temperature). Shear strain rate at each condition was calculated from the screw rotation speed and channel depth; bulk melt temperature was determined from measurements made in the extruder die adaptor using the thermocouple mesh. Using the above models, shear viscosity at each extrusion condition was calculated, allowing specific energy consumption to be plotted against viscosity for each screw geometry and set temperature, as shown in Figure 6.

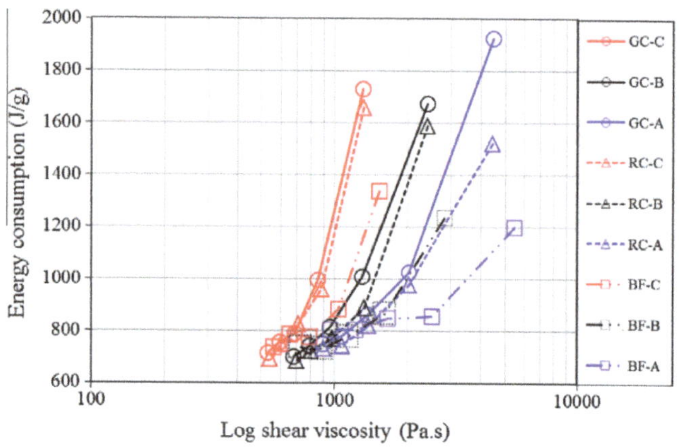

Figure 6. Extruder specific energy consumption vs shear viscosity for Polystyrene.

From Figure 6, it can be seen that specific energy consumption (SEC) ranged from 700 to over 1900 J g^{-1}. The effect of set temperature on melt viscosity was clearly observed, in general lower SEC was observed for lower melt viscosity. However there was not a direct correlation between viscosity and SEC, highlighting the complex nature of melting in polymer extrusion. Highest energy consumption was measured at low extruder screw rotation speeds; here melting is achieved primarily by conduction from the screw and barrel walls. At higher screw speeds, specific energy consumption converged to lower values between 700 and 800 J g^{-1}, due to increased levels of more efficient heating from viscous shearing. The effect of extruder geometry was also apparent, in particular the barrier flighted screw (BF) required lower specific energy than the two single flighted screws (GC and RC) at comparable viscosities. This demonstrates the more efficient melting of the barrier flighted screw for this polymer, which also incorporated a spiral mixer.

Study of the Signal Dynamics and Correlation Coefficients

The experimentally measured signals were observed for any possible correlations of their dynamics. All the signals were normalized to make it easier to observe any possible similarities of their dynamic behavior. The normalized signals of the total extruder power, motor power, melt pressure, T-IR, Trp-0, Trp-14 and Trp-16.5 are shown in Figure 7,Figure 8 and Figure 9 for BF, GC and RC screws (only at set temperature condition A), respectively. From here onwards, T-IR refers to the temperature measured by the IR temperature sensor and Trp refers to the temperature of the radial position, where numbers followed by Trp are in millimeters.

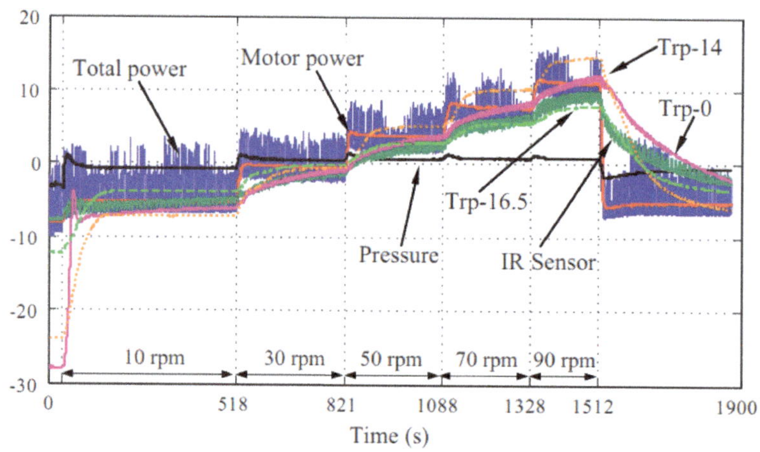

Figure 7. The normalized process signals measured with the BF screw at set temperature condition A.

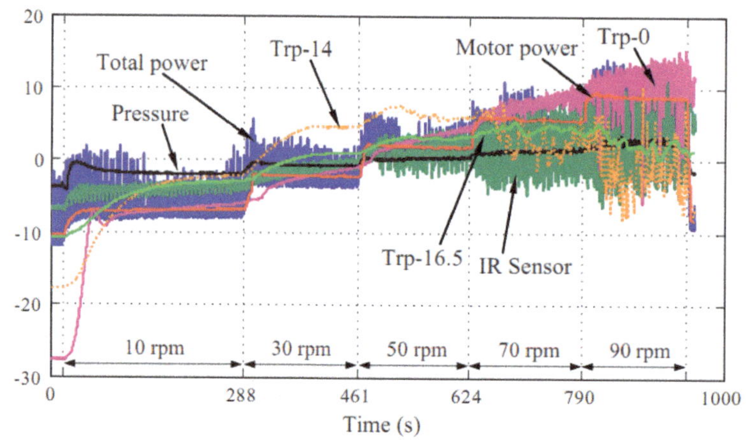

Figure 8. The normalized process signals measured with the GC screw at set temperature condition A.

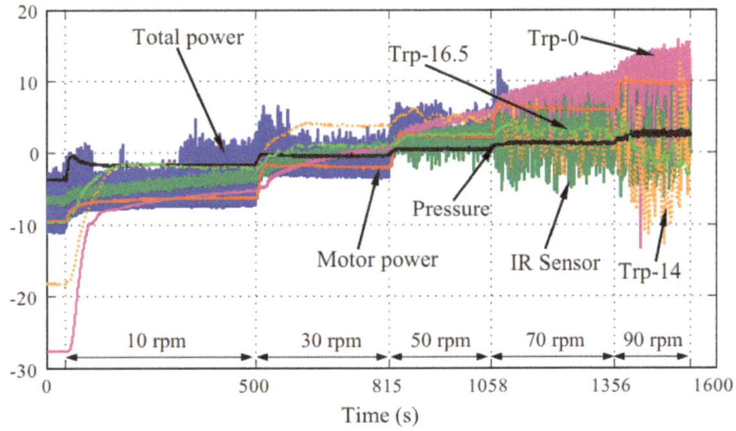

Figure 9. The normalized process signals measured with the RC screw at set temperature condition A.

Table 2. Correlation coefficients between each of the individual signals for the BF screw at set temperature condition A

	T-Power	**M-Power**	**T-IR**	**Pressure**	**Trp-0**	**Trp-14**	**Trp-16.5**
T-Power	1.000						
M-Power	**0.949**	1.000					
T-IR	**0.812**	**0.850**	1.00				
Pressure	0.773	**0.817**	0.569	1.000			
Trp-0	0.670	0.702	**0.902**	0.558	1.000		
Trp-14	**0.800**	**0.846**	**0.944**	0.639	**0.945**	1.000	
Trp-16.5	0.799	**0.846**	**0.944**	0.636	**0.936**	**0.999**	1.000

Table 3. Correlation coefficients between each of the individual signals for the GC screw at set temperature condition A

	T-Power	**M-Power**	**T-IR**	**Pressure**	**Trp-0**	**Trp-14**	**Trp-16.5**
T-Power	1.000						
M-Power	**0.962**	1.000					
T-IR	0.732	0.771	1.00				
Pressure	**0.942**	**0.973**	0.762	1.000			
Trp-0	**0.842**	**0.894**	0.761	**0.873**	1.000		
Trp-14	0.496	0.540	0.453	0.446	0.650	1.000	
Trp-16.5	0.797	**0.848**	0.708	0.777	**0.902**	**0.845**	1.000

Table 4. Correlation coefficients between each of the individual signals for the RC screw at set temperature condition A

	T-Power	**M-Power**	**T-IR**	**Pressure**	**Trp-0**	**Trp-14**	**Trp-16.5**
T-Power	1.000						
M-Power	**0.963**	1.000					
T-IR	**0.833**	**0.867**	1.00				
Pressure	**0.944**	**0.974**	**0.852**	1.000			
Trp-0	**0.862**	**0.904**	**0.818**	**0.900**	1.000		
Trp-14	0.379	0.418	0.431	0.429	0.663	1.000	
Trp-16.5	0.641	0.690	0.633	0.668	**0.850**	**0.837**	1.000

As it is evident, the total power signals vary significantly over time due to the operation of the barrel heaters and cooling fans with on–off action. Otherwise, all the signals show a quite similar trend of dynamic changes with step changes in screw speed.

Correlation coefficients between each signal (under set temperature condition A) were calculated and shown in Table 2, Table

3 and Table 4 for BF, GC and RC screws, respectively. The correlation coefficient (CC) matrix is symmetric and hence only a half of the matrix is shown in the tables. Correlation coefficients higher than 0.8 are shown in bold. Normally, correlation coefficient represents the normalized measure of the strengths and directions of the linear relationship between two variables and this ranges from −1 to 1, where positive values indicate that variables are positively correlated (i.e. variables vary in the same direction) and the strength of the positive correlation increases from 0 to 1. Negative values specify that variables are negatively correlated (i.e. variables vary in the opposite directions) while the strength of the negative correlation between signals increases from 0 to −1. Values close to or equal to 0 suggest there is no linear relationship between the variables.

As was expected, total extruder power and motor power signals show strong correlations with all the screws used. Correlations between the total power and melt pressure are also strong for GC and RC screws (i.e. CC is around 0.94) while it is less strong with the BF screw as well (i.e. CC is around 0.77). Correlations between motor power and melt pressure are stronger than the pressure-total power correlations for all the screws. Correlations between power and melt temperature signals differed depending on the screw geometry. Of the four melt temperature signals, T-IR, Trp-0 and Trp-0 show the highest correlations with the total extruder power or motor power for BF, GC, and RF screws, respectively. Total and motor power signals relating to the GC and RC screws show good correlations with melt temperature signal measured at the middle of the flow (i.e. Trp-0) while these signals relating to the BF screw indicate good correlations with the melt temperature measurements close to the die wall than the middle of the flow.

Comparison of the Power and Melt Temperature Signals

Data collected over the last minute at each screw speed (i.e. after process became stable) were used for plotting the graphs presented in this section and most of the sub-graphs are plotted in the same scale for the ease of comparison.

The average values of the total extruder power (TP), fluctuations of the total extruder power (ΔTP), motor power (MP), fluctuations of the motor power (ΔMP) and the difference between the total power and the motor power (TP–MP) at each screw speed are shown in Figure 10.

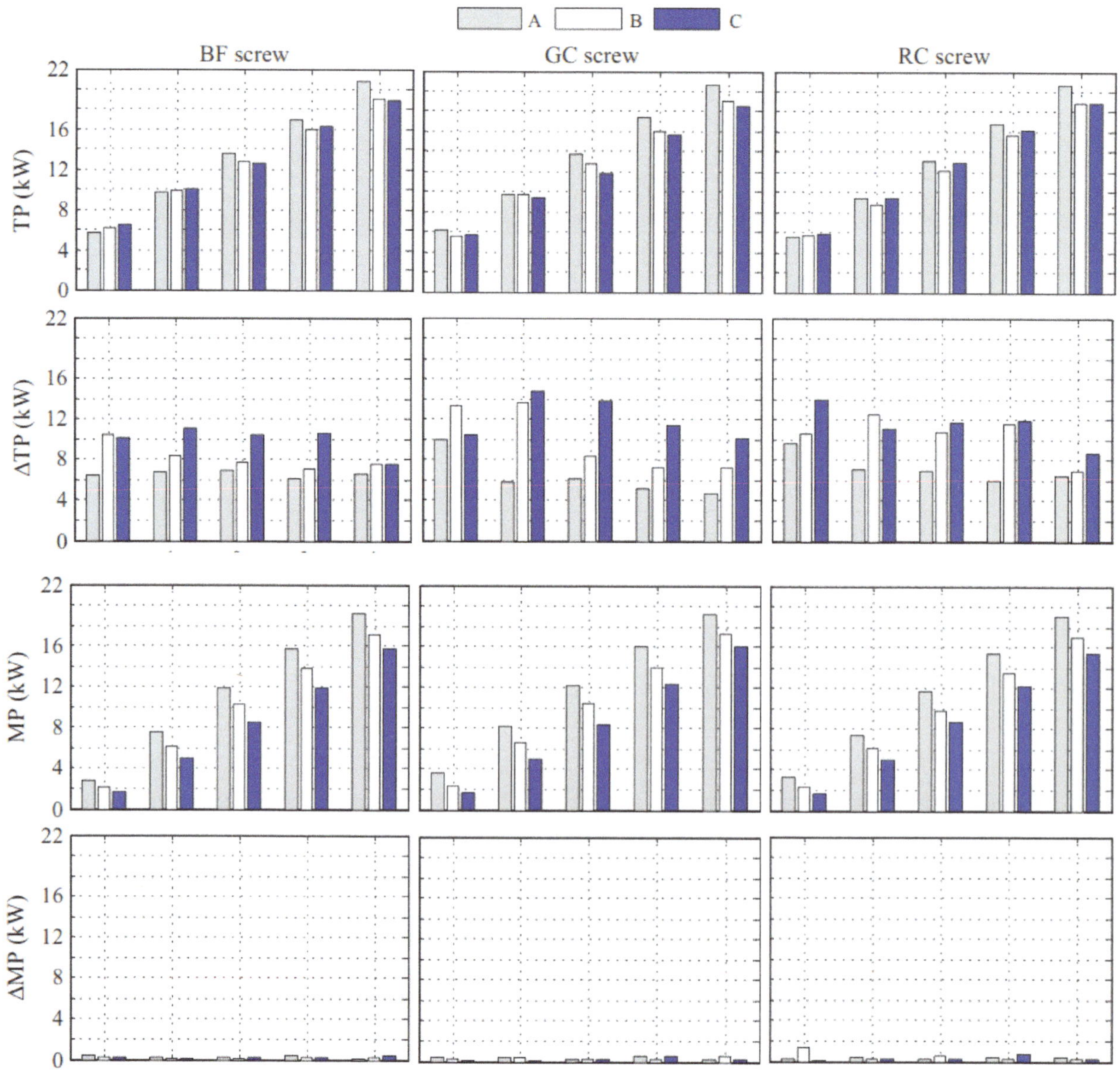

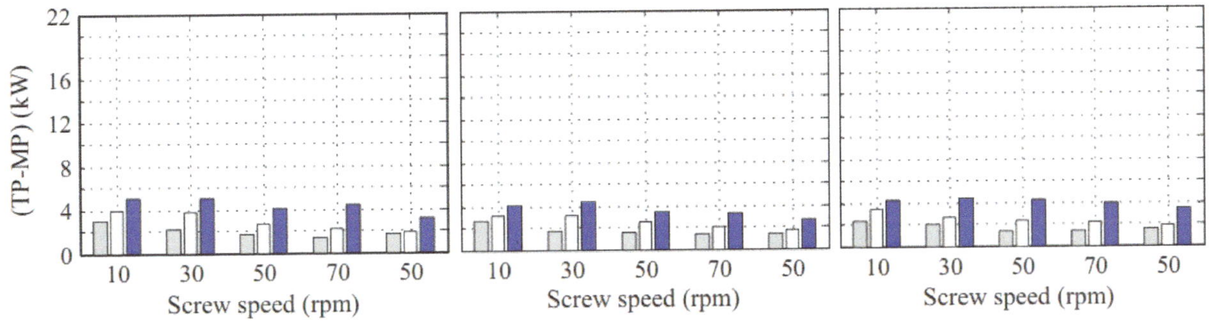

Figure 10. Details of the process energy demand over different processing conditions and screws.

As it is evident, the total extruder power increased with the screw speed quite linearly for all the screws while the power fluctuations increased with increase in barrel set temperature regardless the screw geometry. However, these is no clear link between the total power fluctuations and the screw speed. The motor power also increased linearly with screw speed for all screws but motor power fluctuations are significantly lower compared to the total power variations. Therefore, the majority of variations induced into the total power signal may be due to the on–off action of the barrel heaters and cooling fans of the extruder.

Usually, the difference between the total and motor power signals is the power consumed by the process heating/cooling system, control electronics and other auxiliary equipment. In this study, no auxiliary equipment was employed and hence these values show the power consumed by the process heating/cooling system only (as the typical power consumption of the control electronics is very low at approx. 0.35 kW). Screw speeds corresponding to the highest to lowest value of TP–MP at each set temperature condition for different screws are shown in Table 5.

Table 5. The screw speeds corresponding to the highest to lowest value of TP–MP at each set temperature condition

Screw	**Corresponding screw speeds in descending order of TP–MP**														
	A					B					C				
BF	10	30	50	90	70	10	30	50	70	90	30	10	70	50	90
GC	10	30	50	90	70	10	30	50	70	90	30	10	50	70	90
RC	10	30	90	50	70	10	30	50	70	90	30	50	10	70	90

As it is evident, the heating/cooling system has demanded the highest power at 10 rpm under the set temperature conditions A and B where it is at the 30 rpm for set temperature condition C. Conversely, power demand of the heating/cooling system has gradually reduced in the order of 10–90 rpm at set temperature condition B regardless the screw geometry but this is not true at conditions A and C. Such complex differences of power demand may occur due to the variations of material viscosity, amount of mechanical heat generation, frictional properties of the materials, etc., which may happen with the barrel set temperature changes. An experiment carried out by Cantor [30] with three different materials and screw geometries observed that the difference between the total extruder power and motor power reduces with the screw speed with all the materials and screw geometries tested. However, this experiment was conducted with only a single barrel set temperature condition.

As shown in Figure 10, the difference between the total and motor power signals increased with the increase of the barrel set temperature with all the screws. Actually, this increase of power demand by the heating/cooling system is quite obvious with the increase of barrel set temperatures. Conversely, it is well-known that the heat generated by the process mechanical work increases with the screw speed. Therefore, it can be argued that the amount of heat input by the heating system should reduce as screw speed increased as the heat generated by the screw mechanical work increases, to maintain the process thermal stability. However, experimental results show that the power demand of the heating/cooling system is not always reduced with the increasing screw speed and depend on the level of set temperature and screw geometry as shown in Table 5. This highlights the complex nature of process energy consumption in polymer extrusion.

The mass throughput, specific energy consumption of the extruder (SEC-Extruder) and the specific energy consumption of the motor (SEC-Motor) at each speed (i.e. average values) are shown in Figure 11.

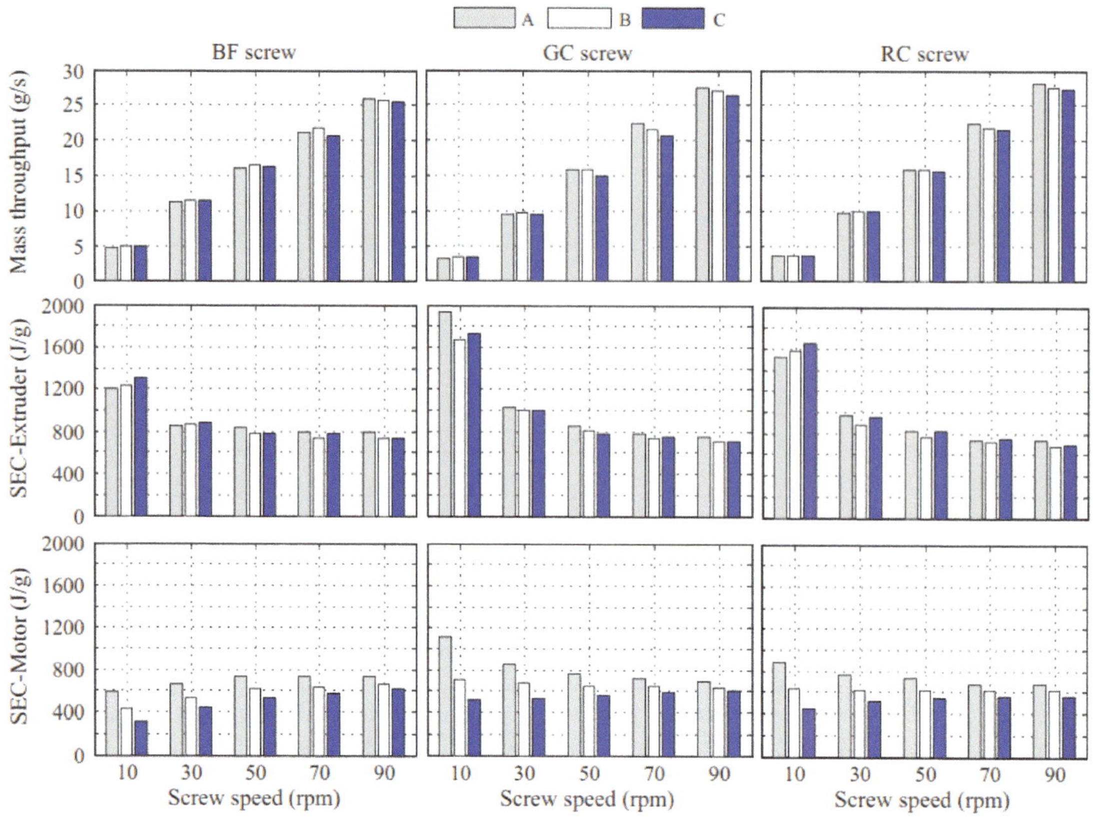

Figure 11. Process mass throughput and SEC of the extruder/motor over different processing conditions.

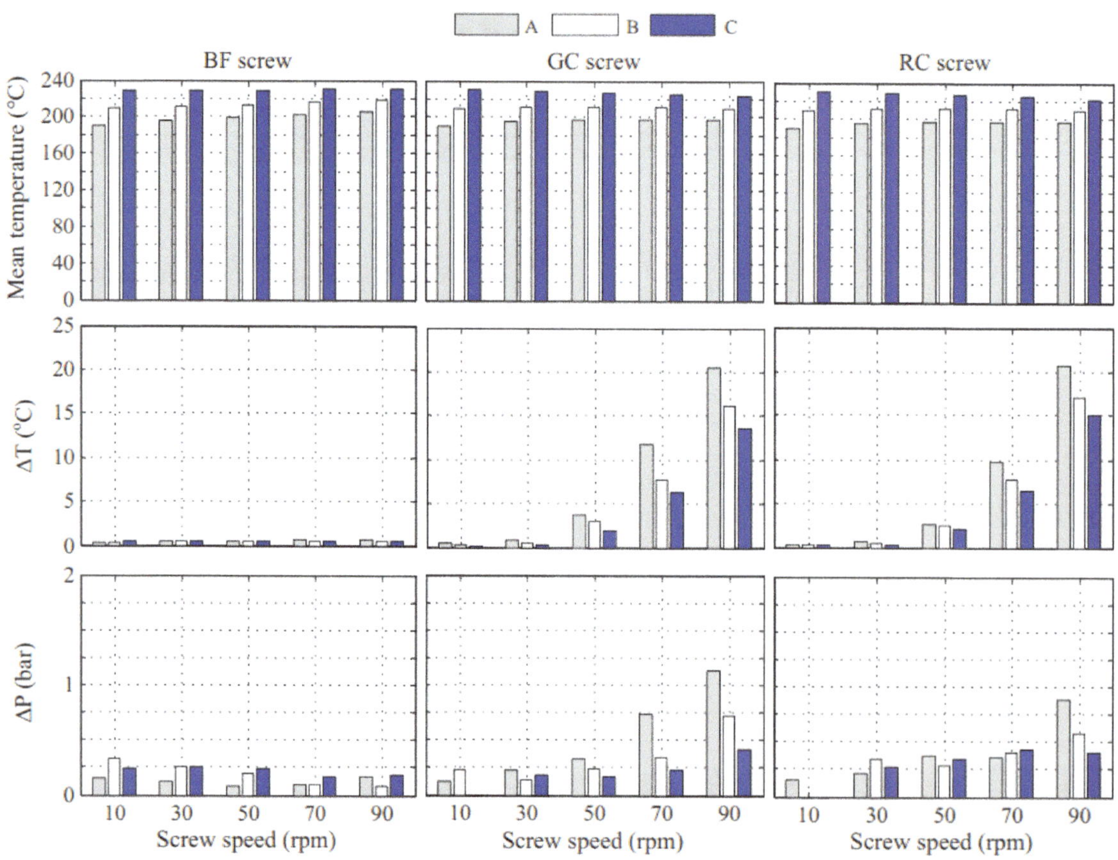

Figure 12. Mean melt temperature across the melt flow, ΔT and ΔP over different processing conditions and screws.

Mass throughput increases with screw speed as expected and slight differences in the rate can be observed with the barrel set temperature and screw geometry at the same screw speed. Also, it is noticeable that the SEC values of the extruder at 10 rpm (where conductive heat is the dominant) are considerably higher than other speeds with all screws. Moreover, it is clear that SEC of the extruder reduces with increase of screw speed regardless of the differences in the screw geometry and barrel set temperatures. However, the SEC of the motor does not reduce with screw speed and shows an increasing trend with screw speed at all set temperature conditions with the BF screw and at the set temperature condition C with the GC and RC screws. Conversely, the SEC of the motor is similar at all screw speeds at the set temperature condition B with the GC and RC screws. A similar trend to this study was observed for the SEC of an extruder with the same BF, GC and RC screws by Kelly et al. [22] for a low density polyethylene (LDPE). Moreover, they observed a reducing trend of the SEC of the motor as well for all the screws with this LDPE. Therefore, it clear that the nature of the motor SEC may depend on the material type, screw geometry and set temperature condition.

The mean temperature across the melt flow (i.e. the average of the mean temperature values at the different radial positions: 0 mm, 2.5 mm, 5 mm, 8 mm, 11 mm, 14 mm, 16.5 mm and 19 mm), the difference between the maximum (i.e. the average of the maximum values of different radial positions) and minimum (i.e. the average of the minimum values of different radial positions) temperatures across the melt flow (ΔT) and the difference between the maximum and minimum melt pressures (ΔP) were calculated and these details are shown in Figure 12.

These figures clearly indicate that the melt temperature fluctuations increase with screw speed (regardless of the screw geometry) although the SEC of the extruder shows a reducing trend with increasing screw speed. As process speed increases, material has less time inside the barrel prior to entering to the die. Since polymers are poor thermal conductors there may not be enough time for good melting and mixing to occur and give a uniform melt at high speeds and this may cause an increase of thermal fluctuations at increasing process speeds. Moreover, the results show that these thermal variations are dependent upon the die radial position and screw geometry as well. With GC and RC screws, melt temperature fluctuations are reduced with increase of the barrel set temperature which is the opposite trend to the total extruder power fluctuations. With the BF screw, thermal fluctuations do not show a clear trend (however fluctuations are very low compared to other two screws) with an increase of the barrel set temperatures while total power fluctuations show an increasing trend. The BF screw shows the lowest thermal fluctuations among the three screws. Therefore, it can be argued that the specific design of the BF screw with solid melt separation and Maddox mixer leads to improved melting/mixing performance of the material. Furthermore, the SEC of the extruder is relatively low with the BF screw compared to the GC and BF screws. Generally, it seems that the BF screw has better thermal and energy efficiency than the other two screws in processing of this particular material. Some of the previous work also reported that the BF screws perform favourably (e.g. efficient melting and mixing) compared to conventional single-flighted screws [22] and [31].

To obtain an extensive understanding of the thermal behaviour across the melt flow cross-section, the mean melt temperatures across the melt flow at each radial position were calculated together with the magnitude of fluctuations (i.e. the difference between the lowest and the highest melt temperatures values) for all the screw geometries at each screw speed and set temperature condition. The corresponding details relevant to the temperature measurements Trp-0, Trp-5, Trp-11, Trp-14 and Trp-16.5 are shown in Figure 13. All the sub-figures are shown in the same scale and hence a comparison can be made on the melt temperature variations at different processing conditions.

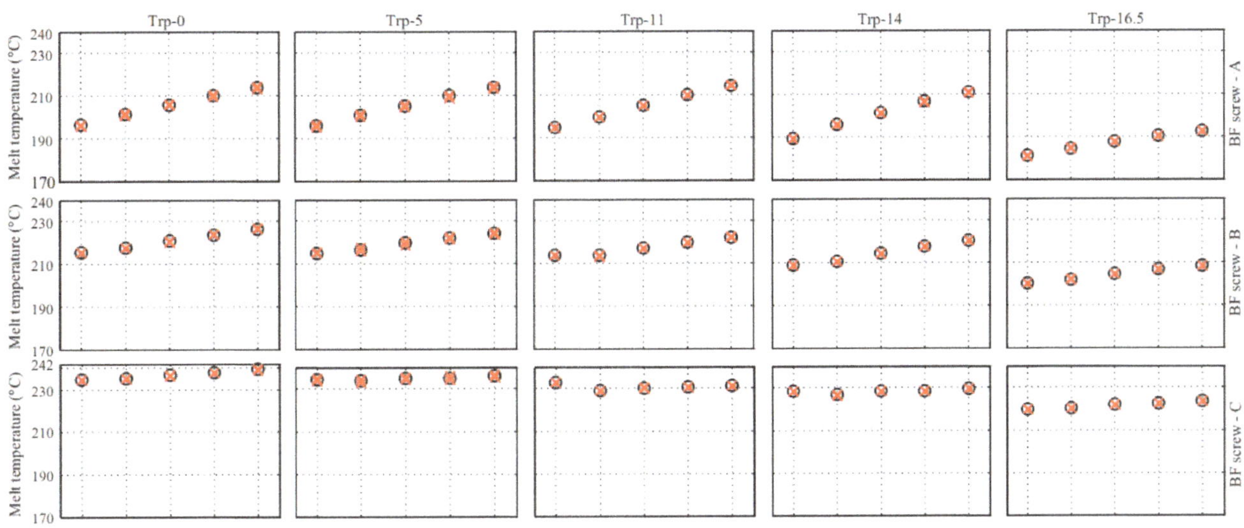

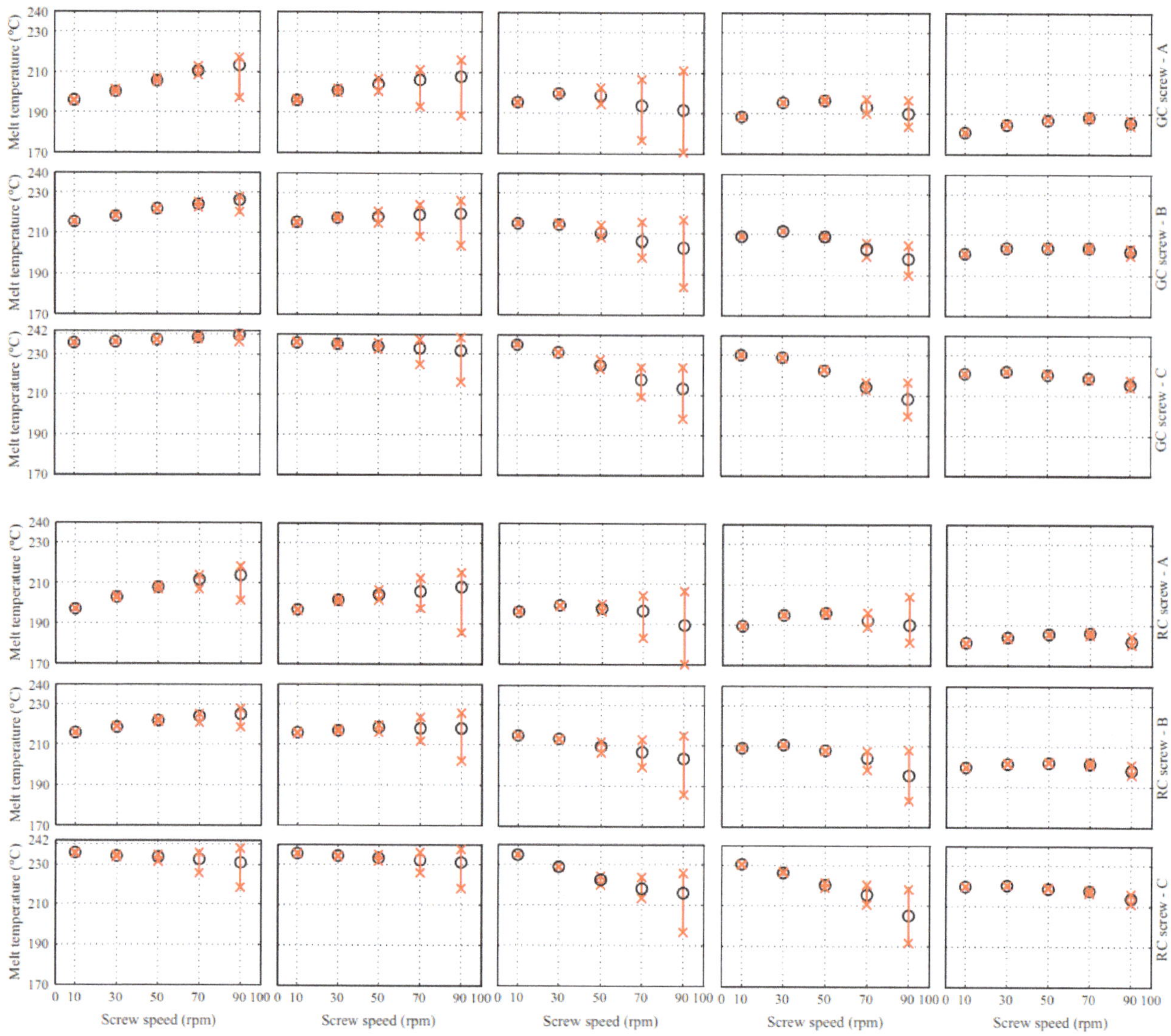

Figure 13. Mean melt temperature observed at each radial position together with the magnitude of fluctuations.

It is clear that melt temperature fluctuations are considerably lower with the BF screw than the GC and RC screws regardless the radial position, processing speed and the barrel set temperature. In general, the mean melt temperature increases with screw speed at the middle of the melt flow (i.e. at the Trp-0) with all the screws but this is not true with the other radial positions across the melt flow. In comparing power and melt temperature behaviours, it is clear that:

- In general, power variations reduce with the screw speed where melt temperature variations show an increasing trend.
- Power variations increase with the increase of barrel set temperature where melt temperature variations show a reducing trend.
- Process energy demand (i.e. the levels of the both total and motor power signals) increases with the screw speed but the melt temperature of some of the radial positions does not increase with the screw speed.
- Changes to the level of mean power demand are quite linear with screw speed but melt temperature shows a non-linear behaviour and this depends on the die radial position, screw geometry and barrel set temperature conditions as well.

Generally, it is clear that the relationship/s between thermal fluctuations and energy consumption differs depending on the screw speed and set temperature conditions, and these are likely to be dependent upon material properties as well (polymer viscosity is dependent on shear stress which in turn depends on the screw configuration. Different polymers will display different rheological responses at a particular shear stress). Therefore, it is better to conduct more research in this area to investigate further about the possible relationship/s between process thermal stability and energy efficiency.

Modelling of total power demand

Eventually, an attempt was made to identify the relationship between the mean total power (*TP*), screw speed (ωsc), mean temperature across the melt flow (T*mean*) and the level of the melt temperature fluctuations across the melt flow (ΔT). Here, the TP relating to all experiential conditions used in this study was modelled as a function of the other variables based on the modelling technique discussed previously [5], [7], [8],[9] and [32]:

$$TP = f(\omega_{sc}, T_{mean}, \Delta T) \tag{11}$$

Of the model examined, a 3rd order model with 5 terms (which shows a root mean square error of 0.1 on an unseen data) which is given in Eq. (12) was selected for the discussion.

$$TP = 1.7981 \times \omega_{sc} + 0.0197 \times T_{mean} - 0.0143 \times \omega_{sc} \times T_{mean} + 3.1224 \times 10^{-5} \times \omega_{sc} \times T_{mean}^2 - 0.0424 \times \Delta T \tag{12}$$

A comparison of the experimental and model predicted mean total power at different speeds is shown in Figure 14.

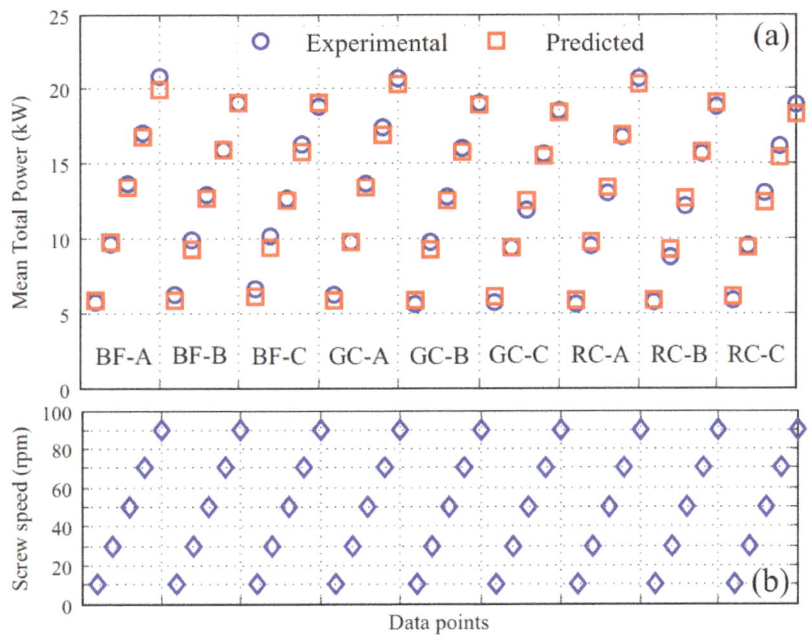

Figure 14. : (a) Experimental and model predicted mean total power at different speeds. (b) The corresponding screw speeds.

Evidently, there is a good fit between the experimental measurements and model predictions. This type of model may be useful for optimizing a particular machine/material combination for thermal and energy efficiency, and it will be considered under future work.

CONCLUSIONS

This study aimed to explore strategies for determining an optimum process operating point by optimizing both the extruder energy and thermal efficiency. An attempt was made to investigate the possible correlation/s between melt temperature stability/fluctuations and energy consumption in polymer extrusion. The results showed that the SEC of the extruder reduces with the screw speed regardless the screw geometry. However, the SEC of the motor was shown to increase with screw speed for a BF screw while it reduced with screw speed for GC and RC screws. Generally, process energy efficiency increases with screw speed. However, process thermal fluctuations increase with screw speed and these variations depend on the screw geometry and the die radial position as well. Conversely, melt temperature shows an increasing trend with screw speed only at the middle of the flow while an opposite trend can be observed at other radial locations away from the melt flow center. Correlation/s between the thermal stability and energy efficiency are highly complex and dependent upon all machine, material and process parameters. According to the results of this work, it is difficult to achieve the energy efficiency achievable at high running speeds without increasing thermal fluctuations and thus reducing thermal efficiency. Further research is highly recommended in this area to determine an accurate generalized model relating process thermal stability, energy efficiency and all other relevant parameters which will allow selection of an optimum operating point for a given machine and material.

ACKNOWLEDGMENTS

This work was funded through an inter-disciplinary research programme (Grant No.EP/G059330/1) by the EPSRC-UK. The technical assistance provided by Ken Howell, Roy Dixon and John Wyborn is greatly appreciated.

REFERENCES

1. Rosato DV, Schott NR, Rosato DV, Rosato MG. Plastics engineering, manufacturing & data handbook: fundamental and processes, vol. 2. Kluwer Academic Publishers; 2001. p. 54–65.
2. Abeykoon C, McAfee M, Thompson S, Li K, Kelly AL, Brown EC. Investigation of torque fluctuations in extrusion through monitoring of motor variables. In: Proceedings of 26th PPS annual Europe/Africa regional meeting, Larnaca, Cyprus. Paper no: 22-O.
3. Abeykoon C, McAfee M, Li K, Martin PJ, Deng J, Kelly AL. Modelling the effects of operating conditions on motor power consumption in single screw extrusion. In:

Proceedings of LSMS 2010 and ICSEE 2010 6329, vol. 2; 2010. p. 9–20.

4. Abeykoon C, McAfee M, Li K, Kelly AL, Brown EC. Monitoring the effect of operating conditions on melt temperature homogeneity in single-screw extrusion. In: SPE ANTEC technical papers; 2010. p. 1799–806.
5. Abeykoon C, Li K, McAfee M, Martin PJ, Niu Q, Kelly AL, et al. A new model based approach for the prediction and optimisation of thermal homogeneity in single screw extrusion. Control Eng Pract 2011;19(8):862–74.
6. Abeykoon C, Martin PJ, Kelly AL, Brown EC. A review and evaluation of melt temperature sensors for polymer extrusion. Sens Actuat A: Phys 2012;182:16–27.
7. Abeykoon C. Modelling and control of melt temperature in polymer extrusion. Ph.D. thesis. United Kingdom: Queen's University Belfast; 2011.
8. Abeykoon C. A novel soft sensor for real-time monitoring of die melt temperature profile in polymer extrusion. IEEE Trans Ind Electron. http:// dx.doi.org/10.1109/ TIE.2014.2321345.
9. Abeykoon C. A novel model-based controller for polymer extrusion. EEE Trans Fuzzy Syst. http://dx.doi. org/10.1109/TFUZZ.2013.2293348.
10. Burmann G, Fischer P, Michels R. High-tech extrusion with barrier-mixing screws and spiral mandrel extrusion dies. In: Proceeding of conference on advances in plastics technology, Katowice, Poland.
11. Rosato DV. Extruding plastics – a practical processing handbook. Springer: Verlag; 1998.
12. Spalding MA, Hyun KS. Troubleshooting mixing problems in single-screw extruders. SPE ANTEC technical papers vol. 1; 2003. p. 229–233.
13. Rauwendaal C. Polymer extrusion. 4th ed. Hanser; 2001.
14. Abeykoon C. Polymer extrusion: a study on thermal monitoring techniques and melting issues. Lap Lambert Academic Publishing: Verlag; 2012.
15. Wahla PR, Trefferb D, Mohra S, Roblegga E, Koschera G, Khinast JG. Inline monitoring and a PAT strategy for pharmaceutical hot melt extrusion. Int J Pharm 2013;455(1–2):159–68.
16. Saerens L, Vervaet C, Remon JP, Beer TD. Process monitoring and visualization solutions for hot-melt extrusion: a review. J Pharm Pharmacol 2014;66(2): 180–203.
17. Drury B. The control techniques drives and controls handbook. London (UK): The Institution of Electrical Engineers; 2010. p 308–12.
18. Vlachopoulos J, Strutt D. Overview: polymer processing. Mater Sci Technol 2003;19:1161–9.
19. Rasid R, Wood AK. Effect of process variables on melt temperature profiles in extrusion process using single screw plastics extruder. Plast Rubb Compos 2003;32(5):193–8.
20. Sorroche JV, Kelly AL, Brown EC, Coates PD, Karnachi N, Harkin-Jones E, et al. Thermal optimisation of polymer extrusion using in-process monitoring techniques. Appl Therm Eng 2012;53(2):405–13.
21. Abeykoon C, McAfee M, Li K, Martin PJ, Kelly AL. The inferential monitoring of screw load torque to predict process fluctuations in polymer extrusion. J Mater Process Technol 2011;211(12):1907–18.
22. Kelly AL, Brown EC, Coates PD. The effect of screw geometry on melt temperature profile in single screw extrusion. Polym Eng Sci 2006;46(12): 1706–14.
23. Brown EC, Kelly AL, Coates PD. Melt temperature field measurement in single screw extrusion using thermocouple meshes. Rev Sci Instrum 2004;75(11): 4742–8.
24. Bagley EB. End corrections in capillary flow of polyethylene. J Appl Phys 1957;28(5):624–7.
25. Bird RB, Armstrong RC, Hassager O. Dynamics of polymeric liquids. Fluid mechanics, 2nd ed., vol. 1. New York: Wiley; 1987.
26. Carreau PJ. Rheological equations from molecular network theories. Ph.D. thesis. Madison (USA): University of Wisconsin; 1968.
27. Carreau PJ. Rheological equations from molecular network theories. Trans Soc Rheol 1972;16(1):99–127.
28. Ferry JD, Landel RF, Williams ML. Extensions of rouse theory of viscoelastic properties to undiluted linear polymers. J Appl Phys 1995;26(4):359–62.
29. Helleloid GT. On the computation of viscosity-shear rate temperature master curves for polymeric liquids. Morehead Electron J Appl Math 2001;1(1):1–11.
30. Cantor KM. Analyzing extruder energy consumption. SPE ANTEC technical papers, vol. 2; 2010. p. 1300–6.
31. Brown EC, Kelly AL, Coates PD. Melt temperature homogeneity in single screw extrusion: effect of material type and screw geometry. SPE ANTEC technical papers; 2004. p. 183–7.
32. Abeykoon C, Martin PJ, Li K, Kelly AL. Dynamic modelling of die melt temperature profile in polymer extrusion: effects of process settings, screw geometry and material. J Appl Math Model 2014;28(4):1224–36.

CHAPTER 18

NANOFLUIDS OF ULTRA-HIGH THERMAL CONDUCTIVITY

Liqiu Wang, Jing Fan

INTRODUCTION

Nanofluids are a new class of fluids engineered by dispersing nanometer-size structures (particles, fibers, tubes, droplets, etc.) in base fluids. The very essence of nanofluids research and development is to enhance fluid macroscopic and system-scale properties through manipulating microscopic physics (structures, properties, and activities) [1, 2]. One of such properties is the thermal conductivity that characterizes the strength of heat conduction and has become a research focus of nanofluid society in the last decade [1–9].

The importance of high-conductivity nanofluids cannot be overemphasized. The success of effectively developing such nanofluids depends very much on our understanding of mechanism responsible for the significant enhancement of thermal conductivity. Both static and dynamic reasons have been proposed for experimental finding of significant conductivity enhancement [1–9]. The former includes the nanoparticle morphology [10, 11] and the liquid layering at the liquid-particle interface [12–17]. The latter contains the coupled (cross) transport [18–20] and the nanoparticle Brownian motion [21–26]. Here, the effect of particle morphology contains those from the particle shape, connectivity among particles (including and generalizing the nanoparticle clustering/aggregating in the literature [10, 11]), and particle distribution in nanofluids. This short review aims for a concise assessment of these contributions, thus identifying the future research needs toward nanofluids of high thermal conductivity. The readers are referred to, for example, [1–9] for state-of-the-art expositions of major advances on the synthesis, characterization, and application of nanofluids.

Citation: Wang and Fan: Nanofluids of ultra-high thermal conductivity. Nanoscale Research Letters 2011 6:153. doi:10.1186/1556-276X-6-153.

STATIC MECHANISMS

Morphology

The nanoparticle morphology in nanofluids can vary from a well-dispersed configuration in base fluids to a continuous phase of interconnected configuration. Such a morphology variation will change nanofluid's effective thermal conductivity significantly [27–32], a phenomenon credited to the particle clustering/aggregating in the literature [1–9]. This appears obvious because the nanofluid's effective conductivity stems mainly from the contribution of continuous phase that constitutes the continuous path for thermal flow [27, 28]. Although particle clustering/aggregating offers a way of changing particle morphology, it is not necessarily an effective means. The research should thus focus not only on the clustering/aggregating, but also on the general ways of varying morphology.

Given that nanofluid thermal conductivity depends heavily on the particle morphology, its lower and upper bounds can be completely determined by the volume fractions and conductivities of the two phases. These bounds have been well developed based on the classical effective-medium theory and termed as the Hashin-Shtrikman (H-S) bounds [33],

$$k_e / k_f = 1 + \frac{3\varphi(k_p / k_f - 1)}{k_p / k_f + 2 - \varphi(k_p / k_f - 1)}, \quad (1)$$

$$k_e / k_f = k_p / k_f \left[1 - \frac{3(1-\varphi)(k_p / k_f - 1)}{3k_p / k_f - \varphi(k_p / k_f - 1)} \right]. \quad (2)$$

Here k_p, k_f, and k_e are the conductivities of particle, base fluid, and nanofluid, respectively, and φ is the particle volume fraction. For the case of $k_p/k_f \geq 1$, Equations (1) and (2) give the lower and the upper bounds for nanofluid effective thermal conductivity, corresponding to the two limiting morphologies where the liquid serves as the continuous phase for the lower bound and the particle disperses the liquid for the upper bound, respectively. When $k_p/k_f \leq 1$, their roles are interchanged, so that Equations (1) and (2) provide the upper and the lower bounds, respectively. Therefore, the upper bound always takes a configuration (morphology) where the continuous phase is made of the higher-conductivity material.

The morphology dependence of nanofluid›s conductivity has been recently examined in detail by either of the two approaches: the constructal approach [1, 2, 29–32] and the scaling-up by the volume average [1, 2, 27, 28]. Such studies not only confirm the features captured in the H-S bounds but also uncover the microscopic mechanism responsible for the morphology dependence of nanofluid›s conductivity. As higher-conductivity particles interconnect each other and disperse the lower-conductivity base fluid into a dispersed phase, the interfacial energy transport between particle and base fluid becomes enhanced significantly such that the nanofluid›s conductivity takes its value of upper H-S bound (Fan J and Wang LQ: Heat conduction in nanofluids: structure-property correlation, submitted).

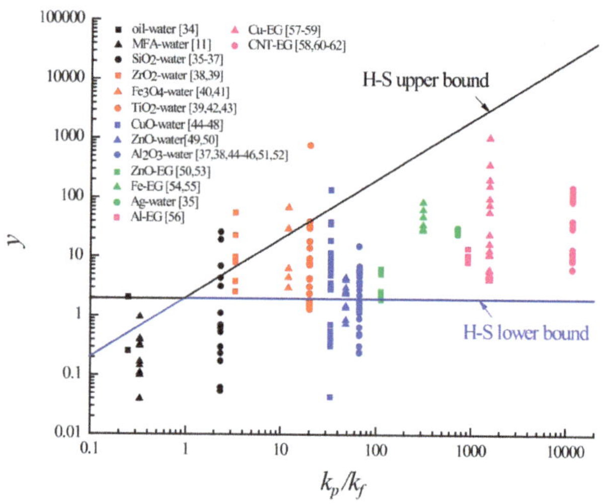

Figure 1: Comparison of experimental data with H-S bounds.

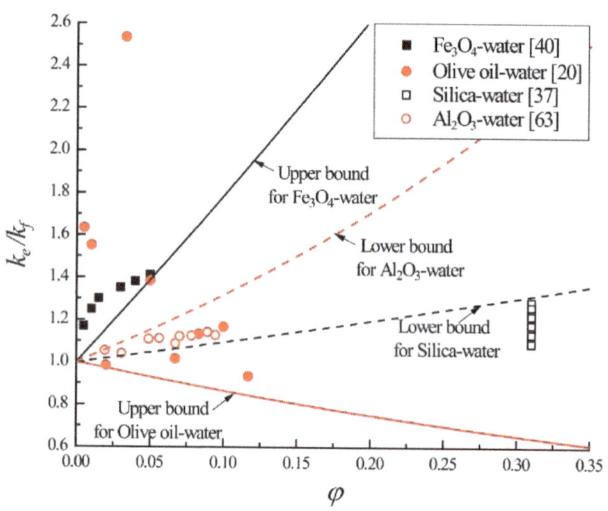

Figure 2: Comparison of effective thermal conductivity between experimental data and H-S bounds.

Figures 1 and 2 compare the experimental data of nanofluid thermal conductivity [11, 20,34–63] with the H-S bounds [33]. For a concise comparison in Figure 1, the H-S bounds (Equations 1 and 2) are rewritten in the form of

$$y=2, \quad (3)$$

and

$$y = 2\frac{k_p}{k_f}, \quad (4)$$

where

$$y \equiv -\frac{\left(k_p / k_f\right)\left(k_e / k_f - 1\right) - \varphi\left(k_p / k_f - 1\right)\left(k_e / k_f\right)}{\left(k_e / k_f - 1\right) - \varphi\left(k_p / k_f - 1\right)}. \quad (5)$$

As k_p/k_f moves away from the unity along both directions, the separation between the upper and lower H-S bounds becomes pronounced (Figures 1 and 2) so that the room for manipulating nanofluid conductivity via changing the particle morphology becomes more spacious. The H-S bounds are respected by some nanofluids for which their thermal conductivity is strongly dependent on particle morphology, such as whether nanoparticles stay well-dispersed in the base fluid, form aggregates, or assume a configuration of continuous phase that disperses the fluid into a dispersed phase (Figure 1). There are thermal conductivity data that fall outside the H-S bounds (Figures 1 and 2).

Ordered Liquid Layer

Both experimental and theoretical evidences have been reported of the presence of ordered liquid layer near a solid surface by which the atomic structure of the liquid layer is significantly more ordered than that of bulk liquid [64–67]. For example, two layers of ice like structures are experimentally observed to be strongly bounded to the crystal surface on a crystal-water interface, followed by two diffusive layers with less significant ordering [65]. Three ordered water layers have also been observed numerically on the Pt (111) surface [64].

The study is very limited regarding why and how these ordered liquid layers are formed. There is also a lack of detailed examination of properties of these layers, such as their thermal conductivity and thickness. Since ordered crystalline solids have normally much higher thermal conductivity than liquids, the thermal conductivity of such liquid layers is believed to be better than that of bulk liquid. The thickness h of such liquid layers around the solid surface can be estimated by [17]

$$h = \frac{1}{\sqrt{3}}\left(\frac{4M_f}{\rho_f N_a}\right)^{1/3}, \quad (6)$$

where N_a is the Avogadro's number, and ρ_f and M_f are the density and the molecular weight of base fluids, respectively. The liquid layer thickness is thus 0.28 nm for water-based nanofluids, which agrees with that from experiments and molecular dynamic simulation on the order of magnitude.

The presence of liquid layers could thus upgrade the nanofluid effective thermal conductivity via augmenting the particle effective volume fraction. For an estimation of an upper limit for this effect, assume that the thickness and the conductivity of the liquid layer are 0.5 nm and the same as that of the solid particle, respectively. For spherical particles of diameter d_p, Equation (1) offers the conductivity ratio with and without this effect:

$$\frac{(k_e)_{\text{with}}}{(k_e)_{\text{without}}} = \frac{1+2\eta\varphi\left(1+2h/d_p\right)^3}{1-\eta\varphi\left(1+2h/d_p\right)^3} \times \frac{1-\eta\varphi}{1+2\eta\varphi}. \tag{7}$$

where $\eta = (k_p - k_f)/(k_p + 2k_f)$. The variation of $(k_e)_{with}/(k_e)_{without}$ with $\eta\varphi$ and $d_p/2h$ is illustrated in Figure 3, showing that the liquid-layering effect is important only when $\eta\varphi$ is large and $d_p/2h$ is small. This is normally not the case for practical nanofluids. For Cu-in-water nanofluids ($\eta \approx 1$), for example, $(k_e)_{with}/(k_e)_{without} \approx 1.005$ with $\varphi = 0.5\%$ and $d_p = 10$ nm.

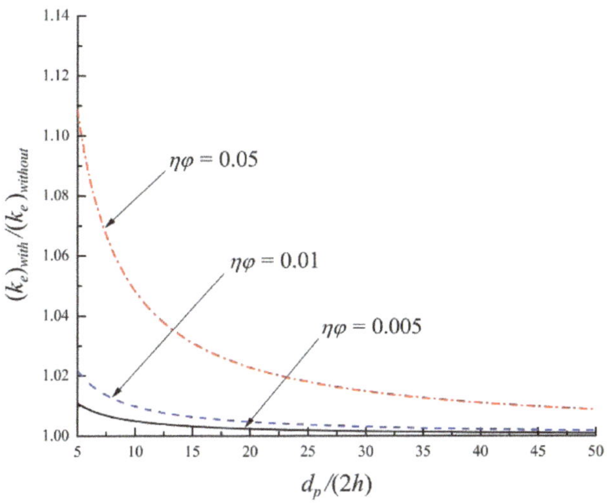

Figure 3: Variations of $(k_e)_{with}/(k_e)_{without}$ with $\eta\varphi$ and $d_p/(2h)$.

Although the liquid layers offer insignificant conductivity enhancement through augmenting the particle volume fraction, their presence do facilitate the formation of particle network by relaxing the requirement of particle physical contact with each other (Figure 4). This will promote the formation of interconnected particle morphology, and thus upgrade the nanofluid thermal conductivity toward its upper bound through the morphology effect.

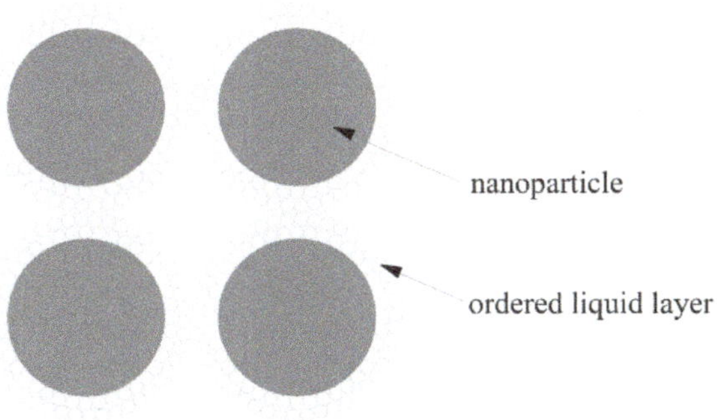

Figure 4: Ordered liquid layer in promoting the formation of interconnected particle morphology.

DYNAMIC MECHANISMS

Coupled Transport

In a nanofluid system, normally, there are two or more transport processes that occur simultaneously. Examples are the heat conduction in dispersed phase, heat conduction in continuous phase, mass transport, and chemical reactions either among the nanoparticles or between the nanoparticles and the base fluid. These processes may couple (interfere) and cause new induced effects of flows occurring without or against its primary thermodynamic driving force, which may be a gradient of temperature, or chemical potential, or reaction affinity. Two classical examples of coupled transport are the Soret effect (also known as thermodiffusion or thermophoresis) in which directed motion of particles or macromolecules is driven by thermal gradient and the Dufour effect that is an induced heat flow caused by the concentration gradient.

While the coupled transport is well recognized to be very important in thermodynamics [68], it has not been well appreciated yet in the nanofluid society. The first attempts of examining the effect of coupled transport on nanofluid heat conduction have been recently made in some studies [1, 2, 9, 18], which are briefly outlined here. With the coupling between the heat conduction in the fluid and particle phases denoted by β and σ- phases, respectively, the temperature T obeys the following energy equations [1, 2]

$$\gamma_\beta \frac{\partial T_\beta}{\partial t} = k_{\beta\beta}\Delta T_\beta + k_{\beta\sigma}\Delta T_\sigma + ha_\upsilon\left(T_\sigma - T_\beta\right) \tag{8}$$

and

$$\gamma_\sigma \frac{\partial T_\sigma}{\partial t} = k_{\sigma\sigma}\Delta T_\sigma + k_{\sigma\beta}\Delta T_\beta - ha_\upsilon\left(T_\sigma - T_\beta\right) \tag{9}$$

where T is the temperature; subscripts β and σ refer to the β and σ- phases, respectively. $\gamma_\beta = (1-\varphi)(\rho c)\beta$ and $\gamma_\sigma = \varphi(\rho c)$ are the effective thermal capacities of β and σ- phases, respectively, with ρ and c as the density and the specific heat. φ is the volume fraction of the σ- phase. h and a_υ come from modeling of the interfacial flux and are the film heat transfer coefficient and the interfacial area per unit volume, respectively. $k_{\beta\beta}$ and $k_{\sigma\sigma}$ are the effective thermal conductivities of the β and σ- phases, respectively; $k_{\beta\sigma}$ and $k_{\sigma\beta}$ are the coupling (cross) effective thermal conductivities between the two phases.

Rewriting Equations (8) and (9) in their operator form, we obtain

$$\begin{bmatrix} \gamma_\beta \frac{\partial}{\partial t} - k_{\beta\beta}\Delta + h & -k_{\beta\sigma}\Delta - ha_\upsilon \\ -k_{\sigma\beta}\Delta - ha_\upsilon & \gamma_\sigma \frac{\partial}{\partial t} - k_{\sigma\sigma}\Delta + ha_\upsilon \end{bmatrix} \begin{bmatrix} T_\beta \\ T_\sigma \end{bmatrix} = 0 \tag{10}$$

An uncoupled form can then be obtained by evaluating the operator determinant such that

$$\left[\left(\gamma_\beta \frac{\partial}{\partial t} - k_{\beta\beta}\Delta + ha_\upsilon\right)\left(\gamma_\sigma \frac{\partial}{\partial t} - k_{\sigma\sigma}\Delta + ha_\upsilon\right) - \left(k_{\beta\sigma}\Delta - ha_\upsilon\right)^2\right]\langle T_i\rangle^i = 0 \tag{11}$$

where the index i can take β or σ. Its explicit form reads, after dividing by $ha_\upsilon(\gamma_\beta + \gamma_\sigma)$

$$\frac{\partial T_i}{\partial t} + \tau_q \frac{\partial^2 T_i}{\partial t^2} = \alpha \Delta T_i + \alpha \tau_T \frac{\partial}{\partial t}(\Delta T_i) + \frac{\alpha}{k}\left[F(\mathbf{r},t) + \tau_q \frac{\partial F(\mathbf{r},t)}{\partial t}\right] \tag{12}$$

where

$$\tau_q = \frac{\gamma_\beta \gamma_\sigma}{ha_\upsilon(\gamma_\beta + \gamma_\sigma)}, \quad \tau_T = \frac{\gamma_\beta k_{\sigma\sigma} + \gamma_\sigma k_{\beta\beta}}{ha_\upsilon(k_{\beta\beta} + k_{\sigma\sigma} + k_{\beta\sigma} + k_{\sigma\beta})},$$

$$k = k_{\beta\beta} + k_{\sigma\sigma} + k_{\beta\sigma} + k_{\sigma\beta}, \quad \alpha = \frac{k_{\beta\beta} + k_{\sigma\sigma} + k_{\beta\sigma} + k_{\sigma\beta}}{\gamma_\beta + \gamma_\sigma},$$

$$F(\mathbf{r},t) + \tau_q \frac{\partial F(\mathbf{r},t)}{\partial t} = \frac{k_{\beta\sigma}k_{\sigma\beta} - k_{\beta\beta}k_{\sigma\sigma}}{ha_\upsilon}\Delta^2 T_i. \tag{13}$$

Equation (12) is not a classical heat-conduction equation, but can be regarded as a dual-phase-lagging (DPL) heat-conduction equation with $((k_{\beta\sigma}k_{\sigma\beta} - k_{\beta\beta}k_{\sigma\sigma})/(ha_\upsilon))\Delta^2 T_i$ as the DPL source-related term $F(\mathbf{r},t) + \tau_q \frac{\partial F(\mathbf{r},t)}{\partial t}$ and with τ_q and τ_T as the phase lags of the heat flux and the temperature gradient, respectively [2, 18, 69]. Here, F(r, t) is the volumetric heat source. k, ρc, and α are the effective thermal conductivity, capacity and diffusivity of nanofluids, respectively.

The computations of $k_{\beta\beta}$, $k_{\sigma\sigma}$, $k_{\beta\sigma}$, and $k_{\sigma\beta}$ are available in [27, 28] for some typical nanofluids. The coupled-transport contribution to the nanofluid thermal conductivity, the term $(k_{\beta\sigma} + k_{\sigma\beta})$, can be as high as 10% of the of the overall thermal conductivity [27, 28]. The more striking effect of the coupled transport on nanofluid heat conduction can be found by considering

$$\frac{\tau_T}{\tau_q} = 1 + \frac{\gamma_\beta^2 k_{\sigma\sigma} + \gamma_\sigma^2 k_{\beta\beta} - 2\gamma_\beta\gamma_\sigma k_{\beta\sigma}}{\gamma_\beta\gamma_\sigma(k_{\beta\beta} + k_{\sigma\sigma} + k_{\beta\sigma} + k_{\sigma\beta})}, \tag{14}$$

which is smaller than 1 when

$$\gamma_\beta^2 k_{\sigma\sigma} + \gamma_\sigma^2 k_{\beta\beta} - 2\gamma_\beta\gamma_\sigma k_{\beta\sigma} = \left(\gamma_\beta\sqrt{k_{\sigma\sigma}} - \gamma_\sigma\sqrt{k_{\beta\beta}}\right)^2 + 2\gamma_\beta\gamma_\sigma\left(\sqrt{k_{\beta\beta}k_{\sigma\sigma}} - k_{\beta\sigma}\right) < 0. \tag{15}$$

Therefore, by the condition for the existence of thermal waves that requires $\tau_T/\tau_q < 1$ [18, 70], thermal waves may be present in nanofluid heat conduction.

Note also that, for heat conduction in nanofluids, there is a time-dependent source term F(r, t) in the DPL heat conduction (Equations (12) and (13)). Therefore, the resonance can also occur. When $k_{\beta\sigma} = k_{\sigma\beta} = 0$ so that τ_T/τ_q is always larger than 1, thermal waves and resonance would not appear. Therefore, the coupled transport could change the nature of heat conduction in nanofluids from a diffusion process to a wave process, thus having a significant effect on nanofluid heat conduction.

Therefore, the cross coupling between the heat conduction in the fluid and particle manifests itself as thermal waves at the macroscale. Depending on factors such as material properties of nanoparticles and base fluids, nanoparticles' geometrical structure and their distribution in the base fluids, and interfacial properties and dynamic processes on particle-fluid interfaces, the cross-coupling-induced thermal waves may either enhance or counteract with the molecular-dynamics-driven heat diffusion. Consequently, the heat conduction may be enhanced or weakened by the presence of nanoparticles. This explains the thermal conductivity data that fall outside the H-S bounds (Figures 1 and 2).

If the coupled transport between heat conduction and particle diffusion is considered, then the temperature T and particle volume fraction φ satisfy the following equations of energy and mass conservation:

$$\gamma_\beta \frac{\partial T_\beta}{\partial t} = k_{\beta\beta}\Delta T_\beta + k_{\beta\sigma}\Delta T_\sigma + \rho_\sigma k_{\beta m}\Delta\varphi + ha_\upsilon\left(T_\sigma - T_\beta\right), \tag{16}$$

$$\gamma_\sigma \frac{\partial T_\sigma}{\partial t} = k_{\sigma\sigma}\Delta T_\sigma + k_{\sigma\beta}\Delta T_\beta + \rho_\sigma k_{\sigma m}\Delta\varphi - ha_\upsilon\left(T_\sigma - T_\beta\right), \tag{17}$$

and

$$\rho_\sigma \frac{\partial \varphi}{\partial t} = \rho_\sigma D_{\sigma\sigma}\Delta\varphi + D_{m\beta}\Delta T_\beta + D_{m\sigma}\Delta T_\sigma + D_{mT}\left(T_\beta - T_\sigma\right), \tag{18}$$

where subscripts m and T stand for mass transport and thermal transport, respectively. $D_{\sigma\sigma}$ is the effective diffusion coefficient for nanoparticles. $k_{\beta m}$, $k_{\sigma m}$, $D_{m\beta}$, $D_{m\sigma}$, and D_{mT} are five transport coefficients for coupled heat and mass transport. By following a similar procedure as that of developing Equation (12), an uncoupled form with u (T_β, T_σ, or φ) as the sole unknown variable is obtained,

$$\frac{\partial(\Delta u)}{\partial t} + \tau_q \frac{\partial^2(\Delta u)}{\partial t^2} = \alpha\Delta(\Delta u) + \alpha\tau_T \frac{\partial}{\partial t}\left(\Delta(\Delta u)\right) + \frac{\alpha}{k}\left[F(\mathbf{r},t) + \tau_q \frac{\partial F(\mathbf{r},t)}{\partial t}\right] \tag{19}$$

where

$$\tau_q = \frac{\gamma_\beta k_{\sigma\sigma} + \gamma_\sigma k_{\beta\beta} + \gamma_\beta\gamma_\sigma D_{\sigma\sigma}}{D_{mT}\left(\gamma_\sigma k_{\beta m} - \gamma_\beta k_{\sigma m}\right) + ha_\upsilon\left(k_{\beta\beta} + k_{\beta\sigma} + k_{\sigma\beta} + k_{\sigma\sigma}\right) + ha_\upsilon D_{\sigma\sigma}\left(\gamma_\beta + \gamma_\sigma\right)}, \tag{20}$$

$$\begin{aligned} k = {} & \frac{\gamma_\beta + \gamma_\sigma}{D_{mT}\left(\gamma_\sigma k_{\beta m} - \gamma_\beta k_{\sigma m}\right) + ha_\upsilon\left(k_{\beta\beta} + k_{\beta\sigma} + k_{\sigma\beta}k_{\sigma\sigma}\right) + ha_\upsilon D_{\sigma\sigma}\left(\gamma_\beta + \gamma_\sigma\right)} \\ & \times\left\{k_{\beta m}\left[D_{mT}\left(k_{\sigma\beta} + k_{\sigma\sigma}\right) - ha_\upsilon\left(D_{m\beta} + D_{mT}\right)\right] - k_{\sigma m}\left[D_{mT}\left(k_{\beta\beta} + k_{\beta\sigma}\right) + ha_\upsilon\left(D_{m\sigma} + D_{m\beta}\right)\right]\right\} \\ & \times\frac{ha_\upsilon D_{\sigma\sigma}\left(\gamma_\beta + \gamma_\sigma\right)\left(k_{\beta\beta} + k_{\beta\sigma} + k_{\sigma\beta} + k_{\sigma\sigma}\right)}{D_{mT}\left(\gamma_\sigma k_{\beta m} - \gamma_\beta k_{\sigma m}\right) + ha_\upsilon\left(k_{\beta\beta} + k_{\beta\sigma} + k_{\sigma\beta}k_{\sigma\sigma}\right) + ha_\upsilon D_{\sigma\sigma}\left(\gamma_\beta + \gamma_\sigma\right)} \end{aligned} \tag{21}$$

$$\alpha = \frac{k}{\gamma_\beta + \gamma_\sigma} \tag{22}$$

$$\begin{aligned} \frac{1}{\tau_T} = {} & \frac{k_{\beta m}\left[D_{mT}\left(k_{\sigma\beta} + k_{\sigma\sigma}\right) - ha_\upsilon\left(D_{m\beta} + D_{mT}\right)\right] - k_{\sigma m}\left[D_{mT}\left(k_{\beta\beta} + k_{\beta\sigma}\right) + ha_\upsilon\left(D_{m\sigma} + D_{m\beta}\right)\right]}{D_{\sigma\sigma}\left(\gamma_\beta k_{\sigma\sigma} - \gamma_\sigma k_{\beta\beta}\right) + k_{\beta\beta}k_{\sigma\sigma} - k_{\beta\sigma}k_{\sigma\beta} - \gamma_\sigma k_{\beta m}D_{m\beta} - \gamma_\beta k_{\sigma m}D_{m\sigma}} \\ & \times\frac{ha_\upsilon D_{\sigma\sigma}\left(k_{\beta\beta} + k_{\beta\sigma} + k_{\sigma\beta} + k_{\sigma\sigma}\right)}{D_{\sigma\sigma}\left(\gamma_\beta k_{\sigma\sigma} - \gamma_\sigma k_{\beta\beta}\right) + k_{\beta\beta}k_{\sigma\sigma} - k_{\beta\sigma}k_{\sigma\beta} - \gamma_\sigma k_{\beta m}D_{m\beta} - \gamma_\beta k_{\sigma m}D_{m\sigma}} \end{aligned} \tag{23}$$

$$\begin{aligned} & F(\mathbf{r},t) + \tau_q \frac{\partial F(\mathbf{r},t)}{\partial t} = \\ & \frac{ha_\upsilon\left(\gamma_\beta + \gamma_\sigma\right)^2}{D_{mT}\left(\gamma_\sigma k_{\beta m} - \gamma_\beta k_{\sigma m}\right) + ha_\upsilon\left(k_{\beta\beta} + k_{\beta\sigma} + k_{\sigma\beta} + k_{\sigma\sigma}\right) + ha_\upsilon D_{\sigma\sigma}\left(\gamma_\beta + \gamma_\sigma\right)}\frac{\partial^2 u}{\partial t^2} \\ & + \frac{\gamma_\beta\gamma_\sigma\left(\gamma_\beta + \gamma_\sigma\right)}{D_{mT}\left(\gamma_\sigma k_{\beta m} - \gamma_\beta k_{\sigma m}\right) + ha_\upsilon\left(k_{\beta\beta} + k_{\beta\sigma} + k_{\sigma\beta} + k_{\sigma\sigma}\right) + ha_\upsilon D_{\sigma\sigma}\left(\gamma_\beta + \gamma_\sigma\right)}\frac{\partial^3 u}{\partial t^3} \\ & + \frac{\left(\gamma_\beta + \gamma_\sigma\right)\left[k_{\beta m}\left(k_{\sigma\sigma}D_{m\beta} - k_{\sigma\beta}D_{m\sigma}\right) + k_{\sigma m}\left(k_{\beta\beta}D_{m\sigma} - k_{\beta\sigma}D_{m\beta}\right)\right]}{D_{mT}\left(\gamma_\sigma k_{\beta m} - \gamma_\beta k_{\sigma m}\right) + ha_\upsilon\left(k_{\beta\beta} + k_{\beta\sigma} + k_{\sigma\beta} + k_{\sigma\sigma}\right) + ha_\upsilon D_{\sigma\sigma}\left(\gamma_\beta + \gamma_\sigma\right)}\Delta^3 u \\ & - \frac{D_{\sigma\sigma}\left(\gamma_\beta + \gamma_\sigma\right)\left(k_{\beta\beta}k_{\sigma\sigma} - k_{\beta\sigma}k_{\sigma\beta}\right)}{D_{mT}\left(\gamma_\sigma k_{\beta m} - \gamma_\beta k_{\sigma m}\right) + ha_\upsilon\left(k_{\beta\beta} + k_{\beta\sigma} + k_{\sigma\beta} + k_{\sigma\sigma}\right) + ha_\upsilon D_{\sigma\sigma}\left(\gamma_\beta + \gamma_\sigma\right)}\Delta^3 u \end{aligned} \tag{24}$$

This can be regarded as a DPL heat-conduction equation regarding Δu with τ_q, τ_T, and $F(\mathbf{r},t) + \tau_q \frac{\partial F(\mathbf{r},t)}{\partial t}$ as the phase lags of the heat flux and the temperature gradient, and the source-related term, respectively. Therefore, the coupled heat and mass transport is capable of varying not only thermal conductivity from that in Equation (13) to the one in Equation (21) but also the nature of heat conduction from that in Equation (12) to the one in Equation (19). As practical nanofluid system always involves many transport processes simultaneously, the coupled transport could play a significant role. For assessing its effect and understanding heat conduction in nanofluids, future research is in great demand on coupling (cross) transport coefficients that are derivable by approaches like the up-scaling with closures [2, 27, 28], the kinetic theory [71, 72], the time-correlation functions [73, 74], and the experiments based on phenomenological flux relations [68]. While the uncoupled form of conservation equations, such as Equations (12) and (19), is very useful for examining nature of heat transport, its coupled form, such as Equations (8), (9), (16)-(18), is normally more readily to be resolved for the temperature or concentration fields after all the transport coefficients are available.

Brownian Motion

In nanofluids, nanoparticles randomly move through liquid and possibly collide. Such a Brownian motion was thus proposed to be one of the possible origins for thermal conductivity enhancement because (i) it enables direct particle-particle transport of heat from one to another, and (ii) it induces surrounding fluid flow and thus so-called microconvection. The ratio of the former contribution to the thermal conductivity (k_{BD}) to the base fluid conductivity (k_f) is estimated based on the kinetic theory [75],

$$\frac{k_{\mathrm{BD}}}{k_{\mathrm{f}}}=\frac{(\rho c)_{\mathrm{p}}\varphi k_{\mathrm{B}}T}{3\pi\mu d_{\mathrm{p}}k_{\mathrm{f}}} \qquad (25)$$

where subscripts p and BD stand for the nanoparticle and the Brownian diffusion, respectively; k_B is the Boltzmann's constant (1.38065×10^{-23}J/K); and μ is the fluid viscosity. The kinetic theory also gives an upper limit for the ratio of the latter›s contribution to the thermal conductivity (k_{BC}) to the base fluid conductivity (k_f) [76],

$$\frac{k_{\mathrm{BC}}}{k_{\mathrm{f}}}=\frac{k_{\mathrm{B}}T}{3\pi\mu d_{\mathrm{p}}\alpha_{\mathrm{f}}} \qquad (26)$$

where subscript BC refers to the Brownian-motion-induced convection, and α_f is the thermal diffusivity of the base fluid.

Consider a 1% volume fraction of d_p = 10 nm copper nanoparticle in water suspension at T = 300 K. $(\rho c)_p = 8900$ kg/m³ × 0.386 kJ/(kg K) = 3435.4 kJ/(m³ K), $\mu = 0.798 \times 10^{-3}$kg/(ms), k_f= 0.615 W/(mK), and $\alpha_f = 1.478 \times 10^{-7}$ m²/s. These yield $k_{BD}/k_f = 3.076 \times 10^{-6}$ and $k_{BC}/k_f = 3.726 \times 10^{-4}$. Therefore, both contributions are negligibly small.

Although the direct contribution of particle Brownian motion to the nanofluid conductivity is negligible, its indirect effect could be significant because it plays an important role in processes of particle aggregating and coupled transport.

CONCLUDING REMARKS

Under the specified volume fractions and thermal conductivities of the two phases in the colloidal state, the interfacial energy transport between the two phases favors a configuration in which the higher-conductivity phase forms a continuous path for thermal flow and disperses the lower-conductivity phase. The effective thermal conductivity is thus bounded by those corresponding to the two limiting morphologies: the well-dispersed configuration of the higher-conductivity phase in the lower-conductivity phase and the well-dispersed configuration of the lower-conductivity phase in the higher-conductivity phase, corresponding to the lower and the upper bounds of thermal conductivity, respectively. Without considering the effect of interfacial resistance and cross coupling among various transport processes, the classical effective-medium theory gives these bounds known as the H-S bounds. A wide separation of these two bounds offers spacious room of manipulating nanofluid thermal conductivity via the morphology effect.

In a nanofluid system, there are normally two or more transport processes that occur simultaneously. The cross coupling among these processes causes new induced effects of flows occurring without or against its primary thermodynamic driving force and is capable of changing the nature of heat conduction via inducing thermal waves and resonance. Depending on the microscale physics (factors like material properties of nanoparticles and base fluids, nanoparticles' morphology in the base fluids, and interfacial properties and dynamic processes on particle-fluid interfaces), the heat diffusion and thermal waves may either enhance or counteract each other. Consequently, the heat conduction may be enhanced or weakened by the presence of nanoparticles.

The direct contributions of ordered liquid layer and particle Brownian motion to the nanofluid conductivity are negligible. Their influence on the particle morphology and/or the coupled transport could, however, offer a strong indirect effect to the nanofluid conductivity.

Therefore, nanofluids with conductivity of upper H-S bound can be obtained by manipulating particles into an interconnected configuration that disperses the base fluid, and thus significantly enhancing the particle-fluid interfacial energy transport. Nanofluids with conductivity higher than the upper H-S bound could also be developed by manipulating the cross coupling among various transport processes and thus the nature of heat conduction in nanofluids.

ACKNOWLEDGEMENTS

The financial support from the Research Grants Council of Hong Kong (GRF718009 and GRF717508) is gratefully acknowledged.

REFERENCES

1. Wang LQ, Fan J: Nanofluids Research: Key Issues. Nanoscale Res Lett 2010, 5: 1241. 10.1007/s11671-010-9638-6
2. Wang LQ, Quintard M: Nanofluids of the future. In Advances in Transport Phenomena. New York: Springer; 2009:179. 2009 2009
3. Choi SUS, Zhang ZG, Keblinski P: Nanofluids. In Encyclopedia of Nanoscience and Nanotechnology. Edited by: Nalwa HS. New York: American Scientific Publishers; 2004:757.
4. Peterson GP, Li CH: Heat and mass transfer in fluids with nanoparticle suspensions. Adv Heat Transfer 2006, 39: 257.

5. Das CH, Choi SUS, Yu W, Pradeep T: Nanofluids: Science and Technology. Hoboken, NJ: John Wiley & Sons, Inc; 2008.
6. Sobhan CB, Peterson GP: Microscale and Nanoscale Heat Transfer: Fundamentals and Engineering Applications. Boca Raton: CRC Press; 2008.
7. Choi SUS: Nanofluids: from vision to reality through research. ASME J Heat Transfer 2009, 131: 033106. 10.1115/1.3056479
8. Fan J, Wang LQ: Review of heat conduction in nanofluids. ASME J Heat Transfer 2011, 133: 040801. 10.1115/1.4002633
9. Wang LQ, Wei XH: Heat conduction in nanofluids. In Handbook of Nanophysics: Nanoparticles and Quantum Dots. Volume Chapter 33. Edited by: Sattler K. Taylor & Francis; 2010:33–1.
10. Prasher R, Phelan PE, Bhattacharya P: Effect of aggregation kinetics on the thermal conductivity of nanoscale colloidal solutions (Nanofluids). Nano Lett 2006, 6: 1529. 10.1021/nl060992s
11. Rusconi R, Rodari E, Piazza R: Optical measurements of the thermal properties of nanofluids. Appl Phys Lett 2006, 89: 261916. 10.1063/1.2425015
12. Yu W, Choi SUS: The role of interfacial layers in the enhanced thermal conductivity of nanofluids: A renovated Hamilton-Crosser model. J Nanopart Res 2004, 6: 355. 10.1007/s11051-004-2601-7
13. Xue L, Keblinski P, Phillpot SR, Choi SUS, Eastman JA: Effect of liquid layering at the liquid-solid interface on thermal transport. Int J Heat Mass Transfer 2004, 47: 4277. 10.1016/j.ijheatmasstransfer.2004.05.016
14. Xie HQ, Fujii M, Zhang X: Effect of interfacial nanolayer on the effective thermal conductivity of nanoparticle-fluid mixture. Int J Heat Mass Transfer 2005, 48: 2926. 10.1016/j.ijheatmasstransfer.2004.10.040
15. Ren Y, Xie H, Cai A: Effective thermal conductivity of nanofluids containing spherical nanoparticles. J Phys D Appl Phys 2005, 38: 3958. 10.1088/0022-3727/38/21/019
16. Leong KC, Yang C, Murshed SMS: A model for the thermal conductivity of nanofluids - the effect of interfacial layer. J Nanopart Res 2006, 8: 245. 10.1007/s11051-005-9018-9
17. Wang BX, Zhou LP, Peng XF: A fractal model for predicting the effective thermal conductivity of liquid with suspension of nanoparticles. Int J Heat Mass Transfer 2003, 46: 2665. 10.1016/S0017-9310(03)00016-4
18. Wang LQ, Zhou XS, Wei XH: Heat Conduction: Mathematical Models and Analytical Solutions. Heidelberg, Berlin: Springer-Verlag; 2008.
19. Wang LQ, Xu MT, Wei XH: Multiscale theorems. Adv Chem Eng 2008, 34: 175. full_text
20. Wang L, Wei X: Nanofluids: synthesis, heat conduction, and extension. ASME J Heat Transfer 2009, 131: 033102. 10.1115/1.3056597
21. Koo J, Kleinstreuer C: A new thermal conductivity model for nanofluids. J Nanopart Res 2004, 6: 577. 10.1007/s11051-004-3170-5
22. Jang SP, Choi SUS: Role of Brownian motion in the enhanced thermal conductivity of nanofluids. Appl Phys Lett 2004, 84: 4316. 10.1063/1.1756684
23. Bhattacharya P, Saha SK, Yadav A, Phelan PE, Prasher RS: Brownian dynamics simulation to determine the effective thermal conductivity of nanofluids. J Appl Phys 2004, 95: 6492. 10.1063/1.1736319
24. Prasher R, Bhattacharya P, Phelan P: Thermal conductivity of nanoscale colloidal solutions (nanofluids). Phys Rev Lett 2005, 94: 025901. 10.1103/PhysRevLett.94.025901
25. Prasher R, Bhattacharya P, Phelan PE: Brownian-motion-based convective-conductive model for the effective thermal conductivity of nanofluids. J Heat Transfer Trans ASME 2006, 128: 588. 10.1115/1.2188509
26. Yu W, Choi SUS: The role of interfacial layers in the enhanced thermal conductivity of nanofluids: A renovated Maxwell model. J Nanopart Res 2003, 5: 167. 10.1023/A:1024438603801
27. Fan J, Wang LQ: Effective thermal conductivity of nanofluids: the effects of microstructure. J Phys D Appl Phys 2010, 43: 165501. 10.1088/0022-3727/43/16/165501
28. Fan J, Wang LQ: Microstructural effects on macroscale thermal properties in nanofluids. NANO 2010, 5: 117. 10.1142/S1793292010002001
29. Fan J, Wang LQ: Constructal design of nanofluids. Int J Heat Mass Transfer 2010, 53: 4238. 10.1016/j.ijheatmasstransfer.2010.05.050
30. Bai C, Wang LQ: Constructal structure of nanofluids. J Appl Phys 2010, 108: 068979.
31. Bai C, Wang LQ: Constructal Allocation of Nanoparticles in Nanofluids. J Heat Transfer Trans ASME 2010, 132: 052404. 10.1115/1.4000473
32. Bai C, Wang LQ: Constructal design of nanofluids for one-dimensional steady heat conduction systems. NANO 2010, 5: 39. 10.1142/S1793292010001895
33. Hashin Z, Shtrikman S: A variational approach to the theory of the effective magnetic permeability of multiphase materials. J Appl Phys 1962, 33: 3125. 10.1063/1.1728579
34. Wei XH, Wang LQ: 1+1 > 2: Extraordinary fluid conductivity enhancement. Curr Nanosci 2009, 5: 527.
35. Kang HU, Kim SH, Oh JM: Estimation of thermal conductivity of nanofluid using experimental effective particle volume. Exp Heat Transfer 2006, 19: 181. 10.1080/08916150600619281
36. Eapen J, Li J, Yip S: Mechanism of thermal transport in dilute nanocolloids. Phys Rev Lett 2007, 98: 028302. 10.1103/PhysRevLett.98.028302
37. Buongiorno J, Venerus DC, Prabhat N, McKrell T, Townsend J, Christianson R, Tolmachev YV, Keblinski P, Hu LW, Alvarado JL, Bang IC, Bishnoi SW, Bonetti M, Botz F, Cecere A, Chang Y, Chen G, Chen HS, Chung SJ, Chyu MK, Das SK, Di Paola R, Ding YL, Dubois F, Dzido G, Eapen J, Escher W, Funfschilling D, Galand Q, Gao JW, Gharagozloo PE, Goodson KE, Gutierrez JG, Hong HP, Horton M, Hwang KS, Iorio CS, Jang SP, Jarzebski AB, Jiang YR, Jin LW, Kabelac S, Kamath

A, Kedzierski MA, Kieng LG, Kim C, Kim JH, Kim S, Lee SH, Leong KC, Manna I, Michel B, Ni R, Patel HE, Philip J, Poulikakos D, Reynaud C, Savino R, Singh PK, Song PX, Sundararajan T, Timofeeva E, Tritcak T, Turanov AN, Van Vaerenbergh S, Wen DS, Witharana S, Yang C, Yeh WH, Zhao WH, Zhou SQ: A benchmark study on the thermal conductivity of nanofluids. J Appl Phys 2009, 106: 094312. 10.1063/1.3245330

38. Williams W, Buongiorno J, Hu LW: Experimental investigation of turbulent convective heat transfer and pressure loss of alumina/water and zirconia/water nanoparticle colloids (nanofluids) in horizontal tubes. J Heat Transfer Trans ASME 2008, 130: 042412. 10.1115/1.2818775

39. Zhang X, Gu H, Fujii M: Experimental study on the effective thermal conductivity and thermal diffusivity of nanofluids. Int J Thermophys 2006, 27: 569. 10.1007/s10765-006-0054-1

40. Zhu HT, Zhang CY, Liu SQ, Tang YM, Yin YS: Effects of nanoparticle clustering and alignment on thermal conductivities of Fe3O4 aqueous nanofluids. Appl Phys Lett 2006, 89: 023123. 10.1063/1.2221905

41. Shima PD, Philip J, Raj B: Role of microconvection induced by Brownian motion of nanoparticles in the enhanced thermal conductivity of stable nanofluids. Appl Phys Lett 2009, 94: 223101. 10.1063/1.3147855

42. Murshed SMS, Leong KC, Yang C: Enhanced thermal conductivity of TiO2 - water based nanofluids. Int J Therm Sci 2005, 44: 367. 10.1016/j.ijthermalsci.2004.12.005

43. Duangthongsuk W, Wongwises S: Measurement of temperature-dependent thermal conductivity and viscosity of TiO2-water nanofluids. Exp Therm Fluid Sci 2009, 33: 706. 10.1016/j.expthermflusci.2009.01.005

44. Mintsa HA, Roy G, Nguyen CT, Doucet D: New temperature dependent thermal conductivity data for water-based nanofluids. Int J Therm Sci 2009, 48: 363. 10.1016/j.ijthermalsci.2008.03.009

45. Lee DY, Vafai K: Analytical characterization and conceptual assessment of solid and fluid temperature differentials in porous media. Int J Heat Mass Transfer 1999, 42: 423. 10.1016/S0017-9310(98)00185-9

46. Das SK, Putra N, Thiesen P, Roetzel W: Temperature dependence of thermal conductivity enhancement for nanofluids. ASME J Heat Transfer 2003, 125: 567. 10.1115/1.1571080

47. Li CH, Peterson GP: Experimental investigation of temperature and volume fraction variations on the effective thermal conductivity of nanoparticle suspensions (nanofluids). J Appl Phys 2006, 99: 084314. 10.1063/1.2191571

48. Zhu HT, Zhang CY, Tang YM, Wang JX: Novel synthesis and thermal conductivity of CuO nanofluid. J Phys Chem C 2007, 111: 1646. 10.1021/jp065926t

49. Hong JG, Kim SH, Kim DS: Effect of laser irradiation on thermal conductivity of ZnO nanofluids. J Phys Conf Ser 2007, 59: 301. 10.1088/1742-6596/59/1/063

50. Kim SH, Choi SR, Kim D: Thermal conductivity of metal-oxide nanofluids: Particle size dependence and effect of laser irradiation. J Heat Transfer Trans ASME 2007, 129: 298. 10.1115/1.2427071

51. Wen DS, Ding YL: Experimental investigation into convective heat transfer of nanofluids at the entrance region under laminar flow conditions. Int J Heat Mass Transfer 2004, 47: 5181. 10.1016/j.ijheatmasstransfer.2004.07.012

52. Gharagozloo PE, Eaton JK, Goodson KE: Diffusion, aggregation, and the thermal conductivity of nanofluids. Appl Phys Lett 2008, 93: 103110. 10.1063/1.2977868

53. Moosavi M, Goharshadi EK, Youssefi A: Fabrication, characterization, and measurement of some physicochemical properties of ZnO nanofluids. Int J Heat Fluid Flow 2010, 31: 599. 10.1016/j.ijheatfluidflow.2010.01.011

54. Hong TK, Yang HS, Choi CJ: Study of the enhanced thermal conductivity of Fe nanofluids. J Appl Phys 2005, 97: 064311. 10.1063/1.1861145

55. Sinha K, Kavlicoglu B, Liu YM, Gordaninejad F, Graeve OA: A comparative study of thermal behavior of iron and copper nanofluids. J Appl Phys 2009, 106: 064307. 10.1063/1.3225574

56. Murshed SMS, Leong KC, Yang C: Determination of the effective thermal diffusivity of nanofluids by the double hot-wire technique. J Phys D Appl Phys 2006, 39: 5316. 10.1088/0022-3727/39/24/033

57. Eastman JA, Choi SUS, Li S, Yu W, Thompson LJ: Anomalously increased effective thermal conductivities of ethylene glycol-based nanofluids containing copper nanoparticles. Appl Phys Lett 2001, 78: 718. 10.1063/1.1341218

58. Assael MJ, Metaxa IN, Kakosimos K, Constantinou D: Thermal conductivity of nanofluids - experimental and theoretical. Int J Thermophys 2006, 27: 999. 10.1007/s10765-006-0078-6

59. Garg J, Poudel B, Chiesa M, Gordon JB, Ma JJ, Wang JB, Ren ZF, Kang YT, Ohtani H, Nanda J, McKinley GH, Chen G: Enhanced thermal conductivity and viscosity of copper nanoparticles in ethylene glycol nanofluid. J Appl Phys 2008, 103: 074301. 10.1063/1.2902483

60. Xie H, Lee H, Youn W, Choi M: Nanofluids containing multiwalled carbon nanotubes and their enhanced thermal conductivities. J Appl Phys 2003, 94: 4967. 10.1063/1.1613374

61. Liu MS, Lin MCC, Huang IT, Wang CC: Enhancement of thermal conductivity with carbon nanotube for nanofluids. Int Commun Heat Mass Transfer 2005, 32: 1202. 10.1016/j.icheatmasstransfer.2005.05.005

62. Choi SUS, Zhang ZG, Yu W, Lockwood FE, Grulke EA: Anomalous thermal conductivity enhancement in nanotube suspensions. Appl Phys Lett 2001, 79: 2252. 10.1063/1.1408272

63. Xie H, Wang J, Xi T, Liu Y, Ai F: Thermal conductivity enhancement of suspensions containing nanosized alumina particles. J Appl Phys 2002, 91: 4568. 10.1063/1.1454184

64. Raghavan K, Foster K, Motakabbir K, Berkowitz M: Structure and Dynamics of Water at the Pt(111) Interface

- Molecular-Dynamics Study. J Chem Phys 1991, 94: 2110. 10.1063/1.459934

65. Reedijk MF, Arsic J, Hollander FFA, de Vries SA, Vlieg E: Liquid order at the interface of KDP crystals with water: Evidence for icelike layers. Phys Rev Lett 2003, 90: 066103. 10.1103/PhysRevLett.90.066103

66. Mo H, Evmenenko G, Dutta P: Ordering of liquid squalane near a solid surface. Chem Phys Lett 2005, 415: 106. 10.1016/j.cplett.2005.08.142

67. Yu CJ, Richter AG, Kmetko J, Dugan SW, Datta A, Dutta P: Structure of interfacial liquids: X-ray scattering studies. Phys Rev E 2001, 63: 021205. 10.1103/PhysRevE.63.021205

68. Demirel Y: Nonequilibrium Thermodynamics: Transport and Rate Processes in Physical, Chemical and Biological Systems. 2nd edition. Amsterdam: Elsevier; 2007.

69. Tzou D: Macro- to Micro-Scale Heat Transfer: The Lagging Behavior. Washington, DC: Taylor & Francis; 1997.

70. Xu MT, Wang LQ: Thermal oscillation and resonance in dual-phase-lagging heat conduction. Int J Heat Mass Transfer 2002, 45: 1055. 10.1016/S0017-9310(01)00199-5

71. Chapman S, Cowling TG: The Mathematical Theory of Non-Uniform Gases: An Account of the Kinetic Theory of Viscosity, Thermal Conduction and Diffusion in Gases. Cambridge: Cambridge University Press; 1991.

72. Hanley HJM: Transport Phenomena in Fluids. New York: Marcel Dekker; 1969.

73. McQuarrie DA: Statistical Mechanics. Sausalito: University Science Books; 2000.

74. Zwanzig RW: Time-correlation functions and transport coefficients in statistical mechanics. Annu Rev Phys Chem 1965, 16: 67. 10.1146/annurev.pc.16.100165.000435

75. Keblinski P, Cahil DG: Comment on “Model for heat conduction in nanofluids. Phys Rev Lett 2005, 95: 209401. 10.1103/PhysRevLett.95.209401

76. Evans W, Fish J, Keblinski P: Role of Brownian motion hydrodynamics on nanofluid thermal conductivity. Appl Phys Lett 2006, 88: 093116. 10.1063/1.2179118

CHAPTER 19

TRANSPORT PHENOMENA IN GEL

Masayuki Tokita

Keywords: Diffusion, friction, scaling theory, colloid gel

INTRODUCTION

Gel has the characteristics and common structure that consists of the three-dimensional polymer network and gel fluid. Because of such a diluted structure, many molecules are transported through gel. Such a characteristic of gel is used in separation technologies, namely, gel electrophoresis, gel permeation chromatography, and so forth. Therefore, information on the transport phenomena in and across gel is of importance in designing a separation system. Moreover, knowledge on transport phenomena in gel has become quite important recently for understanding the volume phase transition of gel and the pattern formation phenomena of gel [1,2,3]. It has been reported that the volume of gel reversibly changes more than a thousandfold in response to the slight environmental change in the vicinity of the volume phase transition point [4]. The huge amount of gel fluid, therefore, flows in and out in accordance with the swelling and shrinking of gel. Kinetic study on the volume change of gel indicates that the characteristic time of the swelling of gel, τ, is determined by the ratio of the longitudinal modulus of gel, E, the friction between gel fluid and the polymer network of gel, f, and the typical size of gel, a, as $\tau=a^2f/E$ [5]. In contrast, the shrining pattern formation phenomena in the shrinking process of gel in the poor solvent is very complicated since the first event that occurs in the beginning of the shrinking process is the diffusion of the poor solvent into gel. Then, the squeeze out of gel fluid came in to create the shrinking patterns. Both the diffusion and the friction compete with each other in pattern formation in the shrinking gels. As a result, various shrinking patterns appear in the shrinking gel [6,7,8]. In addition to this, one of the most striking expectations is deduced from the study of the critical dynamics of the volume phase transition of gel, namely, the friction of gel becomes smaller and it vanishes at the volume phase transition point [9]. The diffusion and the friction, therefore, play important roles in the volume phase transition of gel as well as the pattern formation in gel. The detailed studies on the transport of substances by the diffusion and the flow of gel fluid are still required for the full understanding of the volume phase transition of gel and the pattern formation in gel. Since the polymer network of gel is regarded as an assembly of the molecular mesh, the relationship between the transport properties of gel and the mesh size of the polymer network of gel should first be clarified. The diffusion of small molecules in gel and the frictional properties of gel have been studied by many researchers so far [10,11,12,13]. Although many studies have been conducted, most measurements are made under not ideal conditions. In this review, we will discuss the diffusion and the friction of gel in terms of the structure of gel by which the simple view for the transport phenomena in gel will be deduced.

Citation: Masayuki Tokita, "Transport Phenomena in Gel," Gels 2016, 2(2), 17; doi:10.3390/gels2020017.

GEL

Gel is classified into two classes by the structure of the cross-linking point. One is called chemical gel and the other is known as physical gel. In the case of chemical gel, the cross-linking point consists of a covalent bond. On the other hand, the cross-linking point of physical gel consists of the assembly of polymers that is formed by the short range interactions such as the van der Waals interaction, the hydrophobic interaction, the hydrogen bond, and the Coulomb interaction between charges. The network structure of chemical gel is not varied by the change of the external environment such as the temperature, because the bonding energy of the covalent bond is much higher than the thermal energy. Hence, the cross-linking point of the polymer network is unchanged even if the temperature is changed, say, from 0 to 100°C. Gel is in an equilibrium swelling state [14]. The volume of gel changes in response to environmental changes such as the temperature in the case of chemical gel. If the appropriate chemical structure is given to the polymer network of gel, then gel shows the volume phase transition phenomena. On the other hand, the bonding energy due to short range interactions is almost the same order of magnitude as the thermal energy. Accordingly, most physical gels show the sol-gel transition phenomena when the environmental conditions are changed. The sol-gel transition and the volume phase transition are the characteristic transitions in gel system [15].

In this review, we describe the transport properties of the chemical gels of acrylamide (AAm) and N- (NIPA) cross-linked by N,N-methylenebisacrylamide (BIS). The linear polymer chains of AAm or NIPA are connected by BIS to form a three-dimensional polymer network. The network structure is, therefore, determined both by the concentrations of the linear chain component and the cross-linker. The cross-linking density of gel, which is defined by the mole fraction of the cross-linker to the total amount of the linear component and the cross-linker, is fixed at 0.01 or 0.02 in most studies. The total concentration of gel is changed under the fixed cross-linking density. The gels obtained under these conditions are uniform; hence, gel is transparent and elastic. On the other hand, the cross-linking density is changed at a constant total concentration, usually fixed at 700 mM. The gel prepared at higher concentrations of BIS are opaque and brittle suggesting the formation of a non-uniform structure.

The mesh size of the polymer network is the most important parameter that defines the structure of the polymer network of gel. The gels used here are synthesized by the random copolymerization of the linear chain component and the cross-linker. As a result of this, the network structure of gel becomes more or less random. The size of the polymer mesh is, therefore, expressed by an averaged parameter that is called the correlation length of gel, ξ. The correlation length of gel represents, in a good solvent, the distance between two nearest neighbor contact points of polymers. The correlation length of gel varies with the concentration of the linear chain component, the cross-linking density, and the environmental conditions such as the temperature.

For most parts of the study, poly(acrylamide) gel (PAAm gel) is chosen as the sample gel since this gel is widely used in separation and the purification technologies. Hence, the physical and the chemical properties of PAAm gel are well known. For instance, the concentration of gel can be changed widely from 0.02 to 0.5 g/mL because of a high affinity with water. This is a definite advantage for the present purpose of the study because the correlation length of the polymer network of gel can be changed widely. In addition to this, PAAm gel does not interact strongly with many water soluble molecules. This is another great advantage for the present study since it widens the choice of the probe molecules. On the other hand, poly(n-siopropylacrylamide) gel (PNIPA gel) is known as the thermo-sensitive gel. The gel shows the discontinuous volume change at about 34°C in water. The correlation length of PNIPA gel can be changed reversibly as a function of the temperature. Therefore, the critical behaviors of the frictional property can be revealed by studying the frictional property of PNIPA gel in the vicinity of the volume phase transition point.

PROBE DIFFUSION IN GEL

Many advancements in nuclear magnetic resonance technology have been made at the end of the last century. One of the useful technologies in the study of transport phenomena is the pulsed field gradient nuclear magnetic resonance method (PFG-NMR). The diffusion coefficient of the specific molecules, which will be called the probe molecule hereafter, can be determined by using PFG-NMR invasively if the probe molecule is distinguishable in the NMR spectrum of the whole system. The PGF-NMR method is firstly applied to study the diffusion of the molecules in simple solution. It is then applied to complex systems such as the semi-dilute polymer solution systems, and finally to polymer gels [10,16,17]. Previous studies suggest that it is a requirement to study the diffusion process by changing both the concentration of gel and the size of the probe molecule systematically to obtain the entire aspects of the diffusion process in gel. The diffusion of the probe molecule in gel is, however, extensively affected by short range interactions between the probe molecule and the polymer network. The presence of short range interactions is unfavorable because we firstly focus our attention only on the effects of the polymer network on the diffusion process of the probe molecules. Since the short range interactions depend strongly on the chemical structures of the polymer network and the probe molecules, one can minimize the complex situations by choosing appropriate chemicals for gel and the probe molecules. Then, we can deduce the effects of only the polymer network of gel on the diffusion of the probe molecules.

In this study, the diffusion coefficient of the probe molecules are measured in PAAm gel of the total concentration from 0.02 to 0.5 g/mL at a cross-linking density of 0.02. Although any substance can be a candidate for the probe diffusion experiments in PAAm gel, there are still several requirements for the probe molecule for the present purpose of the study as listed below.

1. Highly soluble in water because water is used as a solvent.
2. Compact molecules to be assumed as a spherical molecule.
3. Molecular weight spreads at least one order of magnitude.
4. Chemical shift does not overlap with that of PAAm gel.
5. Absence of strong interaction with the polymer network of PAAm gel.

Taking into account these requirements, we finally choose water (solvent), ethanol, glycerin, poly(ethylene glycol), and sucrose as the probe molecules. The molecular weight of each probe molecule is, 18, 46, 92, 200, and 342, respectively.

The sample PAAm gel is synthesized in an NMR tube of 10 mm in diameter. The solvent is the mixture of water and the heavy water (H2O:D2O = 9:1) which contains the probe molecule at a concentration of 10 wt % is prepared firstly. Then, an appropriate amount of AAm, BIS, and ammonium persulfate (initiator) are dissolved into the above solvent. The pre-gel solution thus obtained is de-gassed for 20 min. The reaction is initiated by raising the temperature to 60°C for 1 h. The diffusion experiments are made on a JEOL FX-60Q (JEOL, Tokyo, Japan), which is equipped with a field gradient apparatus NMPL-502 (JEOL, Tokyo, Japan), at a frequency of 60 MHz. The temperature is controlled at 30.0 ± 0.5°C. Details of the experimental procedure are given in the previous report [18].

In Figure 1, the diffusion coefficient of the probe molecules in PAAm gels are plotted as a function of the concentration of gel.

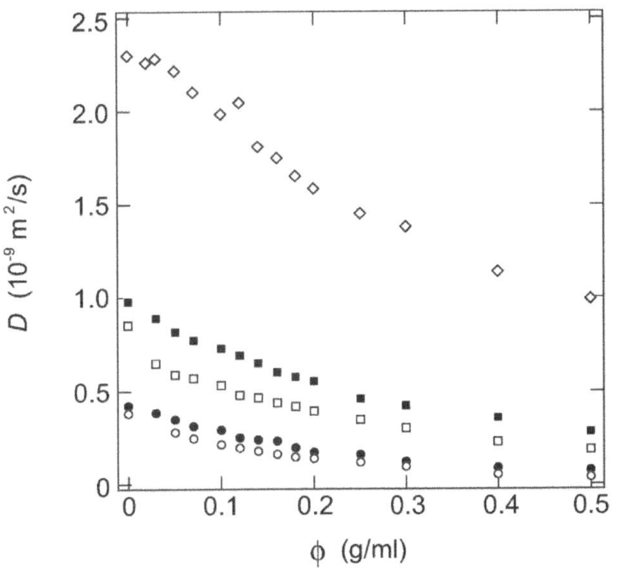

Figure 1. The dependence of the diffusion coefficient of probe molecules on the concentration of acrylamide (AAm) in gel. The probe molecules are water, diamonds, ethanol (filled squares), glycerin (open squares), poly(ethylene glycol) (filled circles), and sucrose (open circles), from top to bottom. The data points at the position of zero concentration represent the diffusion coefficient of each probe molecules in water that are measured by the same experimental setup. The temperature is fixed at 30.0 ± 0.5°C.

The probe molecule, when dissolved in a simple fluid, thermally fluctuates in time and space. The thermal fluctuation and the hydrodynamic friction due to the surrounding fluid, ζ, determine the rate of diffusion. The diffusion coefficient, D_0, of the probe molecule of the hydrodynamic radius, R_h, in a simple fluid of viscosity, η, is expressed by the so-called Stokes–Einstein relationship as follows [19,20]:

$$D_0 = \frac{k_B T}{\zeta} = \frac{k_B T}{6\pi\eta R_h}. \quad (1)$$

Here, kB and T represent the Boltzmann's constant and temperature. It is clear from the above equation that the diffusion coefficient of the probe molecule is essentially determined by the size of the probe molecule, R_h, when it is dissolved in a simple fluid at the constant temperature. However, it is found from Figure 1 that the diffusion coefficients of all probe molecules decrease monotonically with the concentration of AAm in the gel. The diffusion coefficient of the probe molecules decreases even at a very small concentration region of gel, ca., less than 0.1 g/mL. The results obtained here, therefore, indicate that the diffusion of the probe molecule within the gel is far from the free diffusion in a simple fluid even in a tenuous gel. It is clear from Figure 1 that the diffusion coefficient of the probe molecule is a decreasing function of the molecular weight of the probe molecule, and it is also a decreasing function of the concentration of the gel. Furthermore, it may be clear that the concentration dependence of the diffusion coefficients of the probe molecules are similar to each other. It strongly suggests the presence of a universal function for the diffusion coefficient of the probe molecules in the gel. Since both the concentration of the gel and the size of the probe molecule is systematically changed in this study, such a function is written by these two parameters.

A similar phenomena has been also studied in semi-dilute polymer solutions and effort is devoted to finding the universal physical picture of the diffusion in these condensed polymer systems of the semi-dilute solution and gel [21,22]. Among others, one of the hopeful analyses of the diffusion processes in gel is the scaling approach [23]. In modern statistical theory of polymer systems, many physical quantities are expressed in terms of the scaling function, f(x). The diffusion coefficient of the probe molecules in gel that is normalized by D0 is also expected to be expressed by a scaling function as follows:

$$\frac{D}{D_0} = f(x). \quad (2)$$

Here, D and D_0 represent the diffusion coefficient of the probe molecule in the gel and that in the simple fluid, and x is the non-dimensional scaling variable. Since the diffusion coefficient is determined by the linear size of the probe molecule, R_h, one of the candidates for the scaling variable is a ratio of R_h and the correlation length. The correlation length of the polymer network of gel, ξ, is expressed as a function of the concentration of gel, ϕ, as follows [23]:

$$\xi \propto \phi^{-3/4}. \quad (3)$$

On the other hand, the size of the molecules with simple and compact structures is simply assumed as follows using the molecular weight of the probe molecule, M:

$$R_h \propto M^{1/3}. \quad (4)$$

We, therefore, assume the following scaling variable:

$$x = \frac{R_h}{\xi} \propto M^{1/3}\phi^{3/4}. \quad (5)$$

In other words,

$$\frac{D}{D_0} \propto f\left(M^{1/3}\phi^{3/4}\right). \quad (6)$$

The results shown in Figure 1 are, therefore, plotted as a function of the scaling variable, x, in Figure 2.

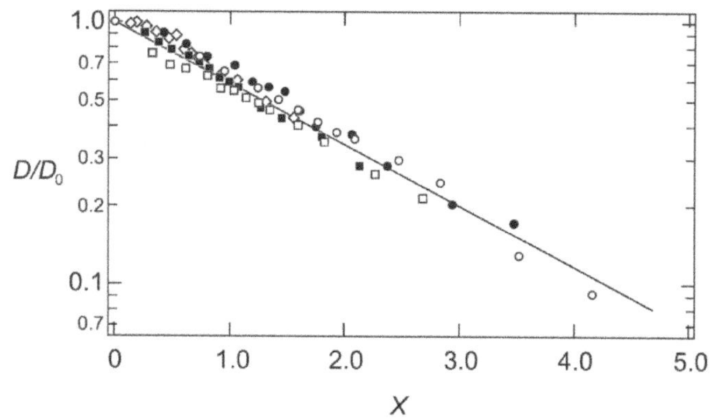

Figure 2. The normalized diffusion coefficient of the probe molecules in gel, D/D_0, is plotted as a function of the scaling variable, $x = M^{1/3}\phi^{3/4}$, in a semi-logarithmic manner. Symbols are the same as that of the Figure 1.

It is found that all data fall onto a single master curve. In addition to this, the results are well explained by a simple functional form as follows:

$$\frac{D}{D_0} = \exp\left(-\frac{R_h}{\xi}\right). \tag{7}$$

The results are in good agreement with the hydrodynamic calculations for the scaling function [21,22]. It is found that the diffusion coefficient of the probe molecules follows the simple scaling theory.

The results obtained here suggest strongly that the probe molecule that diffuses in the polymer network of PAAm gel mainly experiences resistance due to the polymer network. In addition, the effects of the polymer network on the diffusion of the probe molecules are determined by the ratio of the correlation length of the polymer network and the radius of the probe molecule. The results in Equation (7) indicate that the relative rate of the diffusion is equivalent if the ratio of the hydrodynamic radius and the correlation length of the polymer network, Rh/ξ, are equal. The smaller probe molecule diffuses faster within the gel. However, even the size of the probe molecule is larger, it diffuses in the gel at the similar rate as the smaller probe molecules if the mesh size of the gel is large enough. The rate of the diffusion in the gel is, in the simplest case, determined by the geometrical factor of the probe molecule and the polymer network of the gel. The parameter that determines the diffusion of the substance in gel is the ratio of the size of the probe molecule and the correlation length of gel, Rh/ξ. The results obtained here are schematically illustrated in Figure 3.

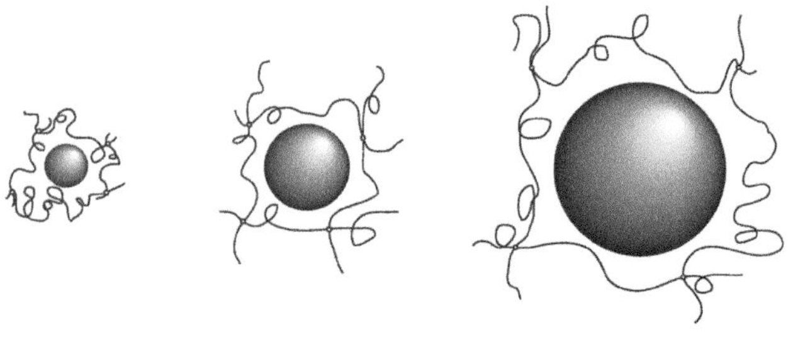

Figure 3. The schematic representation of Equation (7). The size of the probe molecules is different, but the ratio of the size and the correlation length, Rh/ξ, is similar in these three cases. The diffusion coefficient of the probe molecule that is normalized by the diffusion coefficient of the probe molecule in water, D/D0, is almost the same in these cases.

FRICTION OF GEL

Many efforts have been devoted to creating active systems such as artificial muscles after finding the volume phase transition of gel [24]. When gel collapses, a huge amount of gel fluid is squeezed out of gel. Hence, the acceleration of the drainage of gel fluid from gel is the key to success. In other words, an understanding of the friction between the polymer network of gel and gel fluid is required. The study of the kinetics of the swelling of gel indicates that the swelling of gel is determined by the collective diffusion coefficient of gel, Dc [25,26,27]:

$$D_c = \frac{E}{f}. \tag{8}$$

Here, E represents the longitudinal elastic modulus of the polymer network alone and f the friction coefficient between the polymer network of gel and gel fluid. Equation (8) indicates that the polymer network of gel moves under the drag force of the surrounding fluid.

The measurements of the friction between the polymer network of gel and gel fluid is, in principle, simple as schematically shown in Figure 4. However, only a little is known about the frictional properties of gels. The reason is simple—the friction coefficient of the polymer network is huge, while gel is fragile. Detailed study on the frictional properties of the PAAm gels are reported where the strategies to design the apparatus are also reported [28].

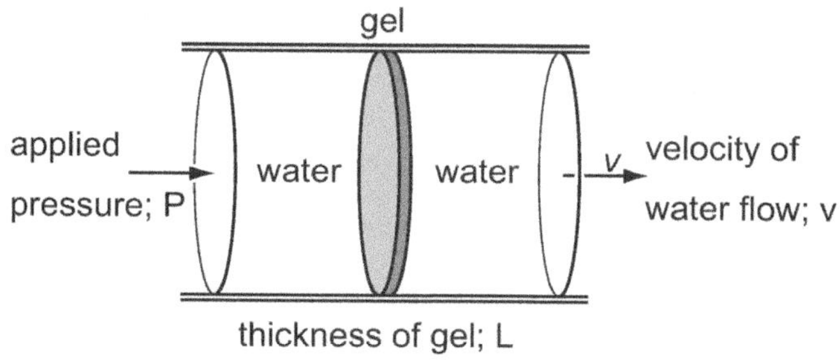

Figure 4. The schematic illustration of the principle of friction measurement. The gel of the thickness L is set in a column with the limb fixed. Then, small pressure, P, is applied to water, which covers the left side of the gel. The velocity, v, of water flowing out of gel is measured in the stationary state. The friction coefficient of gel is defined asf=P/vL.

Friction of Uniform Gel

The first study we should make is to obtain all the aspects of the frictional properties of uniform gel. For this purpose, we use PAAm gel because of its many advantages as described in the previous sections.

First of all, the temperature dependence of the friction coefficient of PAAm gel is studied. The results are given in Figure 5. The friction coefficient of PAAm gel decreases monotonously with the increase in temperature as shown in Figure 5. It may be natural to assume that the friction coefficient itself depends on the viscosity of the flowing fluid.

Thus, the friction coefficient of PAAm gel is normalized by the viscosity of water and the results are also plotted in Figure 5. It is found that the friction coefficient of PAAm gel is independent of the temperature when the friction coefficient is normalized by the viscosity of water, f(T)/η(T). The results obtained here indicate that the structure of the polymer network of PAAm gel is independent of the temperature. The results further indicate that the raw value of the friction coefficient includes the information on the viscosity of the flowing fluid. In the present case, the friction coefficient of PAAm gel is proportional to the viscosity of water, f∝ηwater.

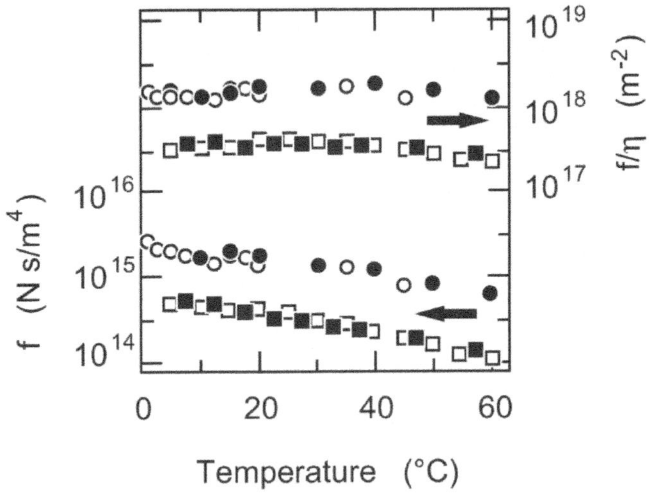

Figure 5. The temperature dependence of the friction coefficient of poly(acrylamide) (PAAm) gel. The concentration of gel is 5 % (squares) and 8% (circles). The friction coefficient that is normalized by the viscosity of water, f(T)/η(T) is also shown in this figure. The solid symbols are used in the increasing of the temperature, and the open symbols are used in the decreasing of the temperature.

Secondly, we study the concentration dependence of the friction coefficient of PAAm gel. The raw value of the friction coefficient of PAAm gel is plotted as a function of the concentration of gel in a double logarithmic manner in Figure 6, since the measurements are made at a constant temperature of 20°C, and hence, the viscosity of water does not show any singular behavior at this temperature.

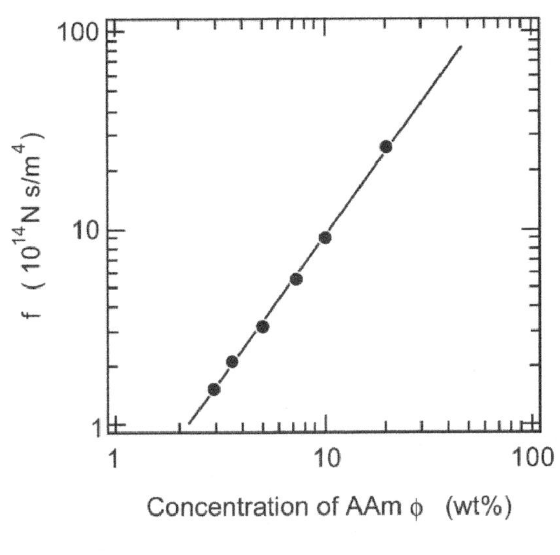

Figure 6. The concentration dependence of the friction coefficient of PAAm gel. The raw value of the friction coefficient is plotted as a function of the concentration of PAAm gel in a double logarithmic manner. The temperature is fixed at 20.0°C in these measurements. The straight line with the slope of 1.5 is the results of the least squares analysis.

The least squares analysis shows that the concentration dependence of the friction coefficient is well expressed by a power law relationship as,

$$f \propto \phi^{1.5}. \tag{9}$$

The scaling suggests that both the collective diffusion coefficient of gel and the elastic modulus of gel can be written as a function of the concentration as follows [23]:

$$D_c \propto \phi^{3/4}, \tag{10}$$

$$E \propto \phi^{9/4}. \tag{11}$$

Equation (8) with the above scaling relationships reveals that the friction coefficient of gel scales with the exponent of 6/4 for the power law relationship. The concentration dependence of the friction coefficient obtained here is in good agreement with the scaling theory. The results are also explained from the molecular structure of gel. When the pressure is applied to the system, water flows through the open space of the polymer network of gel. The gel is, therefore, regarded as the assembly of the pore. The rate of water flow is assumed to be proportional to the viscosity of the fluid, η, and is inversely proportional to the cross section of the pores. The cross section of the pores in gel is reasonably assumed to be proportional to the square of the correlation length of gel, ξ:

$$f \propto \frac{\eta}{\xi^2}. \tag{12}$$

It is suggested by the scaling theory that the correlation length of the polymer network depends on the concentration of gel as follows [23]:

$$\xi \propto \phi^{-3/4}. \tag{13}$$

The scaling again results as $f \propto \phi^{1.5}$. Equation (12) is rather useful when the friction coefficient of gel is discussed in terms of the structure of the gel.

Finally, the friction of PAAm gel is measured as a function of the cross-linking density. It is expected that the size of the pores in the polymer network of gel decreases with the cross-linking density of gel. In other words, the correlation length of the polymer network of gel decreases with the concentration of BIS. Therefore, it may be natural to assume that the friction coefficient of gel increases with the concentration of BIS according to Equation (12). However, the results, as shown in the Figure 7, clearly indicate that the friction of PAAm gel decreases with the concentration of BIS. In addition, it is also found that gel becomes translucent with the concentration of BIS and finally the gel becomes completely opaque when the concentration of BIS exceeds 3 % in mole fraction.

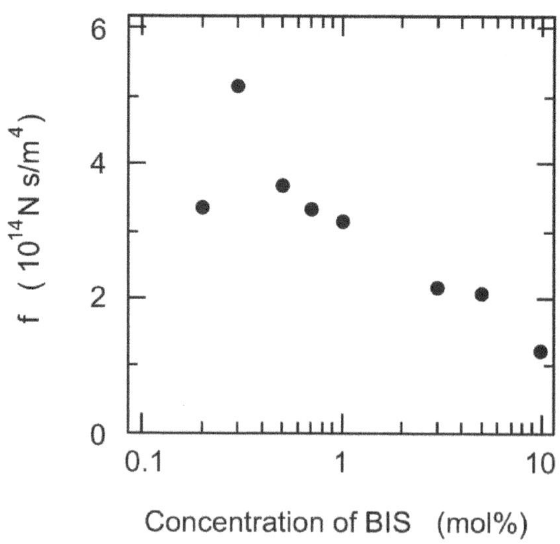

Figure 7. The cross-linker concentration dependence of the friction coefficient of PAAm gel. The total concentration of gel is fixed at 700 mM, and the mole fraction of the cross-linker is varied.

The increase in turbidity of the gel indicates the appearance of the density fluctuations in the polymer network of gel. The density fluctuations that appear in the present system are irreversible and independent of time, and hence, it indicates that the polymer network of gel itself is spatially non-uniform. In contrast, gel is transparent as long as the mole fraction of BIS is less than 3%. The friction of gel, however, clearly decreases with the concentration of BIS even in the concentration region where the transparent gels are obtained. These results thus strongly suggest that the seeds of the density fluctuations emerge in the polymer network of gel at a lower concentration region where gel looks completely transparent. The frictional property of gel, therefore, reflects the spatial homogeneity of gel in an extremely sensitive manner. However, we have to wait for the advancement of new technology that provides us with observations of the structure of the polymer network of the opaque gels in real space, such as the confocal laser scanning microscope (CLSM), for further detailed discussion of the phenomenon. We will return to this point again in the later section.

Friction of Non-Uniform Gel

Many studies have been devoted to clarifying the relationship between the chemical structure of the monomer unit and the biological phenomena [29,30]. Among others, the finding of the volume phase transition in poly(N-isopropylacrylamide) gel (PNIPA gel) is of importance since PNIPA gel shows the volume phase transition in response to the temperature change under water [31]. The transition in PNIPA gel is thought to be a result of the hydrophobic interaction between the N-isopropylacrylamide (NIPA) residues in the polymer chain. According to the fact that PNIPA gel shows the volume phase transition in water, the gel is widely used to study the aspects of the volume phase transition of the gel, for instance, such as the critical kinetics of the volume phase transition of gel [32]. One of the unsolved phenomena currently is the critical behavior of the friction of gel in the vicinity of the volume phase transition point. It is expected in the early stage of the studies of the volume phase transition of gel that the friction of gel becomes smaller in the vicinity of the volume phase transition point of gel [9]. Therefore, the critical behavior of the friction is studied in the vicinity of the volume phase transition temperature of PNIPA gel [33]. The results are given in the Figure 8. The results clearly indicate the drastic decrease of the friction coefficient of gel in the vicinity of the volume phase transition temperature. The friction coefficient of gel decreases more than three orders of magnitude as the temperature approaches the volume phase transition temperature. The change of the friction coefficient of gel is reversible against the temperature change. The results indicate that the reversible decrease of the friction coefficient of gel is caused by the emergence of the critical density fluctuations in the polymer network of gel. The characteristic time scale for the density fluctuations becomes longer in the vicinity of the phase transition point, and hence, the polymer network of gel looks like an assembly of the large pore of the radius that corresponds to the correlation length of the density fluctuations, ξ. Water flows through the open space that is created by the lower density regions of the polymer network, avoiding the denser region that blocks the flow. It is reasonably assumed that the correlation length of the density fluctuations diverge in the vicinity of the volume phase transition point. The friction of the polymer network of gel, therefore, becomes smaller and it practically vanishes at the phase transition point. The critical exponent for the divergence of the correlation length near the critical point is also estimated from these results. We, however, observed a relatively large discrepancy with the expected value. It is clear from the swelling curve of gel that is shown in Figure 8 that the isochore conditions employed here do not hit the exact critical point of gel. Therefore, the critical exponent we obtained here corresponds to the off-critical value. Although the critical slowing down is observed by the friction experiment of PNIPA gel, the critical behaviors of gel is still open to question.

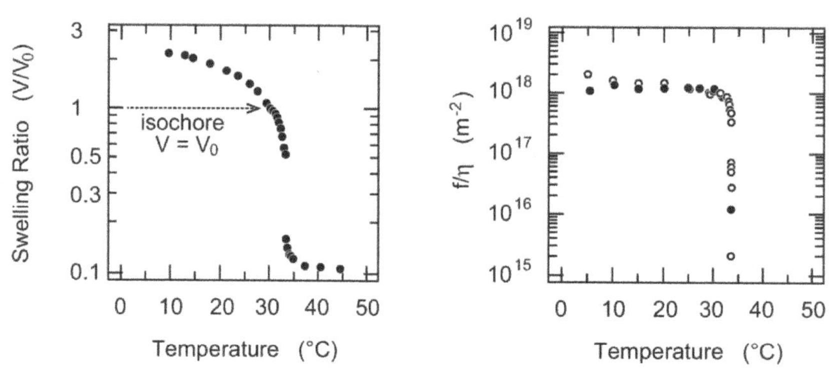

Figure 8. The temperature dependence of the swelling ratio of gel (left; V/V0) and that of the friction coefficient of poly(N-siopropylacrylamide)

(PNIPA) gel that is normalized by the viscosity of water (right; f(T)/η(T)). The friction is measured under a constant volume condition that is shown in the left figure as an arrow. Open circles are used in the increasing of the temperature and solid circles are used in the lowering of the temperature.

Friction of Colloid Gel

It is found that the friction of PAAm gel decreases with the cross-linking density in the previous section. The results suggest strongly that the non-uniform density fluctuations are frozen into the polymer network of gel whenever the cross-link is introduced. The structure of the polymer network of gel, therefore, may change if the concentration of BIS is increased further. The detailed study on the structure of the polymer network of gel, therefore, should be made firstly. Then, the relationship between the structure and the frictional properties of gel is discussed.

The formation of PAAm gel has been studied in detail by changing the concentration of AAm and BIS drastically. The results are expressed as a kind of phase diagram [34]. It has been found that PAAm gel becomes turbid at higher concentration regions of BIS. Therefore, the turbidity change of PAAm gel itself is a well-known phenomenon. The structural study of opaque gels at these concentration regions has been also made by using spectroscopy techniques [35]. Although these studies strongly suggest the formation of the cluster of BIS in the opaque gel, the structure of the opaque PAAm gel in real space could not be obtained.

The idea of the confocal scanning microscopy has been reported in 50's [36]. It became a popular technique when the laser was used as the light source in the 1980s as the confocal laser scanning microscope (CLSM). Firstly, CLSM was used in the research area of biology, and eventually, it spread to the research area of materials science. In the area of gel science, the structure of PNIPA gel was firstly studied using CLSM. It is well known that PNIPA gel also becomes opaque if the reaction temperature is raised to 30°C or more. The network structure of such opaque PNIPA gels is solved by CLSM imaging technique [37]. We also study the structure of the opaque PAAm gel [38]. In Figure 9, CLSM images of the opaque PAAm gel are shown.

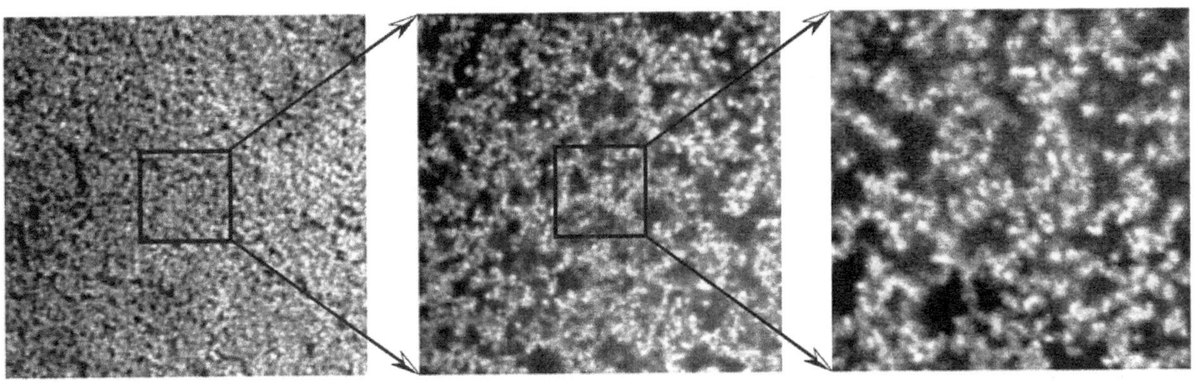

Figure 9. The confocal laser scanning microscope (CLSM) images of opaque PAAm gel. The magnification of the image is 10×, 40×, and 100× from left to right. The length of the side of these images is 189 m, 48.8 m, and 18.9 m from left to right. The concentrations of AAm and N,N-methylenebisacrylamide (BIS) are 350 mM. The total concentration is 700 mM and the mole fraction of BIS is 50%. The squares in the image of 10× is the observation area of 40× observation. The square in the image of 40× is the observation are of 100× observation. These squares only show the relative size of the images and are shown only for a guide to the eye.

The total concentration of PAAm gel observed here is 700 mM and the mole fraction of the BIS is 50%. The images of the same gel are gained at various magnifications. The magnifications of the images are 10×, 40×, and 100×. The brighter portion in the images corresponds to the dense region of AAm and BIS, and the darker region is the dilute region of AAm and BIS. It is clearly shown that the opaque PAAm gel consists of the aggregate of the colloidal particles from these images. The detailed analysis of the confocal images of the opaque PAAm gel yields that the density of the particle is of the order of 1×103 kg/m3. The density of the particle is almost the same as that of AAm and BIS in the solid state. Both AAm and BIS are densely polymerized into a sphere of the colloidal dimension. The particles thus formed in the opaque gels are, therefore, regarded as a hard sphere. In addition to this, the small angle neutron scattering from the opaque gel strongly suggest that gel is a fractal with the fractal dimension of $D_f \simeq 2.7$ [39,40,41,42]. The results suggest that the colloidal particles of AAm and BIS are formed and they aggregate by the process of the diffusion limited aggregation or the cluster-cluster aggregation process. We expect that water can easily flow through the darker portion in the colloidal gel since the mass is mostly localized in the colloidal particles, and, hence, the darker portion of the image serves as the open pore for the fluid flow. The friction experiments are made in the higher concentration region of BIS in gel. The results are given in Figure 10 [43].

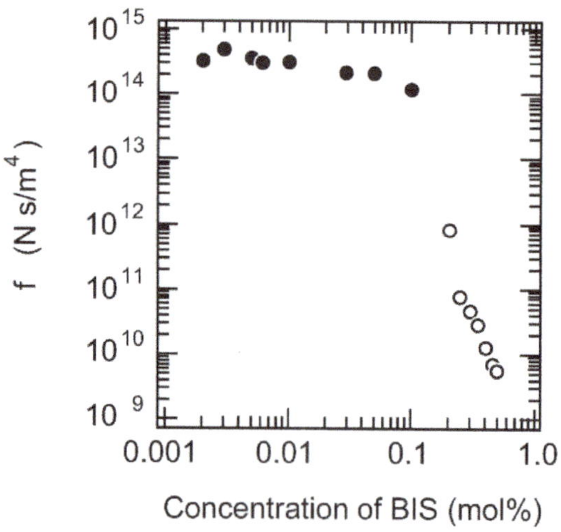

Figure 10. The cross-linker concentration dependence of the friction coefficient of PAAm gel at higher concentration region of BIS (open circles). The results shown inFigure 7 are also plotted in this figure (solid circles). The gels are prepared at a total concentration of 700 mM, and the mole fraction of the cross-linker is changed.

It is found that the friction coefficient of gel decreases more than four orders of magnitude when the mole fraction of BIS is increased to 0.2 or more. The opacity of gel increases at the same concentration of BIS where the drastic decrease of the friction occurs. This suggests strongly that the structure of the polymer network of gel changes from the uniform polymer network to the colloidal aggregate. In accordance with the structural transition of the polymer network of gel, the friction of gel decreases more than four orders of magnitude. The friction of the opaque gel can be explained in terms of the structure of gel because the structural parameters can be deduced from the CLSM images. First of all, the density of the particle is high, and is almost the same as AAm and BIS in a solid state. Secondly, the particle is more or less spherical in shape. The particles are regarded as hard spheres. Water thus flows around the aggregates of hard spheres experiencing the hydrodynamic friction. Since the density of the sphere is almost the same as that of the solid, the aggregates only behave as the fixed obstacles for water flow. If only one solid particle of radius R is suspended in the flowing fluid, the hydrodynamic friction due to the particle is expected to be $f_{obstacle}=6\pi\eta R$. In the colloid gel, however, N particles are included in a unit volume of gel. The friction of the colloid gel thus becomes

$$f_{\text{obstacle}} = N6\pi\eta R. \qquad (14)$$

Here, N represents the number of colloid particle in a unit volume of gel. Both N and R are calculated from the CLSM images of gel that are used as the samples of the friction experiments. Then, the friction of gel is calculated by using Equation (14). The calculated values show the same concentration dependence with the experimental results. The numerical values are, however, about one-half of the experimental values. One of the reasons for this discrepancy arises from the resolution limit of the CLSM that the focal plane of the CLSM has a finite thickness. In the case of the present system, the thickness of the focal plane is about 1 m according to the manufacturer. In addition, the size of the particle in the present case is comparable with the thickness of the focal plane. Therefore, the overlap of the particles in the depth direction in the CLSM image is unavoidable. The number of the particle, N, that is simply calculated from the CLSM image therefore underestimates the actual number of the particle contained in the image. The friction of gel thus calculated becomes smaller than the actual values of the friction. The details are given in the previous reports [43,44].

CONCLUSIONS

We review the probe diffusion of gel and the hydrodynamic friction of the polymer network. It is found that the diffusion coefficient of compact molecules in gel is well described by the scaling theory. The essential parameter that determines the diffusion coefficient is the ratio of the tow sizes, R/ξ, where R is the radius of the probe molecule and ξ the mesh size of the polymer network. In the study, most probe molecules are rather small and compact, and, hence, the scaling results are natural. In addition to this, it is also interesting that the diffusion coefficient of poly(ethylene glycol) follows the same scaling law since poly(ethylene glycol) is a linear polymer. In designing the diffusion experiments, we expected different diffusion mechanisms for poly(ethylene glycol), that is, the diffusion by the reptation [45,46]. The molecular weight of poly(ethylene glycol) used in the diffusion experiments is 200, which indicates that the polymer consists of only about five monomers. The results indicate that polymers of this size can be regarded as compact molecules. Hence, the diffusion experiments using the longer linear polymers is of interest for establishing the entire physical picture of the diffusion in gel. Since we have obtained the simplest view of the diffusion in gel, it may be of interest to study the system where the probe molecule interacts with the polymer network strongly through the typical interactions such as the hydrophobic interaction, the electrostatic interaction and so forth. Further detailed study on the diffusion in gel is required.

The results of the frictional properties of the polymer network of gel is also easily explainable by the scaling theory. The friction of the polymer network is essentially determined by the mesh size of gel. The measure of the mesh size of gel is the correlation length of the polymer network, ξ. In Equation (12), however, we only assume the proportionality between the friction coefficient and the correlation length, $f\propto\eta/\xi^2$. Thus, the numerical coefficient of this equation is not clear at present. It is well known that the Hagen–Poiseuille flow is one of well-defined flow in hydrodynamics:

$$P = 8\eta L\frac{v}{r^2},$$

where P, L, r, and v are the hydrodynamic pressure, the length of the capillary, the radius of the capillary, and the velocity of the flowing fluid. Taking into account Equation (12), one obtains the proportional constant of eight for the friction of the capillary of radius r [47]. It seems that the proportional constant for Equation (12) depends on the structure of the system. Therefore, it will contribute in designing the flow through the porous medium if the numerical coefficient of Equation (12) is clarified. The frictional studies of the well-defined gel or the combined experiments of the structure and the friction as well as the theoretical studies are required for understanding the flow processes in gel.

The diffusion of the molecule, which is dissolved in gel fluid, and the flow of gel fluid are independent phenomena from each other because the origin of the driving force for each process is not the same. Therefore, both transport processes can occur simultaneously in a system as is seen in the pattern formation in the shrinking gel. The well-defined experimental study of such systems is of interest for obtaining information on the general transport phenomena in the soft material. For instance, it is expected that the crossover from the transport by flow to that by diffusion will occur at an extremely low pressure gradient. The detailed study of the crossover phenomena may open new insights into transport phenomena in biological systems.

ACKNOWLEDGMENTS

The author thanks the Ministry of Education, Science, Sports, and Culture of Japan for the financial support (Grant-in-Aid fro Scientific Research, No. 06640523, No. 08640504, and No. 09440153). A part of this work is also supported by the International Joint Research Program of the New Energy and Industrial Technology Development Organization (NEDO).

ABBREVIATIONS

The following abbreviations are used in this manuscript:

PFG-NMR	Pulsed Field Gradient Nuclear Magnetic Resonance
AAm	Acrylamide
PAAm gel	Poly(acrylamide) gel
NIPA	N-isopropylacrylamide
PNIPA gel	Poly(N-isopropylacrylamide) gel
BIS	N, N-methylenebisacrylamide (cross-linker)
CLSM	Confocal laser scanning microscope

REFERENCES

1. Tanaka, T. Collapse of gels and the critical endpoint. Phys. Rev. Lett. **1978**, 40, 820–823.
2. Tanaka, T.; Sun, S.-T.; Hirokawa, Y.; Katayama, S.; Kucera, J.; Hirose, Y.; Amiya, T. Mechanical instability of gels at the phase transition. Nature **1987**, 325, 796–798.
3. Matsuo, E.S.; Tanaka, T. Patterns in shrinking gels. Nature **1992**, 358, 482–485.
4. Tanaka, T. Gels. Sci. Am. **1981**, 244, 124–136.
5. Tanaka, T.; Fillmore, D.J. Kinetics of swelling of gels. J. Chem. Phys. **1979**, 70, 1214–1218.
6. Maskawa, J.; Takeuchi, T.; Maki, K.; Tsujii, K.; Tanaka, T. Theory and numerical calculation of pattern formation in shrinking gels. J. Chem. Phys. **1999**, 110, 10993–10999.
7. Tokita, M.; Suzuki, S.; Miyamoto, K.; Komai, T. Confocal laser scanning microscope imaging of a pattern in shrinking gel. J. Phys. Soc. Jpn. **1999**, 68, 330–333.
8. Tokita, M.; Miyamoto, K.; Komai, T. Polymer network dynamics in shrinking patterns of gels. J. Chem. Phys. **2000**, 113, 1647–1650.
9. Tanaka, T.; Ishiwata, S.; Ishimoto, C. Critical behavior of density fluctuations in gels. Phys. Rev. Lett. **1977**, 38, 771–774.
10. Muhr, A.H.; Blanshard, J.M.V. Diffusion in gels. Polymer **1982**, 23, 1012–1026.
11. Weiss, N.; van Villet, T.; Silberberg, A. Permeability of heterogeneous gels. J. Polym. Sci. Polym. Phys. Ed. **1979**, 17, 2229–2240.
12. Hecht, A.M.; Geissler, E. Gel deswelling under reverse osmosis. J. Chem. Phys. **1980**, 73, 4077–4080.
13. Geissler, E.; Hecht, A.M. Gel deswelling under revers osmosis. II. J. Chem. Phys. **1982**, 77, 1548–1553.
14. Flory, P.J. Principles of Polymer Chemistry; Cornell University Press: Ithaca, NY, USA, 1953; Chapter XIII-3; pp. 576–589.
15. Papon, P.; Leblond, J.; Meijer, P.H.E. The Physics of Phase Transition; Springer-Verlag: Berlin/Heidelberg, Germany, 2002; Chapter 6; pp. 185–209.
16. Park, I.H.; Johnson, C.S., Jr.; Gabriel, D.A. Probediffusion in polyacrylamide gels as observed by means of holographic relaxation methods-search for a universal equation. Macromolecules **1990**, 23, 1548–1553.
17. Gibbs, S.J.; Johnson, C.S., Jr. Pulsed field gradient NMR-study of probe motion in polyacrylamide gels. Macromolecules**1991**, 24, 6110–6113.
18. Tokita, M.; Miyoshi, T.; Takegoshi, K.; Hikichi, K. Probe diffusion in gels. Phys. Rev. E **1996**, 53, 1823–1827.
19. Crank, J. The Mathematics of Diffusion; Oxford University Press: New York, NY, USA, 1989.
20. Einstein, A. On the movement of small particles suspended in a stationary liquid demanded by the molecular-kinetic theory of heat. In Investigations on the Theory of the Brownian Movement; Fürth, R., Ed.; Dover Publications, Inc.: New York, NY, USA, 1956; pp. 1–18.
21. Langevin, D.; Rondelez, F. Sedimentation of large colloidal particles through semidilute polymer-solutions. Polymer**1978**, 53, 875–882.
22. Cukier, R.I. Diffusion of Brownian spheres in semidilute polymer-solutions. Macromolecules **1984**, 17, 252–255.
23. De Gennes, P.G. Scaling Concepts in Polymer Physics; Cornell University Press: Ithaca, NY, USA, 1979; pp. 128–162.
24. Tanaka, T.; Fillmore, J.D.; Sun, S.-T.; Nishio, I.; Swislow, G.; Shah, A. Phase transition in ionic gels. Phys. Rev. Lett.**1980**, 45, 1636–1639.
25. Tanaka, T.; Hocker, L.O.; Benedek, G.B. Spectrum of light scattered from a viscoelastic gel. J. Chem. Phys. **1973**, 59, 5151–5159.
26. Munch, J.P.; Candau, S.; Herz, J.; Hild, G. Inelastic light-scattering by gel modes in semi-dilute polymer solutions and permanent network at equilibrium swollen state. J. Phys. **1977**, 38, 971–976.
27. Munch, J.P.; Lemarechal, P.; Candau, S.; Herz, J. Light-scattering spectroscopy of polydimethylsiloxane-toluene gel. J. Phys. **1977**, 38, 1499–1509.

28. Tokita, M.; Tanaka, T. Friction coefficient of polymer networks of gels. J. Chem. Phys. **1991**, 95, 4613–4619.
29. Ilmain, F.; Tanaka, T.; Kokufuta, E. Volume transition in a gel driven by hydrogen bonding. Nature **1991**, 349, 400–401.
30. Annaka, M.; Tanaka, T. Multiple phases of polymer gels. Nature **1992**, 355, 430–432.
31. Hirokawa, Y.; Tanaka, T. Volume phase transition in a nonionic gel. J. Chem. Phys. **1984**, 81, 6379–6380.
32. Tanaka, T.; Sato, E.; Hirokawa, Y.; Hirotsu, S.; Peetermans, J. Critical kinetics of volume phase transition of gels. Phys. Rev. Lett. **1985**, 55, 2455–2458.
33. Tokita, M.; Tanaka, T. Reversible decrease of gel-solvent friction. Science **1991**, 253, 1121–1123.
34. Richards, E.G.; Temple, C.J. Some properties of polyacrylamide gels. Nature **1971**, 230, 92–96.
35. Bansil, R.; Gupta, M.K. Effects of varying crosslinking density on polyacrylamide gels. Ferroelectrics **1980**, 30, 63–71.
36. Minsky, M.L. Memoir on inventing the confocal scanning microscope. Scanning **1988**, 10, 128–138.
37. Hirokawa, Y.; Jinnai, H.; Nishikawa, Y.; Okamoto, T.; Hashimoto, T. Direct observation of internal structures in poly(N-isoprpylacrylamide) chemical gels. Macromolecues **1999**, 32, 7093–7099.
38. Doi, Y.; Tokita, M. Real space structure of opaque gel. Langmuir **2005**, 21, 5285–5289.
39. Mandelbrot, B.B. The Fractal Geometry of Nature; W. H. Freeman and Company: New York, NY, USA, 1983.
40. Weitz, D.A.; Huang, J.S. Kinetics of Aggregation and Gelation; Family, F., Landau, D.P., Eds.; Elsevier Science Publishing Company, Inc.: New York, NY, USA, 1984; pp. 19–28.
41. Aharony, A. Fractals and Disordered Systems; Bunde, A., Havlin, S., Eds.; Springer-Verlag: Berlin/Heidelberg, Germany, 1996; Chapter 4; pp. 177–199.
42. Mukai, S.; Miki, H.; Garamus, V.; Willmeit, R.; Tokita, M. Structural transition of non-ionic poly(acrylamide) gel.Progr. Colloid Polym. Sci. **2009**, 136, 95–100.
43. Doi, Y.; Tokita, M. Friction coefficient and structural transition in a poly(acrylamide) gel. Langmuir **2005**, 21, 9420–9425.
44. Tokita, M. Structure and frictional properties of colloid gel. Polymers **2014**, 6, 651–666.
45. Doi, M.; Edwards, S.F. The Theory of Polymer Dynamics; Oxford University Press: New York, NY, USA, 1986.
46. Grosberg, A.Y.; Khokhlov, A.R. Giant Molecules—2nd Edition, Here, There, and Everywhere; World Scientific Publishing Company, Co. Pte. Ltd.: Hackensack, NJ, USA, 2011; Chapter 12; pp. 239–259.
47. Tokita, M. Friction between polymer network of gels and solvent. Adv. Polym. Sci. **1993**, 110, 27–47.

INDEX

S

T

V